# Terpenoids: Recent Advances in Extraction, Biochemistry and Biotechnology

Edited by

**Mozaniel Santana de Oliveira**
*Museu Paraense Emilio Goeldi*
*Botanical Coordination*
*Brazil*


**Antônio Pedro da Silva Souza Filho**
*Embrapa Amazônia Oriental*
*Brazil*

# Terpenoids: Recent Advances in Extraction, Biochemistry and Biotechnology

Editors: Mozaniel Santana de Oliveira and Antônio Pedro da Silva Souza Filho

ISBN (Online): 978-1-68108-964-5

ISBN (Print): 978-1-68108-965-2

ISBN (Paperback): 978-1-68108-966-9

need for a court order if at any point you breach any terms of this License Agreement. In no event will any delay or failure by Bentham Science Publishers in enforcing your compliance with this License Agreement constitute a waiver of any of its rights.

3. You acknowledge that you have read this License Agreement, and agree to be bound by its terms and conditions. To the extent that any other terms and conditions presented on any website of Bentham Science Publishers conflict with, or are inconsistent with, the terms and conditions set out in this License Agreement, you acknowledge that the terms and conditions set out in this License Agreement shall prevail.

**Bentham Science Publishers Ltd.**
Executive Suite Y - 2
PO Box 7917, Saif Zone
Sharjah, U.A.E.
Email: subscriptions@benthamscience.net

**BENTHAM SCIENCE**

# CONTENTS

PREFACE ................................................................................................................ i
   ACKNOWLEDGEMENTS .................................................................................... i
   DEDICATION ................................................................................................... ii

LIST OF CONTRIBUTORS ................................................................................... iii

CHAPTER 1  BIOSYNTHESIS OF TERPENOIDS BY PLANTS ............................ 1
*Akemi L. Niitsu, Elesandro Bornhofen* and *Tábata Bergonci*
   INTRODUCTION ............................................................................................ 1
      Mevalonic Acid Pathway ............................................................................. 2
      Methylerythritol Phosphate Pathway ........................................................... 5
      Isomerization of the C5 Building Blocks ...................................................... 8
      Geranyl Pyrophosphate ................................................................................ 9
      Farnesyl Pyrophosphate ............................................................................... 10
      Geranylgeranyl Pyrophosphate .................................................................... 11
      Cytokinin's Biosynthesis ............................................................................. 13
   CONCLUDING REMARKS ............................................................................. 13
   CONSENT FOR PUBLICATION ..................................................................... 13
   CONFLICT OF INTEREST ............................................................................. 13
   ACKNOWLEDGEMENTS ................................................................................ 13
   REFERENCES ................................................................................................. 13

CHAPTER 2  GREEN EXTRACTION TECHNIQUES TO OBTAIN BIOACTIVE
CONCENTRATES RICH IN TERPENOIDS ........................................................... 17
*Ana Carolina de Aguiar, Arthur Luiz Baião Dias* and *Juliane Viganó*
   INTRODUCTION ............................................................................................ 18
   LOW-PRESSURE EXTRACTION METHODS ................................................. 19
      Microwave-Assisted Extraction (MAE) ...................................................... 19
      Ultrasound-Assisted Extraction (UAE) ....................................................... 22
   HIGH-PRESSURE EXTRACTION METHODS ............................................... 27
      Supercritical Fluid Extraction (SFE) ........................................................... 27
      Pressurized Liquid Extraction (PLE) ........................................................... 31
      Liquefied Petroleum Gas (LPG) Extraction ................................................. 32
   CONCLUDING REMARKS ............................................................................. 32
   CONSENT FOR PUBLICATION ..................................................................... 33
   CONFLICT OF INTEREST ............................................................................. 33
   ACKNOWLEDGEMENTS ................................................................................ 33
   REFERENCES ................................................................................................. 34

CHAPTER 3  TERPENOIDS PRODUCED BY PLANT ENDOPHYTIC FUNGI FROM
BRAZIL AND THEIR BIOLOGICAL ACTIVITIES: A REVIEW FROM JANUARY 2015 TO
JUNE 2021 ........................................................................................................... 39
*Lourivaldo Silva Santos, Giselle Skelding Pinheiro Guilhon, Railda Neyva Moreira Araujo,*
*Antonio José Cantanhede Filho, Manoel Leão Lopes Junior, Haroldo da Silva, Ripardo Filho*
and *Kiany Sirley Brandão Cavalcante*
   INTRODUCTION ............................................................................................ 40
      Terpenoids ................................................................................................... 42
      Monoterpenoids ........................................................................................... 43
         *Sesquiterpenoids* ...................................................................................... 44
      Sesterterpenoids .......................................................................................... 54
      Meroterpenoids ........................................................................................... 54

Triterpenoids ........................................................................................... 55
CONCLUDING REMARKS ..................................................................... 58
CONSENT FOR PUBLICATION ............................................................ 59
CONFLICT OF INTEREST ...................................................................... 59
ACKNOWLEDGEMENTS ......................................................................... 59
REFERENCES ............................................................................................ 59

CHAPTER 4 VOLATILE TERPENOIDS IN MYRTACEAE SPECIES: CHEMICAL
STRUCTURES AND APPLICATIONS ........................................................... 67
*Oberdan Oliveira Ferreira, Celeste de Jesus Pereira Franco, Angelo Antônio Barbosa de Moraes,
Giovanna Moraes Siqueira, Lidiane Diniz Nascimento, Márcia Moraes, Cascaes, Mozaniel
Santana de Oliveira* and *Eloisa Helena de Aguiar Andrade*
INTRODUCTION ....................................................................................... 68
Myrtaceae Family and General Aspects ............................................. 70
Essential Oil of Myrtaceae Rich in Terpenoids ................................ 70
APPLICATIONS ......................................................................................... 75
Antioxidant Activity ........................................................................... 75
Anti-Inflammatory Activity ............................................................... 80
Neuroprotective Activity .................................................................... 82
Cytotoxicity Activity .......................................................................... 85
Anti-protozoan Activity ...................................................................... 90
Antidiabetic Activity ........................................................................... 93
Recent Advances in Phytotherapy ..................................................... 93
CONCLUDING REMARKS ..................................................................... 94
CONSENT FOR PUBLICATION ............................................................ 94
CONFLICT OF INTEREST ...................................................................... 95
ACKNOWLEDGEMENTS ......................................................................... 95
REFERENCES ............................................................................................ 95

CHAPTER 5 VOLATILE TERPENOIDS OF ANNONACEAE: OCCURRENCE AND
REPORTED ACTIVITIES ................................................................................. 105
*Márcia M. Cascaes, Giselle M. S. P. Guilhon, Lidiane D. Nascimento, Angelo A. B., de Moraes,
Sebastião G. Silva, Jorddy Neves Cruz, Oberdan O. Ferreira, Mozaniel S Oliveira* and *Eloisa
H. A. Andrade*
INTRODUCTION ....................................................................................... 105
CHEMICAL DIVERSITY OF VOLATILE TERPENOIDS .................. 106
BIOACTIVITY OF THE ESSENTIAL OILS FROM ANNONACEAE SPECIES ............... 112
Acetylcholinesterase Inhibition ......................................................... 112
Antimicrobial Activities ...................................................................... 113
Anti-Inflammatory Activity ............................................................... 115
Antiproliferative and Cytotoxic Activities ....................................... 116
Larvicidal Activity .............................................................................. 118
Trypanocidal and Antimalarial Activities ......................................... 119
Other Activities ................................................................................... 119
ANTIOXIDANT POTENTIAL ................................................................. 121
CONCLUDING REMARKS ..................................................................... 122
CONSENT FOR PUBLICATION ............................................................ 122
CONFLICT OF INTEREST ...................................................................... 122
ACKNOWLEDGEMENTS ......................................................................... 123
REFERENCES ............................................................................................ 123

CHAPTER 6 REPELLENT POTENTIAL OF TERPENOIDS AGAINST TICKS ...... 129

*Tássia L. Vale, Isabella C. Sousa, Caio P. Tavares, Matheus N. Gomes, Geovane F. Silva, Jhone R. S. Costa, Aldilene da Silva Lima, Claudia Q. Rocha* and *Livio Martins Costa-Júnior*

INTRODUCTION ............................................................................................................... 129
    Terpenoids Repellent Against Ticks .............................................................................. 132
    Bioassays to Evaluate Repellent Compounds Against Ticks ......................................... 136
        *Tick Climbing Bioassay* ........................................................................................ 137
        *Olfactometer Bioassay* ......................................................................................... 137
        *Petri Dish Bioassay* ............................................................................................. 138
        *Bioassay of the Falcon Tissue Flask Repellency* ................................................. 138
        *Moving Object Bioassay* ....................................................................................... 138
    Repellent Compound on Ticks of Medical Importance ................................................. 139
    Repellent Compounds on Dogs' Ticks ......................................................................... 140
        *Repellent Compounds on Livestock's Ticks* ........................................................ 141
CONCLUDING REMARKS ................................................................................................ 142
CONSENT FOR PUBLICATION ....................................................................................... 142
CONFLICT OF INTEREST ................................................................................................ 142
ACKNOWLEDGEMENTS .................................................................................................. 142
REFERENCES ..................................................................................................................... 143

**CHAPTER 7 USE OF TERPENOIDS TO CONTROL HELMINTHS IN SMALL RUMINANTS** 148

*Dauana Mesquita-Sousa, Victoria Miro, Carolina R. Silva, Juliana R. F. Pereira1, Livio M. Costa-Júnior, Guillermo Virke* and *Adrian Lifschitz*

INTRODUCTION ............................................................................................................... 149
    Mechanism of Action of the Anthelmintic Compound ................................................. 149
        *Neuromuscular System and Motility Control* ...................................................... 149
        *Terpenoids with Action in GABA* ......................................................................... 150
        *The Action of the Terpenoids on Tubulin* ............................................................ 150
        *Structural Alterations* .......................................................................................... 151
    Combination of Synthetic Anthelmintics and Terpenoids ............................................ 151
    Influence of Pharmacological Properties of Monoterpenes on their Anthelmintic Effect ...... 152
    In Vivo Anthelmintic Effect of Monoterpenes .............................................................. 157
CONCLUDING REMARKS ................................................................................................ 161
CONSENT FOR PUBLICATION ....................................................................................... 161
CONFLICT OF INTEREST ................................................................................................ 162
ACKNOWLEDGEMENT .................................................................................................... 162
REFERENCES ..................................................................................................................... 162

**CHAPTER 8 TERPENES BEHAVIOR IN SOIL** ....................................................... 169

*Marcia M. Mauli, Adriana M. Meneghetti* and *Lúcia H. P. Nóbrega*

INTRODUCTION ............................................................................................................... 169
    Terpenoids ..................................................................................................................... 169
    Biosynthesis of IPP (Isopentyl Diphosphate) and DMAPP (Dimethylallyl Diphosphate) ..... 173
        *The Mevalonic Acid Pathway (MVA)* .................................................................. 173
    Methylerithritol-Phosphate Pathway (MEP) ................................................................. 173
    Hemiterpenes C5 ........................................................................................................... 174
    Monoterpenes C10 ......................................................................................................... 175
    Sesquiterpenes C15 ........................................................................................................ 177
    Diterpenes C20 .............................................................................................................. 178
    Triterpenoids C30 .......................................................................................................... 179
    Tetraterpenoids C40 ....................................................................................................... 181
    Secondary Metabolites .................................................................................................. 181

Allelochemicals Behavior in the Environment ............................................................... 181
Allelochemicals Behavior in Soil ............................................................. 185
Terpenes in Soil ............................................................. 187
**CONSENT FOR PUBLICATION** ............................................................. 191
**CONFLICT OF INTEREST** ............................................................. 191
**ACKNOWLEDGEMENT** ............................................................. 191
**REFERENCES** ............................................................. 192

**CHAPTER 9  POTENTIAL USE OF TERPENOIDS IN WEED MANAGEMENT** ...................... 200
*Mozaniel Santana de Oliveira, Jordd Nevez Cruz, Eloisa Helena de Aguiar Andrade* and *Antônio Pedro da Silva Souza Filho*
**INTRODUCTION** ............................................................. 200
Volatile Terpenoids ............................................................. 202
Monoterpenes with Phytotoxic Potential ............................................................. 204
Sesquiterpenes with Phytotoxic Potential ............................................................. 207
Diterpenes with Phytotoxic Potential ............................................................. 209
**CONCLUDING REMARKS** ............................................................. 213
**CONSENT FOR PUBLICATION** ............................................................. 213
**CONFLICT OF INTEREST** ............................................................. 213
**ACKNOWLEDGMENTS** ............................................................. 213
**REFERENCES** ............................................................. 213

**CHAPTER 10  APPLICATIONS OF NATURAL TERPENOIDS AS FOOD ADDITIVES** ........ 223
*Fernanda Wariss Figueiredo Bezerra, Giselle Cristine Melo Aires, Lucas Cantão,  Freitas Marielba de Los Angeles Rodriguez Salazar, Rafael Henrique Holanda Pinto  Jorddy Neves da Cruz* and *Raul Nunes de Carvalho Junior*
**INTRODUCTION** ............................................................. 224
**DIVERSITY AND CHARACTERISTICS OF TERPENOIDS IN FOOD SYSTEMS** ........... 225
**POSSIBLE APPLICATIONS OF THE TERPENOIDS AS FOOD ADDITIVES** .................. 226
Colorants ............................................................. 226
Flavoring Agent ............................................................. 228
Anti-Oxidants ............................................................. 229
Anti-Microbial ............................................................. 231
Nutraceutical ............................................................. 233
**CONCLUDING REMARKS** ............................................................. 235
**CONSENT FOR PUBLICATION** ............................................................. 235
**CONFLICT OF INTEREST** ............................................................. 235
**ACKNOWLEDGEMENTS** ............................................................. 235
**REFERENCES** ............................................................. 235

**CHAPTER 11  POTENTIAL USE OF TERPENOIDS FOR CONTROL OF INSECT PESTS** ... 246
*Murilo Fazolin, Humberto Ribeiro Bizzo* and *André Fábio Medeiros Monteiro*
**INTRODUCTION** ............................................................. 247
**MECHANISMS OF INSECTICIDAL ACTION OF TERPENOIDS** ............................ 248
Binding of GABA (Gamma-Aminobutyric Acid) Neurotransmitter to Receptors ................ 249
Binding to the Nicotinic Acetylcholine Receptor ............................................................. 249
Inhibition of Transient Receptor Potential (TRP) Channels ............................................................. 249
Activity on Octopamine and Tyramine Receptors ............................................................. 250
Inhibition of Detoxification Enzymes ............................................................. 250
**INSECTICIDE FORMULATIONS USING TERPENOIDS** ............................................................. 254
Synergistic Interactions Between Terpenoids for Insecticide Formulation ............................ 254

Production of Blends From Synergistic Terpenoids Present in Essential Oils and
Development of Commercial Products .................................................................... 259
Essential Oil-Based Products with High Levels of Terpenoids ............................. 262
Challenges to the Production of Commercial Insecticides Based on Essential Oils and
Terpenoids .............................................................................................................. 268
Insect Resistance to Commercial Terpenoid-Based Insecticides .......................... 269
CONCLUDING REMARKS ........................................................................................ 269
CONSENT FOR PUBLICATION .............................................................................. 270
CONFLICT OF INTEREST ........................................................................................ 270
ACKNOWLEDGEMENT ............................................................................................ 270
REFERENCES .............................................................................................................. 270

CHAPTER 12 POTENTIAL ANTIMICROBIAL ACTIVITIES OF TERPENOIDS .................. 279
Hamdy A. Shaaban and Amr Farouk
INTRODUCTION ......................................................................................................... 279
Antibacterial Activity ............................................................................................. 281
Antiviral Effect ...................................................................................................... 283
Terpenoids and Essential Oils as Antimicrobial Agents in Food Preservation ...... 285
The Site of Influence of Terpenoids and Essential Oils ........................................ 287
Factors Affecting Antimicrobial Activity .............................................................. 289
CONCLUDING REMARKS ........................................................................................ 289
CONSENT FOR PUBLICATION .............................................................................. 290
CONFLICT OF INTEREST ........................................................................................ 290
ACKNOWLEDGEMENT ............................................................................................ 291
REFERENCES .............................................................................................................. 291

CHAPTER 13 TERPENOIDS IN PROPOLIS AND GEOPROPOLIS AND APPLICATIONS .... 298
Jorddy Neves Cruz, Mozaniel Santana de Oliveira, Lindalva Maria de Meneses Costa, Ferreira
Daniel Santiago Pereira, João Paulo de Holanda Neto, Aline Carla de Medeiros, Patrício Borges
Maracajá and Antônio Pedro da Silva Souza Filho
INTRODUCTION ......................................................................................................... 298
TERPENOIDS PRESENT IN PROPOLIS FROM APIS MELLIFERA BEES .................. 300
TERPENOIDS PRESENT IN PROPOLIS AND GEOPROPOLIS OF STINGLESS BEES .... 309
RECENT APPLICATIONS OF PROPOLIS AND GEOPROPOLIS IN THE FOOD AND
PHARMACEUTICAL INDUSTRIES ....................................................................... 312
CONCLUDING REMARKS ........................................................................................ 313
CONSENT FOR PUBLICATION .............................................................................. 314
CONFLICT OF INTEREST ........................................................................................ 314
ACKNOWLEDGEMENTS .......................................................................................... 314
REFERENCES .............................................................................................................. 314

CHAPTER 14 TERPENOIDS AND BIOTECHNOLOGY ..................................................... 320
Jorddy Neves Cruz, Fernanda Wariss Figueiredo Bezerra, Renan Campos e Silva, Mozaniel
Santana de Oliveira, Márcia Moraes Cascaes, Jose de Arimateia Rodrigues do Rego, Antônio
Pedro da Silva Souza Filho, Daniel Santiago Pereira and Eloisa Helena de Aguiar Andrade
INTRODUCTION ......................................................................................................... 320
BIOTECHNOLOGY OF TERPENE PRODUCTION IN MICROORGANISMS ............ 323
PHARMACEUTICAL APPLICATIONS OF TERPENOIDS PRODUCED BY
BIOTECHNOLOGICAL METHODS ....................................................................... 327
BIOTECHNOLOGICAL APPLICATIONS OF TERPENES ...................................... 327
CONCLUDING REMARKS ........................................................................................ 330

**CONSENT FOR PUBLICATION** ........................................................................................... 330

**CONFLICT OF INTEREST** ................................................................................................. 330

**ACKNOWLEDGEMENTS** ................................................................................................... 330

**REFERENCES** ..................................................................................................................... 330

**SUBJECT INDEX** ..................................................................................................................... 338

# PREFACE

In natural systems, such as forest areas, and in artificial systems, such as agroecosystems, different types of interactions can occur, promoting changes in the dynamics and density of components, favoring certain components, or harming others. Many of these interactions are due to competition for factors that are essential for the survival of each component, and in others, the interactions are mediated by chemical compounds released by different components in the environment. Plant-plant, plant-fungal, and plant-insect interactions, among others, are good examples of chemical interactions.

Although only in the recent past have significant advances in this area made it possible to understand the real possibilities that this knowledge represented in practical terms for the consolidation of agriculture aimed at meeting the demands of society, the perception of the occurrence of these interactions dates back to a very remote time. Over the years, teams of researchers focused on the subject, and the implementation of properly equipped laboratories enabled the development of research projects that resulted in a substantial accumulation of information on the chemical classes involved in the process and the components of each class.

As new equipment was made available and incorporated into existing laboratories, other laboratories were set up around the world, especially in countries with little tradition in science. In the wake of this process, other researchers were joining the groups already formed, boosting the research even more. As a result of these efforts, several chemicals were isolated and their biological activities identified. Among these studies, the class of terpenoids deserves to be highlighted, representing the group with the largest number of components and the greatest range of activity for the control of weeds, insects, and fungi, among others.

Terpenoids are composed of various chemicals with different polarities, which include both essential oils, formed by monoterpenes, diterpenes, sesquiterpenes, hydrocarbons, and triterpenes, and terpenes and tetranorterpenoids. Transforming available information about this class into a finished product to fight pests and diseases is a challenge that has been overcome.

## ACKNOWLEDGEMENTS

We would like to thank the authors who responded positively, to the challenge that represented the elaboration of the present Book, contributing with its knowledge and experience accumulated over time.

We are grateful to the author, Dr. Mozaniel Santana de Oliveira and would like to give special thanks to PCI-MCTIC/MPEG, as well as CNPq for the scholarship (process number 302050/2021-3).

## DEDICATION

To the dream that inhabits in everyone's heart, those who propose to offer alternatives to the cravings of society.

The author Dr. Mozaniel de Oliveira dedicates this work to his parents and Maria and Manoel de Oliveira and wife Joyce Fontes.

**Antônio Pedro da Silva Souza Filho**
Embrapa – Belém,
Pará, Brazil

# List of Contributors

| | |
|---|---|
| **Adrian Lifschitz** | Laboratorio de Farmacología, Centro de Investigación Veterinaria de Tandil (CIVETAN)(CONICET-CICPBA-UNCPBA), Facultad de Ciencias Veterinarias, Universidad Nacional del Centro, Tandil, Argentina |
| **Adriana M. Meneghetti** | Laboratório de Química e Metabólitos Secundários, Departamento de Biologia, Universidade Tecnológica Federal do Paraná, Santa Helena, Paraná, Brazil |
| **Akemi L. Niitsu** | Department of Biological Sciences, University of Sao Paulo, Piracicaba, Brazil |
| **Aldilene da Silva Lima** | Laboratorio de produtos naturais, Departamento de Química, Universidade Federal do Maranhão (UFMA), São Luis, Maranhão, Brazil |
| **Aline Carla de Medeiros** | Federal University of Campina Grande, , Paraiba, Brazil |
| **Ana Carolina de Aguiar** | Laboratory of High Pressure in Food Engineering, Department of Food Engineering University of Campinas, 13083-862, Campinas, Brazi |
| **André Fábio Medeiros Monteiro** | Agroforestry Research Center of Acre, Brazilian Agricultural Research Corporation (Embrapa/CPAFAC), , |
| **Antônio José Cantanhede Filho** | Instituto Federal de Educação Ciência do Maranhão (IFMA), Departamento Acadêmico de Química, São Luís MA, Brazil |
| **Antônio Pedro da Silva Souza Filho** | Empresa Brasileira de Pesquisa Agropecuária (Embrapa-Amazônia Oriental), Tv. Dr. Eneas Pinheiro, s/n - Marco, Belém – PA, Brazil |
| **Ângelo Antônio Barbosa de Moraes** | Laboratório Adolpho Ducke Laboratory, Botany Coordination, Museu Paraense Emílio Goeldi, Av. Perimetral, 1900, Terra Firme, 66077-830, Belém PA, Brazil |
| **Arthur Luiz Baião Dias** | Laboratory of High Pressure in Food Engineering, Department of Food Engineering, University of Campinas, UNICAMP, 13083-862 Campinas, Brazil |
| **Caio P. Tavares** | Laboratorio de controle de Parasitos, Departamento de Patologia, Universidade Federal do Maranhão (UFMA), São Luis, Maranhão, Brazil |
| **Carolina R. Silva** | Laboratorio de controle de Parasitos, Departamento de Patologia, Universidade Federal do Maranhão (UFMA), São Luis, Maranhão, Brazil |
| **Celeste de Jesus Pereira Franco** | Laboratório Adolpho Ducke Laboratory, Botany Coordination, Museu Paraense Emílio Goeldi, Av. Perimetral 1900, Terra Firme, 66077-830 Belém, PA, Brazil |
| **Claudia Q. Rocha** | Laboratorio de produtos naturais, Departamento de Química, Universidade Federal do Maranhão (UFMA), São Luis, Maranhão, Brazil |
| **Daniel Santiago Pereira** | Laboratorio de produtos naturais, Departamento de Química, Laboratory of Agro-Industry, Embrapa Eastern Amazon, Belem, Pará, Brazil |
| **Dauana Mesquita-Sousa** | Laboratorio de controle de Parasitos, Departamento de Patologia, Universidade Federal do Maranhão (UFMA), São Luis, Belem Pará, Brazil |
| **Elesandro Bornhofen** | Center for Quantitative Genetics and Genomics, Aarhus University, Aarhus, Denmark |
| **Eloisa Helena de Aguiar Andrade** | Laboratório Adolpho Ducke Laboratory, Botany Coordination, Museu Paraense Emílio Goeldi, Av. Perimetral, 1900, Terra Firme, 66077-830, Belém, PA, Brazil |

| | |
|---|---|
| **Eloisa Helena de Aguiar Andrade** | Fernanda Wariss Figueiredo Bezerra LABEX/PPGCTA (Extraction Laboratory / Graduate Program in Food Science and Technology), Federal University of Pará, Rua Augusto Corrêa S/N, 66075-900, Belém, Pará, Brazil |
| **Geovane F. Silva** | Laboratorio de controle de Parasitos, Departamento de Patologia, Universidade Federal do Maranhão (UFMA), São Luis, Maranhão, Brazil |
| **Giovanna Moraes Siqueira** | Laboratório Adolpho Ducke Laboratory, Botany Coordination, Museu Paraense Emílio Goeldi, Belém, PA, Brazil |
| **Giselle Cristine Melo Aires** | LABEX/PPGCTA (Extraction Laboratory / Graduate Program in Food Science and Technology), Federal University of Pará, Rua Augusto Corrêa S/N, 66075-900, Belém, Pará, Brazil |
| **Giselle Skelding Pinheiro Guilhon** | Laboratório de Micro-organismos, Programa de Pós-Graduação em Química, Universidade Federal do Pará-UFPA, 66970-110, Belém, PA, Brazil |
| **Guillermo Virkel** | Laboratorio de Farmacología, Centro de Investigación Veterinaria de Tandil (CIVETAN)(CONICET-CICPBA-UNCPBA), Facultad de Ciencias Veterinarias, Universidad Nacional del Centro, Tandil, Argentina |
| **Hamdy A. Shaaban** | National Research Center, Chemistry of Flavours & Aroma Department, El-Behoose St. Dokki Giza, Egypt |
| **Haroldo da Silva Ripardo Filho** | Instituto Federal do Amapá, Faculdade de Química, Macapá, AP, Brazil |
| **Humberto Ribeiro Bizzo** | National Center for Research on Agroindustrial Food Technology (CTAA), Av. das Américas, nº 29.501, Guaratiba, RJ, CEP 23020-470, Brazil |
| **Isabella C. Sousa** | Laboratorio de controle de Parasitos, Departamento de Patologia, Universidade Federal do Maranhão (UFMA), São Luis, Maranhão, Brazil |
| **Jhone R. S. Costa** | Laboratorio de controle de Parasitos, Departamento de Patologia, Universidade Federal do Maranhão (UFMA), São Luis, Maranhão, Brazil |
| **João Paulo de Holanda Neto** | Federal Institute of education, Science and Technology of Sertão Pernambucano, Oricuri, Pernambuco, Brazil |
| **Jorddy Neves Cruz** | Adolpho Ducke Laboratory, Paraense Emílio Goeldi Museum, Belém, Brazil |
| **Jose de Arimateia Rodrigues do Rego** | Institute of Technology, Federal University of Pará, Belém, Brazil |
| **Juliana R. F. Pereira** | Laboratorio de controle de Parasitos, Departamento de Patologia, Universidade Federal do Maranhão (UFMA), São Luis, Maranhão, Brazil |
| **Juliane Viganó** | Multidisciplinary Laboratory of Food and Health (LabMAS), School of Applied Sciences (FCA), University of Campinas, Rua Pedro Zaccaria 1300, 13484- 350 Limeira, São Paulo, Brazil |
| **Kiany Sirley Brandão Cavalcante** | Instituto Federal de Educação Ciência do Maranhão (IFMA), Departamento Acadêmico de Química, São Luís, MA, Brazil |
| **Lidiane Diniz Nascimento** | Laboratório Adolpho Ducke Laboratory, Botany Coordination, Museu Paraense Emílio Goeldi, Av. Perimetral, 1900, Terra Firme, 66077-830, Belém, PA, Brazil, Brazil |
| **Livio Martins Costa-Júnior** | Laboratorio de controle de Parasitos, Departamento de Patologia, Universidade Federal do Maranhão (UFMA), São Luis, Maranhão, Brazil |

| | |
|---|---|
| **Lourivaldo Silva Santos** | Laboratório de Micro-organismos, Programa de Pós-Graduação em Química, Universidade Federal do Pará-UFPA, 66970-110, Belém, PA, Brazil |
| **Lucas Cantão Freitas** | LABEX/PPGCTA (Extraction Laboratory / Graduate Program in Food Science and Technology), Federal University of Pará, Rua Augusto Corrêa S/N, 66075-900, Belém, Pará, Brazil |
| **Lúcia H. P. Nóbrega** | Laboratório de avaliação de sementes e plantas, Centro de Ciências Exatas e Tecnológicas, Universidade Estadual do Oeste do Paraná – UNIOESTE, Campus de Cascavel, Paraná, Brazil |
| **Lyndalva Maria de Meneses Costa Ferreira** | Laboratório de Nanotecnologia Farmacêutica, Faculdade de Farmácia, Universidade Federal do, Pará, Brazil |
| **Manoel Leão Lopes Junior** | Universidade Federal do Pará, Campus de Cametá, Cametá, PA, Brazil |
| **Marcia M. Mauli** | Secretaria do Estado da Educação do Paraná (Seed), Cascavel, Paraná, Brazil |
| **Marielba de Los Angeles Rodriguez Salazar** | LABEX/PPGCTA (Extraction Laboratory / Graduate Program in Food Science and Technology), Federal University of Pará, Rua Augusto Corrêa S/N, 66075-900, Belém, Pará, Brazil |
| **Márcia Moraes Cascaes** | Program of Post-Graduation in Chemistry, Federal University of Pará, Rua Augusto Corrêa S/N, Guamá, 66075-900 Belém, Pará, Brazil |
| **Marielba de Los Angeles Rodriguez Salazar** | LABEX/PPGCTA (Extraction Laboratory / Graduate Program in Food Science and Technology), Federal University of Pará, Rua Augusto Corrêa S/N, 66075-900, Belém, Pará, Brazil |
| **Matheus N. Gomes** | Laboratorio de controle de Parasitos, Departamento de Patologia, Universidade Federal do Maranhão (UFMA), São Luis, Maranhão, Brazil |
| **Mozaniel Santana de Oliveira** | Laboratório Adolpho Ducke Laboratory, Botany Coordination, Museu Paraense Emílio Goeldi, Av. Perimetral, 1900, Terra Firme, 66077-830, Belém, PA, Brazil |
| **Murilo Fazolin** | Agroforestry Research Center of Acre, Brazilian Agricultural Research Corporation (Embrapa/CPAFAC), , |
| **Oberdan Oliveira Ferreira** | Program of Post-Graduation in biodiversity e biotecnology-Bionorte, Federal University of Pará, Rua Augusto Corrêa S/N, Guamá, 66075-900 Belém, Pará, Brazil |
| **Patrício Borges Maracajá** | , Federal University of Campina Grande, 66075-900 Belém, Pará, Paraiba |
| **Rafael Henrique Holanda Pinto** | LABEX/PPGCTA (Extraction Laboratory / Graduate Program in Food Science and Technology), Federal University of Pará, Rua Augusto Corrêa S/N, 66075-900 Belém, Pará, Paraiba |
| **Railda Neyva Moreira Araujo** | Escola Estadual de Ensino Médio Agostinho Morais de Oliveira, , Inhangapi, PA, Brazil |
| **Raul Nunes de Carvalho Junior** | LABEX/PPGCTA (Extraction Laboratory / Graduate Program in Food Science and Technology), Federal University of Pará, Rua Augusto Corrêa S/N, 66075-900, Belém, Pará, Brazil |

**Renan Campos e Silva**   Program of Post-Graduation in Chemistry, Federal University of Pará, Belém, Brazil

**Sebastião G. Silva**   Adolpho Ducke Laboratory, Paraense Emílio Goeldi Museum, Belém, Brazil

**Tábata Bergonci**   Department of Food Science, Aarhus University, Aarhus, Denmark

**Tássia L. Vale**   Laboratorio de controle de Parasitos, Departamento de Patologia, Universidade Federal do Maranhão (UFMA), São Luis, Maranhão, Denmark

**Victoria Miro**   Laboratorio de Farmacología, Centro de Investigación Veterinaria de Tandil (CIVETAN)(CONICET-CICPBA-UNCPBA), Facultad de Ciencias Veterinarias, Universidad Nacional del Centro, Tandil, Argentina

# CHAPTER 1

# Biosynthesis of Terpenoids By Plants

## Akemi L. Niitsu[1], Elesandro Bornhofen[2] and Tábata Bergonci[3,*]

*[1] Department of Biology Science, University of Sao Paulo, Piracicaba, Brazil*

*[2] Center for Quantitative Genetics and Genomics, Aarhus University, Aarhus, Denmark*

*[3] Department of Food Science, Aarhus University, Aarhus, Denmark*

**Abstract:** Terpenoids are a class of chemicals with over 50,000 individual compounds, highly diverse in chemical structure, founded in all kingdoms of life, and are the largest group of secondary plant metabolites. Also known as isoprenoids, their structure began to be elucidated between the 1940s and 1960s, when their basic isoprenoid building blocks were characterized. They play several basic and specialized physiological functions in plants through direct and indirect interactions. Terpenoids are essential to metabolic processes, including post-translational protein modifications, photosynthesis, and intracellular signaling. All terpenoids are built through $C_5$ units condensed to prenyl diphosphate intermediates. The fusion of these $C_5$ units generates short $C_{15}$-$C_{25}$, medium $C_{30}$-$C_{35}$, and long-chain $C_{40}$-Cn terpenoids. Along with the extension of the chain, the introduction of functional groups, such as ketones, alcohol, esters and, ethers, forms the precursors to hormones, sterols, carotenoids, and ubiquinone synthesis. The biosynthesis of terpenoids is regulated by spatial, temporal, transcriptional, and post-transcriptional factors. This chapter gives an overview of terpenoid biosynthesis, focusing on both cytoplasmic and plastid pathways, and highlights recent advances in the regulation of its metabolic pathways.

**Keywords:** Abscisic Acid, Brassinosteroids, Carotenoids, Dimethylallyl Diphosphate, Gene Regulation, Gene Expression, Glycosylation, Isopentenyl Diphosphate, Isoprene, Isoprenoids, MEP Pathway, MVA Pathway, Plant Hormones, Prenyl Diphosphate, Secondary Metabolites, Sterols, Terpene Synthesis, Terpenoid, Ubiquinone, Volatile Terpenes.

## INTRODUCTION

Terpenoids, also called isoprenoids, are the most diverse class of chemical groups produced by plants. They are the largest category of secondary metabolites derived from the universal 5-carbon compound, isopentenyl diphosphate (IPP),

---

* **Corresponding author Tábata Bergonci:** Department of Food Science, Aarhus University, Aarhus, Denmark; E-mail: tabatab@alumni.usp.br

**Mozaniel Santana de Oliveira & Antônio Pedro da Silva Souza Filho (Eds.)**

and its allylic isomer dimethylallyl diphosphate (DMAPP) [1] (Fig. **1**). The condensation of IPP and/or DMAPP units to prenyl diphosphate intermediates are used as precursors for the biosynthesis of terpenoids.

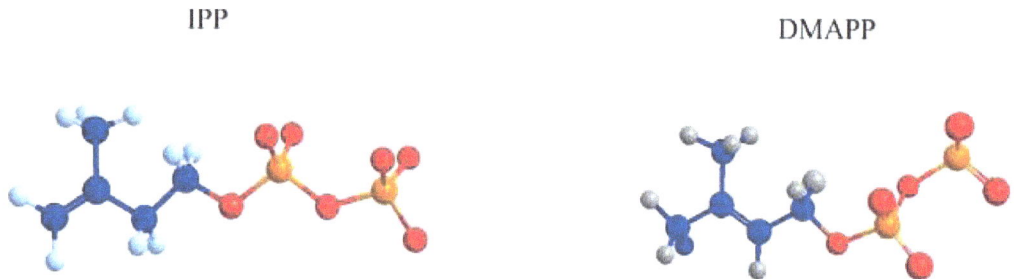

IPP                                                                           DMAPP

**Fig. (1).** Isopentenyl diphosphate (IPP) and dimethylallyl diphosphate (DMAPP) molecules structure.

In plants, IPP and DMAPP are produced by two independent pathways: the mevalonate-dependent pathway, also known as mevalonic acid pathway (MVA), and the methylerythritol phosphate pathway (MEP). Both pathways are regulated at the transcript and protein level and by feedback. The enzyme IPP isomerase is the one responsible to convert IPP to DMAPP, the reaction occurring in both directions [2]. Subsequently, IPP and DMAPP fusion generate short, medium and long-chains of prenyl diphosphates, which then can be modified for many different enzymes downstream in the terpenoid biosynthetic pathways [3]. From the MVA pathway in the cytosol, many compounds are generated, such as brassinosteroid, cytokinin, and protein prenylation. From the MEP pathway in the plastids, we have the generation of carotenoids (and subsequently strigolactones and abscisic acid), gibberellins, cytokinin, ubiquinone, and chlorophyll.

**Mevalonic Acid Pathway**

The MVA pathway is primarily cytosolic and is present in most organisms, including animals, plants, archaebacteria and gram-positive bacteria, and yeasts [4]. It consists of six steps initiated with a condensation reaction of two molecules of acetyl-CoA to acetoacetyl-CoA. This condensation is catalyzed by acetoacetyl-CoA thiolase (AACT) (Fig. **2**). The second step is catalyzed by hydroxymethy-glutaryl-CoA synthase (HMGS), where acetoacetyl-CoA is condensed with another acetyl-CoA molecule to form the C6-compound S-3-hydroxy-3-methylglutaryl-CoA (S-HMG-CoA). In the third step, hydroxymethyglutaryl-CoA reductase (HMGR) catalyzes the conversation of S-HMG-CoA to mevalonate using two NADPH. Mevalonate is phosphorylated to mevalonate-5-phosphate in the 5-OH position in a reaction catalyzed by mevalonate kinase (MK). Mevalonate-5-phosphate produces mevalonate-diphosphate in a reaction catalyzed by phosphomevalonate kinase (PMK). In the last step of the mevalonic acid pathway, mevalonate diphosphate decarboxylase (MPDC) catalyzes the decarboxylative elimination reaction of mevalonate-diphosphate to IPP. The three last steps use one ATP in each reaction.

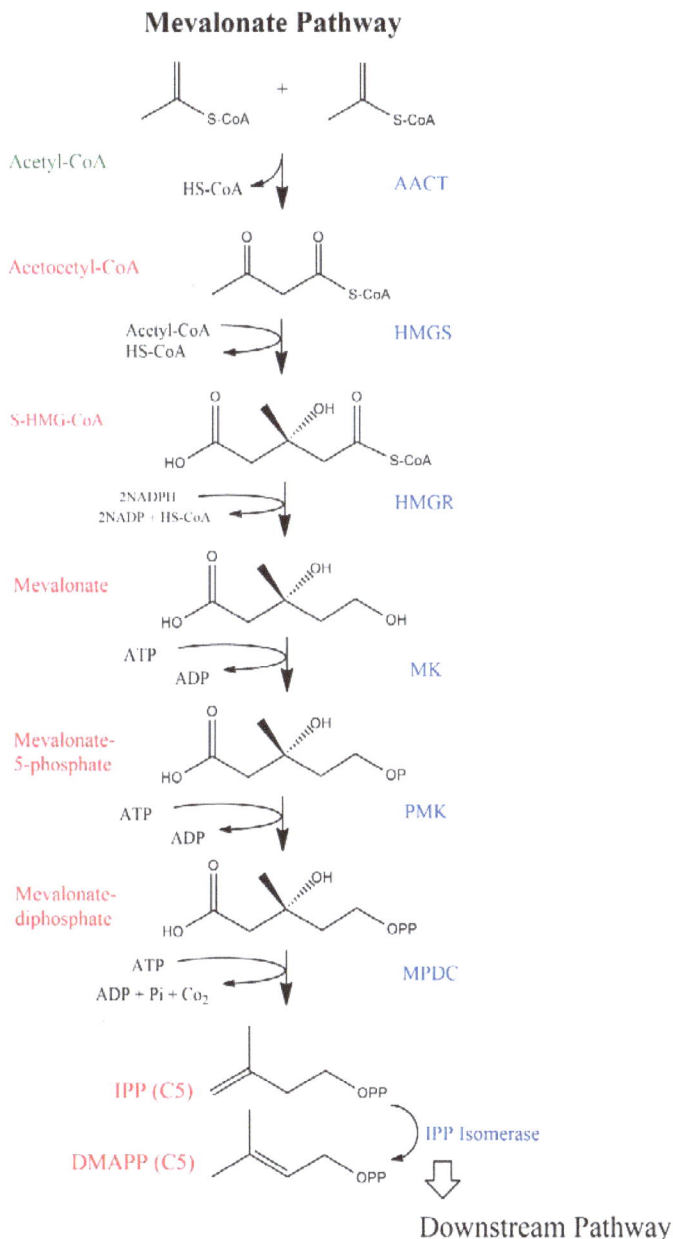

**Fig. (2).** Enzymatic steps of MVA pathway in terpenoid precursor biosynthesis.

The first enzyme in the MVA pathway is AACT, a class II thiolase encoded by a small gene family in plants, with AACT1 localized in cytoplasm and peroxisome, and AACT2 localized only in cytoplasm (Table **1**) [5]. In *Arabidopsis thaliana*, *AACT1* gene has five alternatively spliced isoforms, and the *AACT2* gene has two alternatively spliced isoforms. The second enzyme in the MVA pathway, HMGS, is encoded for gene paralogs in most plant species [6]. In *A. thaliana*, HMGS is encoded by only one gene, and *HMGS* mRNA produces two alternatively spliced variants. HMGS has its subcellular localization at the cytoplasm. In the majority of the plants, several paralog genes encoded the HMGR enzyme [7]. In *A. thaliana*, HMGR is encoded by two genes. HMGR binds to the endoplasmic reticulum with their catalytic site facing the cytosol. The MK enzyme is a cytosolic protein with three alternatively spliced variants in *A. thaliana* [8]. PMK enzyme resides in peroxisomes, encoded by one gene with two alternatively spliced variants in *A. thaliana* [9]. The last enzyme of the MVA pathway, MPDC, is encoded by a single gene in most plants. In *A. thaliana*, two paralogous genes encode the protein [10].

Table 1. Genes encoding enzymes of MVA pathway in terpenoid precursor biosynthesis and its subcellular localization in *Arabidopsis thaliana*.

| Enzyme Name | Abbreviation | Gene Name | Subcellular Localization |
|---|---|---|---|
| Acetoacetyl-CoA thiolase | AACT | *AACT1.1* | Cytosol |
| | | *AACT1.2* | Peroxisome |
| | | *AACT1.3* | Cytosol |
| | | *AACT1.4* | Peroxisome |
| | | *AACT1.5* | Cytosol |
| | | *AACT2.1* | Cytosol |
| | | *AACT2.2* | Cytosol |
| Hydroxymethyglutaryl-CoA synthase | HMGS | *HMGS1.1* | Cytosol |
| | | *HMGS1.2* | Cytosol |
| Hydroxymethyglutaryl-CoA reductase | HMGR | *HMGR1.1* | Endoplasmic reticulum |
| | | *HMGR1.2* | Endoplasmic reticulum |
| | | *HMGR2* | Endoplasmic reticulum |
| Mevalonate kinase | MK | *MK1* | Cytosol |
| | | *MK2* | Cytosol |
| | | *MK3* | Cytosol |
| Phosphomevalonate kinase | PMK | *PMK1.1* | Peroxisome |
| | | *PMK1.2* | Peroxisome |

*(Table 1) cont.....*

| Enzyme Name | Abbreviation | Gene Name | Subcellular Localization |
|---|---|---|---|
| Diphosphate decarboxylase | MPDC | *MPDC1* | Unknown |
| | | *MPDC2* | Unknown |

The regulation of MVA pathway in plants is more complex than in other organisms because plants have two terpenoid biosynthetic pathways in one cell, MVA and MEP. All genes in the MVA pathway are expressed when the demand for IPP/DMAPP is high. Light negatively regulates MVA pathway genes expression [11]. Several other signals regulate the transcriptions of MVA pathway genes, such as drought stress, low and high temperature, herbivory, and wounding [12].

HMGR appears to be the most regulated enzyme in the MVA pathway, with several factors affecting its expression. The phosphorylation of a conserved Ser-5777 in HMGR by a Sucrose nonfermenting1-related kinase 1 (SNRK1) negatively regulates the enzyme by inactivation of it [13]. Lovastatin insensitive 1 (LOI1), a mitochondrial protein participating in the respiration process, also negatively regulates HMGR enzyme [14].

Recently, a study showed that MVA pathway genes increased their expression in two folds when plants of *Tripterygium wilfordii* were treated with methyl jasmonate (MJ) [15]. The accumulation of celastrol, a triterpenoid quinone, is also significantly increased in plants treated with MJ.

**Methylerythritol Phosphate Pathway**

The MEP pathway (also known as non-mevalonate pathway and mevalonate-independent pathway) is used by green algae, gram-negative bacteria, cyanobacteria, and plants [4]. It was discovered in the 1990s and occurred in the plastids. It consists of seven enzymatic steps starting with the condensation of hydroxyethyl thiamine diphosphate, derived from pyruvate, and glyceraldehyde 3-phosphate (GAP) in 1-deoxy-*D*-xylulose-5-phosphate (DXP) (Fig. **3**) [16]. The condensation is catalyzed by 1-deoxy-*D*-xylulose-5-phosphate synthase (DXS) in an irreversible reaction that releases $CO_2$. The reduction of DXP to MEP is catalyzed by 1-deoxy-*D*-xylulose-5-phosphate reductoisomerase (DXR). Further, MEP is converted to 4-cystidine 5-diphospho-2-C-methyl-*D*-erythritol (CDP-ME) by 2-C-methyl-*D*-erythritol 4-phosphate cytidylyltransferase (MCT). CDP-ME is then phosphorylated by 4-cystidine 5-diphospho-2-C-methyl-*D*-erythritol kinase (CMK) and converted to 2-phospho-4-cystidine-5-diphospho-2-C-methyl-*D*-erythritol (CDP-ME2P). CDP-ME2P is converted to 2-C-methyl-*D*-erythritol 2,4-cyclodiphosphate (MEcPP), in a reaction catalyzed by 2-C-methyl-*D*-erythritol 2,4-cyclodiphosphate synthase (MDS). MEcPP is subsequently reduced by 4-

hydroxy-3-methylbut-2-enyldiphosphate synthase (HDS) to 4-hydroxy-3-methylbut-2-enyldiphosphate (HMBPP). In the last step of the MEP pathway, HMBPP is converted into a mixture of IPP and DMAPP in a reaction catalyzed by 4-hydroxy-3-methylbut-2-enyldiphosphate reductase (HDR) [17].

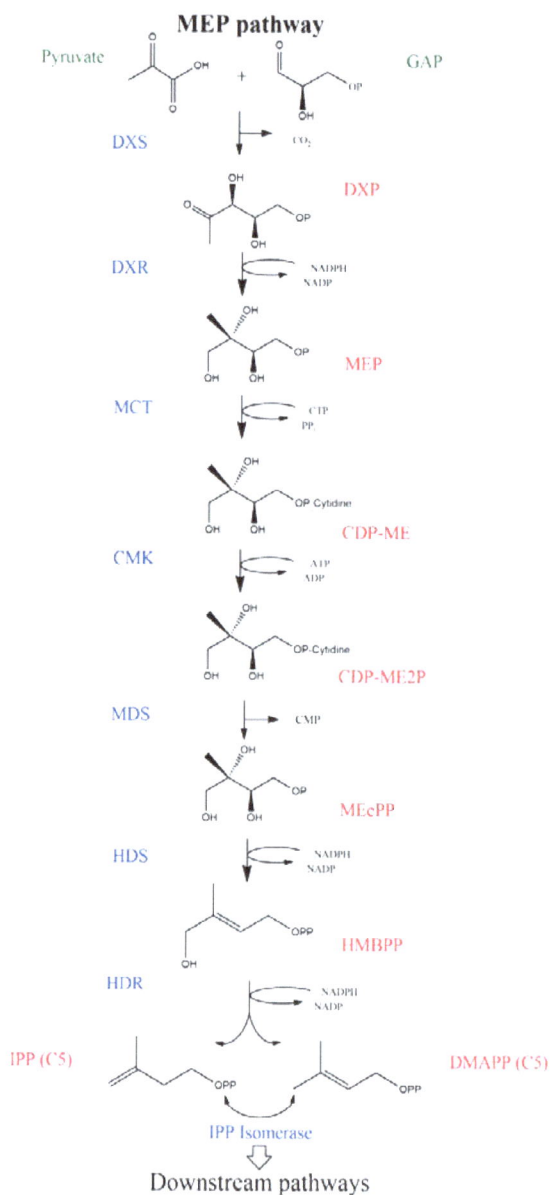

**Fig. (3).** Enzymatic steps of MEP pathway in terpenoid precursor biosynthesis.

All enzymes from MEP pathway reside in the stroma of the plastid (Table **2**) [18]. *A. thaliana* has one *DXS* gene encoding a DXS enzyme, but in most plants, DXS is encoded by multiple gene paralogs [19]. The other enzymes of the MEP pathway are encoded by only one gene each in *A. thaliana* and other species. In *A. thaliana*, *DXR* and *MDS* genes have two alternatively spliced isoforms, and *HDS* gene has three alternatively spliced isoforms.

**Table 2. Genes encoding enzymes of MEP pathway in terpenoid precursor biosynthesis and its subcellular localization in *Arabidopsis thaliana*.**

| Enzyme Name | Abbreviation | Gene Name | Subcellular Localization |
|---|---|---|---|
| 1-deoxy-*D*-xylulose-5-phosphate synthase | DXS | *DXS* | Plastid |
| 1-deoxy-*D*-xylulose-5-phosphate reductoisomerase | DXR | *DXR1.1* | Plastid |
| | | *DXR1.2* | Plastid |
| 2-C-methyl-*D*-erythritol 4-phosphate cytidylyltransferase | MCT | *MCT* | Plastid |
| 5-diphospho-2-C-methyl-*D*-erythritol kinase | CMK | *CMK* | Plastid |
| 2-C-methyl-*D*-erythritol 2,4-cyclodiphosphate synthase | MDS | *MDS1.1* | Plastid |
| | | *MDS1.2* | Plastid |
| 4-hydroxy-3-methylbut-2-enyldiphosphate synthase | HDS | *HDS1.1* | Plastid |
| | | *HDS1.2* | Plastid |
| | | *HDS1.3* | Plastid |
| 4-hydroxy-3-methylbut-2-enyldiphosphate reductase | HDR | *HDR* | Plastid |

MEP pathway genes are regulated by light, and circadian rhythm, *DXS* and *HDR* genes are highly expressed during the day, decreasing their expression during the night [20]. The transcript factors from Phytochrome Interacting Factors (PIFs) family are responsible to regulate MEP pathway. PIFs act negatively regulating *DXS* gene expression in dark [21].

DXS is negatively regulated by mitochondrial respiration at a post-translational level *via* LOI1, similar to HMGR regulation in the MVA pathway [22]. HDR and HDS are also regulated at a post-translational level, being both enzymes' substrates to thioredoxin, which reduces specific disulfide groups to increase their activity [23].

Recently, a study using *Picea glauca* showed that moderate drought treatment reduced photosynthetic rate by 70%, but metabolic flux thought the MEP pathway was reduced only by 37% [24]. The experiments also showed that DXS activity was not decreased even in severe drought, indicating the resilience of the MEP pathway under drought.

## Isomerization of the $C_5$ Building Blocks

In the MVA pathway, IPP needs to be converted to DMAPP by the activity of an isopentenyl diphosphate isomerase (IPP isomerase) in a reversible Mg-dependent reaction (Fig. **4**). In the MEP pathway, both IPP and DMAPP are produced in the reaction catalyzed by HDR (in an 85:15 ratio, IPP/DMAPP, respectively), but IPP isomerase is necessary to produce an optimal ratio of the two molecules [25]. The trafficking of the $C_5$ building blocks between cytosol and plastids has been shown in many studies, although the mechanisms of how it occurs were not elucidated.

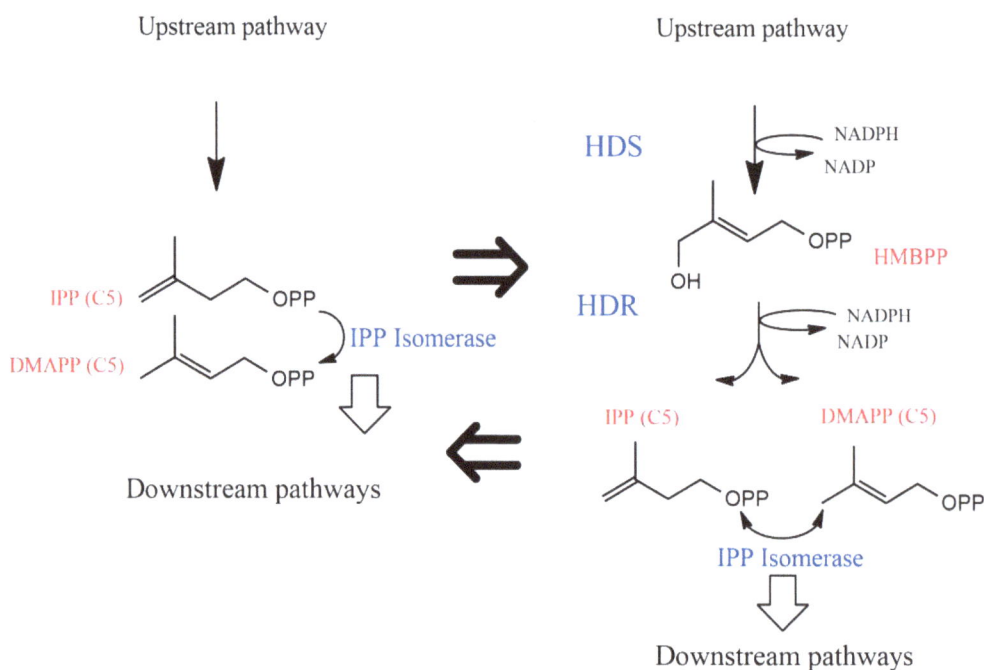

**Fig. (4).** Trafficking of terpenoids precursors, IPP and DMAPP, between the cytosol and plastids.

IPP isomerase is encoded by two paralogs genes in *A. thaliana,* and both of these genes have two alternatively spliced isoforms (Table **3**). In plants, the isoforms are localized in peroxisomes, plastids, and mitochondria [26]. IPP isomerase activity is a limiting step in terpenoid biosynthesis, and the isomerase process regulates the terpenoid flux [27]. In *A. thaliana*, the double mutant *ipp1ipp2* has 50% less ubiquinone than wild-type plants, and a decrease in the incorporation of sterols, showing that IPP is essential for the supply of proper levels of IPP and DMAPP in different subcellular compartments [28].

**Table 3. Genes encoding IPP isomerase enzyme in terpenoid precursor biosynthesis and its subcellular localization in *Arabidopsis thaliana*.**

| Enzyme Name | Abbreviation | Gene Name | Subcellular Localization |
|---|---|---|---|
| Isopentenyl diphosphate isomerase | IPP isomerase | *IPP1.1* | Peroxisome |
| | | *IPP1.2* | Plastid, Mitochondria |
| | | *IPP2.1* | Peroxisome |
| | | *IPP2.2* | Plastid |

The second main step in terpenoid biosynthesis is IPP and DMAPP fusion, which is catalyzed by isoprenyl diphosphate synthases, also called prenyltransferases, to form prenyl diphosphates. Prenyl diphosphates are the precursors of all terpenoids. The first reaction catalyzed by prenyltransferases is the condensation of IPP with DMAPP to produce a $C_{10}$ allylic diphosphate molecule, with the elimination of pyrophosphate. Additional rounds of condensation reactions produce $C_{15}$-Cn chains of prenyl diphosphate [29]. Some main precursors of the biosynthesis of terpenoids are $C_{10}$-geranyl diphosphate, also called geranyl pyrophosphate (GPP), $C_{15}$-farnesyl diphosphate, also known as farnesyl pyrophosphate (FPP), and $C_{20}$-geranylgeranyl diphosphate, also called geranylgeranyl pyrophosphate (GGPP) (Fig. **5**). DMAPP is a precursor for hemiterpenes and isoprenes. GPP is a precursor to monoterpenes. FPP is a precursor to sesquiterpenes and triterpenes. GGPP is a precursor to diterpenes and tetraterpenes.

**Geranyl Pyrophosphate**

GPP is the precursor of $C_{10}$-monoterpenoids, and it is synthesized from IPP and DMAPP by the catalytic activity of geranyl pyrophosphate synthase (GPPS) enzymes, a type of prenyltransferases. GPPS are found in plants as homodimers and heterodimers. Although homodimers enzymes are found in gymnosperms and angiosperms, most studies are focusing on heterodimer GPPS in plants.

Heterodimers GPPS show one large and one small subunit (LSU and SSU, respectively), where LSU is responsible for catalytic activities and SSU is responsible for regulation [30]. The $C_{10}$-monoterpenoids originate different indol alkaloids in plants, as such ibogaine, a psychoactive found in the *Apocynaceae* family, voacangine, an alkaloid found in the roots of *Voacanga africana* tree and other species, and vincamine, an alkaloid found in leaves of *Vinca minor* plants [31].

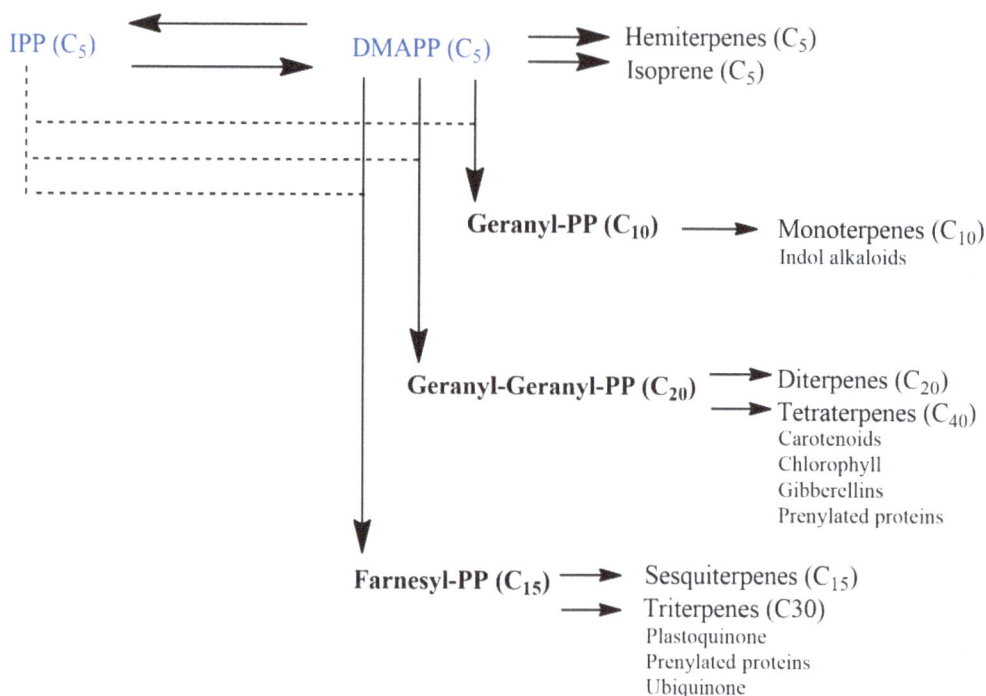

**Fig. (5).** Overview of terpenoids biosynthesis from IPP and DMAPP.

## Farnesyl Pyrophosphate

FPP is the precursor for primary metabolites, specialized metabolites, and protein prenylation (Fig. **6**). FPP is synthesized by the catalytic activity of farnesyl pyrophosphate synthases (FPPS), which are homodimer enzymes localized in cytosol and mitochondria [32]. They are encoded by a small gene family, and it has many different sizes depending on the gene splicing.

Squalene is one of the products in the FPP precursor pathway, and it is an intermediate compound in the synthesis of more than 200 triterpenes [33]. Two molecules of FPP are reduced to form squalene in a reaction catalyzed by the squalene synthase enzyme. Further, phytosterol is generated from the cyclization of squalene, which is catalyzed by cycloartenol synthase. Phytosterol is a key component in brassinosteroid biosynthesis and a substrate for many steps in the pathway [34].

FPP is also a precursor to protein prenylation, a post-translational modification where isoprenoid side chains are added to the carboxyl-terminal of the protein. In plants, several proteins are prenylated. Farnesyl transferase (FTase) is the enzyme that catalyzes the linkage of a farnesyl group to a cysteine residue in the C-terminal protein [35].

A loss of function in the β subunit of FTase enhances the sensitivity and the response of the plant to abscisic acid, leading to stomatal closure and a better drought stress response. Although none of the abscisic acid signaling transduction pathway enzymes are prenylated, a heat shock protein (HSP40) is targeted to FTase and could be involved in abscisic acid responses [35].

**Cytosol**

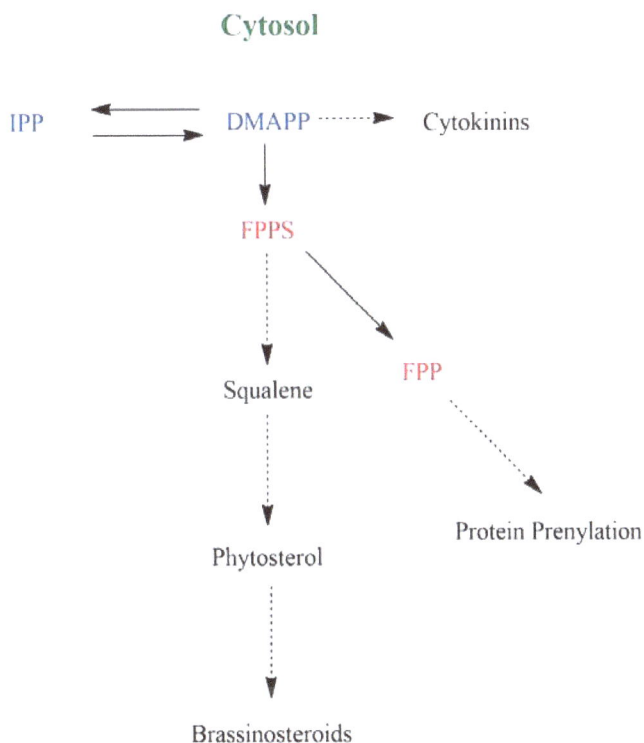

**Fig. (6).** FPP precursor and its products in cytosol.

In the brassinosteroid biosynthesis pathway, a prenylated cytochrome P450 protein catalysis the conversion of castasterone to brassinolide. Mutants lacking prenylation do not have a proper function, showing the importance of prenylation to maintain the normal metabolism in plants [35].

## Geranylgeranyl Pyrophosphate

GGPP is a major precursor to terpenoids biosynthesis in primary and specialized metabolism, with many of the compound's biosynthesis occurring in the plastid (Fig. 7). GGPP is synthesized by the catalytic activity of geranylgeranyl pyrophosphate synthases (GGPPS). In *A. thaliana*, the GGPPS family has 12 paralogs genes, most of them being homodimers.

## Plastid

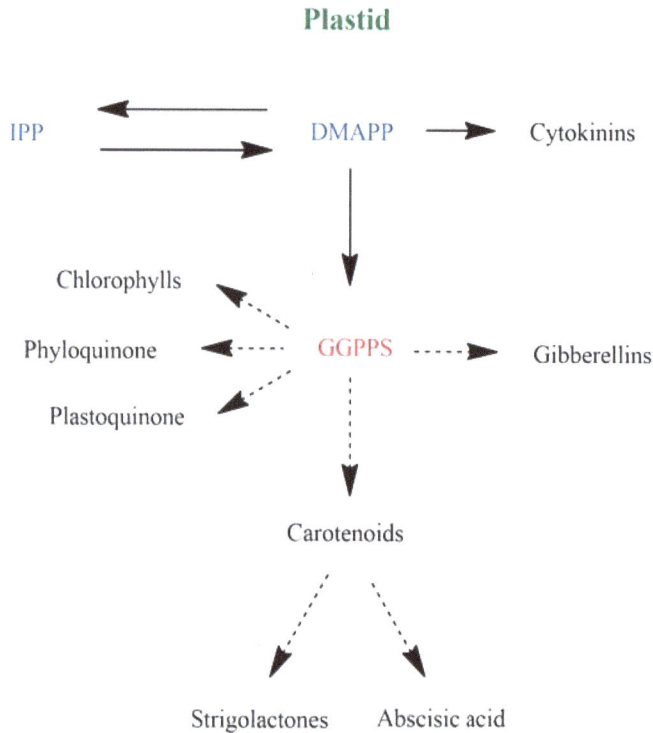

**Fig. (7).**  GGPP precursor and its products in plastid.

In the gibberellin (GA) biosynthesis pathway, GGPP is converted to ent-copalyl diphosphate by ent-copalyl diphosphate synthase enzyme. Further, many steps catalyzed by ent-kaurene oxidase and ent-kaurene acid oxidase enzymes generate GA12. GA12 is then processed to the bioactive GA4 form, by catalytic activity of two enzymes: GA 20-oxidase and GA 3-oxidase [36].

In the carotenoid biosynthesis pathway, the first step is the condensation of two molecules of GGPP to form phytoene, by the catalytic activity of phytoene synthase. Four more reactions generate lycopene, which is then processed by β and ε-cyclases to form α and β-carotene [37].

To form abscisic acid, β-carotene produces zeaxanthin by catalysis of carotene hydroxylases enzymes (BCH1 and BCH2). Zeaxanthin is then converted to violaxanthin by zeaxanthin epoxidase [38]. Three other steps occur in plastids and form xanthoxin, which is then transported to cytosol. Further, two other enzymes, AbscisisAcid2 and AbscisidAcid3 catalysis the last two steps in the production of abscisic acid.

In the strigolactones biosynthesis, β-carotene produces the precursor of strigolactones in the plastid, carlactone (CL) in three steps by catalysis of one isomerase and two dioxygenases' enzymes [39]. CL is then exposed to the cytosol and suffers an oxidation reaction generating many different strigolactones molecules downstream in the pathway.

Chlorophylls, phylloquinone, and plastoquinone are other main compounds derived from GGPP precursor in the terpenoids biosynthesis, and many steps are involved in each of the pathways.

**Cytokinin's Biosynthesis**

Cytokinin is a central hormone in plant metabolism, and it is produced in the cytosol, mitochondria, and plastids (Figs. **6** and **7**). DMAPP is the precursor to cytokinin biosynthesis, and isopentenyl transferases catalyze the first step of the pathway, transferring the isoprenoid moiety to an adenine [40]. Further, the side chain is hydroxylated by cytochrome P450 monooxygenase, and, in the next step, cytokinin nucleotides are hydrolyzed to free bases.

**CONCLUDING REMARKS**

Recent efforts have been made to elucidate the biosynthesis of terpenoids in both MVA and MEP pathways and their regulation. Additional works should be performed to clarify the subcellular localization of several enzymes in both pathways. Further, more efforts to understand the regulation of the biosynthetic terpenoid pathways enzymes will be important to elucidate the regulatory networks coordinating the many several routes in space and time.

**CONSENT FOR PUBLICATION**

Not applicable.

**CONFLICT OF INTEREST**

The author declares no conflict of interest, financial or otherwise.

**ACKNOWLEDGEMENTS**

Declared none.

**REFERENCES**

[1]    Kubeczka, K.H. Handbook of Essential Oil: Science, Technology, and Applications. **2010**, , 3-38.

[2]    Sapir-Mir, M.; Mett, A.; Belausov, E.; Tal-Meshulam, S.; Frydman, A.; Gidoni, D.; Eyal, Y. Peroxisomal localization of Arabidopsis isopentenyl diphosphate isomerases suggests that part of the plant isoprenoid mevalonic acid pathway is compartmentalized to peroxisomes. *Plant Physiol.,* **2008**,

*148*(3), 1219-1228.
[http://dx.doi.org/10.1104/pp.108.127951] [PMID: 18988695]

[3]     Tholl, D. Biosynthesis and Biological Functions of Terpenoids in Plants. **2015**, , 63-106.
[http://dx.doi.org/10.1007/10_2014_295]

[4]     Vranová, E.; Coman, D.; Gruissem, W. Network analysis of the MVA and MEP pathways for isoprenoid synthesis. *Annu. Rev. Plant Biol.,* **2013**, *64*(1), 665-700.
[http://dx.doi.org/10.1146/annurev-arplant-050312-120116] [PMID: 23451776]

[5]     Carrie, C.; Murcha, M.W.; Millar, A.H.; Smith, S.M.; Whelan, J. Nine 3-ketoacyl-CoA thiolases (KATs) and acetoacetyl-CoA thiolases (ACATs) encoded by five genes in Arabidopsis thaliana are targeted either to peroxisomes or cytosol but not to mitochondria. *Plant Mol. Biol.,* **2006**, *63*(1), 97-108.
[http://dx.doi.org/10.1007/s11103-006-9075-1] [PMID: 17120136]

[6]     Montamat, F.; Guilloton, M.; Karst, F.; Delrot, S. Isolation and characterization of a cDNA encoding Arabidopsis thaliana 3-hydroxy-3-methylglutaryl-coenzyme A synthase. *Gene,* **1995**, *167*(1-2), 197-201.
[http://dx.doi.org/10.1016/0378-1119(95)00642-7] [PMID: 8566777]

[7]     Lumbreras, V.; Campos, N.; Boronat, A. The use of an alternative promoter in the Arabidopsis thaliana HMG1 gene generates an mRNA that encodes a novel 3-hydroxy-3-methylglutaryl coenzyme A reductase isoform with an extended *N*-terminal region. *Plant J.,* **1995**, *8*(4), 541-549.
[http://dx.doi.org/10.1046/j.1365-313X.1995.8040541.x] [PMID: 7496400]

[8]     Guirimand, G.; Guihur, A.; Phillips, M.A.; Oudin, A.; Glévarec, G.; Melin, C.; Papon, N.; Clastre, M.; St-Pierre, B.; Rodríguez-Concepción, M.; Burlat, V.; Courdavault, V. A single gene encodes isopentenyl diphosphate isomerase isoforms targeted to plastids, mitochondria and peroxisomes in Catharanthus roseus. *Plant Mol. Biol.,* **2012**, *79*(4-5), 443-459.
[http://dx.doi.org/10.1007/s11103-012-9923-0] [PMID: 22638903]

[9]     Simkin, A.J.; Guirimand, G.; Papon, N.; Courdavault, V.; Thabet, I.; Ginis, O.; Bouzid, S.; Giglioli-Guivarc'h, N.; Clastre, M. Peroxisomal localisation of the final steps of the mevalonic acid pathway in planta. *Planta,* **2011**, *234*(5), 903-914.
[http://dx.doi.org/10.1007/s00425-011-1444-6] [PMID: 21655959]

[10]    Cordier, H.; Karst, F.; Bergès, T. Heterologous expression in Saccharomyces cerevisiae of an Arabidopsis thaliana cDNA encoding mevalonate diphosphate decarboxylase. *Plant Mol. Biol.,* **1999**, *39*(5), 953-967.
[http://dx.doi.org/10.1023/A:1006181720100] [PMID: 10344201]

[11]    Rodríguez-Concepción, M. Early steps in isoprenoid biosynthesis: multilevel regulation of the supply of common precursors in plant cells. *Phytochem. Rev.,* **2006**, *5*(1), 1-15.
[http://dx.doi.org/10.1007/s11101-005-3130-4]

[12]    Stermer, B.A.; Bianchini, G.M.; Korth, K.L. Regulation of HMG-CoA reductase activity in plants. *J. Lipid Res.,* **1994**, *35*(7), 1133-1140.
[http://dx.doi.org/10.1016/S0022-2275(20)39958-2] [PMID: 7964176]

[13]    Douglas, P.; Pigaglio, E.; Ferrer, A.; Halford, N.G.; MacKINTOSH, C. Three spinach leaf nitrate reductase-3-hydroxy-3-methylglutaryl-CoA reductase kinases that are regulated by reversible phosphorylation and/or Ca$^{2-}$ ions. *Biochem. J.,* **1997**, *325*(1), 101-109.
[http://dx.doi.org/10.1042/bj3250101] [PMID: 9245257]

[14]    Kobayashi, K.; Suzuki, M.; Tang, J.; Nagata, N.; Ohyama, K.; Seki, H.; Kiuchi, R.; Kaneko, Y.; Nakazawa, M.; Matsui, M.; Matsumoto, S.; Yoshida, S.; Muranaka, T. Lovastatin insensitive 1, a Novel pentatricopeptide repeat protein, is a potential regulatory factor of isoprenoid biosynthesis in Arabidopsis. *Plant Cell Physiol.,* **2007**, *48*(2), 322-331.
[http://dx.doi.org/10.1093/pcp/pcm005] [PMID: 17213228]

[15]    Liu, Y.J.; Zhao, Y.J.; Su, P.; Zhang, M.; Tong, Y.R.; Hu, T.Y.; Huang, L.Q.; Gao, W. The MVA

pathway genes expressions and accumulation of celastrol in *Tripterygium wilfordii* suspension cells in response to methyl jasmonate treatment. *J. Asian Nat. Prod. Res.,* **2016**, *18*(7), 619-628. [http://dx.doi.org/10.1080/10286020.2015.1134504] [PMID: 26785825]

[16]    Frank, A.; Groll, M. The methylerythritol phosphate pathway to isoprenoids. *Chem. Rev.,* **2017**, *117*(8), 5675-5703. [http://dx.doi.org/10.1021/acs.chemrev.6b00537] [PMID: 27995802]

[17]    Vranová, E.; Hirsch-Hoffmann, M.; Gruissem, W. AtIPD: a curated database of Arabidopsis isoprenoid pathway models and genes for isoprenoid network analysis. *Plant Physiol.,* **2011**, *156*(4), 1655-1660. [http://dx.doi.org/10.1104/pp.111.177758] [PMID: 21617028]

[18]    Phillips, M.A.; D'Auria, J.C.; Gershenzon, J.; Pichersky, E. The Arabidopsis thaliana type I Isopentenyl Diphosphate Isomerases are targeted to multiple subcellular compartments and have overlapping functions in isoprenoid biosynthesis. *Plant Cell,* **2008**, *20*(3), 677-696. [http://dx.doi.org/10.1105/tpc.107.053926] [PMID: 18319397]

[19]    Joyard, J.; Ferro, M.; Masselon, C.; Seigneurin-Berny, D.; Salvi, D.; Garin, J.; Rolland, N. Chloroplast proteomics and the compartmentation of plastidial isoprenoid biosynthetic pathways. *Mol. Plant,* **2009**, *2*(6), 1154-1180. [http://dx.doi.org/10.1093/mp/ssp088] [PMID: 19969518]

[20]    Toledo-Ortiz, G.; Huq, E.; Rodríguez-Concepción, M. Direct regulation of phytoene synthase gene expression and carotenoid biosynthesis by phytochrome-interacting factors. *Proc. Natl. Acad. Sci. USA,* **2010**, *107*(25), 11626-11631. [http://dx.doi.org/10.1073/pnas.0914428107] [PMID: 20534526]

[21]    Toledo-Ortiz, G.; Huq, E.; Quail, P.H. The Arabidopsis basic/helix-loop-helix transcription factor family. *Plant Cell,* **2003**, *15*(8), 1749-1770. [http://dx.doi.org/10.1105/tpc.013839] [PMID: 12897250]

[22]    He, Y.; Yan, Z.; Du, Y.; Ma, Y.; Shen, S. Molecular cloning and expression analysis of two key genes, HDS and HDR, in the MEP pathway in Pyropia haitanenses. *Sci. Rep.,* **2017**, *48*, 322-331.

[23]    Schürmann, P.; Jacquot, J.P. Plant thioredoxin systems revisited. *Annu. Rev. Plant Physiol. Plant Mol. Biol.,* **2000**, *51*(1), 371-400. [http://dx.doi.org/10.1146/annurev.arplant.51.1.371] [PMID: 15012197]

[24]    Perreca, E.; Rohwer, J.; González-Cabanelas, D.; Loreto, F.; Schmidt, A.; Gershenzon, J.; Wright, L.P. Effect of drought on the methylerythritol 4-phosphate (MEP) pathway in the isoprene emitting conifer Picea glauca. *Front. Plant Sci.,* **2020**, *11*546295 [http://dx.doi.org/10.3389/fpls.2020.546295] [PMID: 33163010]

[25]    Phillips, M.A.; D'Auria, J.C.; Gershenzon, J.; Pichersky, E. The *Arabidopsis thaliana* type I Isopentenyl Diphosphate Isomerases are targeted to multiple subcellular compartments and have overlapping functions in isoprenoid biosynthesis. *Plant Cell,* **2008**, *20*(3), 677-696. [http://dx.doi.org/10.1105/tpc.107.053926] [PMID: 18319397]

[26]    Sapir-Mir, M.; Mett, A.; Belausov, E.; Tal-Meshulam, S.; Frydman, A.; Gidoni, D.; Eyal, Y. Peroxisomal localization of *Arabidopsis* isopentenyl diphosphate isomerases suggests that part of the plant isoprenoid mevalonic acid pathway is compartmentalized to peroxisomes. *Plant Physiol.,* **2008**, *148*(3), 1219-1228. [http://dx.doi.org/10.1104/pp.108.127951] [PMID: 18988695]

[27]    Berthelot, k.; Estevez, Y.; Deffieux, A.; Peruch, F. Isopentenyl diphosphate isomerase: a checkpoint to isoprenoid biosynthesis. *Biochimie,* **2012**, *94*(8), 1621-1634.

[28]    Okada, K.; Kasahara, H.; Yamaguchi, S.; Kawaide, H.; Kamiya, Y.; Nojiri, H.; Yamane, H. Genetic evidence for the role of isopentenyl diphosphate isomerases in the mevalonate pathway and plant development in Arabidopsis. *Plant Cell Physiol.,* **2008**, *49*(4), 604-616. [http://dx.doi.org/10.1093/pcp/pcn032] [PMID: 18303110]

[29]    Kharel, Y.; Koyama, T. Molecular analysis of cis-prenyl chain elongating enzymes. *Nat. Prod. Rep.,* **2003**, *20*(1), 111-118.
[http://dx.doi.org/10.1039/b108934j] [PMID: 12636086]

[30]    Chang, T.H.; Hsieh, F.L.; Ko, T.P.; Teng, K.H.; Liang, P.H.; Wang, A.H.J. Structure of a heterotetrameric geranyl pyrophosphate synthase from mint (Mentha piperita) reveals intersubunit regulation. *Plant Cell,* **2010**, *22*(2), 454-467.
[http://dx.doi.org/10.1105/tpc.109.071738] [PMID: 20139160]

[31]    Burke, C.C.; Wildung, M.R.; Croteau, R. Geranyl diphosphate synthase: Cloning, expression, and characterization of this prenyltransferase as a heterodimer. *Proc. Natl. Acad. Sci. USA,* **1999**, *96*(23), 13062-13067.
[http://dx.doi.org/10.1073/pnas.96.23.13062] [PMID: 10557273]

[32]    Cunillera, N.; Boronat, A.; Ferrer, A. The *Arabidopsis thaliana FPS1* gene generates a novel mRNA that encodes a mitochondrial farnesyl-diphosphate synthase isoform. *J. Biol. Chem.,* **1997**, *272*(24), 15381-15388.
[http://dx.doi.org/10.1074/jbc.272.24.15381] [PMID: 9182568]

[33]    Lozano-Grande, M.A.; Gorinstein, S.; Espitia-Rangel, E.; Dávila-Ortiz, G.; Martínez-Ayala, A.L. Plant sources, extraction methods, and uses of squalene. *Int. J. Agron.,* **2018**, *2018*, 1-13.
[http://dx.doi.org/10.1155/2018/1829160]

[34]    Bajguz, A.; Chmur, M.; Gruszka, D. Comprhensive overview of brassinosteroid biosynthesis pathways: substrates, products, inhibitors, and connections. Fron. Plant. Sci. **2020**, ID01034.

[35]    Hála, M.; Žárský, V. Protein prenylation in plant stress responses. *Molecules,* **2019**, *24*(21), 3906.
[http://dx.doi.org/10.3390/molecules24213906] [PMID: 31671559]

[36]    Yamaguchi, S. Gibberellin metabolism and its regulation. *Annu. Rev. Plant Biol.,* **2008**, *59*(1), 225-251.
[http://dx.doi.org/10.1146/annurev.arplant.59.032607.092804] [PMID: 18173378]

[37]    Cazzonelli, C.I.; Pogson, B.J. Source to sink: regulation of carotenoid biosynthesis in plants. *Trends Plant Sci.,* **2010**, *15*(5), 266-274.
[http://dx.doi.org/10.1016/j.tplants.2010.02.003] [PMID: 20303820]

[38]    Finkelstein, R. Abscisic Acid synthesis and response. *Arabidopsis Book,* **2013**, *11*e0166
[http://dx.doi.org/10.1199/tab.0166] [PMID: 24273463]

[39]    Lopez-Obando, M.; Ligerot, Y.; Bonhomme, S.; Boyer, F.D.; Rameau, C. Strigolactone biosynthesis and signaling in plant development. *Development,* **2015**, *142*(21), 3615-3619.
[http://dx.doi.org/10.1242/dev.120006] [PMID: 26534982]

[40]    Frébort, I.; Kowalska, M.; Hluska, T.; Frébortová, J.; Galuszka, P. Evolution of cytokinin biosynthesis and degradation. *J. Exp. Bot.,* **2011**, *62*(8), 2431-2452.
[http://dx.doi.org/10.1093/jxb/err004] [PMID: 21321050]

# CHAPTER 2

# Green Extraction Techniques to Obtain Bioactive Concentrates Rich in Terpenoids

**Ana Carolina de Aguiar**[1,*]**, Arthur Luiz Baião Dias**[1] **and Juliane Viganó**[2]

*[1] Laboratory of High Pressure in Food Engineering, Department of Food Engineering, University of Campinas, UNICAMP, 13083-862, Campinas, Brazil*

*[2] Multidisciplinary Laboratory of Food and Health (LabMAS), School of Applied Sciences (FCA), University of Campinas, Rua Pedro Zaccaria 1300, 13484- 350 Limeira, São Paulo, Brazil*

**Abstract:** Terpenoids, also called isoprenoids or terpenes, are a large class of natural products which display a wide range of biological activities. They are major constituents of essential oils produced by aromatic plants and tree resins. Due to their notable biological activities, these compounds have enormous economic importance, being widely used as bioactive ingredients in the food, cosmetic, and pharmaceutical industries. The growing demand from consumers and regulatory agencies to develop green sustainable industrial processes has resulted in the emergence of new technologies for obtaining bioactive compounds from natural sources. Thus, many works have been reported in the literature regarding the development and application of new methods for obtaining terpenoids from natural sources that meet the demands of green processes, with reduced consumption of solvent and energy, less waste generation, and use of non-toxic solvents. This chapter proposes to present the main methods of green extraction to obtain terpenoids-rich extracts, with an emphasis on low-pressure methods, such as microwave-assisted extraction (MAE) and ultrasound-assisted extraction (UAE); and high-pressure methods (here considered as pressures greater than 5 bar), including extraction with supercritical fluids (SFE), subcritical water (SWE) and liquefied petroleum gas extraction (LPG). In addition, the future perspectives and the main challenges regarding the development of alternative methods for the recovery of terpenoids are presented and discussed.

**Keywords:** Bioactive Compounds, Carbon Dioxide, Conventional Extraction, Green, High-Pressure Extraction, Innovative, Isoprenoids, Low-Pressure Extraction, Microwave-Assisted Extraction, Pressure, Pressurized Liquid Extraction, Supercritical Fluid Extraction, Subcritical Water Extraction, Solvent, Sustainable, Terpenoids, Terpenes, Thermolabile, Temperature, Technologies.

---

*Corresponding author Ana Carolina de Aguiar:** Laboratory of High Pressure in Food Engineering, Department of Food Engineering, University of Campinas, Brazil: E-mail: aguiarea@gmail.com

**Mozaniel Santana de Oliveira & Antônio Pedro da Silva Souza Filho (Eds.)**

## INTRODUCTION

Terpenoids, also called isoprenoids or terpenes, are a large class of natural compounds since over 60,000 structures have already been identified from natural sources [1, 2]. Terpenoids present an extensive range of biological activities, which is often assumed, for certain terpenoids, due to their lipophilicity and ability to partition into cellular membranes, interact with membrane-bounded proteins and disrupt membrane integrity [3]. Terpenoids are major constituents of essential oils produced by aromatic plants and tree resins. Monoterpenes and sesquiterpenes and their oxygenated derivatives are the most abundant groups of chemical substances in essential oils. Although their biological activities have been scientifically proven, many plants and terpenoid-rich extracts were already widely used in traditional medicine for their anti-inflammatory and pain-relieving properties [4 - 6]. Due to their notable sensory aspects and biological activities, these compounds have enormous economic importance, being widely used as bioactive ingredients in the food, cosmetic, and pharmaceutical industries.

Bioactive compounds, including essential oils, carotenoids, fatty acids, phenolic acids, and flavonoids, were conventionally extracted by steam distillation, solvent extraction, Soxhlet extraction, pressing method, and hydro-distillation, mainly due to their equipment and operation simplicity. However, many drawbacks of conventional extraction methods have been recently recognized. For instance, for Soxhlet extraction, the main disadvantages comprise the long extraction time, the use of toxic solvents, usually in large amounts, the necessity of further evaporation or concentration operation to remove the excess of solvent, besides the possibility of thermal degradation of the targeted compounds due to the harsh extraction conditions (high temperature, long time, presence of oxygen and light, *etc.*) [7]. Most of these limitations also apply to other conventional extraction methods, especially a large amount of solvent required.

Regarding the extraction of terpenoids, thermal degradation is notably a major issue. Many terpenoids, such as α-pinene, limonene, camphor, citronellol, carvacrol, camphene, $\Delta^3$-carene, and γ-terpinene are thermolabile at temperatures above 100 °C, under subcritical water conditions [8] and hot air [9]. Large-scale extraction of terpenoids commonly uses organic solvents such as methanol or 2-propanol, ethyl acetate, and light petroleum (1:1:1) at temperatures ranging from 40 °C to 190 °C [10].

The fact that many bioactive compounds are thermolabile, combined with the growing demand from consumers and regulatory agencies to develop green sustainable industrial processes, has resulted in the emergence of new technologies for obtaining bioactive compounds from natural sources [11]. Thus, innovative strategies to extract and isolate bioactive compounds from plant-based materials are gaining attention in the research and development domains.

According to Chemat, Vian and Cravotto [12], green extraction of natural products is based on the discovery and design of extraction processes that will reduce energy consumption, allow the use of alternative solvents and renewable natural products, and ensure a safe and high-quality extract/product. Therefore, microwave-assisted extraction (MAE), ultrasound-assisted extraction (UAE), supercritical fluid extraction (SFE), and pressurized liquid extraction (PLE) [13 - 16], which are readily accessible and environmentally sustainable, can be considered green technologies. Many of these green sustainable extraction methods have already been used to recover different terpenoids from plant matrices. The results obtained so far have demonstrated excellent performance of these processes compared to conventional extraction methods.

In this chapter, we will present the main methods of green extraction techniques to obtain terpenoids-rich extracts, with an emphasis on low-pressure methods, such as MAE and UAE; and high-pressure methods, including SFE, subcritical water (SWE) and liquefied petroleum gas extraction (LPG). In addition, the future perspectives and the main challenges regarding the development of alternative methods for the recovery of terpenoids are presented and discussed.

## LOW-PRESSURE EXTRACTION METHODS

### Microwave-Assisted Extraction (MAE)

Microwaves are radiation of the electromagnetic spectrum ranging in frequency from 300 MHz (radio radiation) to 300 GHz. When applied in chemical processes, the frequencies of 2.45 GHz and 915 MHz are used for laboratory-scale and industrial-scale equipment, respectively [17].

The microwave photon energy corresponding to the frequency used in the microwave heating system ($3.78 \times 10^{-6}$ to $1.01 \times 10^{-5}$ eV) cannot affect the molecular structure since it is lower than the typical ionization energies of chemical bonds (3–8 eV) and hydrogen bonds (0.04-0.44 eV) [18]. As microwave radiation is nonionized, the interaction with materials that absorb the microwave energy occurs by heating. Thus, the efficiency of microwave heating (at a given frequency and temperature) is a function of the capacity of the material to absorb electromagnetic energy and dissipate heat.

Briefly, MAE uses microwave energy to heat solvents containing samples, thereby partitioning analytes from a sample matrix into the solvent. The main advantage of MAE is its capacity to rapidly heat the sample solvent mixture, resulting in its broad applicability for the accelerated extraction of analytes, including thermolabile compounds [19].

Fig. (**1**) shows a schematic diagram of a generic ultrasound-assisted extraction device, composed of a sample vessel, which contains the extraction solvent (although in many applications, the extraction process is conducted without solvents), which is inserted inside an ultrasound oven, and the sample vessel is coupled to a condenser.

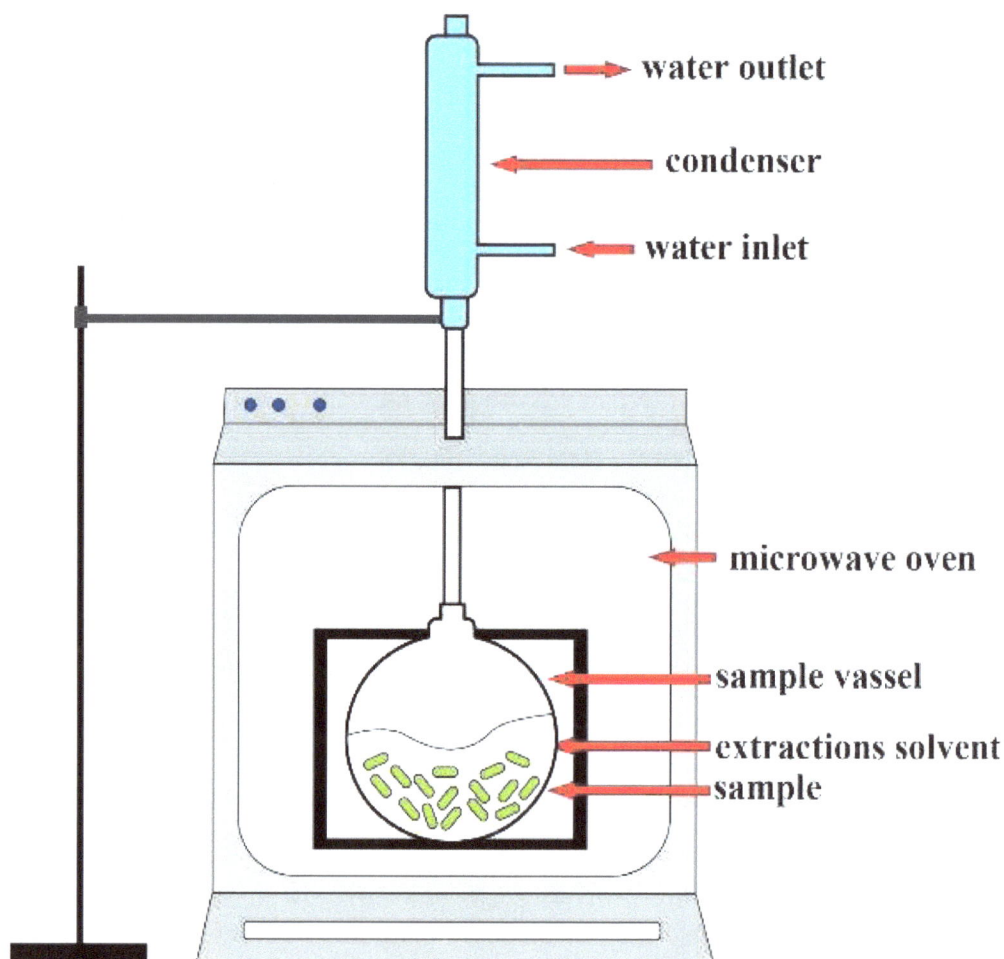

**Fig. (1).** Schematic diagram of the microwave-assisted extraction equipment.

MAE methods can be classified into solvent-free extraction methods (usually for the recovery of volatile compounds) and solvent extraction methods (usually for the recovery of non-volatile compounds). Specifically, for the extraction of oils from natural substrates, the MAE has gained prominence due to its numerous advantages over the conventional heating methods, such as reducing the extraction time, the volume of solvent, and the amount of sample required, plus reaching higher extraction yields [20 - 22].

Specifically, for the extraction of oils from natural substrates, the MAE has gained prominence due to its numerous advantages over the conventional heating methods, such as reducing the extraction time, the volume of solvent and the amount of sample required, plus reaching higher extraction yields [20 - 22].

In these cases, the main factor related to the MAE's better performance is the interaction between the radiation and the matrix water: the microwaves release the essential oil, and the water *in situ* is transferred from the interior to the exterior portion of the vegetable tissue [23]. In addition to the intrinsic characteristics of the plant material, other process parameters affect the efficiency of the MAE, including (i) power level, (ii) duration of the microwave irradiation, (iii) type and volume of the solvent used for extraction (if used), (iv) solvent to feed ratio, and (v) extraction system capacity. Accordingly, the rational optimization of the MAE process parameters is an efficient strategy to achieve better process performance [24]. MAE was successfully applied to recover terpenoids from different plant materials, and in many cases, the response surface methodology was used as a powerful tool to optimize the parameters of the extraction process, as described below.

A green protocol for MAE of volatile oil terpenes from *Pterodon emarginatus* was developed by Vila Verde *et al.* [24]. The process was optimized using experimental design and response surface methodology by evaluating the effect of time, moisture and microwave power on the extraction yield. The MAE had superior performance compared to the conventional method (steam distillation), resulting in a shorter extraction time, and higher energy efficiency, in addition to lower solvent consumption and waste generation. Regarding the terpenes profile in the extracts, the application of the microwave increased the concentration of caryophyllene, γ-muurulene, and γ-elemeno, which have important biological activities (anti-inflammatory and antimicrobial). The results obtained confirmed the efficiency and bio-sustainability of the MAE process.

Response surface methodology was also used to optimize the extraction of essential oil using solvent-free microwave extraction (SFME) from the aerial parts *Limnophila aromatica* [20]. The optimal extraction conditions for the essential oil recovery were 700 W and 25 min. The irradiation time was the most important variable influencing the extraction, followed by the microwave power. Monoterpene hydrocarbons were the major compounds present in oils, with considerable amounts of limonene, perillaldehyde, (*E*)-4-caranone, (*Z*)-4-caranone, and α-pinene.

Microwave-assisted hydro-distillation (MAHD) was applied to extract essential oil from *O. vulgare* L. ssp. *hirtum,* and the results were compared to the conventional hydro-distillation process [21]. MAHD resulted in shorter extraction times, higher extraction yields and concentration of oxygenated compounds in the extracts, besides a lower electrical consumption, when compared to the conventional process. Regarding the essential oil composition, carvacrol was the major compound.

Both MAHD and solvent-free *microwave* extraction (SFME) processes were used to extract terpenoids-rich essential oil from Tunisian *Rosmarinus officinalis* L [22]. Results showed that the SFME was efficient to improve the quality of the essential oil since it provided the best results, that is, lower extraction time, less energy consumption, and better chemical composition of the extracts, mainly consisting of α-pinene (42.57-35.62%), eucalyptol (64.71%), camphor (20.4%), myrtenal (7.39%) and isoborneol (9.8%).

MAHD process was used to recover a volatile hydrodistillate (essential oil) rich in monoterpenes, sesquiterpenes, and a small quantity of phytocannabinoids from *Cannabis sativa* L. inflorescences by Gunjević *et al.* [25]. The optimized extraction procedure had superior performance on obtaining hydrodistillate, reaching 0.35% w/w (in relation to dry inflorescence mass), while the conventional hydrodistillation method yielded 0.12% w/w. Additionally, the hydrodistilled oil was extremely rich in the characteristic *Cannabis* terpenes: α-pinene, β-myrcene, β-ocimene, E-caryophyllene, α-humulene, caryophyllene oxide, and β-selinene. MAHD extraction kinetics showed a progressive enrichment in monoterpenes and a decrease in sesquiterpene. Thus, the MAHD process is a promising alternative for the recovery of active cannabis compounds.

## Ultrasound-Assisted Extraction (UAE)

Ultrasound is a crucial technology in achieving the objective of sustainable extraction. Ultrasound is well known to significantly affect the rate of various processes in the chemical and food industry. Several food components and nutraceuticals, such as aromas, pigments, antioxidants, and other organic compounds have been efficiently extracted from a variety of matrices [26, 27].

The ability of ultrasound to enhance the extraction efficiency of bioactive compounds is mainly due to the cavitation phenomenon produced in the solvent by the passage of an ultrasonic wave, which intensifies mass transfer and close interaction between the solvent and the plant matrix [26, 28]. The collapse of cavitation bubbles near tissue surfaces creates microjets, leading to tissue disruption and extensive solvent penetration into the tissue structure [29].

High-power ultrasound generally can be employed using two types of devices: probe-based ultrasound equipment and an ultrasonic bath. Both systems use a transducer as a source of ultrasound power, and the piezoelectric transducer is the most commonly used in ultrasonic equipment [30].

The ultrasonic bath is the most common type of ultrasonic device consisting of a tank with ultrasonic transducers typically operating at a 40 kHz frequency. The main advantages of these devices are their low cost, availability, and capacity for processing large numbers of samples simultaneously. However, the main drawbacks are the low efficiency in delivering power directly to the sample to be extracted and the low reproducibility compared with probe systems. Additionally, the delivered intensity can be attenuated by the water in the bath and the sample container wall.

High-power ultrasonic probes are more suitable for extraction applications. The probe system is more powerful due to an ultrasonic intensity delivered through a smaller surface (tip of the probe) compared to the ultrasonic bath. These systems generally operate at around 20 kHz. The transducer is bonded to the probe, which is immersed in the extraction vessel, resulting in direct delivery of ultrasound in the solution, minimizing ultrasonic energy loss. As the intensity of ultrasound delivered by the probe to the solution induces a temperature increase, a cooling system (usually a double jacket) is required to conduct the extraction [26]. A schematic diagram of probe-type UAE equipment is presented in Fig. (**2**).

Important UAE parameters to be optimized are the amount and polarity of the extraction solvent, sample mass and particle size, extraction temperature and pressure, sonication time, and the ultrasound source conditions, mainly ultrasound frequency and power [19].

Although methanol and ethanol are the usual extraction solvents, the optimal solvent depends on the chemical characteristic of the bioactive compounds and the extract application. For instance, for sustainable green extraction, toxic solvent as methanol is not allowed. The extraction efficiency can be optimized by using solvents containing 10% to 50% water, depending on the type and characteristics of the samples. Additionally, performing UAE at high temperatures can enhance the extraction process efficiency since high temperatures can increase the number of cavitation bubbles in the extraction media [30].

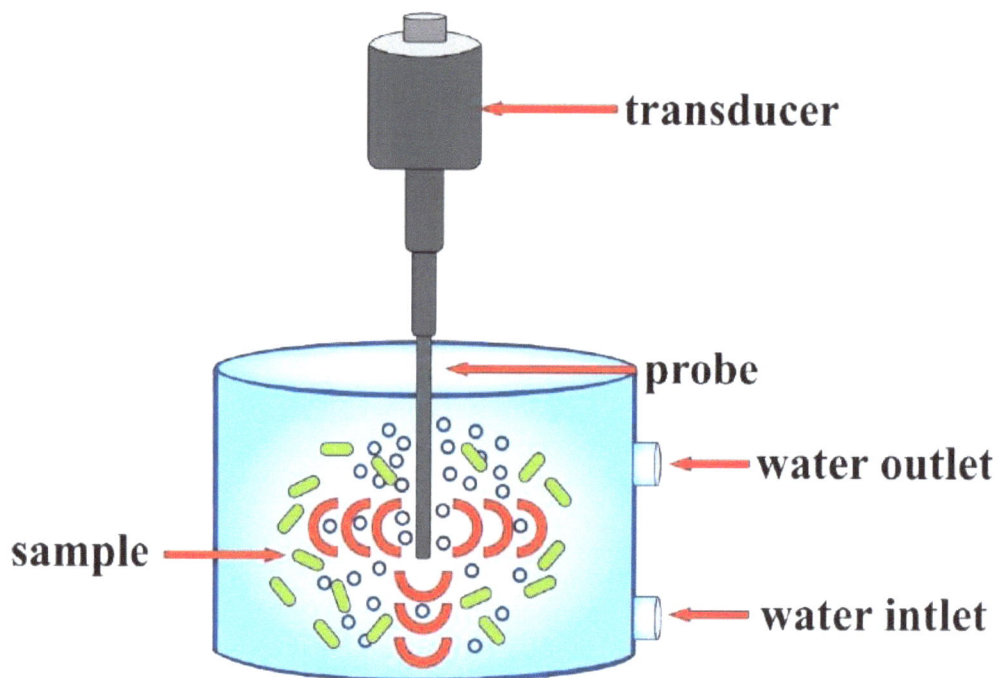

**Fig. (2).** Schematic diagram of a probe-type ultrasound-assisted extraction equipment.

UAE has been widely used for the inexpensive and effective extraction of bioactive compounds, including terpenoids, from natural matrices. Next, we will present and discuss some successful applications of the UAE process to extract terpenoids from plant matrices and their wastes (Table **1**).

**Table 1. Low-pressure green extraction processes for obtaining terpenoids from vegetable matrices.**

| Extraction Method | Sample | Main Bioactive Recovered | Highlights | Refs. |
|---|---|---|---|---|
| MAE[1] | *Pterodon emarginatus* | Caryophyllene γ-muurulene γ-elemeno | Considerable reduction of extraction time and energy required when compared to steam distillation | [24] |
| SFME[2] | *Limnophila aromatica* aerial parts | Limonene Perillaldehyde (*E*)-4-caranone (*Z*)-4-caranone α-pinene | Optimal MAE condition (for essential oil recovery): 700 W for 25 min | [20] |
| MAHD[3] | *Origanum vulgare L.ssp. hirtum* | Carvacrol | Considerable reduction of extraction time, reduction energy required, higher oil recovery when compared to conventional hydro-distillation | [21] |

*(Table 1) cont.....*

| Extraction Method | Sample | Main Bioactive Recovered | Highlights | Refs. |
|---|---|---|---|---|
| MAHD SFME | Tunisian *Rosmarinus officinalis* L. | α-pinene Eucalyptol Camphor Myrtenal Isoborneol | SFME provided the best results regarding the extraction yield, extraction time, energy consumption and extract composition | [22] |
| MAHD | *Cannabis sativa* L. inflorescences | α-pinene β-myrcene β-ocimene E-caryophyllene α-humulene Caryophyllene oxide β-selinene | MAHD extracted hydrodistilled oil was extremely rich in the terpenes Extraction kinetics showed a progressive enrichment in monoterpenes and a decrease in sesquiterpene in the extract | [25] |
| UAE[4] | *Citrus limetta* peel | *d*-limonene | UAE (80 W power and 25 kHz frequency) reduced *d*-limonene extraction time and increased the extraction yield | [31] |
| UAE MAE | *Sambucus nigra* (elder) bark | Ursolic acid *Oleanolic acid* | Optimized extraction conditions: ethanol (solvent), 55 °C (temperature) and ~1h (extraction time) Ursolic and oleanolic acids extraction yields: 86% for UAE and 92% for MAE | [32] |
| UAE | *Ganoderma lucidum* spore powder | Ganoderic acid Lucidenic acid Ganoderenic acid Ganodernoid C Ganoderic acid DM Ganodermic acid TQ Ganoderenic acid D Ganoderiol I Ganoderic acid C2 | Optimized extraction conditions: 95% v/v (ethanol concentration), 50:1 mL/g (solvent to solid ratio), 5.4 min (ultrasound time), 564.7 W (ultrasound power), and 8.2 cm (ultrasound probe distance) | [33] |
| UAE-VD[5] | Spearmint leaves (*Mentha spicata*) | Monoterpene hydrocarbons Oxygenated monoterpenes Sesquiterpene hydrocarbons Oxygenated sesquiterpenes | UAE-VD extracts had strong flavoring capacity, with the concentration of oxygenated compounds about 5x to 8x higher compared with hydrodistillation extract | [34] |

[1]MAE: microwave assisted extraction, [2]SFME: solvent-free microwave extraction, [3]MAHD: microwave-assisted hydro-distillation, [4]UAE: ultrasound assisted extraction, and [5]UAE-VD: ultrasound-assisted extraction coupled to vacuum distillation.

When applying the UAE process to extract *d*-limonene from *Citrus limetta* (sweet lime) peel waste, Khandare and collaborators [31] reported a considerable reduction in the extraction time (20 minutes of total extraction time) compared to the conventional method (185 min). The food-grade hexane solvent (feed to solvent ratio of 1:10) resulted in the best extraction yield of *d*-limonene, reaching 32.9 mg/g (97%) under the following operating conditions: temperature of 60 °C, ultrasound power of 80 W, and frequency of 25 kHz. According to the authors, UAE stands as a fast, efficient, and economical method for extracting *d*-limonene from the fresh sweet lime peel.

Ursolic acid and oleanolic acid (pentacyclic triterpenoids) were obtained from elder (*Sambucus nigra*) bark by UAE and MAE [32]. The use of microwave dielectric heating and ultrasonic cavitation bubbles substantially improved bioactive compounds' extraction compared to conventional heating methods. The optimized extraction conditions were ethanol as solvent, the temperature of 55 °C, and extraction times of at least an hour. Under these conditions, the extraction yield of ursolic and oleanolic acids reached 86% for UAE and 92% for MAE. Both ultrasound- and microwave-assisted extraction yields were superior to the conventional extraction method.

Triterpenoids (ganoderic acid, lucidenic acid, ganoderenic acid, ganodernoid C, ganoderic acid DM, ganodermic acid TQ, ganoderenic acid D, ganoderiol I, and ganoderic acid C2) were recovery from *Ganoderma lucidum* spore powder (GLSP) using UAE [33]. The extraction parameters (ultrasound power, ultrasound time, and ultrasound source distance) were optimized by response surface methodology. The extraction yields were compared to those obtained using the non-assisted extraction method. The optimum extraction conditions were ethanol concentration of 95% (v/v), solvent to solid ratio of 50:1 (mL/g), ultrasound time of 5.4 min, ultrasound power of 564.7 W, and ultrasound probe distance of 8.2 cm. In addition, SEM images suggested that the verified increase in triterpenoid extraction rate when using ultrasound may be due to the disruption of the matrix structure. The reported findings demonstrate that UAE is a more efficient method for extracting triterpenoid from GLSP than the convention method.

An innovative ultrasound-assisted extraction and vacuum distillation coupling strategy has been proposed to obtain aroma compounds from fresh leaves of two varieties of spearmint (*Mentha spicata*) [34]. The coupled process aimed to remove and subsequently concentrate and separate (without overheating) the aromatic compounds. The ultrasound-assisted extraction coupled to vacuum distillation was an effective strategy since it provided extracts with higher flavoring capacity due to the increased concentration of oxygenated compounds (from 5 to 8 times) compared with hydrodistillation (conventional extraction process). Flavor volatiles extraction yields reached 0.04–0.13% for the ultrasound coupled to vacuum distillation method and 0.01–0.02% for the hydrodistillation method.

# HIGH-PRESSURE EXTRACTION METHODS

## Supercritical Fluid Extraction (SFE)

Over the last decades, supercritical technology has been applied to obtain valuable natural compounds from numerous raw materials, *i.e.*, herbs, seaweed, and industrial by-products, among others [35]. Many industrial plants have been designed and installed to substitute conventional extraction plants. This market phenomenon has a precise motivation for the food and pharmaceutical industry to obtain natural products with clean and efficient technologies [36].

Before discussing supercritical extraction, a process using $CO_2$ as a solvent, it is necessary to register the environmental importance of industrial processes based on $CO_2$ consumption without returning it to the environment. Its use in industrial processes generates demand for $CO_2$ and therefore brings opportunities for the development of efficient $CO_2$ capture and storage techniques, leading to a reduction of the $CO_2$ concentration in the atmosphere [37].

$CO_2$ is widely used as a solvent and anti-solvent for sub and supercritical applications. It easily reaches supercritical conditions ($T_c$ = 31.1 °C and $P_c$ = 7.4 MPa), therefore being appropriate for processing thermolabile compounds. It has definite advantages (low toxicity, inflammability and cost, and high purity) over other fluids [38]. Moreover, the solute separation from the solvent can efficiently be bypassed by depressurization of supercritical fluid, and the recycling and reuse of it are possible, thus minimizing waste generation [39]. Supercritical $CO_2$ (sc-$CO_2$) processes present an important advantage over low-pressure methods based on the tunability properties, *i.e.*, the sc-$CO_2$ selectivity can be adjusted by varying temperature and pressure to solubilize fractions of different compounds [40]. A disadvantage of using $CO_2$ as a solvent is its low polarity, which makes it ideal for lipid, fat, and non-polar substance, but unsuitable for many pharmaceuticals and drug samples. Such limitation can be overcome by the use of chemical modifiers as a co-solvent such as ethanol and water [39, 41].

The operation of an extractor applying sub or supercritical $CO_2$ starts from a solvent reservoir, from which the $CO_2$ is released and, generally, passes through a heat exchanger to liquefy the fluid before being pressurized at the pressure required for the process. Then, the $CO_2$ may be preheated at the temperature of the extraction process, and then it enters the extraction vessel, containing the raw material with the target compound. The vessel needs a heat source to keep the extraction temperature constant throughout the extraction period. In the extraction vessel, $CO_2$ solubilizes the soluble compounds for the set temperature and pressure. Afterward, it is common to depressurize the system by stages in components called separators, which operate with pressure drop until reaching the last stage, at atmospheric pressure. The separators have a technical function that goes beyond just dropping the pressure, in which the fractionation of the extract can also be obtained to concentrate on target compounds. The selection of pressure and temperature in the separator allows the precipitation of different classes of compounds, thus being able to concentrate the extract from the extraction vessel in different fractions. After that, $CO_2$ is recycled, undergoes a compression step, and is re-injected into the supply line [42]. A schematic diagram of a general sc-$CO_2$ extraction equipment is showed in Fig. (3).

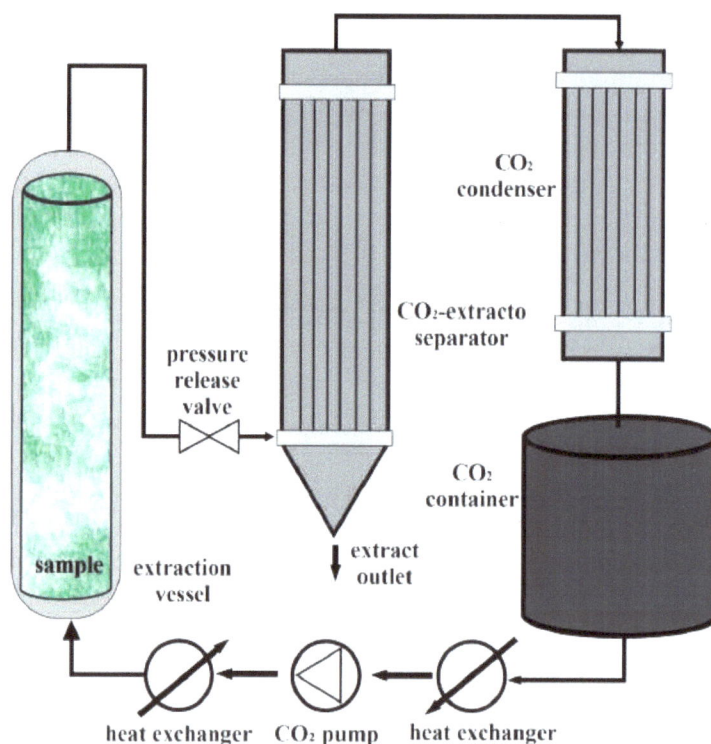

**Fig. (3).** Schematic diagram of a supercritical carbon dioxide extraction equipment.

Temperature and pressure are fundamental variables in the SFE process. Based on both critical temperature and pressure of $CO_2$, the processes are commonly developed for temperatures between 40 and 70 °C and pressures from 10 to 50 MPa. Other variables are important in this process, such as sample particle size, solvent flow rate, mass ratio solvent-to-feed, and properties of the extraction bed

[43, 44]. Different terpenoids have been extracted using sc-$CO_2$ from plant material, as shown in Table **2**.

**Table 2. High-pressure green extraction processes for obtaining terpenoids from vegetable matrices.**

| Extraction Method | Sample | Main Bioactive Recovered | Highlights | Refs. |
|---|---|---|---|---|
| sc-$CO_2^1$ | *Salvia officinalis* L. (sage) leaves | Camphor<br>α-thujone<br>β-thujone<br>1,8 cineole<br>α-humulene<br>Viridiflorol<br>Manool | Optimized extraction conditions: 13 MPa (pressure), 40 °C (temperature) and 3 kg/h ($CO_2$ flow rate) | [47] |
| sc-$CO_2$ | Green coffee | Kahweol<br>Cafestol | sc-$CO_2$ with or without ethanol provided extracts with higher free fatty acids and diterpene contents | [59] |
| sc-$CO_2$ | *Piper klotzschianum* leaves | Germacrene D<br>Pipercallosidine<br>14-oxy-α muurolene<br>Bicyclogermacrene<br>(E)-caryophyllene | Highest extraction yield, of 1.36% was obtained at 353 K and 220 bar using carbon dioxide as solvent | [48] |
| sc-$CO_2$ | 15 carotenoid-rich vegetable samples | β-carotene<br>α-carotene<br>Lutein<br>Lycopene | SFE at 59 °C, 350 bar, 15.5% ethanol (as co-solvent), 15 g/min $CO_2$ flow rate and 30 min od extraction recovered more than 95% of samples β-carotene and α-carotene | [45] |
| sc-$CO_2$ | *Echinophora platyloba* DC | Linalool<br>*Trans*-β-ocimene<br>α-pinene<br>Spathulenol | sc-$CO_2$ extraction in optimized conditions had a superior extraction yield compared to the hydrodistillation method | [60] |
| sc-$CO_2$ | *Rosmarinus officinalis* (rosemary) | Camphor<br>Eucalyptol<br>β -caryophyllene<br>Borneol acetate | SFE resulted in higher yield of essential oil and higher antioxidant activity (~14 times higher) than conventional extraction methods (steam distillation and hydrodistillation). | [49] |
| sc-$CO_2$ | Spearmint leaves (*Mentha spicata*) | Limonene<br>1,8-cineole<br>Dihydrocarvone<br>Pulegone<br>Carvone<br>Piperitenone<br>β-caryophyllene | Cross over pressure of 100 bar and cross over temperature of 45 °C for extraction pressure higher than 100 bar was reported | [61] |

*(Table 2) cont.....*

| Extraction Method | Sample | Main Bioactive Recovered | Highlights | Refs. |
|---|---|---|---|---|
| SWE[2] | Basil and oregano leaves | α-pinene Limonene Camphor Citronellol Carvacrol | Terpenoids concentrations decreased with the increase water temperature due to poorer stability of terpenes at the evaluated temperature conditions (100 to 250 °C). | [8] |
| PLE[3] | *Piper gaudichaudianum* Kunth leaves | Nerolidol Phytol Squalene | PLE decreased total extraction time and amount of solvent required compared to Soxhlet extraction | [56] |
| LPG[4] | Orange and apple pomace Pine needle and pine wood shavings | Limonene α-pinene β-pinene | LPG was superior to conventional organic solvent to extract terpenoids, promoting less decomposition of the extracted compounds | [10] |

Footnotes: [1]sc-$CO_2$: supercritical carbon dioxide extraction, [2]SWE: subcritical water extraction, [3]PLE: pressurized liquid extraction, and [4]LPG: gas extraction with liquefied petroleum gas.

Extraction with supercritical carbon dioxide and ethanol as a co-solvent was evaluated (under the fixed operational conditions of 59 °C, 350 bar, 15.5% ethanol, $CO_2$ flow rate of 15 g/min and 30 min of extraction) as an alternative for the recovery of carotenoids (β-carotene, α-carotene, lutein, and lycopene) from 15 different materials of plant origin: flesh and peels of sweet potato, tomato, apricot, pumpkin and peach, as well as flesh and wastes of green, yellow and red peppers [45]. The SFE process was successfully applied, recovering over 95% of β-carotene and α-carotene present in the samples. According to the authors, in cases where the fruit and vegetable matrix were rich in complex polysaccharides, the SFE process could possibly benefit from a long extraction time to guarantee the solubilization of the remaining carotenoid molecules trapped in locations that are not readily accessible by the solvent; this occurs in the later stages of the extraction and is primarily governed by diffusive mass transfer phenomena [46].

The optimization of the sc-$CO_2$ extraction parameters (pressure, temperature, and $CO_2$ flow rate) using response surface methodology of *Salvia officinalis* L. (sage) leaves was performed by Jokić *et al.* [47]. The higher pressures (within the range of 15 or 20 MPa) significantly influenced the concentrations of 1,8-cineole, α-/--thujone, and camphor. In contrast, the pressure of 10 MPa resulted in higher concentrations of α-humulene, viridiflorol, and manool, which indicates the possibility of selectively extracting terpenoids with sc-$CO_2$.

A factorial experiment was used to optimize the extraction of secondary metabolites from dry leaves of *Piper klotzschianum* [48]. Among the tested conditions, the pressure of 220 bar and temperature of 353 K resulted in the highest extraction yield (1.36%) when using pure $CO_2$ as solvent. Germacrene D,

14-oxy-α-muuroleno, bicyclogermacrene and (E)-caryophyllene were the major terpenoids found at the SFE extract.

The sc-$CO_2$ extraction was also applied to recover essential oil of rosemary and the results were compared to the hydrodistillation and steam distillation conventional methods [49]. Different temperatures (40 and 50 °C) and pressures (10.34 and 17.24 MPa) were evaluated. SFE extraction yield varied from 1.41 to 2.53 g essential oil/100 g of dry rosemary (%, w/w). Camphor, eucalyptol, b-caryophyllene, and borneol acetate were the main terpenoids found in the extracts.

Based on the studies on terpenoid extraction with supercritical carbon dioxide, the experimental conditions for the recovery of this class of compounds vary significantly with the type of material extracted and the operational characteristics of the extraction device. However, the results reported in the literature suggest that the extraction process should be conducted at a temperature around 40 °C and the pressure around 250 bar to reach the highest yields [50].

## Pressurized Liquid Extraction (PLE)

PLE has been consolidated as a green extraction method for the sustainable extraction of bioactive compounds from natural sources and the determination of a wide variety of important analytes in food and environmental matrices [51, 52]. Briefly, PLE employs solvent at high temperatures and pressures but below their respective critical points. Consequently, the solvent is kept in the liquid state during the entire extraction procedure, which leads to enhanced mass transfer rates, a decrease in solvent surface tension and viscosity, besides an increase in solubility of analytes [53]. PLE is a versatile method due to the possibility of choosing a wide range of solvents. When water is used as the extraction solvent, this technique is commonly called subcritical water extraction (SWE) [54]. The main advantages of PLE are faster extraction processes, lower solvent consumption, and the possibility of scale-up and automation [55]. Below, we will present some studies that used pressurized liquids to extract terpenoids from plant sources.

Pressurized liquid extraction with petroleum ether and ethanol as solvents was applied to obtain terpenes (terpenic alcohols and phytosterols), fatty acids and vitamin E from leaves of *Piper gaudichaudianum* Kunth [56]. PLE significantly decreased the total extraction time and the amount of solvent required compared to Soxhlet extraction. In addition, PLE was more effective for the extractions of terpenes (terpenic alcohols and phytosterols), fatty acids and vitamin E.

A study of the stability of five terpenes (α-pinene, limonene, camphor, citronellol, and carvacrol) under subcritical water conditions at four different temperatures

(100, 150, 200, and 250 °C) with two different heating times (30 and 300 min) reveled that degradation of terpenes increased with the increase of water temperature and heating time [8]. Additionally, the terpenoids recoveries from basil and oregano leaves were around 70 to 80% for SWE at 100 °C, decreasing with increasing water temperature due to poorer stability of terpenes. The results indicate that the SWE presents considerable limitations for the extraction of terpenoids.

## Liquefied Petroleum Gas (LPG) Extraction

Gas extraction with liquefied petroleum gas (butane/propane) (LPG) was developed as a clean, fast and cheap method for the extraction of terpenes and their derivatives. According to the authors, LPG extraction offers an industrial and laboratory-scale solution for the recovery of terpenes [10].

The extraction process with LPG is similar to that with supercritical fluids, except for the solvent state, which is in subcritical conditions. The LPG is placed in contact with the material to be extracted, solubilizing the soluble compounds present. For the batch process, after the defined contact time, the solvent is removed by depressurization, and the extract is recovered. The advantages of LPG extraction are the same as those described for SFE, except that LPG is explosive, which represents a considerable limitation to its industrial application. This extraction process was previously applied to improve the enzymatic hydrolysis of sugarcane bagasse [57] and extract steroids from pepper fruits [58].

LPG was applied to extract terpenes from agro-industrial (orange and apple pomace) and forest (pine needle and pine wood shavings) wastes [10]. The extraction process was successfully applied to recover terpenoids, presenting superior performance compared to conventional organic solvent extraction (using hexane and dichloromethane: *n*-pentane). Orange waste extracts showed the highest extraction yield and concentration of limonene, α-pinene, β-pinene, among all the solvents used. Additionally, as expected, the LPG extraction displayed less decomposition of the extracted compounds.

## CONCLUDING REMARKS

The development of green sustainable extraction processes to obtain terpenoids from natural matrices is a field in increasing expansion and with promising application on a large scale to produce ingredients for the food, chemical, and pharmaceutical industries. The use of extraction intensification strategies, such as the application of ultrasound and microwaves, the possibility of carrying out the process at high pressures (with solvents in sub and supercritical states), are the primary tools to optimize solvent extraction processes, reducing the extraction time, the generation of residues and the consumption of solvents, in addition to resulting in extracts with greater purity and quality. The MAE, UAE, SFE, SWE, and PLE techniques are some examples of modern processes that can be considered green and have been successfully used to produce terpenoid-rich (mainly monoterpenoids, sesquiterpenoids, diterpenoids, and tetraterpenoids) extracts from plants.

However, the literature is still scarce regarding the cost analysis and scale-up of many green sustainable extraction processes. The industrial-scale process is still challenging for many companies, especially those using ultrasound, microwave, and high-pressure, mainly due to equipment design, high expenses with facilities, and process optimization. Moreover, the lack of expert laborers and weak competition among supplier companies can also be highlighted as obstacles to the scale-up of these techniques. However, as these technologies are improved, these drawbacks tend to be limited and later overcome.

The combination of different extraction techniques (when the intensification processes co-occur in the extraction), for instance, the high-pressure extraction assisted by ultrasound, has shown promising results for the recovery of many bioactive compounds [30]. Thus, it may represent a significant advance in optimizing the terpenoid extraction processes, with even shorter extraction times and solvent consumption, in addition to providing greater selectivity and quality of the extracts obtained. Finally, the use of solvents as ionic liquids [62, 63] and deep eutectic solvents [64, 65] in the development of extraction processes for bioactive compounds are also areas of increasing interest and can be used for the recovery of terpenoids.

## CONSENT FOR PUBLICATION

Not applicable.

## CONFLICT OF INTEREST

The author declares no conflict of interest, financial or otherwise.

## ACKNOWLEDGEMENTS

Ana Carolina would like to thank the Coordenação de Aperfeiçoamento de Pessoal de Nível Superior – Brasil (CAPES) – Financial code 001. Arthur L.B. Dias would like to thank the Conselho Nacional de Desenvolvimento Científico e Tecnológico (CNPq) (151005/2019-2). Juliane Viganó is grateful to FAPESP (2020/15774-5).

# REFERENCES

[1]     Köksal, M.; Hu, H.; Coates, R.M.; Peters, R.J.; Christianson, D.W. Structure and mechanism of the diterpene cyclase ent-copalyl diphosphate synthase. *Nat. Chem. Biol.,* **2011**, *7*(7), 431-433.
[http://dx.doi.org/10.1038/nchembio.578] [PMID: 21602811]

[2]     Berthelot, K.; Estevez, Y.; Deffieux, A.; Peruch, F. Isopentenyl diphosphate isomerase: A checkpoint to isoprenoid biosynthesis. *Biochimie,* **2012**, *94*(8), 1621-1634.
[http://dx.doi.org/10.1016/j.biochi.2012.03.021] [PMID: 22503704]

[3]     Agatonovic-Kustrin, S.; Morton, D.W. Agatonovic-Kustrin, S.; Morton, D. W. The Current and Potential Therapeutic Uses of Parthenolide. *In Studies in Natural Products Chemistry,* **2018**, 61-91.
[http://dx.doi.org/10.1016/B978-0-444-64056-7.00003-9]

[4]     Salminen, A.; Lehtonen, M.; Suuronen, T.; Kaarniranta, K.; Huuskonen, J. Terpenoids: natural inhibitors of NF-κB signaling with anti-inflammatory and anticancer potential. *Cell. Mol. Life Sci.,* **2008**, *65*(19), 2979-2999.
[http://dx.doi.org/10.1007/s00018-008-8103-5] [PMID: 18516495]

[5]     Adlard, E.R. Handbook of Essential Oils. Science, Technology and Applications. *Chromatographia,* **2010**, *72*(9-10), 1021-1021.
[http://dx.doi.org/10.1365/s10337-010-1680-0]

[6]     Ni, Z.J.; Wang, X.; Shen, Y.; Thakur, K.; Han, J.; Zhang, J.G.; Hu, F.; Wei, Z.J. Recent updates on the chemistry, bioactivities, mode of action, and industrial applications of plant essential oils. *Trends Food Sci. Technol.,* **2021**, *110*, 78-89.
[http://dx.doi.org/10.1016/j.tifs.2021.01.070]

[7]     Wang, L.; Weller, C.L. Recent advances in extraction of nutraceuticals from plants. *Trends Food Sci. Technol.,* **2006**, *17*(6), 300-312.
[http://dx.doi.org/10.1016/j.tifs.2005.12.004]

[8]     Yang, Y.; Kayan, B.; Bozer, N.; Pate, B.; Baker, C.; Gizir, A.M. Terpene degradation and extraction from basil and oregano leaves using subcritical water. *J. Chromatogr. A,* **2007**, *1152*(1-2), 262-267.
[http://dx.doi.org/10.1016/j.chroma.2006.11.037] [PMID: 17126345]

[9]     McGraw, G.W.; Hemingway, R.W.; Ingram, L.L.; Canady, C.S.; McGraw, W.B. Thermal Degradation of Terpenes: Camphene, $\Delta^3$-Carene, Limonene, and α-Terpinene. *Environ. Sci. Technol.,* **1999**, *33*(22), 4029-4033.
[http://dx.doi.org/10.1021/es9810641]

[10]    Bier, M.C.J.; Medeiros, A.B.P.; de Oliveira, J.S.; Côcco, L.C.; da Luz Costa, J.; de Carvalho, J.C.; Soccol, C.R. Liquefied gas extraction: A new method for the recovery of terpenoids from agroindustrial and forest wastes. *J. Supercrit. Fluids,* **2016**, *110*, 97-102.
[http://dx.doi.org/10.1016/j.supflu.2015.12.016]

[11]    Martínez, J.; de Aguiar, A.C.; Machado, A.P. da F.; Barrales, F.M.; Viganó, J.; dos Santos, P. Process Integration and Intensification.*Comprehensive Foodomics.,* **2021**, , 786-807.
[http://dx.doi.org/10.1016/B978-0-08-100596-5.22819-9]

[12]    Chemat, F.; Vian, M.A.; Cravotto, G. Green extraction of natural products: concept and principles. *Int. J. Mol. Sci.,* **2012**, *13*(7), 8615-8627.
[http://dx.doi.org/10.3390/ijms13078615] [PMID: 22942724]

[13]    Zhang, Q.W.; Lin, L.G.; Ye, W.C. Techniques for extraction and isolation of natural products: a comprehensive review. *Chin. Med.,* **2018**, *13*(1), 20.
[http://dx.doi.org/10.1186/s13020-018-0177-x] [PMID: 29692864]

[14]    Chemat, F.; Rombaut, N.; Sicaire, A.G.; Meullemiestre, A.; Fabiano-Tixier, A.S.; Abert-Vian, M. Ultrasound assisted extraction of food and natural products. Mechanisms, techniques, combinations, protocols and applications. A review. *Ultrason. Sonochem.,* **2017**, *34*, 540-560.
[http://dx.doi.org/10.1016/j.ultsonch.2016.06.035] [PMID: 27773280]

[15] Ghasemi, E.; Raofie, F.; Najafi, N.M. Application of response surface methodology and central composite design for the optimisation of supercritical fluid extraction of essential oils from Myrtus communis L. leaves. *Food Chem.,* **2011,** *126*(3), 1449-1453.
[http://dx.doi.org/10.1016/j.foodchem.2010.11.135]

[16] Ae, K.; Singh, A.; Nc, S.; Jp, P. Novel Eco-Friendly Techniques for Extraction of Food Based Lipophilic Compounds from Biological Materials. *Nat. Prod. Chem. Res.,* **2016,** *4*(5)
[http://dx.doi.org/10.4172/2329-6836.1000231]

[17] Vinatoru, M.; Mason, T.J.; Calinescu, I. Ultrasonically assisted extraction (UAE) and microwave assisted extraction (MAE) of functional compounds from plant materials. *Trends Analyt. Chem.,* **2017,** *97*, 159-178.
[http://dx.doi.org/10.1016/j.trac.2017.09.002]

[18] *Microwaves in Organic Synthesis,* 3rd ed.; Hoz, A. de la, Loupy, A., Eds.; Wiley-VCH. **2013**.

[19] Kataoka, H. Pharmaceutical Analysis | Sample Preparation. *In Encyclopedia of Analytical Science,* Worsfold, P., Townshend, A., Poole, C., Eds.; Elsevier,. **2018**.
[http://dx.doi.org/10.1016/B978-0-12-409547-2.14358-6]

[20] Yingngam, B.; Brantner, A.; Treichler, M.; Brugger, N.; Navabhatra, A.; Nakonrat, P. Optimization of the eco-friendly solvent-free microwave extraction of Limnophila aromatica essential oil. *Ind. Crops Prod.,* **2021,** *165*113443
[http://dx.doi.org/10.1016/j.indcrop.2021.113443]

[21] Drinić, Z.; Pljevljakušić, D.; Živković, J.; Bigović, D.; Šavikin, K. Microwave-assisted extraction of O. vulgare L. spp. hirtum essential oil: Comparison with conventional hydro-distillation. *Food Bioprod. Process.,* **2020,** *120*, 158-165.
[http://dx.doi.org/10.1016/j.fbp.2020.01.011]

[22] Farhat, A.; Benmoussa, H.; Bachoual, R.; Nasfi, Z.; Elfalleh, W.; Romdhane, M.; Bouajila, J. Efficiency of the optimized microwave assisted extractions on the yield, chemical composition and biological activities of Tunisian Rosmarinus officinalis L. essential oil. *Food Bioprod. Process.,* **2017,** *105*, 224-233.
[http://dx.doi.org/10.1016/j.fbp.2017.07.011]

[23] Vian, M.A.; Fernandez, X.; Visinoni, F.; Chemat, F. Microwave hydrodiffusion and gravity, a new technique for extraction of essential oils. *J. Chromatogr. A,* **2008,** *1190*(1-2), 14-17.
[http://dx.doi.org/10.1016/j.chroma.2008.02.086] [PMID: 18343393]

[24] Vila Verde, G.M.; Barros, D.A.; Oliveira, M.S.; Aquino, G.L.B.; Santos, D.M.; de Paula, J.R.; Dias, L.D.; Piñeiro, M.; Pereira, M.M. A Green Protocol for Microwave-Assisted Extraction of Volatile Oil Terpenes from Pterodon emarginatus Vogel. (Fabaceae). *Molecules,* **2018,** *23*(3), 651.
[http://dx.doi.org/10.3390/molecules23030651] [PMID: 29534046]

[25] Gunjević, V.; Grillo, G.; Carnaroglio, D.; Binello, A.; Barge, A.; Cravotto, G. Selective recovery of terpenes, polyphenols and cannabinoids from Cannabis sativa L. inflorescences under microwaves. *Ind. Crops Prod.,* **2021,** *162*113247
[http://dx.doi.org/10.1016/j.indcrop.2021.113247]

[26] Vilkhu, K.; Mawson, R.; Simons, L.; Bates, D. Applications and opportunities for ultrasound assisted extraction in the food industry — A review. *Innov. Food Sci. Emerg. Technol.,* **2008,** *9*(2), 161-169.
[http://dx.doi.org/10.1016/j.ifset.2007.04.014]

[27] Kumar, K.; Srivastav, S.; Sharanagat, V.S. Ultrasound assisted extraction (UAE) of bioactive compounds from fruit and vegetable processing by-products: A review. *Ultrason. Sonochem.,* **2021,** *70*105325
[http://dx.doi.org/10.1016/j.ultsonch.2020.105325] [PMID: 32920300]

[28] Vinatoru, M. Ultrasonically assisted extraction (UAE) of natural products some guidelines for good practice and reporting. *Ultrason. Sonochem.,* **2015,** *25*, 94-95.

[http://dx.doi.org/10.1016/j.ultsonch.2014.10.003] [PMID: 25454822]

[29]   Both, S.; Chemat, F.; Strube, J. Extraction of polyphenols from black tea – Conventional and ultrasound assisted extraction. *Ultrason. Sonochem.*, **2014**, *21*(3), 1030-1034.
[http://dx.doi.org/10.1016/j.ultsonch.2013.11.005] [PMID: 24315029]

[30]   Dias, A.L.B.; de Aguiar, A.C.; Rostagno, M.A. Extraction of natural products using supercritical fluids and pressurized liquids assisted by ultrasound: Current status and trends. *Ultrason. Sonochem.*, **2021**, *74*105584
[http://dx.doi.org/10.1016/j.ultsonch.2021.105584] [PMID: 33975187]

[31]   Khandare, R.D.; Tomke, P.D.; Rathod, V.K. Kinetic modeling and process intensification of ultrasound-assisted extraction of d-limonene using citrus industry waste. *Chem. Eng. Process.*, **2021**, *159*108181
[http://dx.doi.org/10.1016/j.cep.2020.108181]

[32]   Bachtler, S.; Bart, H.J. Increase the yield of bioactive compounds from elder bark and annatto seeds using ultrasound and microwave assisted extraction technologies. *Food Bioprod. Process.*, **2021**, *125*, 1-13.
[http://dx.doi.org/10.1016/j.fbp.2020.10.009]

[33]   Shen, S.F.; Zhu, L.F.; Wu, Z.; Wang, G.; Ahmad, Z.; Chang, M.W. Production of triterpenoid compounds from *Ganoderma lucidum* spore powder using ultrasound-assisted extraction. *Prep. Biochem. Biotechnol.*, **2020**, *50*(3), 302-315.
[http://dx.doi.org/10.1080/10826068.2019.1692218] [PMID: 31755817]

[34]   Da Porto, C.; Decorti, D. Ultrasound-assisted extraction coupled with under vacuum distillation of flavour compounds from spearmint (carvone-rich) plants: Comparison with conventional hydrodistillation. *Ultrason. Sonochem.*, **2009**, *16*(6), 795-799.
[http://dx.doi.org/10.1016/j.ultsonch.2009.03.010] [PMID: 19406680]

[35]   de Aguiar, A.C.; da Fonseca Machado, A.P.; Figueiredo Angolini, C.F.; de Morais, D.R.; Baseggio, A.M.; Eberlin, M.N.; Maróstica Junior, M.R.; Martínez, J. Sequential high-pressure extraction to obtain capsinoids and phenolic compounds from biquinho pepper (Capsicum chinense). *J. Supercrit. Fluids,* **2019**, *150*(150), 112-121.
[http://dx.doi.org/10.1016/j.supflu.2019.04.016]

[36]   del Valle, J.M. Extraction of natural compounds using supercritical $CO_2$: Going from the laboratory to the industrial application. *J. Supercrit. Fluids,* **2015**, *96*, 180-199.
[http://dx.doi.org/10.1016/j.supflu.2014.10.001]

[37]   Davarazar, M.; Jahanianfard, D.; Sheikhnejad, Y.; Nemati, B.; Mostafaie, A.; Zandi, S.; Khalaj, M.; Kamali, M.; Aminabhavi, T.M. **2019**.
[http://dx.doi.org/10.1016/j.jcou.2019.05.022]

[38]   Zougagh, M.; Valcárcel, M.; Ríos, A. Supercritical fluid extraction: a critical review of its analytical usefulness. *Trends Analyt. Chem.*, **2004**, *23*(5), 399-405.
[http://dx.doi.org/10.1016/S0165-9936(04)00524-2]

[39]   Azmir, J.; Zaidul, I.S.M.; Rahman, M.M.; Sharif, K.M.; Mohamed, A.; Sahena, F.; Jahurul, M.H.A.; Ghafoor, K.; Norulaini, N.A.N.; Omar, A.K.M. Techniques for extraction of bioactive compounds from plant materials: A review. *J. Food Eng.*, **2013**, *117*(4), 426-436.
[http://dx.doi.org/10.1016/j.jfoodeng.2013.01.014]

[40]   Viganó, J.; Zabot, G.L.; Martínez, J. Supercritical fluid and pressurized liquid extractions of phytonutrients from passion fruit by-products: Economic evaluation of sequential multi-stage and single-stage processes. *J. Supercrit. Fluids,* **2017**, *122*, 88-98.
[http://dx.doi.org/10.1016/j.supflu.2016.12.006]

[41]   Wijngaard, H.; Hossain, M.B.; Rai, D.K.; Brunton, N. Techniques to extract bioactive compounds from food by-products of plant origin. *Food Res. Int.*, **2012**, *46*(2), 505-513.
[http://dx.doi.org/10.1016/j.foodres.2011.09.027]

[42]   de Aguiar, A.C.; Osorio-Tobón, J.F.; Silva, L.P.S.; Barbero, G.F.; Martínez, J. Economic Analysis of Oleoresin Production from *Malagueta Peppers* (Capsicum Frutescens) by Supercritical Fluid Extraction. *J. Supercrit. Fluids,* **2018,** *133.*
       [http://dx.doi.org/10.1016/j.supflu.2017.09.031]

[43]   Zabot, G.L.; Moraes, M.N.; Meireles, M.A.A. Influence of the bed geometry on the kinetics of rosemary compounds extraction with supercritical $CO_2$. *J. Supercrit. Fluids,* **2014,** *94,* 234-244.
       [http://dx.doi.org/10.1016/j.supflu.2014.07.020]

[44]   Reverchon, E.; De Marco, I. Supercritical fluid extraction and fractionation of natural matter. *J. Supercrit. Fluids,* **2006,** *38*(2), 146-166.
       [http://dx.doi.org/10.1016/j.supflu.2006.03.020]

[45]   de Andrade Lima, M.; Kestekoglou, I.; Charalampopoulos, D.; Chatzifragkou, A. Supercritical Fluid Extraction of Carotenoids from Vegetable Waste Matrices. *Molecules,* **2019,** *24*(3), 466.
       [http://dx.doi.org/10.3390/molecules24030466] [PMID: 30696092]

[46]   Sovová, H. Rate of the vegetable oil extraction with supercritical $CO_2$. Modelling of extraction curves. *Chem. Eng. Sci.,* **1994,** *49*(3), 409-414.
       [http://dx.doi.org/10.1016/0009-2509(94)87012-8]

[47]   Jokić, S.; Molnar, M.; Jakovljević, M.; Aladić, K.; Jerković, I. Optimization of supercritical $CO_2$ extraction of Salvia officinalis L. leaves targeted on Oxygenated monoterpenes, α-humulene, viridiflorol and manool. *J. Supercrit. Fluids,* **2018,** *133,* 253-262.
       [http://dx.doi.org/10.1016/j.supflu.2017.10.022]

[48]   Lima, R.N.; Ribeiro, A.S.; Cardozo-Filho, L.; Vedoy, D.; Alves, P.B. Extraction from Leaves of Piper klotzschianum using Supercritical Carbon Dioxide and Co-Solvents. *J. Supercrit. Fluids,* **2019,** *147,* 205-212.
       [http://dx.doi.org/10.1016/j.supflu.2018.11.006]

[49]   Conde-Hernández, L.A.; Espinosa-Victoria, J.R.; Trejo, A.; Guerrero-Beltrán, J.Á. CO 2 -supercritical extraction, hydrodistillation and steam distillation of essential oil of rosemary ( Rosmarinus officinalis ). *J. Food Eng.,* **2017,** *200,* 81-86.
       [http://dx.doi.org/10.1016/j.jfoodeng.2016.12.022]

[50]   Uwineza, P.A.; Waśkiewicz, A. Recent Advances in Supercritical Fluid Extraction of Natural Bioactive Compounds from Natural Plant Materials. *Molecules,* **2020,** *25*(17), 3847.
       [http://dx.doi.org/10.3390/molecules25173847] [PMID: 32847101]

[51]   Herrero, M.; Sánchez-Camargo, A.P.; Cifuentes, A.; Ibáñez, E. Plants, seaweeds, microalgae and food by-products as natural sources of functional ingredients obtained using pressurized liquid extraction and supercritical fluid extraction. *Trends Analyt. Chem.,* **2015,** *71,* 26-38.
       [http://dx.doi.org/10.1016/j.trac.2015.01.018]

[52]   Alvarez-Rivera, G.; Bueno, M.; Ballesteros-Vivas, D.; Mendiola, J.A.; Ibañez, E. Pressurized Liquid Extraction. In: *Liquid-Phase Extraction*; Elsevier, **2020**; pp. 375-398.
       [http://dx.doi.org/10.1016/B978-0-12-816911-7.00013-X]

[53]   Herrero, M.; Castro-Puyana, M.; Mendiola, J.A.; Ibañez, E. Compressed fluids for the extraction of bioactive compounds. *Trends Analyt. Chem.,* **2013,** *43,* 67-83.
       [http://dx.doi.org/10.1016/j.trac.2012.12.008]

[54]   Carr, A.G.; Mammucari, R.; Foster, N.R. A review of subcritical water as a solvent and its utilisation for the processing of hydrophobic organic compounds. *Chem. Eng. J.,* **2011,** *172*(1), 1-17.
       [http://dx.doi.org/10.1016/j.cej.2011.06.007]

[55]   Mustafa, A.; Turner, C. Pressurized liquid extraction as a green approach in food and herbal plants extraction: A review. *Anal. Chim. Acta,* **2011,** *703*(1), 8-18.
       [http://dx.doi.org/10.1016/j.aca.2011.07.018] [PMID: 21843670]

[56]   Péres, V.F.; Saffi, J.; Melecchi, M.I.S.; Abad, F.C.; de Assis Jacques, R.; Martinez, M.M.; Oliveira,

E.C.; Caramão, E.B. Comparison of soxhlet, ultrasound-assisted and pressurized liquid extraction of terpenes, fatty acids and Vitamin E from Piper gaudichaudianum Kunth. *J. Chromatogr. A,* **2006,** *1105*(1-2), 115-118.
[http://dx.doi.org/10.1016/j.chroma.2005.07.113] [PMID: 16439256]

[57]     Silva, J.R.F.; Cantelli, K.C.; Soares, M.B.A.; Tres, M.V.; Oliveira, D.; Meireles, M.A.A.; Oliveira, J.V.; Treichel, H.; Mazutti, M.A. Enzymatic hydrolysis of non-treated sugarcane bagasse using pressurized liquefied petroleum gas with and without ultrasound assistance. *Renew. Energy,* **2015,** *83*, 674-679.
[http://dx.doi.org/10.1016/j.renene.2015.04.065]

[58]     Cunico, M.M.; Miguel, O.G.; Miguel, M.D.; Carvalho, J.L.S.; Montrucchio, D.P.; Ferreira, J.L.; Oliveira, J.S. Extração de esteróides em frutos de Ottonia martiana Miq., Piperaceae, com gás liqüefeito. *Quim. Nova,* **2003,** *26*(6), 803-806.
[http://dx.doi.org/10.1590/S0100-40422003000600004]

[59]     Bitencourt, R.G.; Ferreira, N.J.; Oliveira, A.L.; Cabral, F.A.; Meirelles, A.J.A. High pressure phase equilibrium of the crude green coffee oil – $CO_2$ – ethanol system and the oil bioactive compounds. *J. Supercrit. Fluids,* **2018,** *133*, 49-57.
[http://dx.doi.org/10.1016/j.supflu.2017.09.017]

[60]     Sodeifian, G.; Sajadian, S.A. Investigation of essential oil extraction and antioxidant activity of Echinophora platyloba DC. using supercritical carbon dioxide. *J. Supercrit. Fluids,* **2017,** *121*, 52-62.
[http://dx.doi.org/10.1016/j.supflu.2016.11.014]

[61]     Shahsavarpour, M.; Lashkarbolooki, M.; Eftekhari, M.J.; Esmaeilzadeh, F. Extraction of essential oils from Mentha spicata L. ( Labiatae) *via* optimized supercritical carbon dioxide process. *J. Supercrit. Fluids,* **2017,** *130*, 253-260.
[http://dx.doi.org/10.1016/j.supflu.2017.02.004]

[62]     Xiao, J.; Chen, G.; Li, N. Ionic Liquid Solutions as a Green Tool for the Extraction and Isolation of Natural Products. *Molecules,* **2018,** *23*(7), 1765.
[http://dx.doi.org/10.3390/molecules23071765] [PMID: 30021998]

[63]     Ventura, S.P.M.; e Silva, F.A.; Quental, M.V.; Mondal, D.; Freire, M.G.; Coutinho, J.A.P. Ionic-Liquid-Mediated Extraction and Separation Processes for Bioactive Compounds: Past, Present, and Future Trends. *Chem. Rev.,* **2017,** *117*(10), 6984-7052.
[http://dx.doi.org/10.1021/acs.chemrev.6b00550] [PMID: 28151648]

[64]     Ivanović, M.; Islamčević Razboršek, M.; Kolar, M. Innovative Extraction Techniques Using Deep Eutectic Solvents and Analytical Methods for the Isolation and Characterization of Natural Bioactive Compounds from Plant Material. *Plants,* **2020,** *9*(11), 1428.
[http://dx.doi.org/10.3390/plants9111428] [PMID: 33114332]

[65]     Zainal-Abidin, M.H.; Hayyan, M.; Ngoh, G.C.; Wong, W.F.; Hayyan, A. Extraction of Bioactive Compounds. In: *Deep Eutectic Solvents*; Wiley, **2019**; pp. 217-233.
[http://dx.doi.org/10.1002/9783527818488.ch11]

**CHAPTER 3**

# Terpenoids Produced by Plant Endophytic Fungi from Brazil and their Biological Activities: A Review from January 2015 To June 2021

**Lourivaldo Silva Santos[1,\*], Giselle Skelding Pinheiro Guilhon[1], Railda Neyva Moreira Araujo[2], Antonio José Cantanhede Filho[3], Manoel Leão Lopes Junior[4], Haroldo da Silva Ripardo Filho[5]** and **Kiany Sirley Brandão Cavalcante[3]**

[1] *Laboratório de Micro-organismos, Programa de Pós-Graduação em Química, Universidade Federal do Pará-UFPA, 66970-110, Belém, PA, Brasil*

[2] *Escola Estadual de Ensino Médio Agostinho Morais de Oliveira, Inhangapi, PA, Brasil*

[3] *Instituto Federal de Educação Ciência do Maranhão (IFMA), Departamento Acadêmico de Química, São Luís, MA, Brasil*

[4] *Universidade Federal do Pará, Campus de Cametá, Cametá, PA, Brasil*

[5] *Instituto Federal do Amapá, Faculdade de Química, Macapá, AP, Brasil*

**Abstract:** Endophytic fungi are fungi that live inside plant tissues at any moment of their life cycle without causing damage or disease symptoms to their hosts. These microorganisms are producers of important substances with several biological activities. Terpenoids are one of the main classes of natural products produced by endophytic fungi, and have a wide range of biological activities, such as anti-inflammatory, anticancer, antioxidant, antifungal, antimicrobial, anticholinesterase, antidepressant, antipyretic, antimalarial, among others. Brazil has one of the largest plant reserves on the planet, consisting of an almost untapped source of endophytic fungi. Thus, in this review chapter, we present the results of the research work of Brazilian researchers, with a focus on the isolation and identification of secondary metabolites of the terpenoid class produced by endophytic fungi and their biological activities. The review period includes January 2015 and June 2021.

**Keywords:** Bioactive Compounds, Diterpenoids, Endophytic Fungi, Isoprenoids, Meroterpenes, Microorganism, Monoterpenes, Monoterpenoids Diterpenes, Sesquiterpenes, Sesquiterpenoids, Terpenoids, Terpenes, Triterpenes, Triterpenoids.

---
\* **Corresponding author Lourivaldo Santos:** Laboratório de Micro-organismos, Programa de Pós-Graduação em Química, Universidade Federal do Pará-UFPA, 66970-110, Belém, PA, Brasil; E-mail: lss@ufpa.br

**Mozaniel Santana de Oliveira & Antônio Pedro da Silva Souza Filho (Eds.)**

## INTRODUCTION

Natural products are compounds isolated from different natural sources such as plants, animals, microbes, insects, plant pathogens, and endophytes and marines [1]. Microorganisms are very versatile and found everywhere, even in inhospitable habitats, in all ecosystems around the globe. It is preconized that less than 1% of all bacteria species and less than 5% of all fungi species are described, suggesting at least 10 million microbial species are unknown, remaining hidden in nature [2]. Besides, based on genetic research, 90% of biosynthetic skills of microorganisms keep unattainable, which ratifies the significance of microbial natural products research for drug discovery and, even for complete biodiversity knowledge and ecological relationships understanding [3].

Endophytic fungi are fungi that live inside plant tissues at any moment of their life cycle, without causing damage or disease symptoms to their hosts [4 - 8]. These microorganisms are producers of important substances with several biological activities, such as anti-inflammatory, anticancer, antioxidante, antifungal, antimicrobial, anticholinesterase, antidepressant, antipyretic, antimalarial, among others [9 - 12].

Endophytic fungi are one of the most important elements in plant micro-ecosystems and have relevant influences on the growth and development of host plants. Basic knowledge about the relationships between endophytic fungi and their host plants is of significant importance [13, 14]. Any plant-fungal interaction is preceded by a physical encounter between a plant and a fungus, followed by several physical and chemical barriers that must be overcome to successfully establish an association (Fig. **1**) [15 - 17]. To know how an endophyte avoids activating the host defenses, ensures self-resistance before being in capacitated by the toxic metabolites of the host, and manages to grow within its host without causing visible manifestations of infection or disease was initially proposed by the *balanced antagonism* hypothesis [13, 14]. This hypothesis proposed that asymptomatic colonization is a balance of antagonisms between the host and the endophyte. Endophytes and pathogens both possess many virulence factors that are countered by plant defense mechanisms. If fungal virulence and plant defense are balanced, the association remains apparently asymptomatic and avirulent (Fig. **1B**). This phase is only a transitory period where environmental factors play a major role to destabilize the delicate balance of antagonisms. If the plant defense mechanisms completely counteract the fungal virulence factors, the fungus will perish (Fig. **1C**). Conversely, if the plant succumbs to the virulence of the fungus, a plant- pathogen relationship would lead to plant disease (Fig. **1A**). They might be influenced by certain intrinsic or environmental factors to express factors that lead to pathogenicity because many endophytes could possibly be latent pathogens [15]. The plant-endophyte interaction might not be just an equilibrium between virulence and defense, but a much more complex and precisely controlled interaction. Endophytes might protect host plants by creating a heterogeneous chemical composition within and among plant organs that are otherwise genetically uniform [16], according to the *mosaic effect* theory.

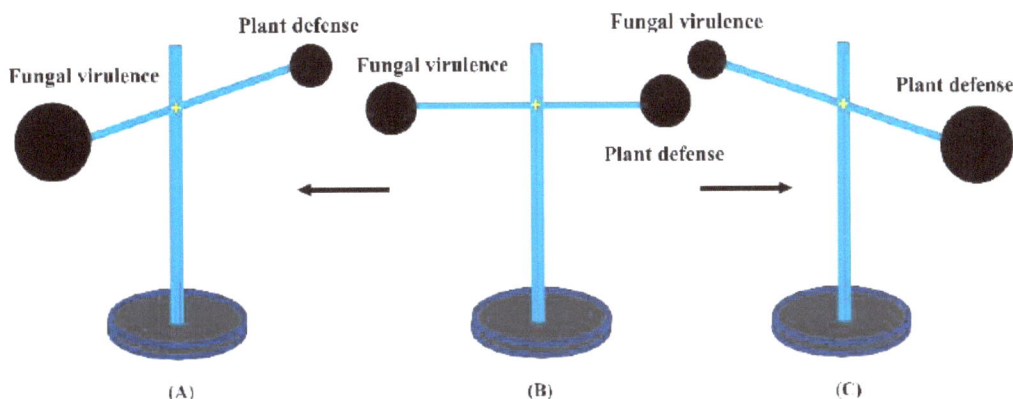

**Fig. (1).**  Balanced antagonism hypothesis: **A)** equilibrium situation; **B)** phytopathogenicity; **C)** healthy plant.

A study based on the report of several authors indicated some benefits promoted by endophytic fungi to their host plants after colonization [18]. Three different beneficial aspects after an interaction are listed: a) First, some endophytic fungi could produce different plant hormones to enhance the growth of their host plants [19]. For example, the growth of wheat (*Triticum aestivum* L.) could be enhanced by *Azospirillum* sp. under drought stresses [20]; b) Second, some endophytic fungi would produce different bioactive compounds, such as alkaloids, diterpenes, flavonoids, and isoflavonoids, to increase the resistance to biotic and abiotic stresses of their host plants [21, 22], and c) Third, some endophytic fungi could promote the accumulation of secondary metabolites (including important medicinal components or drugs) originally produced by plants. These metabolites may be produced by both the host plants or/and endophytic fungi, according to the references surveyed [23].

De Bary described the first endophytic fungi in 1866 [24], but the greatest attention given to these microorganisms as producers of biologically active substances occurred with the isolation of the diterpenoid Taxol (**1**) in 1993, from the endophytic fungus named *Taxomyces andreanae* associated to the phloem of *Taxus brevifalia* Nutt [25]. From then on, endophytes have been recognized as important sources of secondary metabolites such as terpenoids, polyketides, alkaloids, benzopyranones, benzoquinones, naphthoquinones, phenols, steroids, tetralones and xanthones [7, 26 - 30]. The selection of the host plant is an important factor in the production of endophytic fungi and the isolation of their bioactive secondary metabolites [26 - 28, 31 - 38].

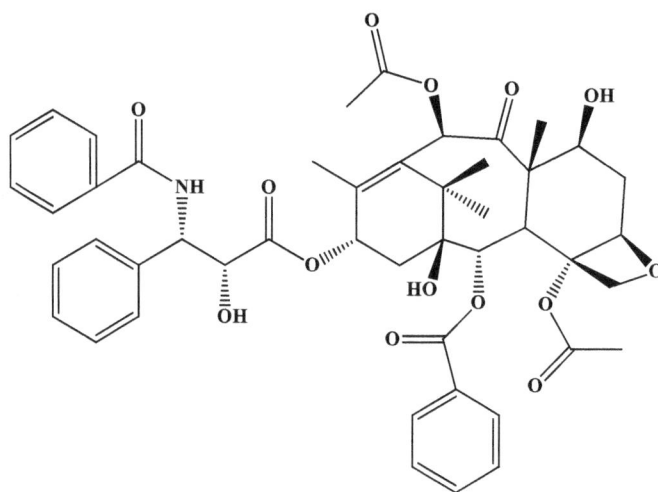

**Taxol (1).**

Endophytic fungi have been associated with plants for over 400 million years [33]. Tropical and temperate forests are considered to be the most diverse terrestrial ecosystems, with the greatest number and diversity of endophytic fungi since the plants residing in these regions are in a constant evolutionary race to survive [34, 35]. The constant innovation present in ecosystems where the evolutionary race to survive is the most active may result in the production of a plethora of chemical molecules [36, 37].

Brazil has one of the largest plant reserves on the planet, consisting of an almost untapped source of endophytic fungi. Despite the efforts of Brazilian researchers, research on the production of secondary metabolites of endophytic fungi is still incipient, when it is taken into account the vast biodiversity existing in Brazilian territory.

This paper summarized the research progress on terpenoids and biological activities from plant endophytic fungi between January 2015 and June 2021 through a literature analysis of studies reported by Brazilian researchers.

## Terpenoids

Terpenoids are widely found in nature and are referred to as terpenes or as isoprenoids. According to the number of building blocks, terpenoids are classified as monoterpenes, sequiterpenes, diterpenes, sesterpenes, meroterpenes, triterpenes, tetraterpenes and polyterpenes [38]. Originally the term terpene was resricted to hydrocarbons, it is now used to include substituted derivatives too. The fundamental building block of terpenes is the isoprene unit, $C_5H_8$. The larger structures are assembled from several isoprene units, usually by head-to-tail linked isoprene units. Terpenes can be cyclic or acyclic, with a large range of structural variations. The chemical class of terpenes is prevalent in fungi. Numerous compounds of class terpenes have been isolated and identified and are of great scientific interest because of their biochemical properties and huge potential for drug development [36, 37].

All terpenes and terpenoids of fungal origin are derived from five-carbon intermediate isopentenyl diphosphate (IPP) and dimethylallyl diphosphate (DMAPP), synthesized from acetyl coenzyme A (CoA) through the mevalonate pathway. Further it is followed by a head-to-tail condensation of IPP units to DMAPP to give isoprenyl diphosphates (10C-geranyl pyrophosphate, 15C-farnesyl pyrophosphate, or 20C-geranylgeranyl pyrophosphate); this reaction is catalyzed by isoprenyldiphosphate synthases [38]. However, longer chains ($C_{30}$, $C_{40}$) are synthesized by head-to-tail condensation of two farnesyl pyrophosphate or two geranyl-geranyl pyrophosphates, and the reaction is catalyzed by squalene synthase and phytoene synthase, respectively [39, 40]. These linear compounds are substrates of many different enzymes that either transfer the prenyl residue to another molecule, mostly aromatic compounds, or initiate the cyclization of prenyl chain, giving rise to tens of thousands of different isoprenoids-derived secondary metabolites. There are several classes of prenyl transferases and cyclases in fungi related to the production of different natural products. Terpenoids of different classes can be distinguished on the basis of scaffolds, either derived solely from IPP units or from mixed origin. Farmer group is divided into mono-, di-, sesqui-, or triterpenoids containing 2-6 IPP units; this group is also characterized by the presence of carotenoids and some rare sesquiterpenoids, and the latter group includes prenylated aromatic secondary metabolites [41 - 43].

## Monoterpenoids

*Monoterpenes* are a class of terpenes that consist of two isoprene units and have the molecular formula $C_{10}H_{16}$. Monoterpenes may be linear (acyclic) or contain rings (monocyclic and bicyclic). Modified terpenes, such as those containing oxygen functionality or missing a methyl group, are called monoterpenoids. Some monoterpenes are acyclic hydrocarbons (such as myrcene and ocimene) or their derivatives (such as linalool and geraniol), but cyclic monoterpenes and their derivatives (such as limonene, pinenes, camphor, alpha-terpineol, perillyl alcohol, carveol, carvone, and menthols) are more abundant in nature and have more industrial applications [44]. Many monoterpenoids and products of their transformation are of practical importance in the pharmaceutical and cosmetic industries, for the production of flavor additives and pesticides [45, 46], and for fine chemical industry [47]. In addition, monoterpenoids and their derivatives have a variety of biological activities, such as cytotoxic, antimicrobial, and anti-inflammatory, which have potential application value in clinical medicine [48].

From the leaves of *Caesalpinia pyramidalis* Tul (Fabaceae) collected in northeastern Brazil was isolated an uncharacterized fungal endophyte (coded BR109), identified as *Diaporthe* (*Phomopsis*) sp. by ITS sequencing [49]. Phylogenetic analyses showed that fungal isolate BR109 was most closely related to *Diaporthe anacardii* and *Diaporthe foeniculaceae* [50]. The isolated endophyte produced four $C_{10}$ monoterpenoids: *trans*-piperitol (**2**), piperitone epoxide (**3**), $\gamma$-terpineol (**4**) and 4-terpineol (**5**) (Fig. **2**).

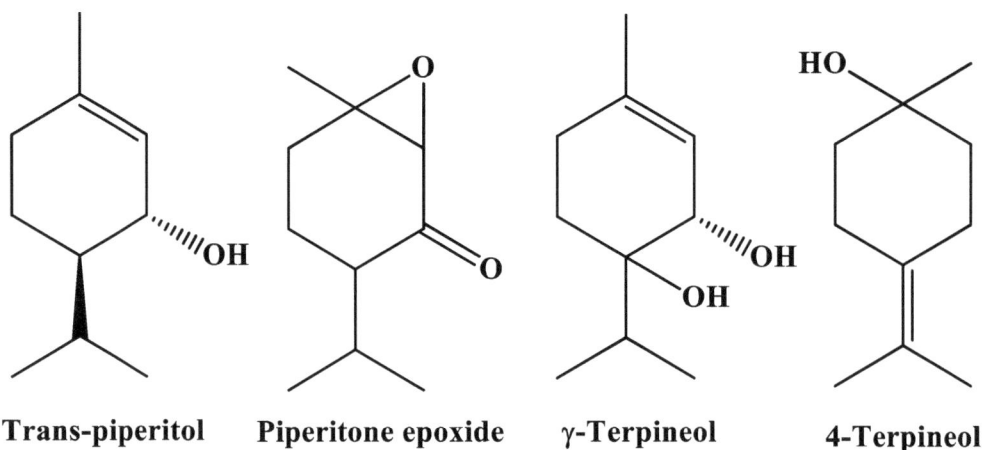

**Trans-piperitol**     **Piperitone epoxide**     **$\gamma$-Terpineol**     **4-Terpineol**

**Fig. (2).** Monoterpenoids from the leaves of *Caesalpinia pyramidalis* Tul (Fabaceae).

## Sesquiterpenoids

*Sesquiterpenoids* are 15-carbon-length structurally diverse natural compounds isolated from plants, fungi, and bacteria. They are synthesized from farnesyl pyrophosphate catalyzed by the enzyme sesquiterpene synthase. The enzyme sesquiterpene synthases have a conserved active site structure, aspartate-rich motifs coordinated by $Mg^{2+}$ cluster and the enzyme bind to the pyrophosphate group of farnesyls *via* $Mg^{2+}$ cluster and after binding to the pyrophosphate group the orientation of prenyl chain changes towards hydrophobic cavity of the enzyme [51]. This orientation results in conformational changes and causes active site closure followed by concomitant cleavage of pyrophosphate to give rise to transoid allylic carbocation [52 - 55]. This carbocation gets transferred to the isoprenyl chain and is ultimately quenched by proton abstraction or by a water molecule. Folding of isoprenyl chain is determined by binding pocket of enzyme, thereby defining the product profile of a particular sesquiterpene synthase. Enzymes catalyze different initial cyclization reactions to produce cyclic carbocation intermediates, for example, *trans*-humulylcarbocation, which is a 1,11-cyclization product. Secondary carbocation produced can undergo additional cyclization and rearrangement until quenching in the active site occurs [56].

Mandavid and co-workers [57], isolated the endophytic fungus *Diaporthe* sp. SNB-GSS10 from a medicinal plant *Sabicea cinerea* that produced the known sesquiterpene altiloxin A (**7**) and an unknown sesquiterpene derivative eremofortin F (**6**) (Fig. **3**). The cytotoxic activities of isolated compounds were evaluted against KB, MDA-MB-435, and MRC5 cell lines. Altiloxin A (**7**) showed no effect ($IC_{50} > 30$ μM) on all tested cell lines, and eremofortin F (**6**) was cytotoxic on KB and MRC5 cells ($IC_{50}$ = 13.9 and 12.2 μM, respectively). On KB and MDA-MB-435 cell lines, the positive control was docetaxel (0.2 and 0.5 nM, respectively). On the MRC5 cell line, the positive control was doxorubicin (20 nM).

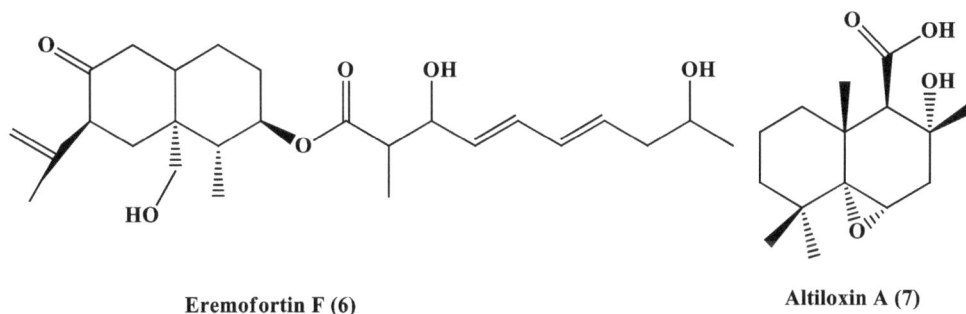

**Eremofortin F (6)**                                              **Altiloxin A (7)**

**Fig. (3).** Sesquiterpenoids isolated from the endophytic fungus *Diaporthe* sp. SNB-GSS10.

Volatile organic compounds (VOCs) were produced by ten species of endophytic fungi: *Lasiodiplodia theobromae* (strain 71), *L. pseudotheobromae* (strain 277),*L. citricola* (strain 258), *L. gonubiensis* (strain 474), *L. parva* (strain 511), *Neofusicoccum cordaticola* (strain 434), *N. parvum* (strain 600), *N. ribis* (strain 683), *Botryosphaeria mamane* (strain 20), and *Pseudofusicoccum stromaticum* (strain 477) from the Botryosphaeriaceae Family, associated with plants from the

Caatinga biome (state of Ceará, Brazil) [58]. A total of 24 volatile compounds were identified as being produced by the fungal strains; between them, 14 non-oxygenated (sesquiterpenes) and 10 oxygenated (sesquiterpenoids) are displayed in Figs. (**4 and 5**), respectively. Most of the VOCs were bicyclic sesquiterpenoids with eudesmane [2 compounds: juniper camphor (**8**) and α-selinene (**9**)], guaiane [3 compounds: globulol (**10**), guaiol acetate (**11**) and palustrol (**12**)], cadinane [6 compounds: δ-amorfane (**13**), *trans*-cadina-1(2)4-diene (**14**), γ-cadinene (**15**), δ-cadinene (**16**), α-cadinol (**17**) and zonarene (**18**)], cedrane [3 compounds: 5-ne-cedranol (**19**), α-cedrene epoxide (**20**) and β-cedren-2-one (**21**)], aristolane [2 compounds: aristolene (**22**) and calarene (**23**)], valencane [three compounds: aristolochene (**24**), 13-hydroxyvalencene (**25**) and valencene (**26**)], copaane [one compound: α-copaene (**27**)] and eremophylane [one compound: eremophylene (**28**)] skeletons. Only three monocyclic sesquiterpenoids [α-bisabolol (**29**), β-elemene (**30**) and germacrene D (**31**)] were found. The non-oxygenated compounds belong mostly to the cadinene group, while the oxygenated ones possess the guaiane and cedrane skeleton [58]. Fifteen sesquiterpenoids were reported for the first time as fungal VOCs: juniper camphor (**8**), globulol (**10**), guaiol acetate (**11**), palustrol (**12**), *trans*-cadina-1(2)4-diene (**14**), γ-cadinene (**15**), δ-cadinene (**16**), α-cadinol (**17**), 5-neo-cedranol (**19**), α-cedrene epoxide (**20**), β-cedren-2-one (**21**), aristolene (**22**), aristolochene (**24**), 13-hydroxyvalencene (**25**), eremophylene (**28**). The authors used multivariate data analysis (PCA and HCA) which allowed the differentiation of all investigated species, and proved to be efficient for the differentiation of *Neofusicocum parvum* and *N. ribis*, which are considered very similar species. Sesquiterpenoids α-bisabolol (**29**), α-selinene (**9**), α-cedrene epoxide (**20**) and guaiol acetate (**11**) were suggested as biomarkers [58].

Two new rearranged sesquiterpenes, 3,5,9-trihydroxy presilphiperfolane (**32**) and 4-deoxy-10-oxodihydrobotrydial (**33**), along with two eremophilane sesquiterpenes xylarenone C (**34**) and xylarenone D (**35**), (Fig. **6**), were isolated from solid substrate cultures of *Camarops* sp., an endophyte of *Alibertia macrophylla* (Rubiaceae) [59]. Xylarenone C (**34**) and xylarenone D (**35**) exhibited cytotoxic activity against leukemia (HL-60), melanoma (MDA/MB-435), colon (HCT-8) and glioblastoma (SF-295) human tumor lines, following 72 h of treatment, with $IC_{50}$ values of 2.4 mg mL$^{-1}$ for **34** in MDA/MB-435 cells and $IC_{50}$ of 1.2 mg mL$^{-1}$ for **35** in HL-60 cells. Compounds **34** and **35** were tested for their acetylcholinesterase (AChE) inhibitory activity using a bioautography assay. Preliminary TLC results suggested weak activity for compounds **35**. Compound **34** had a minimum AChE inhibitory concentration of 6.25 mg mL$^{-1}$. Galantamine (positive control) was active at 1mg mL$^{-1}$ [59].

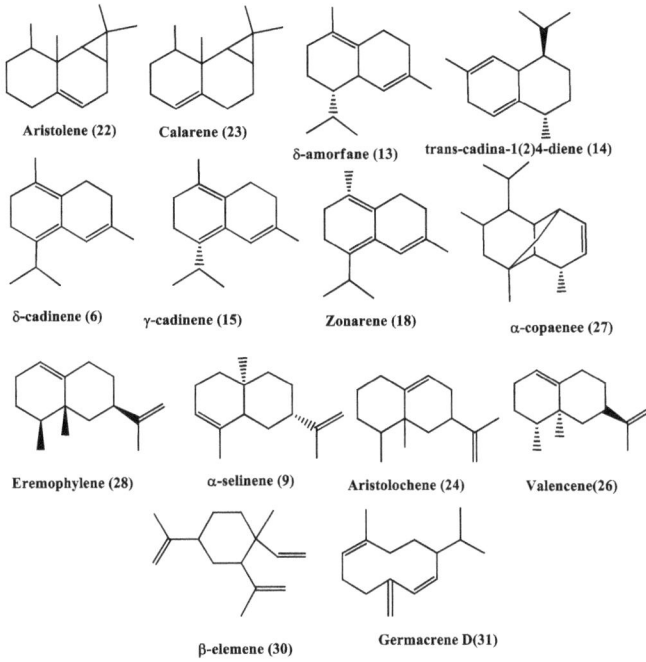

**Fig. (4).** Sesquiterpenes identifieds in ten species of Botryosphaeriaceae.

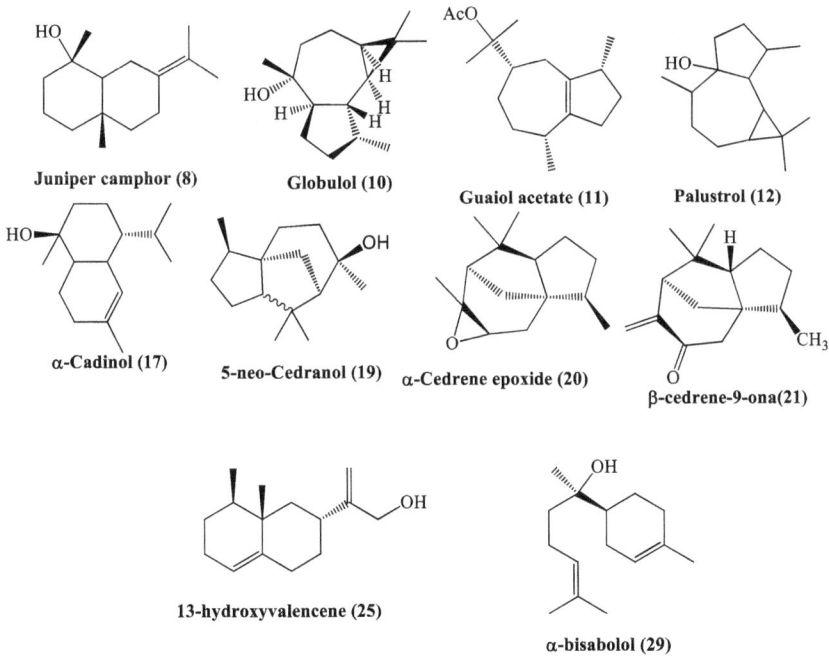

**Fig. (5).** Sesquiterpenoids identifieds in ten species of Botryosphaeriaceae.

**3,5,9-trihydroxy presilphiperfolane (32)**          **4-deoxy-10-oxodihydrobotrydial (33)**

**xylarenone C (34)**                                        **xylarenone D (35)**

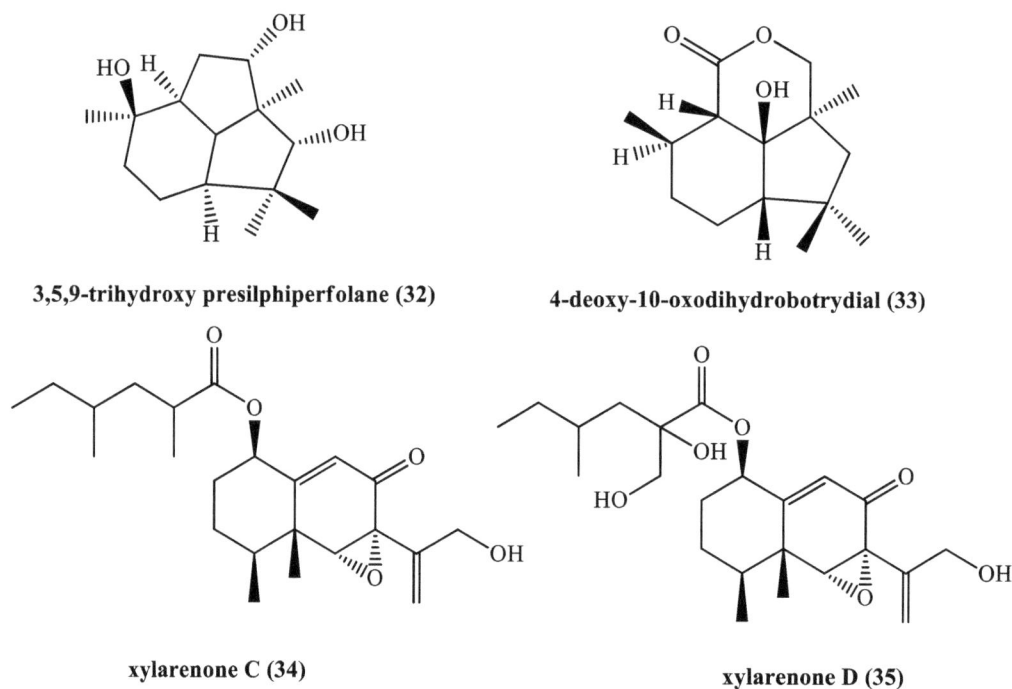

**Fig. (6).** Structures of sesquiterpenoids isolated from the cultures of *Camarops* sp.

To explore the untapped potential of fungi belonging to the division Ascomycota in producing the sesquiterpenoide Δ6-protoilludene (**36**), the precursor to the cytotoxic illudins, which are pharmaceutically relevant as anticancer therapeutics [60], De Sena Filho and co-workers [49] isolated the fungal endophyte *Diaporthe* sp. BR109 and show that it produces a diversity of terpenoids including Δ6-protoilludene (**36**). The endophytic fungus BR109 was isolated from the leaves of *Caesalpinia pyramidalis* and produced the sesquiterpenoids neotrifaradiene (**37**), Δ6-protoilludene (**36**), 3,7-di-epi-trifara-9,14-diene (**38**), β-elemene (**30**), thujopsene (**39**), nardosina-7,9,11-triene (**40**), aristolene (**22**), α-muurolene (**41**), α-bulnesene (**42**), guaia-9,11-diene (**43**), norpatchulenol (**44**), and maaliol (**45**) (Fig. 7). The authors described the first characterization of a Δ6-protoilludene synthase from an Ascomycete, and provides insights into the evolutionary pathway of individual genes and potentially minimal sesquiterpenoid secondary metabolic pathways across fungal phyla.

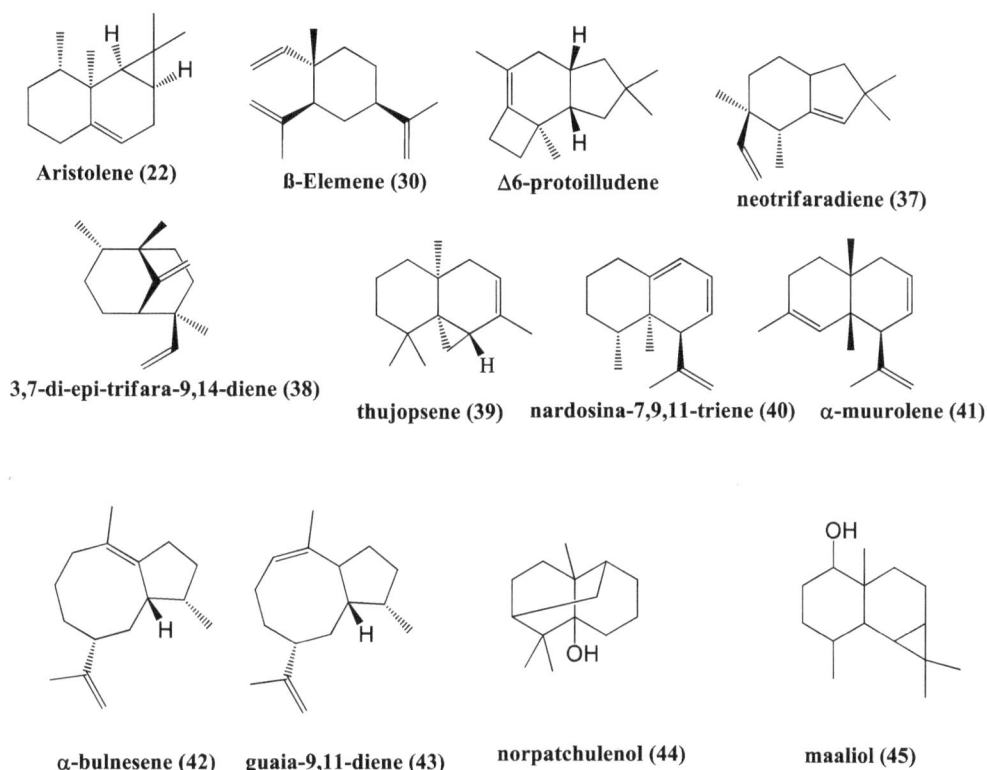

**Aristolene (22)**　　**ß-Elemene (30)**　　**Δ6-protoilludene**　　**neotrifaradiene (37)**

**3,7-di-epi-trifara-9,14-diene (38)**　　**thujopsene (39)**　**nardosina-7,9,11-triene (40)**　**α-muurolene (41)**

**α-bulnesene (42)**　**guaia-9,11-diene (43)**　**norpatchulenol (44)**　　**maaliol (45)**

**Fig. (7).** Chemical structures of sesquiterpenoids prodeced by endophytic fungus *Diaporthe* sp. BR109 isolated from the leaves of *Caesalpinia pyramidalis.*

Fractionation of the ethyl acetate extract (EtOAc) from the culture of *Nectria pseudotrichia* (Nectriaceae), an ascomycete isolated from the tree *Caesalpinia echinata* (Fabaceae), yielded a new sesquiterpenoid 10-acetyl trichoderonic acid A (**46**), and two know sesquiterpenoids hydroheptelidic acid (**47**) and xylaric acid D (**48**) (Fig. **8**) [61]. All three sesquiterpenoids were tested against amastigotes forms of *Leishmania (Viannia) braziliensis*. Sesquiterpenoid **46** was more active, with $IC_{50}$ values of 21.4 μM, **47** presented 24.8 μM, and **48** less active ($IC_{50}$ > 200 μM). All sesquiterpenoids showed low toxicity to Vero and THP-1 cells ($IC_{50}$ values greater than 200 μM). The $IC_{50}$ value of **47** was ~ 10 times lower than the $IC_{50}$ values determined for meglumine antimoniate (178.2-330.2 μM) and paromomycin (233.6-344.4 μM), but ~ 5 times higher than the $IC_{50}$ values determined for miltefosine (0.8-5.4 μM) and azithromycin (5.7 μM), and ~ 1000 times higher than the $IC_{50}$ values obtained for amphotericin B (0.02-0.06 μM) in the assay with amastigote forms of *L. (V.) braziliensis* using different time periods of drug exposure (three, five, and seven days) [62]. Moreover, the $IC_{50}$ value for amphotericin B (0.12 μM), used as a positive control in the assays, was also ~ 200 times lower than the $IC_{50}$ determined for compound **47**. Considering compounds **46-48** (Fig. **8**), and heptelidic acid (Fig. **9**), the authors hypothesise that they originate from a putative common precursor that could cyclise in two different ways to form five and seven-membered lactone rings (Fig. **9**, routes *a* and *b*, respectively). The new compound **46** could be formed by the opening of the epoxy ring of heptelidic acid by an acetoxy group (Fig. **9**) [61].

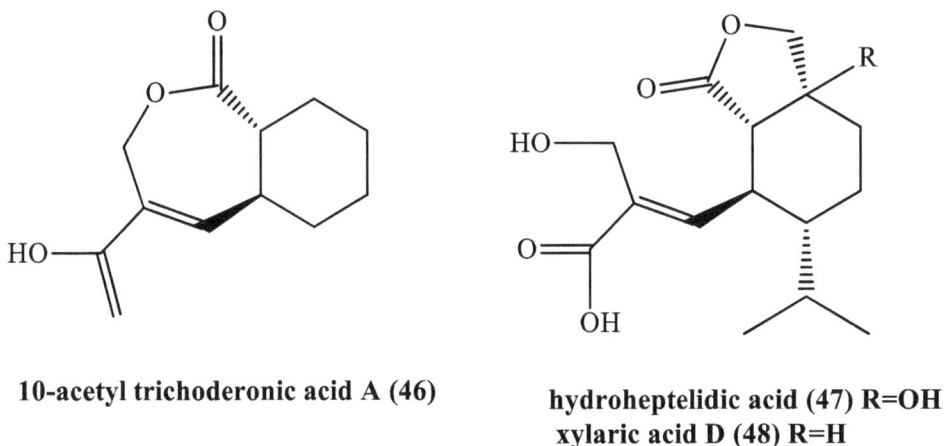

**10-acetyl trichoderonic acid A (46)**

**hydroheptelidic acid (47) R=OH**
**xylaric acid D (48) R=H**

**Fig. (8).** Chemical structures of sesquiterpenoids produced by endophytic fungus *Nectria pseudotrichia* (Nectriaceae) isolated from the leaves of *Caesalpinia echinata* (Fabaceae).

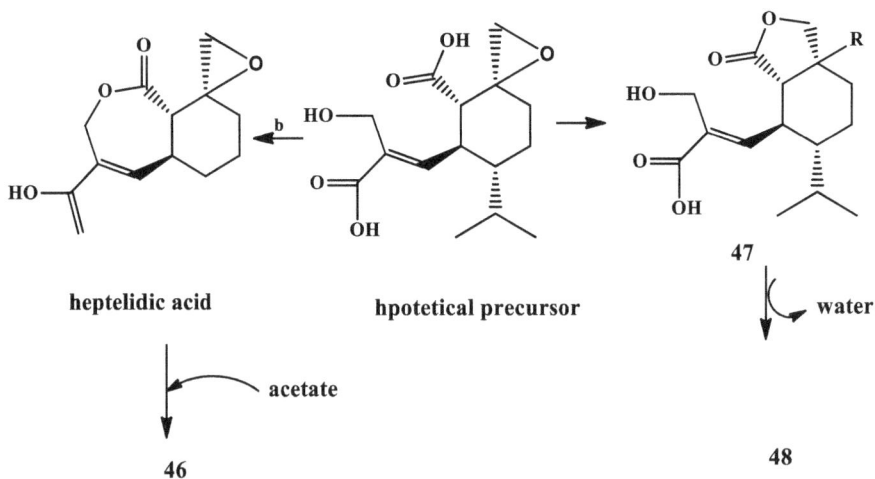

heptelidic acid        hpotetical precursor                           47

                                                                       water

                acetate

        46                                                            48

**Fig. (9).** Hypothetical biosynthetic precursor to explain the formation of compounds **46-48**, and heptelidic acid [61].

A strategy for the rapid identification of purified fungal metabolites by HPLC-UV-MS and $^1$H NMR analyses and, less often, by $^{13}$C NMR analysis, enabled the identification of several compounds. The yet unidentified fungal strain P2AF2F3, isolated from the plant *Anthurium loefgrenii*, was inoculated in 2% malt extract medium, producing the tremulane-type sesquiterpenoid conocenol B (**49**) [63].

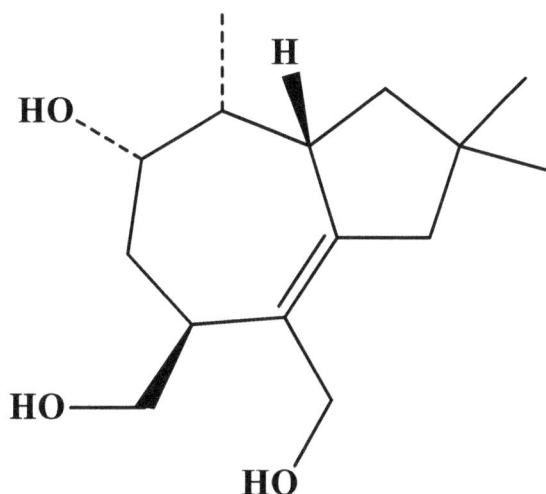

Conocenol B (**49**).

A chemical study of the EtOAc extract of *Nemania bipapillata* (AT-05), an endophytic fungus isolated from the marine red alga *Asparagopsis taxiformis* - Falkenbergia stage, led to the isolation of five new botryane sesquiterpenoids, including the diastereomeric pair (+)-(2R,4S,5R,8S)-(**50**) and (+)-(2R,4R,5R,8S--4-deacetyl-5-hydroxy-botryenalol (**51**), (+)-(2R,4S,5R,8R)-4-deacetyl-botryenalol (**52**), one pair of diastereomeric botryane norsesquiterpenoids bearing an unprecedented degraded carbon skeleton, (+)-(2R,4R,8R)-(**53**) and (+)-(2R,4S,8S)-(**54**), which were named nemenonediol A and nemenonediol B, respectively, in addition to the known sesquiterpenoid **55**, (Fig. **10**) [64]. The authors presented a proposal for the biosynthetic formation of compounds **50-55**, based on the fact that the botryane derivatives have farnesyldiphosphate (FPP) as a precursor and the key intermediate A is generated after cyclization, rearrangement, and hydroxylation (Fig. **11**) [64]. Subsequent hydroxylation and acetylation reactions could transform A to 4β-acetoxy-9β,10β,15α-trihydroxy-probotrydial (**55**), an intermediate in the biosynthesis of botrydial and its derivatives [65 - 68]. Further steps in the biogenetic proposal include cleavage of the vicinal diol in A to give the dialdehyde B, followed by reduction, dehydration, and hydroxylation to generate **50** and **51**. A retro Aldol reaction might epimerize C-$_8$ to give intermediate C$_{21}$ [66]. Reduction, dehydration, and hydroxylation can convert intermediate C into **52** and intermediate D. Allylic rearrangement and oxidative decarboxylation would convert **50** into **54** and D into **53**, the two new *N. bipapillata* metabolites with unprecedented rearranged and degraded terpenoid carbon skeletons (Fig. **11**). The isolated metabolites **50-55** were tested for their ability to inhibit cholinesterase (ChEIs). The inhibition assays were carried out using an immobilized capillary enzyme reactor (ICER) based on acetyl-cholinesterase human recombinant and/or butyrylcholinesterase from human serum, respectively huAChE-ICER and huBChE-ICER (in accordance with the published procedure) [69 - 71]. Compounds **51-55** were more active towards huAChE than huBChE, indicating a selective cholinesterase inhibition, and **53** was the most active compound, with 27.7% (100 M) inhibition against huAChE. Compound **50** was considered a non-selective inhibitor, as it showed similar inhibitory potentials against both huAChE and huBChE (19.9 and 14.1%, respectively), while its stereoisomer (compound **51**) inhibited only huAChE (18.3%). The results represented only modest inhibition of huAChE or huBChE, but they suggest that botryane sesquiterpenoids could act as lead compounds for synthetic development of more potent selective cholinesterase ligands. Compounds **50, 52, 54** and **55** were tested against colorectal carcinoma HCT-116 and breast adenocarcinoma MCF-7 cell lines using the 3-(4,5-dimethylthiazo--2-yl)-2,5-diphenyltetrazolium bromide tetrazolium reduction (MTT) assay, and presented no significant toxicixity at tested concentrations ($IC_{50}$ μM) against both cell lines.

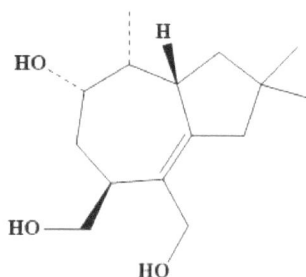

Conocenol B (49)

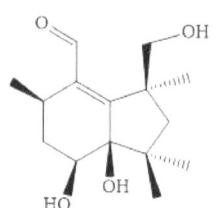

diastereomeric pair
(+)-(2R,4S,5R,8S)-(50)

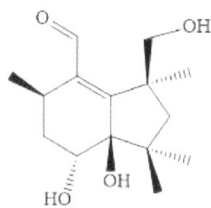

(+)-(2R,4R,5R,8S)-
4-deacetyl-5-hydroxy
-botryenalol (51)

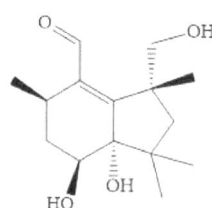

(+)-(2R,4S,5R,8R)
-4-deacetyl-botryenalol
(52)

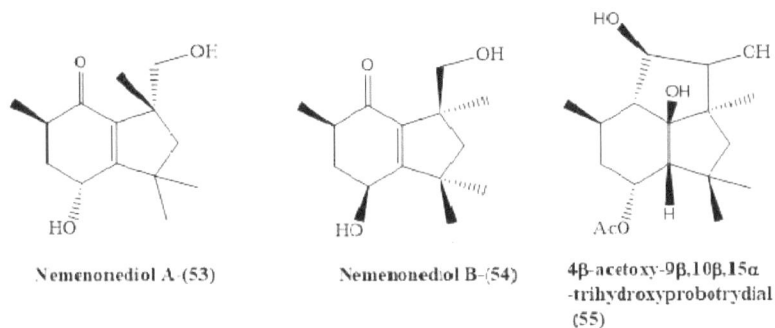

Nemenonediol A (53)          Nemenonediol B-(54)          4β-acetoxy-9β,10β,15α
                                                         -trihydroxyprobotrydial
                                                         (55)

**Fig. (10).** Botryane sesquiterpenoids isolated from EtOAc extract of *Nemania bipapillata* cultures.

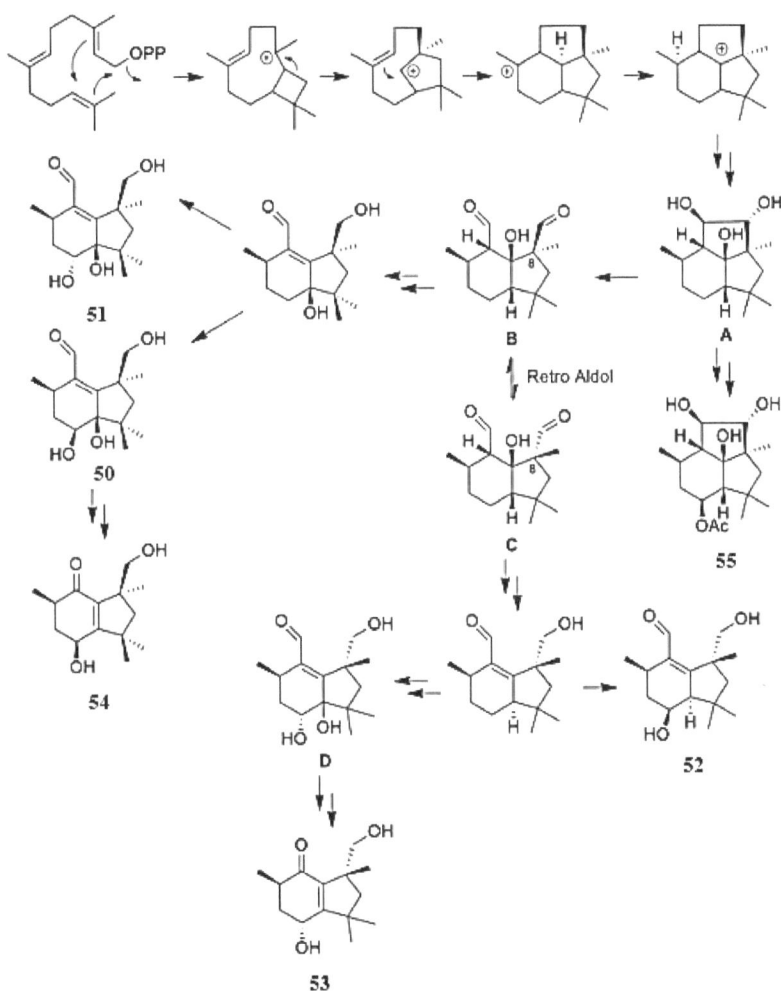

**Fig. (11).** Biogenetic proposal for formation of the compounds **50-55** [64].

## Sesterterpenoids

*Sesterterpenoids* are ubiquitous secondary metabolites in fungi, marine organisms, and plants. Their structural diversity encompasses carbotricyclic ophiobolanes, polycyclic anthracenones, polycyclic furan-2-ones, polycyclic hydroquinones, among many other carbon skeletons [72].

Sesterterpenoids originate from geranylfarnesylpyrophosphate and are found primarily in fungi and marine organisms [73 - 77]. Sesterterpenoids have been a subject of a lot of recent investigations aimed at structural, chemical, and biological characterizations and were found to show a broad spectrum of biological activities against bacteria, fungi, and nematodes [76 - 78]. Many of them exhibit strong anticancer activity *via* novel mode(s) of action [73].

The screening for antimicrobial activity of twenty-five endophytic fungi isolated from the aerial parts of *Paepalanthus chiquitensis* (Eriocaulaceae) was assayed against the bacteria Gram-positive *Staphylococcus aureus*, Gram-negative *Escherichia coli* and *Salmonella setubal*, and the yeast fluconazole-resistant *Candida albicans* [79]. The ethyl acetate extract produced by *Fusarium fujikuroi* was the most bioactive and this fungus was chosen for the chemical study, affording the sesterterpenoid terpestacin (**56**), among others compounds. The small amount of isolated terpestacin did not allow the antimicrobial tests with the substance.

## Meroterpenoids

*Meroterpenoids* are hybrid natural products partially derived from terpenoid pathways as the prefix "mero-" means "part, partial, and fragment" [80]. On the basis of their definition, numerous compounds derived from animals, plants, bacteria, and fungi can be grouped into this class of natural products, which range from widely distributed molecules to species-specific secondary metabolites. Based on their different biosynthetic origins, meroterpenoids can be grouped into 2 major classes: polyketide-terpenoids and nonpolyketide-terpenoids [81]. According to their biosynthetic pathway, meroterpenoids are consisted of at least two parts: a terpenoid fragment (mainly mevalonate pathway) and a nonterpenoid fragment (including polyketide-, amino acid- (typically indole-diterpenes), shikimate-, and miscellaneous-biogenetic pathways). The different nonterpenoid moiety, diverse terpenoid portion (the length of the terpenoid chain and its cyclization mode), and the tailoring reactions make the chemical diversity of meroterpenoids [81]. Therefore, based on their nonterpenoid starting scaffolds, meroterpenoids can be divided into four classes meroterpenoids containing polyketide-terpenoid moiety (polyketide-terpenoids), containing indole-terpenoid moiety (indole-terpenoids), containing shikimate-terpenoid moiety (shikimate-terpenoids), and miscellaneous meroterpenoids [81].

The meroterpenoid austin (**57**) was produced in PDA solid media (on Petri dish) from co-culture (mixture) of two endophytic strains isolated from *Handroanthus impetiginosus* leaves, identified as *Talaromyces purpurogenus* H4 and *Phanerochaete* sp. H2 [82]. The stress caused by the presence of the two fungi species in the same Petri dish seems to be efficient in triggering biosynthetic pathways. Although fungal cultures in agar plates are not ideal substrates in ecological terms, this culture method allows the sampling of different regions of the mycelia as well as the observation of inhibition zones [83].

The bioactivity of austin (**57**) toward *Trypanosoma cruzi* epimastigotes were assessed, and inhibition rates of 100% and 96.39% were observed at concentrations of 200 μg/mL and 100 μg mL$^{-1}$, respectively [82]. The IC$_{50}$ value for austin (**57**) against the epimastigote form was $36.60 \pm 1.20$ μg mL$^{-1}$, while the reference drug benznidazole showed an IC$_{50}$ value of $8.01 \pm 1.31$ μg mL$^{-1}$. In the cytotoxicity assay against uninfected H9c2 cells, austin (**57**) showed a CC$_{50}$ value of $175.65 \pm 1.20$ μg mL$^{-1}$, while in comparison, the CC$_{50}$ value for benznidazole was $187.85 \pm 2.15$ μg mL$^{-1}$ [82]. The selectivity indexes for austin (**57**) and for benznidazole were calculated as 4.79 and 23.45, respectively. Although the selectivity index of austin (SI = 4.79) was less favorable than that observed for benznidazole (SI = 23.45), the meroterpenoid was about fivefold less toxic to the host cell than to the epimastigote. The results demonstrate that **57** shows good potential for use in further studies on the development of trypanocidal drugs [82].

## Triterpenoids

*Triterpenoids* are common terpenoids with a 30-carbon skeleton based on 6 isoprene units. Triterpen/oids include a large group of chemical compounds and are predominantly present in the animal and plant kingdoms. The essential role of triterpenoids is to function as membrane sterols in eukaryotes to provide membrane fluidity. The majority of fungal-derived triterpen/oids natural products have been isolated from basidiomycetes; however, certain ascomycetes are reported having triterpenoids [51]. Animal, plant, and fungal kingdoms are reported to have three kinds of sterols: phytosterols (sitosterol, stigmasterol, campestrol) in plants, cholesterol in animals, and ergosterol in fungi [83, 84]. The majority of fungal-derived triterpenoids natural products have been isolated from basidiomycetes; however, certain ascomycetes are reported producing triterpenoids [85 - 87].

The endophyte *Scleroderma* UFSM Sc1 (Persoon) Fries associated with *Eucalyptus grandis* led to the isolation of two new lanostane triterpenoids, sclerodols A (**58**) and B (**59**), and a known related lanostane triterpenoid (**3S\*, 5R\*,10S\*,13R\*,14R\*,17R\*,20R)-lanosta-8,23-diene-3β,25-diol (60)**, (Fig. **12**) [88]. The new compounds were evaluated for their anti-candidal potential against *Candida albicans, C. tropicalis, C. crusei, C. parapsiosis* activities. Sclerodal B (**59**) showed good anticandidal activity against *C. albicans, C. tropicalis, C. crusei, C. parapsiosis* with the MIC (minimum inhibitory concentration) of 25.0, 25.0, 6.25 and 12.5 µg mL$^{-1}$ and MLC (minimal lethal concentration) of 25.0, 25.0, 12.5 and 25.0 µg mL$^{-1}$ respectively [88]. These results show that the sclerodal B is fungicidal and fungistatic at these concentrations. Sclerodal A (**58**) and triterpenoid **60** were less active against the tested strain than sclerodal B, with the MIC in the range of 12.5-100 µg mL$^{-1}$ and MLC ≥ 50.0 µg mL$^{-1}$. Control nystatin exhibited anti-candidal activities against tested strains with the MIC in the range of 0.77-1.52 µg mL$^{-1}$ and MLC in the range of 3.12-6.25 µg mL$^{-1}$ [88]. Although compounds **58-60** show less activity than the antimicrobial control nystatin, phytochemical products that produce MIC in the range of 100-1000 µg mL$^{-1}$*In vitro* susceptibility tests can be classified as antimicrobials [89].

The ethyl acetate extract from the fermented broth of *Talaromyces* sp. VrTrb2 1.1 isolated of plant *Victoria amazonica* (Nymphaeaceae) yielded the known sterols ergosterol (**61**) and ergosterol peroxide (**62**) [90]. The antifungal potential of **61** was evaluated against wild hospital strains of *Candida albicans* and *C. tropicalis*. Ketoconazole (4 µg mL$^{-1}$) was used as positive antifungal control, and dimethyl sulfoxide (DMSO) 10% was used as a negative control. However, compound **61** exhibited week antifungal activity with inhibition in concentrations (MIC) above 250 µg mL$^{-1}$.

Continuous studies on the chemistry of the endophytic fungus associated with the plant *Bauhinia guianensis* (Fabaceae) have led to the isolation of the fungi *Aspergillus* sp. EJC 04 [91], *Exserohilum rostratum* [92], *Scedosporium apiospermum* [93] and *Pestalotiopsis* sp. EJC07 [94]. From the cultures of the fungi *Aspergillus* sp. EJC 04, *Pestalotiopsis* sp. EJC07 and *Exserohilum rostratum* were produced ergosterol (**61**) and ergosterol peroxide (**62**) [91, 92, 94]. The lethality of the ergosterol (**61**) produced by *Aspergillus* sp. EJC 04 was tested against *Artemia salina*, and activity was observed at concentrações above 250 µg mL$^{-1}$ [91]. Sterol cerevisterol (**63**) was produced by *Exserohilum rostratum* [92] and *Pestalotiopsis* sp. EJC07 [94].

The fungus *Colletotrichum gloeosporioides* was isolated from the leaves of *Virola michelli* (Myristicaceae) [95], a typical Amazon plant, used in folk medicine against skin infection [96]. From the culture of *C. gloeosporioides,* sitosterol (**64**) and stigmasterol (**65**) (in mixture) were identified, and sitostenone (**66**), squalene (**67**), ergosterol (**61**) and ergosterol peroxide (**62**) were isolated (Fig. **13**).

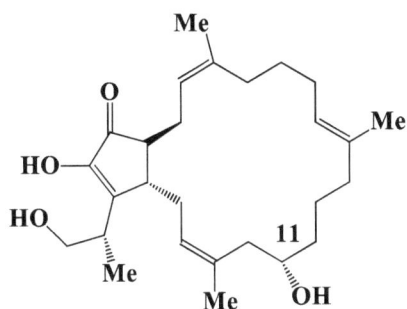

**Terpestacin (56)**

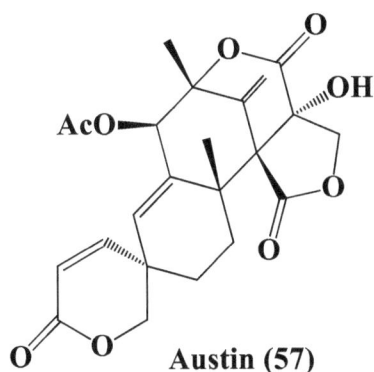

**Austin (57)**

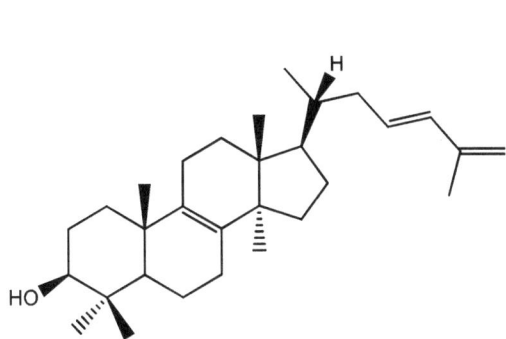

**Sclerodol A (58)**

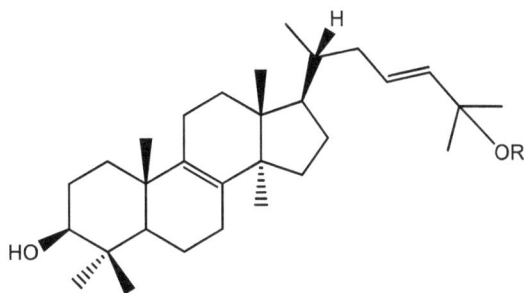

**R=CH3 Sclerodol B (59)**
**R=H  Triterpenoid**

**Fig. (12).** Chemical structures of triterpenoids from *Scleroderma* UFSM Sc1 (Persoon) Fries an endophyte associated with *Eucalyptus grandis*.

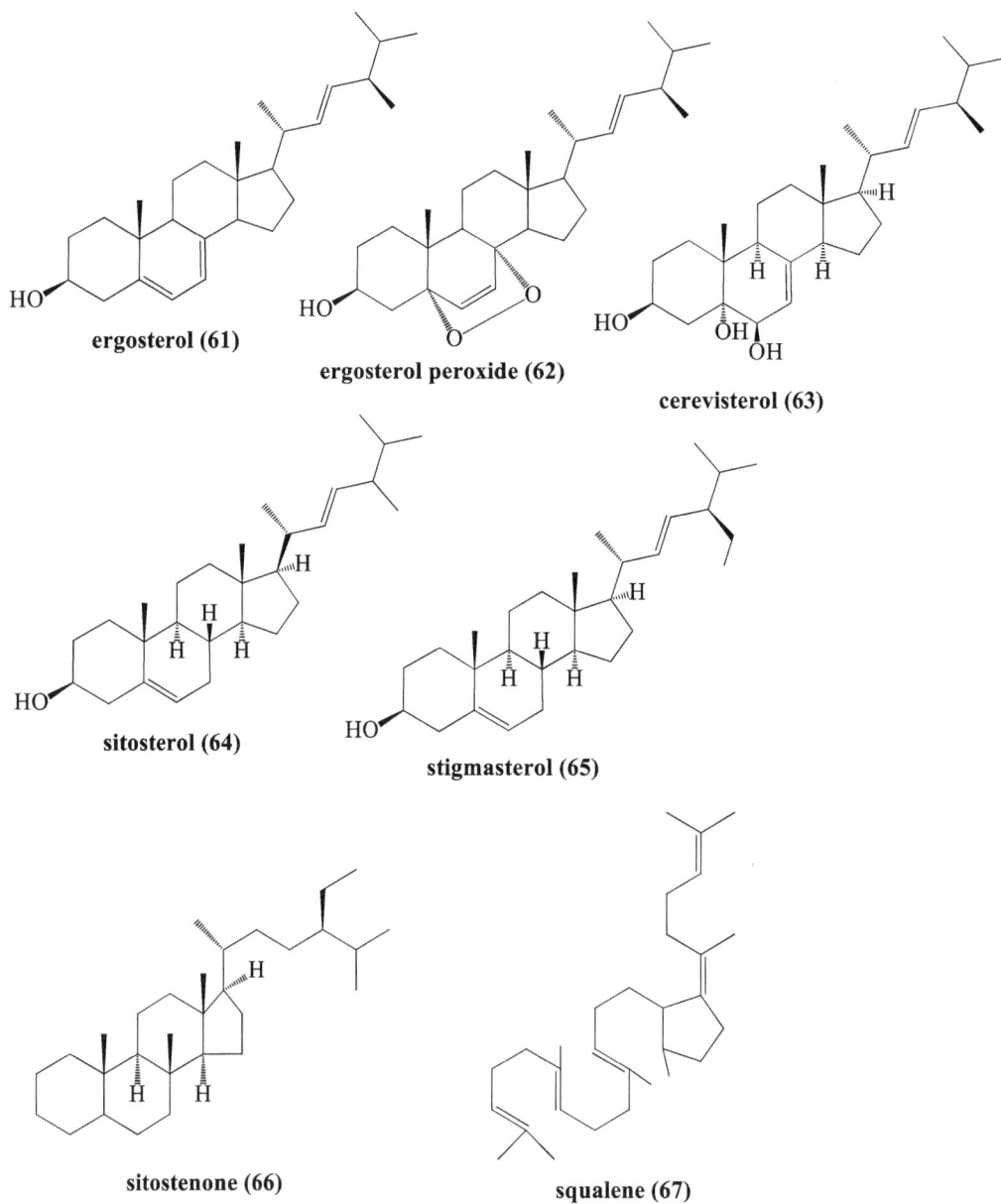

**Fig. (13).**  Chemical structures of triterpenoids produced by endophytic fungi associated with Amazonian plants.

## CONCLUDING REMARKS

Endophytic fungi are of great importance in the production of bioactive substances from different classes of natural products. In this review, we focus on the studies of Brazilian researchers on the terpenoids produced by endophytic fungi and their biological activities published between January 2015 and June 2021. Twenty-six fungi associated with twelve plants from Brazilian were cited. Sixty-seven terpenoids have been described, including four monoterpenoids, fifty-one sesquiterpenoids, one sesterterpene, one meroterpene and ten triterpenoids. None diterpenoid was mentioned during this period. The biological activities of isolated substances have been described, with emphasis on the production of Δ6-protoilludene, the precursor to the cytotoxic illudins, which are pharmaceutically relevant as anticancer therapeutics.

## CONSENT FOR PUBLICATION

Not applicable.

## CONFLICT OF INTEREST

The author declares no conflict of interest, financial or otherwise.

## ACKNOWLEDGEMENTS

The authors would like to thank the Coordenação de Aperfeiçoamento de Pessoal de Nível Superior – Brasil (CAPES) for financial support to the project Proc. 8887.200524/2018-00.

## REFERENCES

[1]    Perveen, S. *Terpenes and Terpenoids*; Perveen, S.; Al-Taweel, A., Eds.; Intech Open: London, UK, **2018**, pp. 1-12.
[http://dx.doi.org/10.5772/intechopen.71175]

[2]    Bérdy, J. Thoughts and facts about antibiotics: Where we are now and where we are heading. *J. Antibiot. (Tokyo)*, **2012**, *65*(8), 385-395.
[http://dx.doi.org/10.1038/ja.2012.27] [PMID: 22511224]

[3]    Walsh, C.T.; Fischbach, M.A. Natural products version 2.0: connecting genes to molecules. *J. Am. Chem. Soc.*, **2010**, *132*(8), 2469-2493.
[http://dx.doi.org/10.1021/ja909118a] [PMID: 20121095]

[4]    Li, L-Y.; Ding, Y.; Groth, I.; Menzel, K-D.; Peschel, G.; Voigt, K.; Deng, Z-W.; Sattler, I.; Lin, W-H. Pyrrole and indole alkaloids from an endophytic *Fusarium incarnatum* (HKI00504) isolated from the mangrove plant *Aegiceras corniculatum. J. Asian Nat. Prod. Res.*, **2008**, *10*(7-8), 775-780.
[PMID: 18696331]

[5]    Azevedo, J.L. Botânica: uma ciência básica ou aplicada? *Rev. Bras. Bot.*, **1999**, *22*, 225-229.
[http://dx.doi.org/10.1590/S0100-84041999000500002]

[6]    Petrini, O.; Sieber, T.N.; Toti, L.; Viret, O. Ecology, metabolite production, and substrate utilization in endophytic fungi. *Nat. Toxins*, **1993**, *1*(3), 185-196.
[http://dx.doi.org/10.1002/nt.2620010306] [PMID: 1344919]

[7]    Qin, J.C.; Zhang, Y.M.; Gao, J.M.; Bai, M.S.; Yang, S.X.; Laatsch, H.; Zhang, A.L. Bioactive metabolites produced by *Chaetomium globosum*, an endophytic fungus isolated from *Ginkgo biloba. Bioorg. Med. Chem. Lett.*, **2009**, *19*(6), 1572-1574.
[http://dx.doi.org/10.1016/j.bmcl.2009.02.025] [PMID: 19246197]

[8]     Macías-Rubalcava, M.L.; Hernández-Bautista, B.E.; Jiménez-Estrada, M.; González, M.C.; Glenn, A.E.; Hanlin, R.T.; Hernández-Ortega, S.; Saucedo-García, A.; Muria-González, J.M.; Anaya, A.L. Naphthoquinone spiroketal with allelochemical activity from the newly discovered endophytic fungus *Edenia gomezpompae*. *Phytochemistry,* **2008**, *69*(5), 1185-1196.
        [http://dx.doi.org/10.1016/j.phytochem.2007.12.006] [PMID: 18234248]

[9]     Soares, D.A.; Rosa, L.H.; da Silva, J.F.M.; Pimenta, R.S. A review of bioactive compounds produced by endophytic fungi associated with medicinal plants: Uma revisão de compostos bioativos produzidos por fungos endofíticos associados a plantas medicinais. *Bol. Mus. Para. Emílio Goeldi Ciênc. Nat.,* **2017**, *12*, 331-352.
        [http://dx.doi.org/10.46357/bcnaturais.v12i3.83]

[10]    Sudha, V.; Govindaraj, R.; Baskar, K.; Al-Dhabi, N.A.; Duraipandiyan, V. Biological properties of endophytic fungi. *Braz. Arch. Biol. Technol.,* **2016**, *59*(0)e16150436
        [http://dx.doi.org/10.1590/1678-4324-2016150436]

[11]    Zheng, R.; Li, S.; Zhang, X.; Zhao, C. Biological activities of some new secondary metabolites isolated from endophytic fungi: A Review Study. *Int. J. Mol. Sci.,* **2021**, *22*(2), 959.
        [http://dx.doi.org/10.3390/ijms22020959] [PMID: 33478038]

[12]    Kumar, G.; Chandra, P.; Choudhary, M. Endophytic fungi: A potential source of bioactive compounds. *Chem. Sci. Ver. Lett.,* **2017**, *6*, 2373-2381.

[13]    Schulz, B.; Römmert, A.K.; Dammann, U.; Aust, H.J.Ü.R.; Strack, D. The endophyte-host interaction: a balanced antagonism? *Mycol. Res.,* **1999**, *103*(10), 1275-1283.
        [http://dx.doi.org/10.1017/S0953756299008540]

[14]    Schulz, B.; Boyle, C. The endophytic continuum. *Mycol. Res.,* **2005**, *109*(6), 661-686.
        [http://dx.doi.org/10.1017/S095375620500273X] [PMID: 16080390]

[15]    Arnold, A.E. *Tropical Forest Community Ecology*; Carson, W.P.; Schnitzer, S.A., Eds.; Wiley-Blackwell: West Sussex, UK, **2008**, pp. 254-271.

[16]    Carroll, G.C. *Microbial ccology of leaves*; Andrews, J.H.; Hirano, S.S., Eds.; Springer-Verlag: New York, USA, **1991**, pp. 358-375.
        [http://dx.doi.org/10.1007/978-1-4612-3168-4_18]

[17]    Santos, L.S.; Ripardo Filho, H.S.; Araujo, R.N.M.; Guilhon, G.M.S.P. *Souza Filho, A.P.S. (Org.)*; Barnhart, P.O.A.C.E.A.E., Ed.; Marques: Belém, **2017**, pp. 123-143.

[18]    Jia, M.; Chen, L.; Xin, H.L.; Zheng, C.J.; Rahman, K.; Han, T.; Qin, L.P. A Friendly relationship between endophytic fungi and medicinal plants: A systematic Review. *Front. Microbiol.,* **2016**, *7*, 906.
        [http://dx.doi.org/10.3389/fmicb.2016.00906] [PMID: 27375610]

[19]    Waqas, M.; Khan, A.L.; Kamran, M.; Hamayun, M.; Kang, S.M.; Kim, Y.H.; Lee, I.J. Endophytic fungi produce gibberellins and indoleacetic acid and promotes host-plant growth during stress. *Molecules,* **2012**, *17*(9), 10754-10773.
        [http://dx.doi.org/10.3390/molecules170910754] [PMID: 22960869]

[20]    Dingle, J.; Mcgee, P.A. Some endophytic fungi reduce the density of pustules of *Puccinia recondita* f. sp. tritici in wheat. *Mycol. Res.,* **2003**, *107*(3), 310-316.
        [http://dx.doi.org/10.1017/S0953756203007512] [PMID: 12825500]

[21]    Firáková, S.; Šturdíková, M.; Múčková, M. Bioactive secondary metabolites produced by microorganisms associated with plants. *Biologia (Bratisl.),* **2007**, *62*(3), 251-257.
        [http://dx.doi.org/10.2478/s11756-007-0044-1]

[22]    Rodriguez, R.J.; White, J.F., Jr; Arnold, A.E.; Redman, R.S. Fungal endophytes: diversity and functional roles. *New Phytol.,* **2009**, *182*(2), 314-330.
        [http://dx.doi.org/10.1111/j.1469-8137.2009.02773.x] [PMID: 19236579]

[23]    Shwab, E.K.; Keller, N.P. Regulation of secondary metabolite production in filamentous ascomycetes.

*Mycol. Res.,* **2008**, *112*(2), 225-230.
[http://dx.doi.org/10.1016/j.mycres.2007.08.021] [PMID: 18280128]

[24]    De Bary, A. *Hofmeister's handbook of physiological botany*; Wilhelm Engelmann: Leipzig, Germany, **1866**, pp. 1831-1888.

[25]    Stierle, A.; Strobel, G.; Stierle, D. Taxol and taxane production by *Taxomyces andreanae*, an endophytic fungus of Pacific yew. *Science,* **1993**, *260*(5105), 214-216.
[http://dx.doi.org/10.1126/science.8097061] [PMID: 8097061]

[26]    Guo, B.; Wang, Y.; Sun, X.; Tang, K. Bioactive natural products from endophytes: a review. *Prikl. Biokhim. Mikrobiol.,* **2008**, *44*(2), 153-158.
[PMID: 18669256]

[27]    Ribeiro, B.A.; da Mata, T.B.; Canuto, G.A.B.; Silva, E.O. Chemical diversity of secondary metabolites produced by brazilian endophytic fungi. *Curr. Microbiol.,* **2021**, *78*(1), 33-54.
[http://dx.doi.org/10.1007/s00284-020-02264-0] [PMID: 33108493]

[28]    Zhao, Y.; Cui, J.; Liu, M.; Zhao, L. Progress on erpenoids with biological activities produced by plant endophytic fungi in China between 2017 and 2019. *Nat. Prod. Commun.,* **2017**, *15*(7), 1-18.

[29]    Souza, J.J.; Vieira, I.J.C.; Rodrigues-Filho, E.; Braz-Filho, R. Terpenoids from endophytic fungi. *Molecules,* **2011**, *16*(12), 10604-10618.
[http://dx.doi.org/10.3390/molecules161210604] [PMID: 22183885]

[30]    Oliveira, M.N.; Santos, L.S.; Guilhon, G.M.S.P.; Santos, A.S.; Ferreira, I.C.S.; Lopes-Junior, M.L.; Arruda, M.S.P.; Marinho, A.M.R.; Silva, M.N.; Rodrigues-Filho, E.; Oliveira, M.C.F. Novel anthraquinone derivatives produced by Pestalotiopsis guepinii, an endophytic of the medicinal plant Virola michelii (Myristicaceae). *J. Braz. Chem. Soc.,* **2011**, *22*(5), 993-996.
[http://dx.doi.org/10.1590/S0103-50532011000500025]

[31]    Verma, V.C.; Kharwar, R.N.; Strobel, G.A. Chemical and functional diversity of natural products from plant associated endophytic fungi. *Nat. Prod. Commun.,* **2009**, *4*(11)1934578X0900401
[http://dx.doi.org/10.1177/1934578X0900401114] [PMID: 19967984]

[32]    Strobel, G.; Daisy, B.; Castillo, U.; Harper, J. Natural products from endophytic microorganisms. *J. Nat. Prod.,* **2004**, *67*(2), 257-268.
[http://dx.doi.org/10.1021/np030397v] [PMID: 14987067]

[33]    Krings, M.; Taylor, T.N.; Hass, H.; Kerp, H.; Dotzler, N.; Hermsen, E.J. Fungal endophytes in a 400-million-yr-old land plant: infection pathways, spatial distribution, and host responses. *New Phytol.,* **2007**, *174*(3), 648-657.
[http://dx.doi.org/10.1111/j.1469-8137.2007.02008.x] [PMID: 17447919]

[34]    Strobel, G.A. Microbial gifts from rain forests. *Can. J. Plant Pathol.,* **2002**, *24*(1), 14-20.
[http://dx.doi.org/10.1080/07060660109506965]

[35]    Strobel, G.A. Rainforest endophytes and bioactive products. *Crit. Rev. Biotechnol.,* **2002**, *22*(4), 315-333.
[http://dx.doi.org/10.1080/07388550290789531] [PMID: 12487423]

[36]    Strobel, G. Harnessing endophytes for industrial microbiology. *Curr. Opin. Microbiol.,* **2006**, *9*(3), 240-244.
[http://dx.doi.org/10.1016/j.mib.2006.04.001] [PMID: 16647289]

[37]    Redell, P.; Gordon, V. *Biodiversity: New Leads for Pharmaceutical and Agrochemical Industries; Wrigley, S.K.; Hayes, M.A.; Thomas, R.; Chrystal; E.J.T*; Nicholson, N., Ed.; The Royal Society of Chemistry: Cambridge, UK, **2000**, pp. 205-212.

[38]    Rabi, T.; Bishayee, A. Terpenoids and breast cancer chemoprevention. *Breast Cancer Res. Treat.,* **2009**, *115*(2), 223-239.
[http://dx.doi.org/10.1007/s10549-008-0118-y] [PMID: 18636327]

[39] Lindequist, U.; Niedermeyer, T.H.J.; Jülich, W.D. The pharmacological potential of mushrooms. *Evid. Based Complement. Alternat. Med.,* **2005**, *2*(3), 285-299.
[http://dx.doi.org/10.1093/ecam/neh107] [PMID: 16136207]

[40] Xu, J.W.; Zhao, W.; Zhong, J.J. Biotechnological production and application of ganoderic acids. *Appl. Microbiol. Biotechnol.,* **2010**, *87*(2), 457-466.
[http://dx.doi.org/10.1007/s00253-010-2576-5] [PMID: 20437236]

[41] Abraham, W.R. Bioactive sesquiterpenes produced by fungi: are they useful for humans as well? *Curr. Med. Chem.,* **2001**, *8*(6), 583-606.
[http://dx.doi.org/10.2174/0929867013373147] [PMID: 11281843]

[42] Elisashvili, V. Submerged cultivation of medicinal mushrooms: bioprocesses and products (review). *Int. J. Med. Mushrooms,* **2012**, *14*(3), 211-239.
[http://dx.doi.org/10.1615/IntJMedMushr.v14.i3.10] [PMID: 22577974]

[43] Wasser, S.P. Current findings, future trends, and unsolved problems in studies of medicinal mushrooms. *Appl. Microbiol. Biotechnol.,* **2011**, *89*(5), 1323-1332.
[http://dx.doi.org/10.1007/s00253-010-3067-4] [PMID: 21190105]

[44] Aram, K.; Taek, S.L. *In: Biotechnology for Biofuel Production and Optimization; Carrie, A.E.; Cong, T.T., Eds.; Elsevier B.V: Amsterdam,* **2016**, 35-71.

[45] Breitmaier, E. *Terpenes: Flavors, Fragrances, Pharmaca, Pheromones*; Wiley-VCH: Weinheim, **2006**.
[http://dx.doi.org/10.1002/9783527609949]

[46] Dehsheikh, A.B.; Sourestani, M.M.; Dehsheikh, P.B.; Mottaghipisheh, J.; Vitalini, S.; Iriti, M. Monoterpenes: Essential oil components with valuable features. *Mini Rev. Med. Chem.,* **2020**, *20*(11), 958-974.
[http://dx.doi.org/10.2174/1389557520666200122144703] [PMID: 31969098]

[47] Schwab, W.; Fuchs, C.; Huang, F.C. Transformation of terpenes into fine chemicals. *Eur. J. Lipid Sci. Technol.,* **2013**, *115*(1), 3-8.
[http://dx.doi.org/10.1002/ejlt.201200157]

[48] Zielińska-Błajet, M.; Feder-Kubis, J. Monoterpenes and their derivatives: Recent development in biological and medical pplications. *Int. J. Mol. Sci.,* **2020**, *21*(19), 7078.
[http://dx.doi.org/10.3390/ijms21197078] [PMID: 32992914]

[49] de Sena Filho, J.G.; Quin, M.B.; Spakowicz, D.J.; Shaw, J.J.; Kucera, K.; Dunican, B.; Strobel, S.A.; Schmidt-Dannert, C. Genome of *Diaporthe* sp. provides insights into the potential inter-phylum transfer of a fungal sesquiterpenoid biosynthetic pathway. *Fungal Biol.,* **2016**, *120*(8), 1050-1063.
[http://dx.doi.org/10.1016/j.funbio.2016.04.001] [PMID: 27521636]

[50] Gomes, R.R.; Glienke, C.; Videira, S.I.R.; Lombard, L.; Groenewald, J.Z.; Crous, P.W. <I>Diaporthe</I>: a genus of endophytic, saprobic and plant pathogenic fungi. *Persoonia,* **2013**, *31*(1), 1-41.
[http://dx.doi.org/10.3767/003158513X666844] [PMID: 24761033]

[51] Singh, A.K.; Rana, H.K.; Pandey, A.K. *Advancement in White Biotechnology Through Fungi, Fungal Biology.,* **2019**, , 229-248.
[http://dx.doi.org/10.1007/978-3-030-14846-1_8]

[52] Davis, E.M.; Croteau, R. Cyclization enzymes in the biosynthesis of monoterpenes, sesquiterpenes, and diterpenes. *Top. Curr. Chem.,* **2000**, *209*, 53-95.
[http://dx.doi.org/10.1007/3-540-48146-X_2]

[53] Cane, D.E.; Kang, I. Aristolochene synthase: purification, molecular cloning, high-level expression in *Escherichia coli*, and characterization of the *Aspergillus terreus* cyclase. *Arch. Biochem. Biophys.,* **2000**, *376*(2), 354-364.
[http://dx.doi.org/10.1006/abbi.2000.1734] [PMID: 10775423]

[54]   Christianson, D.W. Structural biology and chemistry of the terpenoid cyclases. *Chem. Rev.,* **2006**, *106*(8), 3412-3442.
[http://dx.doi.org/10.1021/cr050286w] [PMID: 16895335]

[55]   Vedula, L.S.; Jiang, J.; Zakharian, T.; Cane, D.E.; Christianson, D.W. Structural and mechanistic analysis of trichodiene synthase using site-directed mutagenesis: probing the catalytic function of tyrosine-295 and the asparagine-225/serine-229/glutamate-233-Mg$^{2+}$B motif. *Arch. Biochem. Biophys.,* **2008**, *469*(2), 184-194.
[http://dx.doi.org/10.1016/j.abb.2007.10.015] [PMID: 17996718]

[56]   Lesburg, C.A.; Caruthers, J.M.; Paschall, C.M.; Christianson, D.W. Managing and manipulating carbocations in biology: terpenoid cyclase structure and mechanism. *Curr. Opin. Struct. Biol.,* **1998**, *8*(6), 695-703.
[http://dx.doi.org/10.1016/S0959-440X(98)80088-2] [PMID: 9914250]

[57]   Mandavid, H.; Rodrigues, A.M.S.; Espindola, L.S.; Eparvier, V.; Stien, D. Secondary Metabolites isolated from the Amazonian endophytic fungus *Diaporthe* sp. SNB-GSS10. *J. Nat. Prod.,* **2015**, *78*(7), 1735-1739.
[http://dx.doi.org/10.1021/np501029s] [PMID: 26149922]

[58]   Oliveira, F.C.; Barbosa, F.G.; Mafezoli, J.; Oliveira, M.C.F.; Camelo, A.L.M.; Longhinotti, E.; Lima, A.C.A.; Câmara, M.P.S.; Gonçalves, F.J.T.; Freire, F.C.O. Volatile organic compounds from filamentous fungi: a chemotaxonomic tool of the botryosphaeriaceae family. *J. Braz. Chem. Soc.,* **2015**, *26*, 2189-2194.
[http://dx.doi.org/10.5935/0103-5053.20150204]

[59]   Gubiani, J.R.; Nogueira, C.R.; Pereira, M.D.P.; Young, M.C.M.; Ferreira, P.M.P.; de Moraes, M.O.; Pessoa, C.; Bolzani, V.S.; Araujo, A.R. Rearranged Sesquiterpenes and branched polyketides produced by the endophyte *Camarops* sp. *Phytochem. Lett.,* **2016**, *17*, 251-257.
[http://dx.doi.org/10.1016/j.phytol.2016.08.007]

[60]   Schobert, R.; Knauer, S.; Seibt, S.; Biersack, B. Anticancer active illudins: recent developments of a potent alkylating compound class. *Curr. Med. Chem.,* **2011**, *18*(6), 790-807.
[http://dx.doi.org/10.2174/092986711794927766] [PMID: 21182482]

[61]   Cota, B.B.; Tunes, L.G.; Maia, D.N.B.; Ramos, J.P.; Oliveira, D.M.; Kohlhoff, M.; Alves, T.M.A.; Souza-Fagundes, E.M.; Campos, F.F.; Zani, C.L. Leishmanicidal compounds of Nectria pseudotrichia, an endophytic fungus isolated from the plant Caesalpinia echinata (Brazilwood). *Mem. Inst. Oswaldo Cruz,* **2018**, *113*(2), 102-110.
[http://dx.doi.org/10.1590/0074-02760170217] [PMID: 29236928]

[62]   de Morais-Teixeira, E.; Gallupo, M.K.; Rodrigues, L.F.; Romanha, A.J.; Rabello, A. *In vitro* interaction between paromomycin sulphate and four drugs with leishmanicidal activity against three New World *Leishmania* species. *J. Antimicrob. Chemother.,* **2014**, *69*(1), 150-154.
[http://dx.doi.org/10.1093/jac/dkt318] [PMID: 23970484]

[63]   Iócaa, L.P.; Romminger, S.; Santos, M.F.C.; Bandeira, K.F.; Rodrigues, F.T.; Kossuga, M.H.; Nicacio, K.J.; Ferreira, E.L.F.; Morais-Uranoa, R.P.; Passosa, M.S.; Kohn, L.K.; Arns, C.W.; Sette, L.D.; Berlinck, R.G.S. A strategy for the rapid identification of fungal metabolites and the discovery of the antiviral activity of pyrenocine a and harzianopyridone. *Quim. Nova,* **2016**, *39*, 720-731.

[64]   Medina, R.P.; Araujo, A.R.; Batista, J.M., Jr; Cardoso, C.L.; Seidl, C.; Vilela, A.F.L.; Domingos, H.V.; Costa-Lotufo, L.V.; Andersen, R.J.; Silva, D.H.S. Botryane terpenoids produced by *Nemania bipapillata*, an endophytic fungus isolated from red alga *Asparagopsis taxiformis* - Falkenbergia stage. *Sci. Rep.,* **2019**, *9*(1), 12318.
[http://dx.doi.org/10.1038/s41598-019-48655-7] [PMID: 31444403]

[65]   Durán-Patrón, R.; Colmenares, A.J.; Hernández-Galán, R.; Collado, I.G. Some key metabolic intermediates in the biosynthesis of botrydial and related compounds. *Tetrahedron,* **2001**, *57*(10), 1929-1933.

[http://dx.doi.org/10.1016/S0040-4020(01)00016-3]

[66]     Collado, I.G.; Sánchez, A.J.M.; Hanson, J.R. Fungal terpene metabolites: biosynthetic relationships and the control of the phytopathogenic fungus *Botrytis cinerea. Nat. Prod. Rep.,* **2007**, *24*(4), 674-686.
[http://dx.doi.org/10.1039/b603085h] [PMID: 17653354]

[67]     Yuan, Y.; Feng, Y.; Ren, F.; Niu, S.; Liu, X.; Che, Y. A botryane metabolite with a new hexacyclic skeleton from an entomogenous fungus *Hypocrea* sp. *Org. Lett.,* **2013**, *15*(23), 6050-6053.
[http://dx.doi.org/10.1021/ol402953k] [PMID: 24490837]

[68]     Moraga, J.; Dalmais, B.; Izquierdo-Bueno, I.; Aleu, J.; Hanson, J.R.; Hernández-Galán, R.; Viaud, M.; Collado, I.G. Genetic and molecular basis of botrydial biosynthesis: Connecting cytochrome P450-encoding genes to biosynthetic intermediates. *ACS Chem. Biol.,* **2016**, *11*(10), 2838-2846.
[http://dx.doi.org/10.1021/acschembio.6b00581] [PMID: 27529428]

[69]     Vanzolini, K.L.; Vieira, L.C.C.; Corrêa, A.G.; Cardoso, C.L.; Cass, Q.B. Acetylcholinesterase immobilized capillary reactors-tandem mass spectrometry: an on-flow tool for ligand screening. *J. Med. Chem.,* **2013**, *56*(5), 2038-2044.
[http://dx.doi.org/10.1021/jm301732a] [PMID: 23330848]

[70]     Vilela, A.F.L.; Seidl, C.; Lima, J.M.; Cardoso, C.L. An improved immobilized enzyme reactor-mass spectrometry-based label free assay for butyrylcholinesterase ligand screening. *Anal. Biochem.,* **2018**, *549*, 53-57.
[http://dx.doi.org/10.1016/j.ab.2018.03.012] [PMID: 29550345]

[71]     Seidl, C.; Vilela, A.F.L.; Lima, J.M.; Leme, G.M.; Cardoso, C.L. A novel on-flow mass spectrometry-based dual enzyme assay. *Anal. Chim. Acta,* **2019**, *1072*, 81-86.
[http://dx.doi.org/10.1016/j.aca.2019.04.057] [PMID: 31146868]

[72]     Evidente, A.; Kornienko, A.; Lefranc, F.; Cimmino, A.; Dasari, R.; Evidente, M.; Mathieu, V.; Kiss, R. Sesterterpenoids with anticancer activity. *Curr. Med. Chem.,* **2015**, *22*(30), 3502-3522.
[http://dx.doi.org/10.2174/0929867322666150821101047] [PMID: 26295461]

[73]     Lorigooini, Z.; Jamshidi-kia, F.; Dodman, S. *Recent Advances in Natural Products Analysis*; Silva, A.S.; Fazel, S.; Mina, N.; Seyed, S.; Nabavi, M., Eds.; Elsevier Inc.: Amsterdam, **2020**, pp. 289-312.
[http://dx.doi.org/10.1016/B978-0-12-816455-6.00008-1]

[74]     Dewick, P.M. *Medicinal Natural Products - A Biosynthetic Approach.,* **2009**, , 197-306.
[http://dx.doi.org/10.1002/9780470742761]

[75]     Osbourn, A.E.; Lanzotti, V. Plant-derived Natural Products - Synthesis, Function and Application; Springer: Dortdrecht, **2009**, 3-50.

[76]     Cimmino, A.; Andolfi, A.; Evidente, A. Phytotoxic terpenes produced by phytopathogenic fungi and allelopathic plants. *Nat. Prod. Commun.,* **2014**, *9*(3)1934578X1400900
[http://dx.doi.org/10.1177/1934578X1400900330] [PMID: 24689226]

[77]     González, M.A. Scalarane sesterterpenoids. *Curr. Bioact. Compd.,* **2010**, *6*, 178-210.
[http://dx.doi.org/10.2174/157340710793237362]

[78]     Wang, L.; Yang, B.; Lin, X.P.; Zhou, X.F.; Liu, Y. Sesterterpenoids. *Nat. Prod. Rep.,* **2013**, *30*(3), 455-473.
[http://dx.doi.org/10.1039/c3np20089b] [PMID: 23385977]

[79]     Hilário, F.; Chapla, V.M.; Araujo, A.R.; Sano, P.T.; Bauaba, T.M.; Dos Santos, L.C. Antimicrobial screening of endophytic fungi isolated from the aerial parts of *Paepalanthus chiquitensis* (Eriocaulaceae) led to the isolation of secondary metabolites produced by *Fusarium fujikuroi. J. Braz. Chem. Soc.,* **2017**, *28*, 1389-1395.

[80]     Geris, R.; Simpson, T.J. Meroterpenoids produced by fungi. *Nat. Prod. Rep.,* **2009**, *26*(8), 1063-1094.
[http://dx.doi.org/10.1039/b820413f] [PMID: 19636450]

[81]     Jiang, M.; Wu, Z.; Liu, L.; Chen, S. The chemistry and biology of fungal meroterpenoids

(2009–2019). *Org. Biomol. Chem.,* **2021**, *19*(8), 1644-1704.
[http://dx.doi.org/10.1039/D0OB02162H] [PMID: 33320161]

[82]   Do nascimento, J. S.; Silva, F.M.; Magallanes-Noguera, C. A.; Kurina-Sanz, M.; Dos Santos, E. G.;
       Caldas, I. S.; Luiz, J. H. H.; Silva, E. O. Natural trypanocidal product produced by endophytic fungi
       through co-culturing. *Folia Microbiol. (Praha),* **2020**, *65*, 323-328.

[83]   Dupont, S.; Lemetais, G.; Ferreira, T.; Cayot, P.; Gervais, P.; Beney, L. Ergosterol biosynthesis: a
       fungal pathway for life on land? *Evolution,* **2012**, *66*(9), 2961-2968.
       [http://dx.doi.org/10.1111/j.1558-5646.2012.01667.x] [PMID: 22946816]

[84]   Yadav, A.N.; Verma, P.; Kumar, V.; Sangwan, P.; Mishra, S.; Panjiar, N.; Gupta, V.K.; Saxena, A.K.
       *New and Future Developments in Microbial Biotechnology and Bioengineering: Microbial Secondary
       Metabolites Biochemistry and Applications.,* **2018**, , 3-18.
       [http://dx.doi.org/10.1016/B978-0-444-63501-3.00001-6]

[85]   Zhao, M.; Gödecke, T.; Gunn, J.; Duan, J.A.; Che, C.T. Protostane and fusidane triterpenes: a mini-
       review. *Molecules,* **2013**, *18*(4), 4054-4080.
       [http://dx.doi.org/10.3390/molecules18044054] [PMID: 23563857]

[86]   Ríos, J.L.; Andújar, I.; Recio, M.C.; Giner, R.M. Lanostanoids from fungi: a group of potential
       anticancer compounds. *J. Nat. Prod.,* **2012**, *75*(11), 2016-2044.
       [http://dx.doi.org/10.1021/np300412h] [PMID: 23092389]

[87]   Hiscox, J.; Baldrian, P.; Rogers, H.J.; Boddy, L. Changes in oxidative enzyme activity during
       interspecific mycelial interactions involving the white-rot fungus *Trametes versicolor. Fungal Genet.
       Biol.,* **2010**, *47*(6), 562-571.
       [http://dx.doi.org/10.1016/j.fgb.2010.03.007] [PMID: 20371297]

[88]   Morandini, L. M. B.; Neto, A. T.; Pedroso, M.; Antoniolli, Z. I.; Burrow, R.; Bortoluzzi, A. J.;
       Mostardeiro, M. A.; Silva, U.; Dalcol, I. I.; Morel, A. F. Lanostane-type triterpenes from the fungal
       endophyte Scleroderma UFSMSc1 (Persoon) Fries. *Bioorg. Med. Chem. Lett.,* **2016**, *26*, 1173-1176.

[89]   Simões, M.; Bennett, R.N.; Rosa, E.A.S. Understanding antimicrobial activities of phytochemicals
       against multidrug resistant bacteria and biofilms. *Nat. Prod. Rep.,* **2009**, *26*(6), 746-757.
       [http://dx.doi.org/10.1039/b821648g] [PMID: 19471683]

[90]   Da Silva, P.H.F.; De Souza, M.P.; Bianco, E.A.; Da Silva, S.R.S.; Soares, L.N.; Costa, E.V.; Da Silva,
       F.M.A.; Barison, A.; Forim, M.R.; Cass, Q.B.; De Souza, A.D.L.; Koolen, H.H.F.; De Souza, A.Q.L.
       Antifungal polyketides and other compounds from Amazonian endophytic *Talaromyces* fungi. *J. Braz.
       Chem. Soc.,* **2018**, *29*, 622-630.

[91]   Feitosa, A.O.; Dias, A.C.S.; Ramos, G.C.; Bitencourt, H.R.; Siqueira, J.E.S.; Marinho, P.S.B.; Barison,
       A.; Ocampos, F.M.M.; Marinho, A.M.R. Lethality of cytochalasin B and other compounds isolated
       from fungus Aspergillus sp. (Trichocomaceae) endophyte of Bauhinia guianensis (Fabaceae). *Rev.
       Argent. Microbiol.,* **2016**, *48*(3), 259-263.
       [http://dx.doi.org/10.1016/j.ram.2016.04.002] [PMID: 27567521]

[92]   Pinheiro, E.A.A.; Borges, F.C.; Pina, J.R.S.; Ferreira, L.R.S.; Cordeiro, J.S.; Carvalho, J.M.; Feitosa,
       A.O.; Campos, F.R.; Barison, A.; Souza, A.D.L.; Marinho, P.S.B.; Marinho, A.M.R. Annularins I and
       J: New Metabolites Isolated from Endophytic Fungus *Exserohilum rostratum. J. Braz. Chem. Soc.,*
       **2016**, *27*, 1432-1436.
       [http://dx.doi.org/10.5935/0103-5053.20160140]

[93]   Cordeiro, J.S.; Carvalho, J.M.; Feitosa, A.O.; Pinheiro, E.A.A.; Marinho, P.S.B.; Marinho, A.M.R.
       Brefeldin A and other chemical constituents from endophytic fungus. Rev. Virt. *Quím,* **2019**, *11*, 210-
       217.

[94]   Souza, E.M.C.; Silva, E.L.D.; Marinho, A.M.R.; Marinho, P.S.B. (4S)-4,8-dihydroxy-1-tetralone and
       other chemical constituents from *Pestalotiopsis* sp. EJC07, endophytic from*Bauhinia guianensis. An.
       Acad. Bras. Cienc.,* **2016**, *88*(1), 29-33.
       [http://dx.doi.org/10.1590/0001-3765201620140378] [PMID: 26871501]

[95]   Carvalho, J.M.; Paixão, L.K.O.; Dolabela, M.F.; Marinho, P.S.B.; Marinho, A.M.R. Phytosterols isolated from endophytic fungus Colletotrichum gloeosporioides(Melanconiaceae). *Acta Amazon.,* **2016**, *46*(1), 69-72.
[http://dx.doi.org/10.1590/1809-4392201500072]

[96]   Santos, L.S.; Borges, F.C.; Oliveira, M.N.; Souza Filho, A.P.S.; Guilhon, G.M.P.; Arruda, M.S.P.; Muller, A.H.; Santos, A.S.; Arruda, A.C. Allelochemicals isolated from the leaves of *Virola michelli. Allelopathy J.,* **2007**, *20*, 235-243.

# Volatile Terpenoids in Myrtaceae Species: Chemical Structures and Applications

**Oberdan Oliveira Ferreira**[1,2]**, Celeste de Jesus Pereira Franco**[2]**, Angelo Antônio Barbosa de Moraes**[2]**, Giovanna Moraes Siqueira**[2]**, Lidiane Diniz Nascimento**[3]**, Márcia Moraes Cascaes**[2]**, Mozaniel Santana de Oliveira**[1,2,*] **and Eloisa Helena de Aguiar Andrade**[3]

[1] *Program of Post-Graduation in Biodiversity e Biotecnology-Bionorte, Federal University of Para, Rua Augusto Corrêa S/N, Guamá, 66075-900 Belém, Pará, Brazil*

[2] *Laboratório Adolpho Ducke Laboratory, Botany Coordination, Museu Paraense Emílio Goeldi, Av. Perimetral, 1900, Terra Firme, 66077-830, Belém, PA, Brasil, Brazil*

[3] *Program of Post-Graduation in Chemistry, Federal University of Para, Rua Augusto Corrêa S/N, Guamá, 66075-900 Belém, Pará, Brazil*

**Abstract:** Terpenes are compounds derived from the secondary metabolism of plants, which act biologically in several functionalities, fighting several predators such as fungi and bacteria. Monoterpenes and sesquiterpenes are some of the main compounds that characterize the chemical composition of essential oils. However, this concentration depends on several factors, such as the type of ecosystem, climate, temperature, and other circumstances that can directly impact the chemical composition of essential oil. The Myrtaceae family is considered one of the main families of Brazilian flora and presents a wide diversity of species. Within this family, some species produce essential oils rich in terpenoids, which, besides being responsible for some biological activities, have contributed to the expansion and search for new natural bioactive substances present in such volatile substances. Given the above, this chapter presents a literature search with current studies that prove the biological and antioxidant activities of terpenoids present in essential oils of species of the Myrtaceae family.

**Keywords:** Biological Activities, Bioactive Compounds, Volatile Compounds.

* **Corresponding author Mozaniel Oliveira:** Program of Post-Graduation in Biodiversity e Biotecnology-Bionorte, Federal University of Para, Rua Augusto Corrêa S/N, Guamá, 66075-900 Belém, Pará, Brazil and Laboratório Adolpho Ducke Laboratory, Botany Coordination, Museu Paraense Emílio Goeldi, Av. Perimetral, 1900, Terra Firme, 66077-830, Belém, PA, Brasil, Brazil; E-mail: mozaniel. oliveira@yahoo.com.br

# INTRODUCTION

Essential oils are complex, highly volatile mixtures of low molecular weight [1], originating from the secondary metabolism of plants, which present in their chemical composition several organic compounds such as terpenes (monoterpenes and sesquiterpenes), alcohols, ethers, esters, ketones, aldehydes, phenols, lactones, and phenolic ethers (oxygenated groups) [2].

However, it is important to mention that terpenes are also responsible for the application of essential oils in several sectors. For instance, menthol is used in the preparation of perfumes and fragrances; limonene and citronella are used in the manufacture of repellents, while pinene and limonene are used as air purifiers. Others are applied as expectorants, diuretics, and in the production of ointments for itching and pain relief [3].

Terpenoids are biologically active compounds produced by plants, which can be classified according to the number of carbon atoms: hemiterpenes ($_5$C), monoterpenes ($_{10}$C), sesquiterpenes ($_{15}$C), diterpenes ($_{20}$C), sesterterpenes ($_{25}$C), triterpenes ($_{30}$C), tetraterpenes ($_{40}$C), and polyterpenes ($_n$C) [4]. However, it is important to note that in the chemical composition of essential oils, the strong presence of monoterpenes and sesquiterpenes is peculiar, and monoterpenes can represent about 90% of the essential oil, depending on the type of species studied [5].

In the biosynthesis of terpenic compounds, there are two universal precursors: isopentenyl pyrophosphate (IPP) and dimethylallyl diphosphate (DMAPP). In plants, IPP is biosynthesized through two pathways: *via* mevalonate (MVA) and non-mevalonate (mevalonate-independent), or the deoxyxylulose phosphate pathway, as shown in Fig. (**1**). Through the mevalonate pathway, the IPP intermediate is formed through melavonic acid, which condenses three parts of acetyl coenzyme-A. The non-mevalonate pathway involves 2-C-methyl-*D*-erythritol-4-phosphate (MEP) and 1-deoxy-*D*-xylulose-5-phosphate (DOXP), and results in the condensation of glyceraldehyde phosphate and pyruvate [6]. The first pathway occurs in the cytoplasm, in which most sesquiterpenes are formed, while the second one occurs in chloroplasts, in which there is the formation mainly of monoterpenes and diterpenes [6 - 7].

Monoterpenes are widely distributed in the plant kingdom, especially in some plant species of the Myrtaceae family [5]. They are defined as natural constituents present in the essential oils of plants, which are mostly presented as unsaturated hydrocarbons ($C_{10}$). Moreover, these compounds have some functions, such as antibacterial, analgesic, stimulant, and expectorant properties [2].

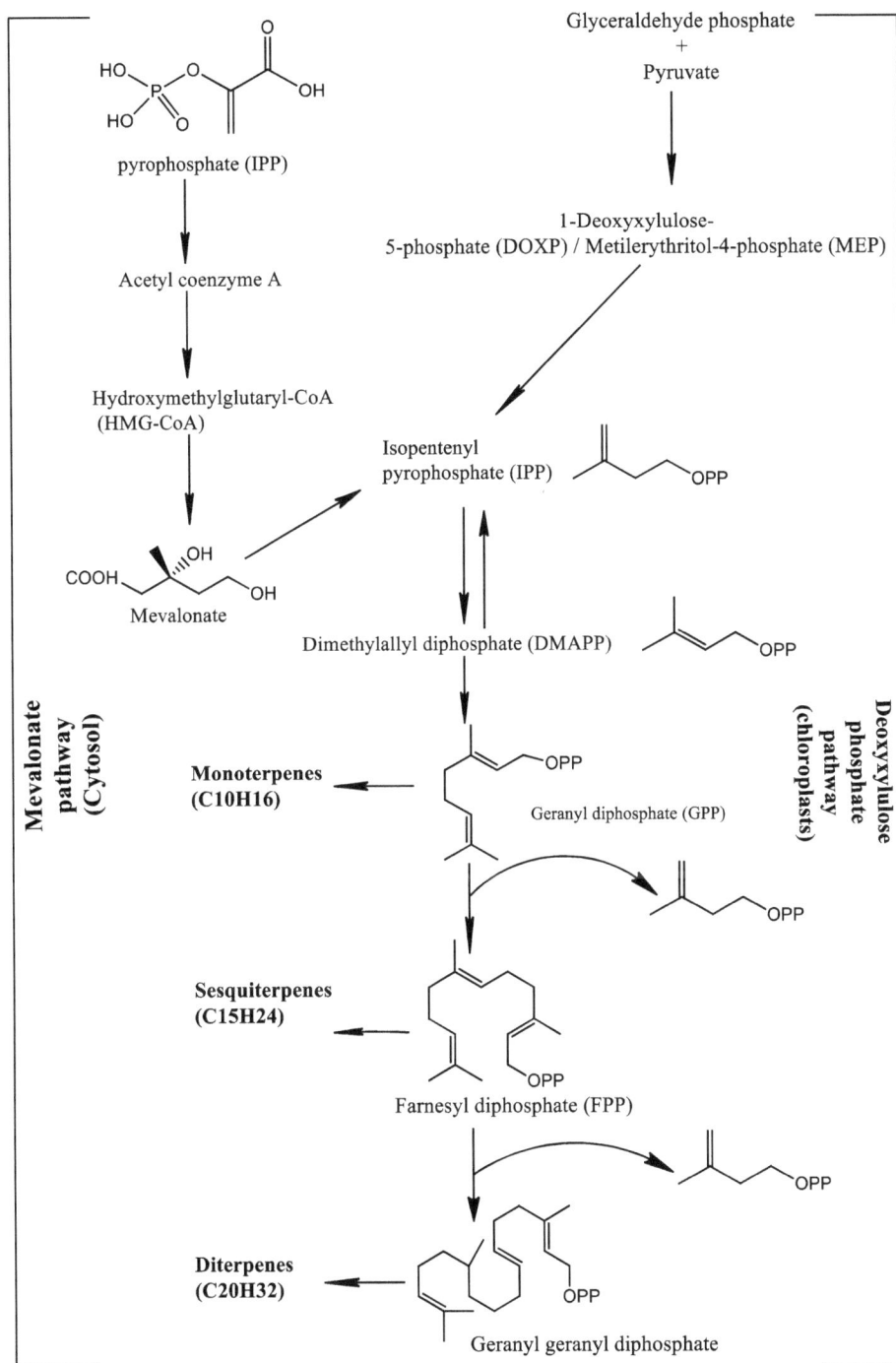

**Fig. (1).** Biosynthesis of terpenes [6].

Sesquiterpenes, in turn, comprise a group of compounds frequently identified in plants, mainly in aerial parts, such as flowers and leaves. Moreover, it is important to note that sesquiterpenes have the α-methylene-γ-lactone group in their structure, which is basically responsible for numerous biological functions, such as cardiovascular disease regulation, antimalarial, analgesic, antineurodegenerative, gastroprotective, and antioxidant activities [5]. Diterpenes, on the other hand, are compounds that have two isoprene units in their structure. They are found in resins, but can characterize some essential oils. In addition, diterpenes are considered heavier than monoterpenes and sesquiterpenes, so their obtaining requires long distillations [6].

## Myrtaceae Family and General Aspects

Myrtaceae family is considered an important botanical family with a wide distribution worldwide. Myrtaceae occurs mainly in tropical and subtropical regions of the world, and South America is the center of its diversity. This family is subdivided into two subfamilies: *Psiloxyloideae* and *Myrtoideae*. The first subfamily has unisexual flowers, and the second one, bisexual flowers. In Brazil, most species belong to the *Myrtoideae* subfamily and *Myrteae* tribe [8], which represent one of the most frequent vegetation formations in the Brazilian flora [9].

Myrtaceae is a family of edible fruits [10], comprising about 140 genera and 6,000 species in the form of shrubs, sub-shrubs and trees. Many representatives of this family have great economic and cultural importance in Brazil, such as eucalyptus (*Corymbia* spp., *Eucalyptus* spp.), which is used in paper and cellulose industries. *Eucalyptus* and *Melaleuca* species are used as antiseptics, natural pesticides, and in the perfume industry. In addition, the Myrtaceae family is widely used for food purposes and in the manufacture of juices, such as guava (*Psidium guajava*); and regarding afforestation and urban landscaping, the weeping bottle brush (*Melaleuca viminalis*) stands out [11].

Essential oils occur naturally throughout the plant kingdom and are not restricted to a specific/taxonomic group. Thus, Myrtaceae family stands out, which is known for presenting some species with essential oils rich in terpenes [12].

## Essential Oil of Myrtaceae Rich in Terpenoids

Species in the Myrtaceae family are rich in essential oils that are produced from aerial parts of plants: leaves, flowers, branches, and fruits. In this family, the diversity of terpenes can be seen in species of different genera, such as *Eucalyptus* [12, 13], *Eugenia* [14, 15], *Myrcia* [17], *Psidium* [18, 19], and others. Table **1** shows the main constituents (≥ 6%) found in essential oils of Myrtaceae species.

**Table 1. Main constituents (≥ 6%) found in essential oils of Myrtaceae species produced from different plant fractions.**

| Species | Plant Fraction (Yield) | Major Components (≥ 6%) | Extraction Method | References |
|---|---|---|---|---|
| *Algrizea minor* Sobral, Faria & Proença | Fresh leaves (n.i) | *β*-Pinene (59.9), *α*-Pinene (16.57) | HD | [22] |
| *Baeckea frutescens* L. | Fresh leaves (2.23%) | *β*-Pinene (19.0), *γ*-Terpinene (11.7), *α*-Pinene (11.1), 1,8-cineole (10.1), *α*-Humulene (9.9), *p*-Cymene (8.9), *β*-Caryophyllene (7.1) | HD | [23] |
| *Callistemon citrinus* (Curtis) Skeels | Fresh leaves (0.62%) | 1,8-Cineole (56.32), *α*-Pinene (18.1), *α*-Terpineol (11.2) | HD | [23] |
| | Fruits (0.34%) | *α*-Pinene (35.1), 1,8-Cineole (32.4), *α*-Terpineol (5.8), Limonene (8.2) | HD | [23] |
| *Eucalyptus citriodora* Hook. | Fresh leaves (2.9%) | Citronellal (29.31), Geraniol (27.63), *β*-Citronellol (14.88), *δ*-Cadinene (6.32) | HD | [13] |
| *Eucalyptus globulus* Labill. | Fresh leaves (0.40%) | *γ*-Terpinene (44.60), *p* –Cymene (28.75), *α*- Pinene (8.57) | HD | [14] |
| *Eugenia brejoensis* Mazine | Dry leaves (n. i) | *δ*-Cadinene (22.6), *β*-Caryophyllene (14.4), *α*-Muurolol (9.34), *α*-Cadinol (8.49), Bicyclogermacrene (7.93) | HD | [15] |
| *Eugenia dysenterica* DC. | Dry leaves (1.5%) | Limonene (16), Caryophyllene oxide (15), Citral (9.0), *β*-Caryophyllene (8), 1,8-Cineole (7.3) | HD | [24] |
| *Eugenia dysenterica* DC. | Fresh leaves (0.34%) | *cis*-*β* –ocimene (19.14), *β*-Caryophyllene (15.36), Caryophyllene oxide (8.23), *α*-Humulene (8.07) | HD | [25] |
| *Eugenia candolleana* DC. | Fresh leaves (0.82%) | *β*-Elemene (35.87), *β*-Caryophyllene (8.15), *δ*-Elemene (8.28), Viridiflorene (6.96) | HD | [16] |
| *Eugenia luschnathiana* Klotzsch ex B.D. Jacks | Fresh leaves (0.12%) | Spathulenol (21.9), Caryophyllene oxide (20.4) | HD | [26] |
| *Eugenia luschnathiana* Klotzsch ex B.D. Jacks | Fresh leaves (0.06%) | *β*-Caryophyllene (33.3), δ-Cadinene (22.38), | HD | [26] |

*(Table 1) cont.....*

| Species | Plant Fraction (Yield) | Major Components (≥ 6%) | Extraction Method | References |
|---|---|---|---|---|
| *Eugenia pyriformis* Cambess | Fresh leaves (n.i) | Limonene (14.8), Nerolidol (11.0), α-cadinol (10.3), Caryophyllene oxide (9.9), β-Pinene (7.1), Spathulenol (6.8) | HD | [27] |
| *Eugenia uniflora* L. | Leaves (0.51%) | Germacrone (8.52), Spathulenol (8.20), α-Selinene (7.50) | HD | [28] |
| *Melaleuca cajuputi* Powel | Fresh Leaves (0.31%) | Terpinolene (21.61), γ-Terpinene (17.81), β-Caryophyllene (12.22), β-Pinene (8.03), Platyphyllos (7.50), o-Cymene (7.34) | HD | [14] |
| *Melaleuca citrina* (Curtis) Dum. | Fresh Leaves (0.18%) | 1,8-Cineole (58.06), α- Pinene (10.98), o-Cymene (7.56), Limonene (6.23) | HD | [14] |
| *Melaleuca leucadendra* (L.) L. | Young leaves (1%) | α-Eudesmol (21.2), Guaiol (12.5) | HD | [23] |
| | Old leaves (1%) | α-Eudesmol (17.6), Guaiol (10.9), β-Caryophyllene (7.0) | HD | |
| | Stem bark (1%) | α-Eudesmol (24.1), Guaiol (11.3) | HD | |
| | Fruits (1%) | α-Eudesmol (30.7), Guaiol (10.4) | HD | |
| | Branch tips (1%) | α-Eudesmol (13.7), Guaiol (7.3), p-Cymene (8.7) | HD | |
| *Myrcia rostrata* DC. | Fresh leaves (0.42%) | Carotol (17.68), Germacrene B (7.28), β-Caryophyllene (6.45) | HD | [19] |
| *Myrcia eximia* DC. | Dry leaves (0.36%) | β-Caryophyllene (15.71), Caryophyllene oxide (10.25), 14-Hydroxy-9-*epi*-(*E*) -caryophyllene (7.02) | HD | [29] |
| | Dry leaves (0.01%) | Hexanal (26.1), β-Caryophyllene (15.0), Caryophyllene oxide (22.16), 14-Hydroxy- 9-*epi*-(*E*)-caryophyllene (7.84) | HD | |
| | Dry leaves (0.08%) | β-Caryophyllene (20.3), Caryophyllene oxide (16.3) | SD | |
| *Myrciaria floribunda* (H. West ex Willd.) O. Berg | Fresh leaves (0.37%) | 1,8-Cineole (10.4), β-Selinene (8.4), α-Selinene (7.4) | HD | [30] |

*(Table 1) cont.....*

| Species | Plant Fraction (Yield) | Major Components (≥ 6%) | Extraction Method | References |
|---|---|---|---|---|
| *Myrciaria tenella* (DC.) O. Berg | Fresh leaves (0.39%) | α- Pinene (25.1), β-Pinene (20.9), β-Caryophyllene (9.99), Platiphyllol (8.87) | HD | [26] |
| | Fresh leaves (0.36%) | α- Pinene (45.7) | HD | |
| *Pimenta Dioica* (L.) Merr | Dry leaves (0.8%) | Eugenol (85.33) | HD | [31] |
| *Plinia cauliflora* (Mart.) | Leaves (0.05%) | β-Caryophyllene (14.69), β-Bisabolene (9.36), (*E, E*) -α-farnecene (8.07), Globulol (7, 86) | HD | [28] |
| *Psidium cattleianum* Sabine | Fresh leaves (0.3%) | 1,8-Cineole (10.8), β-Caryophyllene (11.8), β-Selinene (8.6), α-Humulene (6.0) | HD | [32] |
| *Psidium guajava* Linn. | Fresh leaves (0.19%) | Limonene (36.55), α- Pinene (14.70), 1,8-Cineole (7.37), β- Caryophyllene (7.22) | HD | [14] |
| *Psidium guajava* Linn. | Flowers (1.0%) | α-Cadinol (37.8), β-Caryophyllene (12.2), Nerolidol (9.1), α-Selinene (8.8), β-Selinene (7.4), Caryophyllene oxide (7.2) | HD | [21] |
| *Psidium laruotteanum* Cambess | Dry leaves (0.3) | *p*-Cymene (24.8), α-Pinene (13.4), 1,8-Cineole (19.2), Terpinen-4-ol (6.3) | HD | [20] |
| | Dry leaves (0.4%) | *p*-Cymene (19.4), α-Pinene (11.6), γ-terpinene (14.0), 1,8-Cineole (6.9), | HD | |
| *Psidium laruotteanum* Cambess | Dry leaves (0.3) | *p*-Cymene (34.8), 1,8-Cineole (12.5), α-Pinene (9.2), γ-terpinene (6.9), α-terpineol (6.0) | HD | [20] |
| *Psidium myrtoides* O. Berg | Fresh leaves (n.i) | *trans*-β-Caryophyllene (30.9), α-Humulene (15.9), α-Copaene (7.8), Caryophyllene oxide (7.3), α-Bisabolol (5.3) | HD | [33] |
| *Psidium myrsinites* DC. | Dry leaves and inflorescences (0.1%) | Caryophyllene oxide (26.1), Humulene epoxide II (8.8), β-Caryophyllene (7.4) | HD | [34] |
| *Syzygium cumini* (L.) | Fresh leaves (0.24%) | Terpinolene (19.08), γ-Terpinene (16.33), β-Caryophyllene (12.25), Platyphyllos (9.69), *o*-Cymene (7.34) | HD | [14] |
| *Syzygium cumini* (L.) | Leaves (0.03%) | α-Pinene (21.20), Globulol (15.30), Eugenol (11.20), α-Terpineol (8.88), Aromadendrene (6.79) | HD | [28] |

*(Table 1) cont.....*

| Species | Plant Fraction (Yield) | Major Components (≥ 6%) | Extraction Method | References |
|---------|------------------------|-------------------------|-------------------|------------|
| *Syzygium lanceolatum* (Lam.) Wt. & Arn. | Fresh Leaves (6.3 ml/kg) | Phenyl propanal (18.3), Caryophyllene oxide (15.7), α-Humulene (14.5), β-Caryophyllene (12.8), Humulene epoxide II (6.5) | HD | [35] |
| *Syzygium nervosum* DC. | Fresh Leaves (0.2%) | (Z)-β-ocimene (20.3), Caryophyllene oxide (13.2), β-Caryophyllene (12.1) | HD | [23] |
| *Syzygium samarangense* (Blume) Merr. and L. M. Perry | Fresh Leaves (0.08%) | o-Cymene (54.33), α- Pinene (7.51), | HD | [14] |

HD= Hydrodistillation, SD= Steam Distillation.

Veras *et al.* [22] show that the essential oils from the fresh leaves of *Algrizea minor* are rich in hydrocarbon monoterpenes, with α- and β-pinene as the main components. *Eucalyptus citriodora* oils have been shown to be rich in oxygenated monoterpenes [13], whereas *Eucalyptus globulus* is rich in hydrocarbon monoterpenes [14].

Giang An *et al.* [23] studied the chemical composition of four species (*Baeckea frutescens, Callistemon citrinus, Melaleuca leucadendra, and Syzygium nervosum*) from the Myrtaceae family and demonstrated the chemical diversity among them. Different parts (young leaves, old leaves, stem barks, fruits, and branch tips) of *Melaleuca leucadendra* have been shown to be rich in oxygenated mono and sesquiterpenes, with α-eudesmol and guaiol as the main constituents. The essential oil of *Syzygium nervosum* proved to be rich in oxygenated monoterpenes and sesquiterpenes and sesquiterpene hydrocarbons, having as main constituents (Z)-β-ocimene, caryophyllene oxide, and β-caryophyllene.

The authors also demonstrated that fresh leaves and fruits of *Callistemon citrinus* are rich in hydrocarbon monoterpenes and oxygenated monoterpenes, presenting as main compounds 1,8-cineole, α-pinene, α-erpineol. The compounds α-pinene and 1,8-cineole were also two of the main substances found in the essential oil of *Baeckea frutescens*. Moreover, β-pinene, γ-terpinene, α-humulene, p-cymene, β-caryophyllene were also found.

The essential oils of the genus *Eugenia* have shown to present a great diversity of terpenes [15, 16, 24 - 28]. Studies performed by Santos *et al.* [24] with the essential oils of the dried leaves of *Eugenia dysenterica* proved to be rich in hydrocarbon and oxygenated monoterpenes and sesquiterpenes, with predominance of limonene, caryophyllene oxide, citral, β-caryophyllene, and 1,8-cineole. On the other hand, a study performed with the fresh leaves of this species showed the predominance of *cis-β–ocimene*, *β*-caryophyllene, caryophyllene oxide, and *α*-humulene [25].

Monteiro *et al.* [26] showed the differences in the chemical composition of *Eugenia luschnathiana* leaf essential oil. The specimen collected in São Paulo - Brazil presented spathulenol and caryophyllene oxide as major constituents, and the specimen collected in Rio de Janeiro - Brazil presented β-caryophyllene and δ-cadinene as the main substances. The authors also studied the variation in the chemical composition of *Myrciaria tenella*. The specimen collected in São Paulo-Brazil presented as major constituents α- pinene, β-pinene, β-caryophyllene, and platiphyllol; and the specimen collected in Rio de Janeiro - Brazil, presented α-pinene as the main compound. These results indicated that the collection site may influence the chemical composition.

*Myrcia* is one of the main genera of the Myrtaceae family, and its species are sources of essential oils with the predominance of monoterpenes and sesquiterpenes [16, 26, 27]. The leaf essential oil of *Myrcia splendens* is rich in oxygenated sesquiterpenes, mainly *trans*-nerolidol and α-bisabolol [18].

Ferreira *et al.* [29] conducted the first report on the chemical composition of *Myrcia eximia* essential oil from specimens collected in different years (2017 and 2018) and extracted by different extraction methods (hydrodistillation - HD and steam distillation - SD) [30 - 33]. In the specimens collected in 2017 (HD) and 2018 (SD), the presence of sesquiterpenes is remarkable, mainly β-caryophyllene and caryophyllene oxide. Fig. (**2**) shows the structures of the major constituents found in Myrtaceae essential oils.

## APPLICATIONS

### Antioxidant Activity

Antioxidants can protect an organism from free radical damage as they can inhibit or interrupt the formation and propagation of these radicals [34 - 37]. Various *in vitro* trials are dedicated to evaluating the antioxidant potential of essential oils and their components. Some methodologies assess the radical scavenging activity of isolated substances or mixtures, such as the methodologies that use the radicals 2,2-diphenyl-1-picrylhydrazyl (DPPH) and 2,2'-azino-bis (3-ethylbenzothiazo-line-6-sulfonic acid) (ABTS); or focus on assessing the reducing power of the analyzed compounds, such as TBARS, FRAP, and CUPRAC [38]. The antioxidant activity of aromatic plants is constantly investigated since their composition is characterized by the presence of many antioxidant substances.

Species of Myrtaceae family have been widely studied regarding their chemical composition and antioxidant character, as presented in Table **2**.

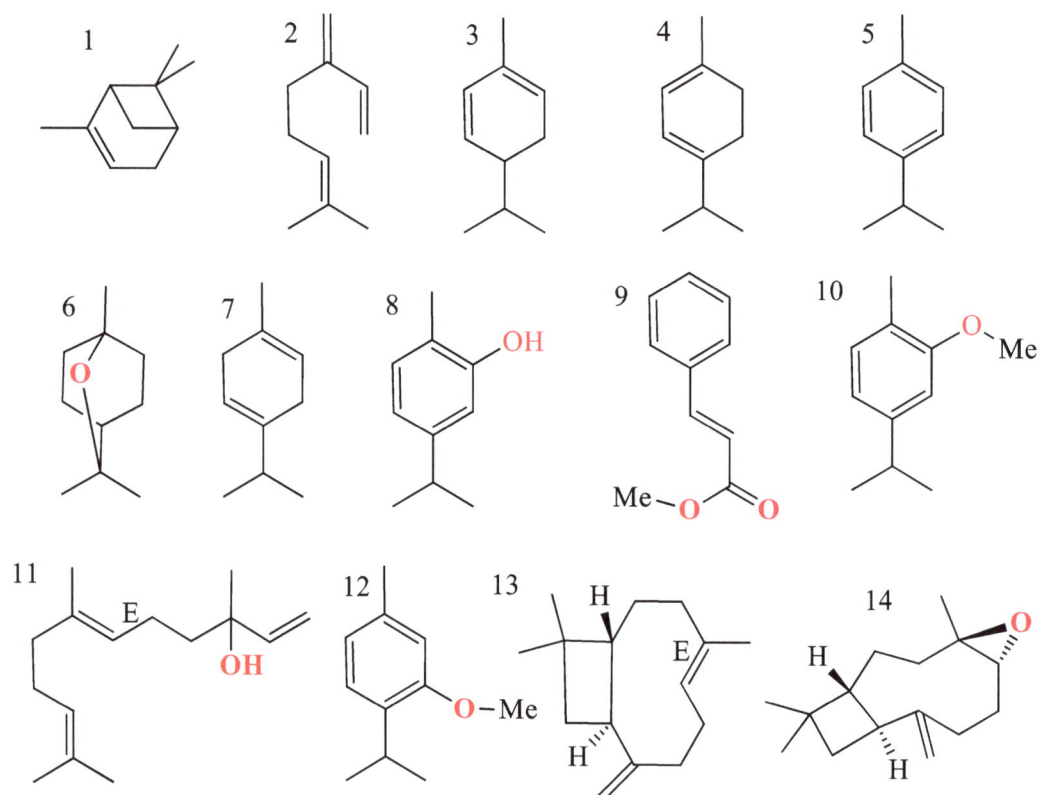

**Fig. (2).** 2D chemical structures of the main compounds present in *Lippia origanoides* essential oil. 1 = (α-pinene), 2 = (myrcene), 3 = (α-phellandrene), 4 = (α-terpinene), 5 = (*p*-cymene), 6 = (1,8-cineole), 7 = (γ-terpinene), 8 = (carvacrol), 9 = ((*E*)-methyl cinnamate), 10 = (carvacrol methyl ether), 11= ((*E*)-nerolidol), 12 = (thymol methyl ether), 13 = ((*E*)-caryophyllene), 14 = (caryophyllene oxide).

Essential oils of different species of the genus *Eucalyptus* were evaluated through DPPH and ABTS assays. The antioxidant profile varied strongly at 10 mg/mL concentration (DPPH test). The inhibition percentage (IP) of DPPH ranged from 10.75 ±1.05% to 52.69 ± 4.59% in *Eucalyptus angulosa* and *E. camaldulensis,* respectively. From the results of the reducing power test, *E. gomphocornuta* showed the highest reducing capacity ($EC_{50}$ = 2.09 ± 0.05 mg/mL), followed by *E. camaldulensis* and *E. paniculata*. As can be seen in Table **2**, *E. angulosa* presented the monoterpene 1.8-cineole (60.48%) as the major constituent, whereas *E. camaldulensis* was characterized by 32.88% of sesquiterpene spathulenol, 19.41% of o-cymene, and 16.31% of 1.8-cineole [40].

**Table 2. Major compounds, methodologies for antioxidant activity determination, and results obtained for essential oils from Myrtaceae species.**

| Species (Plant Fraction) | Major Compound (%) | Methodology | Results Obtained | References |
|---|---|---|---|---|
| *Algrizea minor* (leaves) | α-Pinene (16.57%); β-pinene (56.99%); germacrene D (4.67%) | **DPPH** **ABTS** Phosphomolybdenum Reducing Power | $IC_{50}$ = 114.25 µg/mL (DPPH) $IC_{50}$ = 163.75 µg/mL (ABTS) $IC_{50}$ = 626.1 µg/mL (phosphomolybdenum) $IC_{50}$ = 2738.0 µg/mL (Reducing Power) | [22] |
| *Callistemon citrinus* (flowers) | (-)-Bornylacetate (10.02%); E-caryophyllene (11.89%); α-eudesmol (12.93%). | DPPH ABTS | $IC_{50}$ = 1.13 mg/mL (DPPH) $IC_{50}$ = 0.03 mg/mL (ABTS) | [39] |
| *Callistemon citrinus* (leaves) | α-Pinene (20.02%); eucalyptol (48.98%); α-Terpineol (8.01%). | DPPH ABTS | $IC_{50}$ = 1.49 mg/mL (DPPH) $IC_{50}$ = 0.14 mg/mL ABTS | [39] |
| *Eucalyptus astringens* (leaves) | α-Pinene (28.24%); 1.8-cineole (48.44%). | DPPH FRAP | IP = 19.89 ± 1.05% (DPPH) $EC_{50}$ = 7.46 ± 0.26 mg/mL (FRAP) | [40] |
| *Eucalyptus camaldulensis* (leaves) | o-Cymene (19.14%); 1.8-cineole (16.31%); spathulenol (32.88%) | DPPH FRAP | IP = 52.69 ± 4.59% (DPPH) $EC_{50}$ = 2.96 ± 0.11 mg/mL (FRAP) | [40] |
| *Eucalyptus angulosa* (leaves) | α-Pinene (17.67%); 1.8-cineole (60.48%); pinocarveol (8.04%) | DPPH FRAP | IP = 10.75 ± 1.05% (DPPH) $EC_{50}$ = 12.59 ± 0.34 mg/mL (FRAP) | [40] |
| *Eucalyptus erythrocorys* (leaves) | α-Pinene (9.94%); 1.8-cineole (60,74%); globulol (6.29%) | DPPH FRAP | IP = 29.57 ± 1.05% (DPPH) $EC_{50}$ = 8.45 ± 0.75 mg/mL (FRAP) | [40] |
| *Eucalyptus globulus* (leaves) | α-Pinene (17.15%); 1.8-cineole (70.15%). | DPPH FRAP | IP = 23.66 ± 1.05% (DPPH) $EC_{50}$ =18.79 ± 0.70 mg/mL (FRAP) | [40] |
| *Eucalyptus gomphocornuta* (leaves) | o-Cymene (12.03%); globulol (19.87%); spathulenol (11.98%). | DPPH FRAP | IP = 32.8 ± 1.05% (DPPH) $EC_{50}$ = 2.09 ± 0.05 mg/mL (FRAP) | [40] |

*(Table 2) cont.....*

| Species (Plant Fraction) | Major Compound (%) | Methodology | Results Obtained | References |
|---|---|---|---|---|
| *Eucalyptus . lehmannii* (leaves) | α-pinene (18.21%); 1.8-cineole (68.46%) | DPPH FRAP | IP = 15.05 ± 1.05% (DPPH) $EC_{50}$ = 20.14 ± 0.49 mg/mL (FRAP) | [40] |
| *Eucalyptus maidenii* (leaves) | α-Pinene (10.00%); 1.8-cineole (71.16%). | DPPH FRAP | IP = 24.73 ± 1.05% (DPPH) $EC_{50}$ = 17.61 ± 1.12 mg/mL (FRAP) | [40] |
| *Eucalyptus melliodora* (leaves) | α-Pinene (17.06%); 1.8-cineole (72.91%) | DPPH FRAP | IP = 19.35 ± 1.83% (DPPH) $EC_{50}$ = 43.1 ± 1.74 mg/mL (FRAP) | [40] |
| *Eucalyptus microcarpa* (leaves) | 1.8-Cineole (82.64%); α-pinene (4.27%); pinocarveol (3.19%). | DPPH FRAP | IP = 12.37 ± 1.05% (DPPH) $EC_{50}$ = 41.6 ± 1.73 mg/mL (FRAP) | [40] |
| *Eucalyptus oxidantes* (leaves) | 1.8-Cineole (64.40%); globulol (14.84%); epiglobulol (3.03%). | DPPH FRAP | IP =16.13 ± 1.83% (DPPH) $EC_{50}$ = 4.57 ± 0.07 mg/mL (FRAP) | [40] |
| *Eucalyptus paniculata* (leaves) | o-Cymene (13.53%); α-pinene (8.21%); 1.8-cineole (54.89%); | DPPH FRAP | IP = 23.66 ± 1.05% (DPPH) $EC_{50}$ = 4 ± 0.18 mg/ mL (FRAP) | [40] |
| *Eugenia caryophyllata* (bud flowers) | Eugenol (89.8%); E-caryophyllene (5.4%) | DPPH | IP = 15.4–60.4% | [41] |
| *Syzygium aromaticum* | Eugenol (71.56%) and eugenol acetate (8.99%) | DPPH | Effects on DPPH (10 to 93%) | [42] |
| *Eugenia involucrata* (leaves) | Elixene (26.53%), (*E*)-caryophyllene (13.16%), α-copaene (8.41%) | DPPH | $IC_{50}$ = 38.61±1.11 mg/mL | [43] |
| *Eugenia klotzschiana* (fresh leaves) | α-Copaene (10.6%); spathulenol (8.7%); caryophyllene oxide (7.4%) | DPPH ABTS | $EC_{50}$ = 29.77±2.25 µg/mL (DPPH) 57.81 ± 4.80 µM trolox/g (ABTS) | [44] |
| *Eugenia klotzschiana* (flowers) | Germacrene-D (12.1%); α-(E--bergamotene (29,9%); | DPPH ABTS | $EC_{50}$ = 5.70±0.37 µg/mL (DPPH) 104.61±4.46 µM trolox/g (ABTS) | [44] |
| *Eugenia klotzschiana* (dried leaves) | Germacrene-D (13.3%); β-bisabolene (17.4%) | DPPH ABTS | $EC_{50}$ = 7.61±0.10 (DPPH) 106.27±4.75 µM trolox/g (ABTS) | [44] |

*(Table 2) cont.....*

| Species (Plant Fraction) | Major Compound (%) | Methodology | Results Obtained | References |
|---|---|---|---|---|
| *Eugenia patrisii* (leaves) | (2E,6E)-Farnesol (34.50%); (2E,6Z)-farnesol (23.20%); caryophylla-4(12)-8(13)-dien-5-ol (15.60%). | DPPH | TEAC =111.2 ± 12.4 mg TE/mL | [42] |
| *Eugenia uniflora* (leaves) | Curzerene (35.8-53.1%); germacrene B (3.50-7.50%) | DPPH | $IC_{50}$ = 42.6 ± 0.3 to 64.2 ± 0.3% | [45] |
| *Melaleuca alternifolia* | α-Pinene (13.05%); α-carene (17.41%); terpinen-4-ol (13.17%). | DPPH β-carotene-linoleic acid | $IC_{50}$ = not detected (DPPH) IP = 82.44% (β-carotene system) | [46] |
| *Myrcia oblongata* (leaves) | *Trans*-verbenol (11.94%); caryophyllene oxide (22.03%); (-) − espatulenol (4.22%). | DPPH | IP = 88.33% | [47] |
| *Myrcia splendens* (leaves) | *Trans*-nerolidol (67.81%); α-bisabolol (17.81%). | DPPH | $IC_{50}$ = 43,537.00 ± 15 µg/mL | [18] |
| *Syzygium aromaticum* | 3-Allylguaiacol (42.64%), Caryophyllene (15.51%); Eugenol acetate (15.91%). | DPPH β-carotene-linoleic acid | $IC_{50}$ = 0.26 ± 0.7 mg/mL (DPPH) IP = 98% (β-carotene system) | [46] |
| *Syzgium guineense* (leaves) | α-Cadinol (6.68%); aromadendrene (6.98%); germacrene B (5.52%); τ-cadinol (6.64%). | DPPH | IP = 70% | [48] |

$IC_{50}$: inhibitory concentration for 50%; $EC_{50}$: efficacy concentration for 50%; IP: inhibition percentage; TEAC: Results expressed in milligrams of Trolox.

Veras *et al.* [22] applied four different methodologies (DPPH, ABTS, phosphomolybdenum, and reducing power) to evaluate the antioxidant activity of *Algrizea minor* essential oil collected in Pernambuco (Brazil). According to the authors, the samples demonstrated relevant antioxidant activity, which could be related to its chemical composition, since the oil showed high levels of the monoterpenes β-pinene (56.99%) and α-pinene (16.57%).

The DPPH assay has been applied to evaluate the radical scavenging activity of essential oils from *Eugenia* species. Hadidi *et al.* [41] studied the antioxidant activity of *Eugenia caryophyllata* before and after the encapsulation in chitosan nanoparticles. The inhibition percentage of the oil ranged from 15.4–60.4%, and the encapsulated oil varied from 15.9 to 71.8%, indicating that encapsulation favored the antioxidant character. It presented 89.8% of the monoterpene eugenol, which was reported as strongly related to the antioxidant profile [41].

## Anti-Inflammatory Activity

Natural products are considered a significant source of anti-inflammatory substances, such as terpenes, which have effects on acute and chronic conditions [49]. Also, essential oils are reported in the literature to exhibit anti-inflammatory activity [49, 50].

Most Myrtaceae species contain a variety of volatile substances in their leaves, making them attractive sources with potential biological activities [51, 52]. Among the biological properties of the essential oils of Myrtaceae, the anti-inflammatory activity can be highlighted, since it is derived from phenolic compounds. Good examples are: *Eugenia stipitata, Psidium guineense, Calyptranthes restingae*, as well as the genus *Callistemon*.

In the study performed by Costa *et al.* [53], the leaf essential oil of *Eugenia stipitata* was used to treat paw edema and peritonitis in mice, both induced by carrageenans. *E. stipitata* essential oil was able to reduce paw edema at doses of 40, 100, and 250 mg/kg, showing 88.66%, 91.3%, and 96.94% of inhibition, respectively. Regarding peritonitis, the essential oil showed control action and reduction of leukocyte and neutrophil migration, using doses of 40, 100, and 250 mg/kg. A migratory decrease in the total number of leukocytes (76.9%, 81.7%, and 86.5%, respectively) and neutrophils (74.5%, 79.6%, and 77.9%, respectively) was noted. The major constituents identified were guaiol (13.77%), (E)-caryophyllene (11.36%), β-eudesmol (8.13%), and γ-eudesmol (6.55%). In a general aspect, the use of *E. stipitata* essential oil was indicated as a natural source of numerous chemical compounds, with anti-inflammatory potential, combined with low side effects.

*Psidium guineense* leaf essential oil had its biological activity tested against carrageenan-induced paw edema and pleurisy in mice. In the first hour, the essential oil orally applied (300 mg/kg dose) significantly reduced edema formation, as well as in the second and fourth hours, in which the inhibition values were $59.46 \pm 3\%$, $51.03 \pm 5\%$, and $48.48 \pm 2\%$, respectively. Regarding the pleurisy test, there was a reduction of leukocytes, previously increased due to the action of carrageenans. After the application of the essential oil, the inhibition of leukocytes was verified in the first, second and fourth hours, and resulted in inhibition of $45.33 \pm 3\%$, $77.70 \pm 5\%$, and $75.20 \pm 1\%$, respectively. The major constituent present was spathulenol (80.71%), which is responsible for the anti-inflammatory action [54]. These results corroborate the traditional use of this specimen to treat inflammation.

The leaf essential oil extracted from *Calyptranthes restingae* was tested in mice against leukocyte migration induced by carrageenan injection. The essential oil was applied intraperitoneally in doses of 25, 50, and 100 mg/kg, and half an hour later, a dose of carrageenans was also applied to induce leukocyte migration. However, this migration was inhibited, indicating the action of *C. restingae* essential oil. One of the possible causes of its anti-inflammatory action is the inhibition of the synthesis of inflammatory mediators. This result shows the potential of *C. restingae* essential oil in helping fight inflammatory pain. Also, its major constituent was calyptrantone (81.03%) [55].

Leaves of three species of *Callistemon* genus had their essential oils analyzed for anti-inflammatory potential: *C. citrinus, C. rigidus,* and *C. viminalis*. The anti-inflammatory activity was evaluated by *in vitro* assays using erythrocyte hemolysis induced by a hypotonic solution. The major constituent found in the three essential oils studied was eucalyptol (1,8-cineol), in the range of 71.27% - 81.70%, 69.15% - 81.70%, 64.63% - 79.39%, and 55.69% - 80% from specimens collected in spring, summer, fall, and winter seasons, respectively. As a result, it was noted that *C. viminalis* presented high membrane-stabilization activity, showing $IC_{50}$ of 25.6 µg/mL, being compared to indomethacin (17.02 µg/mL) - the standard anti-inflammatory drug used. *C. citrinus* showed moderate activity with an $IC_{50}$ value of 39.9 µg/mL. Finally, *C. rigidus* showed weak activity with $IC_{50}$ of 217.1 µg/mL [56]. It is concluded then, that the genus *Callistemon* has strong anti-inflammatory potential, especially *C. viminalis*, the species that showed the greatest response in the study.

*Pimenta dioica* essential oil, which was characterized by the major constituents eugenol (90%) and α-terpineol (2%), was tested against human macrophage THP-1cells, and showed the following results in the anti-inflammatory test: 50 pg/mL (IL-6) and 20 pg/mL (TNF-α). The levels of TNF-α and IL-6 were significantly different from lipopolysaccharide (positive control), which was 150 pg/mL and 800 pg/mL, respectively. These results showed that the essential oil of *Pimenta dioica* did not present this sort of biological activity [57].

(*E*)-caryophyllene (14%) and caryophyllene oxide (6.9%) were the major constituents in *Campomanesia phaea* essential oil. In this study, it was used for quantification of cytokines IL-6 and TNF-α, and inhibited 41 and 46% of IL-6 production at concentrations of 50 and 100 µg/mL, respectively. It also inhibited 32.3%, 49.5%, 74.7%, and 76.4% of TNF-α production at concentrations of 1, 10, 50, and 100 µg/mL, respectively. In the luciferase assay, the essential oil was quantified at 20 µg/mL for NF-κB inhibition in HEK 293 cells (human embryonic kidney), which showed inhibition potential equal to 36.2% ± 1.8 µg/mL. According to this result, it was possible to infer that the essential oil of *C. phaea* may indicate a possible molecular mechanism on the NF- κB pathway, in which it inhibits oxidative stress and the expression of pro-inflammatory mediators [58].

In the study performed with the commercial *Eugenia caryophyllata* essential oil, whose major compound is eugenol, its anti-inflammatory potential was evaluated in primary human neonatal fibroblasts stimulated with a mixture of interleukin (IL)-1β, tumor necrosis factor (TNF)-α, interferon (IFN)-γ, basic fibroblast growth factor (bFGF), epidermal growth factor (EGF), and platelet-derived growth factor (PDGF), *i.e.*, this study was based on a dermal fibroblast system, HDF3CGF, that simulates the microenvironment of human inflamed skin cell disease. *E. caryophyllata* essential oil significantly reduced the levels of inflammatory biomarkers such as vascular cell adhesion molecule-1 (VCAM-1), interferon gamma-induced protein 10 (IP-10), interferon-inducible alpha T-cell chemoattractant (I-TAC), and monokine induced by interferon-gamma (MIG). The authors pointed out that the anti-inflammatory activity presented in this essential oil is related to its major compound, eugenol [59].

**Neuroprotective Activity**

Neurodegenerative diseases are characterized by a progressive loss of the structure or function of specific cells. Within this branch of diseases are alzheimer's, parkinson's, stroke, and glaucoma. And, to curb the damaging effects of these diseases, there is a constant search for alternative treatments that prioritize mainly natural products such as terpenoids, which can be considered a promising source as neuroprotectors [60], the neuroprotective potential and presented in Table **3**.

The essential oils from six Myrtaceae species were studied, to evaluate the acetylcholinesterase inhibitory activity (AChEI): *Eucalyptus globulus, Melaleuca cajuputi, Melaleuca citrina, Psidium guajava, Syzygium cumini,* and *Syzygium samarangense*. The results showed that the essential oil of *M. citrina* presented the highest percentage of inhibition (71.77 ± 2.11%) followed by *E. globulus* (47.65 ± 2.26%), *P. guajava* (24.96 ± 2.38%), *M. cajuputi* (21.18 ± 0,54%), *S. cumini* (19.97 ± 1.10%), and *S. samarangense* (13.78 ± 1.52%). α-Pinene was found in the essential oils of all species, whereas 1,8-cineole was the major constituent present in the essential oil of *M. Citrina*, which proved to be an active compound in AChEI test. However, there are other compounds that may have shown synergism, and thus contributed to this high inhibition potential [14].

*Eugenia dysenterica* essential oil presented $IC_{50}$ = 0.92 µg.ml$^{-1}$ (AChEI test). This result was better compared to the drug rivastigmine ($IC_{50}$ = 1.87 µg.ml$^{-1}$), used in the treatment of Alzheimer's disease. Caryophyllene oxide was the major compound of this essential oil, which after purification, was tested for AChEI and showed $IC_{50}$ = 0.31 µg.ml$^{-1}$. These results were considered encouraging, according to the authors, and suggest the continuation of *in vivo* tests [61]. Acetylcholinesterase (AChE) is an enzyme involved in the final step of nerve impulse transmission by hydrolyzing a neurotransmitter called acetylcholine (ACh). This neurotransmitter is reduced by AChE inhibitors, *i.e.*, they reduce the level of ACh in the brain and increase its concentration at synapses. Thus, these inhibitors are the main compounds approved for clinical use in the management of neurodegenerative diseases such as Alzheimer's [62].

The essential oil extracted from the peel of *Myrciaria floribunda* fruits was tested in acetylcholinesterase extracted from gills of *Crassostrea rhizophorae,* and in the commercial AChE obtained from the electric organ of *Electrophorus electricus* (Sigma). The results showed a percentage of average inhibition with $IC_{50}$ equal to 0.08 μg/mL for *C. rhizophorae,* and $IC_{50}$ equal to 23 μg/mL for *E. electricus*. Regarding neostigmine (standard), $IC_{50}$ values were 23.3 μg/mL and 6.2 μg/mL, respectively. The inhibition of AChE may be related to δ-cadinene and γ-cadinene sesquiterpenes [63]. On the other hand, in the study performed with *Pimenta dioica* essential oil, $IC_{50}$ was equal to 1320 μg mL$^{-1}$ for AChE, and 3340 μg mL$^{-1}$ for Butyrylcholinesterase (BChE) [64].

In a study carried out with nine samples (OE1 – OE9) of essential oil extracted from fresh leaves of *Myrtus communis*, it was shown that the essential oil EO1 had greater inhibition potential (BChE) with $IC_{50}$ value equal to 77.8 μg/mL. OE5 also showed significant inhibition, with $IC_{50}$ = 87.9 μg/mL for AChE and $IC_{50}$ = 96.0 μg/mL for BChE. According to the authors, these high $IC_{50}$ values may be related to the complex interactions between the chemical constituents present in the composition of EO1 and EO5 [65]. In contrast, *Eugenia brasiliensis* essential oil showed a weak AChE activity with $IC_{50}$ > 1,000μg.mL$^{-1}$ [66]. Similarly, *Eugenia hiemalis* essential oil showed no activity in the acetylcholinesterase assay ($IC_{50}$> 1000 μg mL$^{-1}$) [67].

The leaf essential oil of *Eucalyptus globulus* was tested in Wistar rats to determine the AChE activity in the hippocampus region of their brains. Concentrations of 1000 mg/kg and 2000 mg/kg for 21 days were significant to decrease AChE activity in the mentioned region. However, a dose of 500 mg/kg was ineffective. In addition, *Eucalyptus globulus* essential oil, at the same concentrations mentioned above, demonstrated an increase in the level of neurotransmitter GABA in the cortical region of their brains. Nevertheless, the level of dopamine in the cortex showed a significant decrease. Based on this study, it can be noted that eucalyptus essential oil contains neuroprotective and therapeutic compounds, which probably work by facilitating GABA neurotransmission and blocking dopamine and AchE activity [68]. The essential oils of two species of Myrtaceae, *Myrcianthes myrsinoides* and *Myrcia mollis,* were evaluated regarding their potential against the enzymes acethylcholinesterase

(AChE) and butyrylcholinesterase (BChE). *Myrcianthes myrsinoides* essential oil showed IC$_{50}$ of 78.6 µg/mL for AChE and 18.4 µg/mL for BChE, whereas the essential oil of *Myrcia mollis* presented IC$_{50}$ > 50 µg/mL for both enzymes. IC$_{50}$ of donepezil (positive control) was 0.04 µg/mL (AChE) and 3.6 µg/mL (BChE). These results, according to the authors, demonstrate great potential for the treatment of Alzheimer's disease [69].

Table 3. Neuroprotective activity in some essential oils from the Myrtaceae family.

| Species | Method Applied | IC$_{50}$ Value | References |
|---|---|---|---|
| *Eucalyptus globulus* | AChE | 47.65 ± 2.26% | [14] |
| *E. globulus* | AChE | 45.13 ± 3.51-Keta + EO (1000 mg/kg) 41.05 ± 3.38- Keta + EO (2000 mg/kg) | [68] |
| | GABA | 256.83 ± 20.77- Keta +EO (1000 mg/kg) 273.33 ± 19.33- Keta + EO (2000 mg/kg) | |
| *Eugenia brasiliensis* | AChE | > 1,000µg/mL | [66] |
| *E. dysenterica* | AChE | 0.92 mg/ml | [61] |
| *E. hiemalis* | AChE | > 1,000µg/mL | [67] |
| *Melaleuca citrina* | AChE | 71.77 ± 2.11% | [14] |
| *M. cajuputi* | AChE | 21.18 ± 0,54% | |
| *Myrcia mollis* | AChE | > 50 µg/mL | [69] |
| | BchE | > 50 µg/mL | |
| *Myrciaria floribunda* | AChE of *C. rhizophorae* | 0.08 µg/mL | [63] |
| | AChE of *E. electricus* | 23 µg/mL | |
| *Myrcianthes myrsinoides* | AchE | 78.6 µg/mL | [69] |
| | BChE | 18.4 µg/mL | |
| *Myrtus communis* | BChE (OE1) | 77.8 µg/mL | [65] |
| | AChE (EO5) | 87.9 µg/mL | |
| | BChE (OE5) | 96.0 µg/mL | |
| *Pimenta dioica* | AChE | 1320 µg/mL | [64] |
| | BChE | 3340 µg/mL | |
| *Psidium guajava* | AChE | 24.96 ± 2.38% | [14] |
| *Syzygium cumini* | AChE | 19.97 ± 1.10% | |
| *Syzygium samarangense* | AChE | 13.78 ± 1.52% | |

**Abbreviations**: AChE: Acetylcholinesterase; BChE: Butyrylcholinesterase; GABA: Gamma-aminobutyric acid.

## Cytotoxicity Activity

The extensive use of synthetic insecticides and herbicides has caused an increase in environmental pollution, generating several risks to human and animal health, in addition to increasing the risk of weed proliferation, interfering harmfully in agriculture and in several other plant species [70]. As a result, there is a growing search for environmentally friendly and sustainable natural products that possess cytotoxic activity against the most diverse types of existing pests [71 - 73]. Table 4 summarizes information on the cytotoxicity of essential oils from Myrtaceae.

Essential oils are potential candidates for environmentally friendly bioinsecticides because they present cytotoxic activities against various pathogens [74 - 76]. In addition, several studies indicate that essential oils also have cytotoxic effects against tumor and non-tumor cells and can be used as alternatives for the production of anticancer drugs [77 - 79].

Natural products from the Myrtaceae family are already known for their promising antitumor, anticancer, bioinsecticidal, and larvicidal potentials [80 - 83]. Therefore, the essential oils of this family may be promising in the treatment of the most diverse types of cancer and parasitic agents [62, 84 - 87].

**Table 4. Cytotoxic activity present in the essential oils of Myrtaceae.**

| Species (Plant Fraction) | Methodology | Results Described | References |
|---|---|---|---|
| *Myrcia sylvatica* (dry leaves) | MTT colorimetric method | *Myrcia sylvatica* oil (Msyl) did not show activity against breast cells, but only against melanoma ($IC_{50}$ 20.0 µg.mL$^{-1}$) and gastric cancer cells ($IC_{50}$ 17.3 µg.mL$^{-1}$). | [88] |
| *Psidium guineense 1* (dry leaves) | MTT colorimetric method | Pgui-1 sample was more active than Pgui-2 against all cell lines; It showed activity against breast ($IC_{50}$ 12.6 and 18.2 µg.mL$^{-1}$), melanoma ($IC_{50}$ 11.1 and 19.1 µg.mL$^{-1}$), gastric ($IC_{50}$ 8.2 and 15.7 µg.mL$^{-1}$), and normal human fibroblast ($IC_{50}$ of 8.27 and 24.0 µg.mL$^{-1}$). | [88] |
| *Psidium guineense 2* (dry leaves) | MTT colorimetric method | | [88] |
| *Eugenia patrisii* (dry leaves) | MTT colorimetric method | It showed activity against melanoma (SKMEL-19) and colon (HCT116) cells with $IC_{50}$ values of 5.8 and 6.7 µg.mL$^{-1}$, respectively. Likewise, gastric ascites (AGP-01) were the most sensitive human cancer cells, showing $IC_{50}$ value of 3.2 µg.mL$^{-1}$ and SI of 1.1. | [88] |

*(Table 4) cont.....*

| Species (Plant Fraction) | Methodology | Results Described | References |
|---|---|---|---|
| *Psidium guajava* (dry leaves) | MTT colorimetric method | It showed activity against melanoma ($IC_{50}$ 15.3 $\mu g.mL^{-1}$), gastric ($IC_{50}$ 16.3 $\mu g.mL^{-1}$), and breast ($IC_{50}$ 12.4 $\mu g.mL^{-1}$) human cancer cells, with SI of 1.7 against MCF7 cells. It also presented activity against breast ($IC_{50}$ 12.6 and 18.2 $\mu g.mL^{-1}$), melanoma ($IC_{50}$ 11.1 and 19.1 $\mu g.mL^{-1}$), gastric cancer cells ($IC_{50}$ 8.2 and 15.7 $\mu g.mL^{-1}$), and normal human fibroblast ($IC_{50}$ 8.27 and 24.0 $\mu g.mL^{-1}$). | [88] |
| *Eugenia stipitata* (dry leaves) | MTT colorimetric method | It displayed cytotoxic activity against melanoma ($IC_{50}$ 17.2 $\mu g.mL^{-1}$), breast cancer cells ($IC_{50}$ 19.1 $\mu g.mL^{-1}$), and normal human fibroblast ($IC_{50}$ 13.8 $\mu g.mL^{-1}$). | [88] |
| *Myrcia splendens* (dry leaves) | MTT colorimetric method | It showed significant cytotoxic activity against gastric ($IC_{50}$ 4.7 $\mu g.mL^{-1}$), melanoma ($IC_{50}$ 8.6 $\mu g.mL^{-1}$), and colon ($IC_{50}$ 8.8 $\mu g.mL^{-1}$) human cancer cells; it was more selective against AGP01 (SI 1.4). | [88] |
| *Cassia alata* and *Psidium guajava* (dry leaves) | MTT colorimetric method | The $IC_{50}$ values were 160 $\mu g/ml$ for C. alata and 200 $\mu g/ml$ for P guajava | [89] |
| *Psidium myrtoides* (leaves) | PM-EO and DXR | GM07492a ($IC_{50}$ PM-EO) 359.8 ± 6.3 and $IC_{50}$ DXR 0.5 ± 0.2), MCF-7 ($IC_{50}$ PM-EO) 254.5 ± 6.3 and $IC_{50}$ DXR 62.1 ± 2.0), HeLa ($IC_{50}$ PM-EO 324.2 ± 41.4 and $IC_{50}$ DXR 5.3 ± 1.3) and M059J ($IC_{50}$ PM-EO 289.3 ± 10.9, and $IC_{50}$ DXR 16.2 ± 2.5). | [33] |
| *Eugenia egensis* (aerial parts) | The MTT colorimetric method | It presented cytotoxicity against HCT-116 ($IC_{50}$ > 25 $\mu g.mL^{-1}$), AGP-01 (> 25 $\mu g.mL^{-1}$), SKMEL-19 (> 25 $\mu g.mL^{-1}$), and hemolysis (> 200 $\mu g.mL^{-1}$). | [42] |
| *Eugenia flavescens* (aerial parts) | The MTT colorimetric method | It showed cytotoxicity against HCT-116 ($IC_{50}$ 13.9 $\mu g.mL^{-1}$), MCRC5 ($IC_{50}$ 14.0 $\mu g.mL^{-1}$), AGP-01 (> 25 $\mu g.mL^{-1}$), SKMEL-19 (> 25 $\mu g.mL^{-1}$), and hemolysis (> 200 $\mu g.mL^{-1}$). | [42] |
| *Eugenia patrisii* (aerial parts) | The MTT colorimetric method | It showed cytotoxicity activity against HCT-116 ($IC_{50}$ 16.4 $\mu g.mL^{-1}$), MCRC5 ($IC_{50}$ 18.1 $\mu g.mL^{-1}$), AGP-01 (> 25 $\mu g.mL^{-1}$), SKMEL-19 (> 25 $\mu g.mL^{-1}$), and hemolysis (> 200 $\mu g.mL^{-1}$). | [42] |
| *Eugenia polystachya* (aerial parts) | The MTT colorimetric method | It presented cytotoxicity activity against HCT-116 ($IC_{50}$ 10.3 $\mu g.mL^{-1}$), AGP-01 (> 25 $\mu g.mL^{-1}$), AGP-01 (> 25 $\mu g.mL^{-1}$), SKMEL-19 (> 25 $\mu g.mL^{-1}$), and hemolysis (> 200 $\mu g.mL^{-1}$). | [42] |

*(Table 4) cont.....*

| Species (Plant Fraction) | Methodology | Results Described | References |
|---|---|---|---|
| *Eugenia cuspidifolia* (dry leaves) | AB assay | HCT-116 (IC$_{50}$ 15.25 µg.mL$^{-1}$), MCF-7 (IC$_{50}$ 18.11 µg.mL$^{-1}$), SKMEL-19 (IC$_{50}$ 26.17 µg.mL$^{-1}$), ACP02 (IC$_{50}$ > 50 µg.mL$^{-1}$), MCR-5 (IC$_{50}$ 25.51 µg.mL$^{-1}$). | [90] |
| *Eugenia tapacumensis* (dry leaves) | AB assay | HCT-116 (IC$_{50}$ 12.37 µg.mL$^{-1}$), MCF-7 (IC$_{50}$ 24.35 µg.mL$^{-1}$), SKMEL-19 (IC$_{50}$ > 50 µg.mL$^{-1}$), ACP02 (IC$_{50}$ > 50 µg.mL$^{-1}$), MCR-5 (IC$_{50}$ 36.12 µg.mL$^{-1}$). | [90] |
| *Eugenia dysenterica* (leaves) | The MTT colorimetric method | L929 (IC$_{50}$ 542.2 µg.mL$^{-1}$) and RAW (IC$_{50}$ 290.0 µg.mL$^{-1}$) | [91] |
| *Eugenia calycina* (leaves) | The MTT colorimetric method | HeLa (137.4 ± 9.6) | [92] |
| *Eucalyptus grandis* (fresh leaves) | Cell cycle analysis | It was less effective in inducing cytotoxic effects in meristematic *L. sativa* cells. | [93] |
| *Eucalyptus citriodora* (fresh leaves) | Cell cycle analysis | It was more effective in inducing cytotoxic effects in meristematic *L. sativa* cells. | [93] |
| *Eucalyptus globulus* (fresh leaves) | MTT colorimetric method | Possible application in the treatment of lung cancer. | [94] |
| *Leptospermum citratum* (blossoms) | Fumigant toxicity and contact tests with different species of flies | Fumigant - male: LC$_{50}$ 2.39 mg.L$^{-1}$ and female: LC$_{50}$ 3.24 mg.L$^{-1}$ Contact - male: LD$_{50}$ 3,31 mg.L$^{-1}$ and female: LD$_{50}$ 5.22 mg.L$^{-1}$ | [95] |
| *Leptospermum ericoides* (leaves) | Contact tests with different species of flies | Male LD$_{50}$ 0.71 mg.L$^{-1}$ and female LD$_{50}$ 1.23 mg.L$^{-1}$ | [95] |
| *Leptospermum scoparium* (leaves) | Contact tests with different species of flies | Male LD$_{50}$ 0.60 mg.L$^{-1}$ and female LD$_{50}$ 1.10 mg.L$^{-1}$ | [95] |
| *Eugenia calycina* (leaves) | Microplate dilution method | HeLa: CC$_{50}$ 266.8 ± 46.5 (µg mL$^{-1}$) 48 h Vero: CC$_{50}$ 312.1 ± 42.5 (µg mL$^{-1}$) 48 h | [96] |
| *Myrcia splendens* (leaves) | MTT colorimetric method | A549: IC$_{50}$ 100.99 ± 2.32 µg mL$^{-1}$ MCF-7: IC$_{50}$ 5.59 ± 0.13 µg mL$^{-1}$ HaCaT: IC$_{50}$ 21.58 ± 1.26 µg mL$^{-1}$ | [18] |

According to Jerônimo *et al.* [88], essential oils from species of the Myrtaceae family showed cytotoxic and antiproliferative activity against the following cancer cell lines: breast (MCF7), melanoma (SKMEL-19), gastric (AGP01), colon (HCT116), and a non-human lung fibroblast malignant cell (MRC5), at concentrations below 100 µg.mL$^{-1}$. The authors also pointed out that *Eugenia patrisii* essential oil showed the greatest activity against melanoma (SKMEL-19) and colon cells (HCT116), with $IC_{50}$ values of 5.8 and 6.7 µg.mL$^{-1}$, respectively. This result was linked to the presence of (E)-caryophyllene (32.0%) and bicyclogermacrene (10.0%) in their chemical composition.

Regarding other species, *Eugenia stipitata* showed cytotoxic activity against melanoma ($IC_{50}$ 17.2 µg.mL$^{-1}$), breast cancer cells ($IC_{50}$ 19.1 µg.mL$^{-1}$), and a normal human fibroblast ($IC_{50}$ 13.8 µg.mL$^{-1}$). *Myrcia splendens* showed significant cytotoxic activity against gastric ($IC_{50}$ 4.7 µg.mL$^{-1}$), melanoma ($IC_{50}$ 8.6 µg.mL$^{-1}$), and colon cancer cells ($IC_{50}$ 8.8 µg.mL-1). *Myrcia silvatica* showed activity against melanoma ($IC_{50}$ 20.0 µg.mL-1) and gastric cancer cells ($IC_{50}$ 17.3 µg.mL$^{-1}$). And *Psidium guajava* showed cytotoxic potential against melanoma, gastric, and breast cancer cells, with specimen 1 being more active against all cell types, despite the similarity in their chemical compositions.

According to Silva *et al.* [89], the essential oil of *Eugenia uniflora* showed cytotoxic potential when used at concentrations below 100 mg.mL$^{-1}$, for the following cancer cells: GM07492A ($IC_{50}$ 39.8 ± 1.4 µg.mL$^{-1}$), M059J ($IC_{50}$ 84.5 ± 5.7 µg.mL$^{-1}$), and MCF-7 ($IC_{50}$ 76.5 ± 8.4 µg.mL$^{-1}$). According to the authors, *Plinia cauliflora* showed significant cytotoxic activity against M059J ($IC_{50}$ 79.1 ± 6.5 µg.mL$^{-1}$), MCF-7 ($IC_{50}$ 76.6 ± 8.2 µg.mL$^{-1}$), and *Syzygium cumini* showed activity only against GM07492A ($IC_{50}$ 679.0 ± 48.0 µg.mL$^{-1}$). The authors also pointed out that the essential oils of these Myrtaceae species, which occur in the Cerrado biome (Brazil), can be used for the development of new drugs.

According to Dias *et al.* [33], *Psidium myrtoides* essential oil may be a potential source for the search for new antitumor agents, possessing cytotoxic and antiproliferative activities against the following cancer cells: GM07492a, MCF-7, HeLa, and M059J. According to the authors, such results may be due to the presence of the hydrocarbon sesquiterpene *trans*-β-caryophyllene, which is the major constituent of this natural product.

Silva *et al.* [42] studied the essential oil of *Eugenia egensis,* which showed cytotoxic potential against tumor cells AGP-01, HCT-116, and SKMEL19, only when used at concentrations above 25 mg.mL$^{-1}$. Regarding the other essential oils, the authors pointed out that *E. polystachya* showed better activity against HCT-116, the cause of human colon cancer (10.3 mg.mL$^{-1}$); and *E. flavescens* showed better activity against MRC5 (14.0 mg.mL$^{-1}$), the human fibroblast.

Aranha *et al.* [90] noted that *Eugenia cuspidifolia* EO showed better cytotoxic activity than *E. tapacumensis* against tumor cells MCF-7, SKMEL-19, and MRC-5. However, for both species, the best inhibitory potential was achieved against HCT116 cells with $IC_{50}$ equal to 15.25 and 12.37 mg.mL$^{-1}$, respectively.

Regarding the ACP02 cell, none of the species showed cytotoxic activity at concentrations below 50 mg.mL$^{-1}$.

Silva *et al.* [91], in their work, stated that the essential oil of *Eugenia dysenterica* leaves presented an IC$_{50}$ value equal to 542.2 µg.mL$^{-1}$ for cell L929, a fibroblast cell line, whereas in the macrophage test (RAW), IC$_{50}$ was 290.0 µg.mL$^{-1}$. These values were associated with the presence of α-humulene at high levels.

Sousa *et al.* [92] conducted a study that identified the *Eugenia calycina* essential oil as a potential inhibitor of the growth of HeLa-type cells (IC$_{50}$ 137.4 ± 9.6). Aragão *et al.* [93] evaluated the cytotoxicity of the essential oil from the dried leaves of lemon eucalyptus (*Eucalyptus citriodora*) and eucalyptus (*Eucalyptus grandis*) on lettuce (*Lactuca sativa*). According to these authors, *E. citriodora* is more effective in inducing cytotoxic effects in meristematic cells of lettuce, acting as aneugen, and on a lesser scale, as a clastogenic agent.

Adnan [94] conducted a study with commercial *Eucalyptus globulus* essential oil. According to him, this product has a cytotoxic effect against A549 cells, indicating its possible application in the treatment of lung cancer. Moreover, the author pointed out that despite the promising results, it has less activity than the standard chemotherapy drug Fluorouracil.

Park *et al.* [95] evaluated the cytotoxicity of essential oils from species of the Myrtaceae family occurring in Oceania and South Africa. According to these authors, such oils and their components are quite promising for the development of natural insecticides, as they have excellent contact toxicity against *Drosophila* flies.

Silva *et al.* [96] evaluated the cytotoxicity of *Eugenia calycina* essential oil in mammalian HeLa and Vero cells. According to the authors, the cytotoxic concentrations (266.8 ± 46.5 and 312.1 ± 42.5 g.mL$^{-1}$, respectively) in 48 h of exposure were higher than CL$_{50}$, besides exhibiting larvicidal activity, resulting in a positive selectivity index.

According to Scalvenzi *et al.* [18], the EO obtained from the leaves of *Myrcia splendens* presented potential against the MCF-7 cell line (IC$_{50}$ 5.59 ± 0.13 µg.mL$^{-1}$), very similar to the result generated by the synthetic drug doxorubicin (IC$_{50}$ 2.10 ± 0.42 µg.mL$^{-1}$). This value was associated with the presence of α-bisabolol, which individually exhibited IC$_{50}$ equal to 1.24 ± 0.03.

## Anti-protozoan Activity

The leaf essential oils of *Calyptranthes grandifolia*, *Calyptranthes tricona*, *Eugenia anomala*, *Eugenia arenosa*, *Eugenia pyriformis*, *Myrrhinium atropurpureum,* and *Psidium salutare* were analyzed (*in vitro*) to determine their antileishmanial potential against promastigotes of *Leishmania amazonenses*. They presented $IC_{50}$ values of 31.27 ± 6.40μg/mL, 26.13 ± 8.60 μg/mL, 62.88 ± 3.19 μg/mL, 13.72 ± 8.65 μg/mL, 19.73 ± 5.40 μg/mL, 154.1 ± 8.14 μg/mL, and 69.71 ± 2.30 μg/mL, respectively. The results found for *C. grandifolia, C. tricona, E. arenosa,* and *E. pyriformis* were not different from those found in the reference drug, pentamidine ($IC_{50}$ = 23.22 ± 9.04 μg/mL). According to the authors, these results demonstrated both the antileishmanial potential and the importance of continuing the studies on the search and development of new phytotherapeutic medicines from the Brazilian flora [97], the antiprotozoal potential is shown in Table **5**.

**Table 5. Antiprotozoan activity in Myrtaceae essential oils.**

| Plant Species | Protozoan | Methodology | $IC_{50}$ Value | References |
|---|---|---|---|---|
| *Calyptranthes grandifolia* | *Leishmania amazonensis* | MTT assay | 31.27 ± 6.40 μg/mL | |
| *C. tricona* | *L. amazonensis* | MTT assay | 26.13 ± 8.60 μg/mL | [97] |
| *Eugenia anômala* | *L. amazonensis* | MTT assay | 62.88 ± 3.19 μg/mL | |
| *E. arenosa* | *L. amazonensis* | MTT assay | 13.72 ± 8.65 μg/mL | |
| *E. brejoensis* | *Trypanosoma cruzi* | Optical microscopy | 29 ± 4.5 μg/ml (Epimastigote form) | [51] |
| | | | 0.62 μg/ml (Trypomastigote form) | |
| | | | 12.5 ± 1,74 μg/ml (Amastigote form) | |
| | | | 251.4 ± 44.3 μg/ml (Macrophage form) | |
| *E. dysenterica* | *T. cruzi* | MTT colorimetric assay | 9.5 μg / mL | [24] |
| *E. pitanga* | *L. amazonensis* | MTT assay | 6.10 ± 1.80 μg/mL | [100] |
| *E. pyriformis* | *L. amazonensis* | MTT assay | 19.73 ± 5.40 μg/mL | [97] |
| *E. uniflora* | *L. amazonensis* | MTT assay | 0.99 μg / mL | [28] |
| *Melaleuca styphelioides* | *Acanthamoeba* | Colorimetric 96-well microtiter plate assay | 69.03 ± 9.17 mg/mL | [102] |

| Plant Species | Protozoan | Methodology | IC$_{50}$ Value | References |
|---|---|---|---|---|
| *Myrciaria plinioides* | *L. amazonensis* | MTT assay | 14.16 ± 7.40 µg / mL | [99] |
| | *L. infantum* | | 101.50 ± 5.78 µg / mL | |
| *Myrrhinium atropurpureum* | *L. amazonensis* | MTT assay | 154.1 ± 8.14 µg/mL | [97] |
| *Myrtus communis* | *L. tropica* | MTT assay | 8.4 µg/ml (Promastigote form) | [101] |
| | | | 11.6 µg/mL (Amastigote form) | |
| *Plinia cauliflora* | *L. amazonensis* | MTT assay | 0.46 µg/mL | [28] |
| *Psidium salutare* | *L. amazonensis* | MTT assay | 69.71 ± 2.30 µg/mL | [97] |
| *P. guajava* | *T. cruzi* | MTT colorimetric assay | 14.6 µg/mL | [21] |
| *Syzygium cumini* | *L. amazonensis* | MTT assay | 8.78 µg/mL | [28] |
| *S. cumini* | *L. amazonensis* | MTT colorimetric assay | 43.9 mg/mL (axenic amastigotes) | [98] |
| | | | 38.1 mg/mL (intracellular amastigotes) | |

**Abbreviations:** MTT assay: 3-(4,5-dimethylthiazol-2-yl)-2,5-diphenyl tetrazolium bromide.

Three Myrtaceae species - *Eugenia uniflora, Plinia cauliflora,* and *Syzygium cumini* were tested against the growth of *Leishmania amazonenses* promastigotes. The results were IC$_{50}$ 0.99 µg/mL for *E. uniflora*, IC$_{50}$ 0.46 µg/mL for *Plinia cauliflora,* and IC$_{50}$ 8.78 µg/mL for *Syzygium cumini*, while amphotericin B had IC$_{50}$ equal to 0.60 µg/mL. Compared to amphotericin B, the results showed that the essential oils of *E. uniflora* and *P. cauliflora,* were promising for leishmanicidal activity [26]. The antileishmanial activity was evidenced in a study performed with the essential oil of another species of *S. cumini* against the protozoan *L. amazonenses*. This essential oil showed IC$_{50}$ of 43.9 mg/mL against axenic amastigotes and 38.1 mg/mL against intracellular amastigotes; and the major constituent α-pinene was more active, presenting IC$_{50}$ values of 16.1 mg/mL against axenic amastigotes and 15.6 mg/mL against intracellular amastigotes. These results demonstrated that both *S. cumini* essential oil and its major constituent α-pinene have antileishmanial activity [98].

The leaf essential oil of *Myrciaria plinioides* was tested *in vitro* to determine its antileishmanial potential against promastigotes of *Leishmania amazonenses* and *Leishmania infantum*. It showed IC$_{50}$ equal to 14.16 ± 7.40 µg/mL against *L. amazonenses* and IC$_{50}$ of 101.50 ± 5.78 µg/mL against *L. infantum*. These results demonstrated that the essential oil of *M. plinioides* is a promising source for new antileishmanial agents against *L. amazonenses* [99]. *Eugenia pitanga* essential oil was also tested *in vitro* against *L. amazonenses*, and presented $IC_{50}$ equal to $6.10 \pm 1.80$ µg/mL. Despite this result, the authors suggested new studies to explain the antileishmanial activity present in this essential oil [100].

The essential oil from the aerial parts of *Myrtus communis* was used to determine the antileishmanial activity against *Leishmania tropica in vitro*. It showed $IC_{50}$ of 8.4 µg/mL against *L. tropica* promastigotes, whereas the control drug (meglumine antimoniate) showed $IC_{50}$ of 88.3 µg/mL. Regarding the anti-amastigote assay, the essential oil showed $IC_{50}$ 11.6 µg/mL, and the control drug, $IC_{50}$ 44.6 µg/mL. These results demonstrated that the essential oil of *M. communis* has antileishmanial potential, however it was more effective in the anti-amastigote assay [101]. The *in vitro* trypanocidal activity was demonstrated by the essential oil of *Eugenia brejoensis* leaves against some evolutionary forms of *Trypanosoma cruzi*. It showed the following results: $IC_{50}$ of $29 \pm 4.5$ µg/mL (epimastigote form), $CL_{50}$ of $17.39 \pm 0.62$ µg/mL (trypomastigote form), $IC_{50}$ of $12.5 \pm 1.74$ µg/mL (amastigote form), and $CC_{50}$ of $251.4 \pm 44.3$ µg/mL (macrophages form).

These results demonstrated that the essential oil of *E. brejoensis* was more effective against the amastigote form, and it is a strong, promising agent for the treatment of parasitic diseases, which can bring perspectives to the development of new drugs against Chagas disease [51]. The leaf essential oil of *Eugenia dysenterica* was investigated against trypomastigote forms of *Trypanosoma cruzi*. It showed $IC_{50}$ of 9.5 µg/mL, while benznidazole (positive control) presented $IC_{50}$ of 9.8 µg/mL, revealing a remarkable trypanocidal action [24].

The essential oil from fresh *Psidium guajava* flowers was studied for its trypanocidal potential against *T. cruzi*. In this study, it proved to be active with reduction of trypomastigote cells of this protozoan, exhibiting satisfactory trypanocidal activity with $IC_{50} = 14.6$ µg/mL, compared to benznidazole (positive control) with $IC_{50} = 9.8$ µg /mL [21].

In the study carried out with the leaf essential oil of *Melaleuca styphelioides,* it was possible to determine its amoebicidal activity through a colorimetric 96-well microtiter plate assay. The essential oil was characterized by caryophyllene oxide (23.42%), spathulenol (20.5%), isoaromadendrene epoxide (7.45%), ledol (5.98%), α-pinene (3.82%), isopinocarveol (2.18%), and showed growth inhibition against *Acanthamoeba* equal to $69.03 \pm 9.17$ mg/mL. This result shows that *M. styphelioides* essential oil is promising for the development of anti-ameba drugs [102].

## Antidiabetic Activity

Natural products have an extensive field of investigation for antidiabetic activity, for many of these products have direct or indirect inhibition effects on the enzymatic pathways of diabetes [103]. Within this line of research, the Myrtaceae family can be noted, since many of its species are reported in popular medicine as promising antidiabetic agents [104]. Nevertheless, there are few reports of such activities in essential oils of this family involving the terpenoid class.

*Syzygium aromaticum* essential oil was evaluated for its antidiabetic activity through the α-amylase inhibition assay. In this study, five emulsions of the essential oil (A1, A2, A3, A4, and A5) were tested, and presented the following inhibition potentials: 68.14%, 68.76%, 70.60%, 88.07%, and 95.30%, respectively. The emulsion A5 presented maximum antidiabetic activity with 95.30% of α-amylase inhibition. This result may be associated with eugenol, which was the major constituent of this essential oil (18.7%) [105].

The leaf essential oils of *Corymbia citriodora*, *Eucalyptus globulus*, *E. radiata*, *Melaleuca alternifolia*, *M. viridiflora, Myrtus communis*, and *Syzygium aromaticum* (in this case, buds were evaluated as well) were tested *in vitro* to determine their inhibitory activity against α-amylase. This test was performed *via* a spectrophotometric assay, and the essential oils showed the following percentages of α-amylase inhibition: *Corymbia citriodora* (44%), *Eucalyptus globulus* (34%), *E. radiata* (65%), *Melaleuca alternifolia* (14%), *M. viridiflora* (28%), *Myrtus communis* (20%), and *Syzygium aromaticum* essential oil showed no activity. These results demonstrated that the essential oil of *Eucalyptus radiata* leaves is a promising source of α-amylase inhibitors [106].

The essential oil from leaves and stems of *Myrtus communis* was also evaluated regarding its antidiabetic potential against α-amylase enzyme. In this study [107], the essential oil from the leaves showed inhibition potential ($IC_{50}$) equal to 29.94 μg/mL, whereas the essential oil from stems presented $IC_{50}$ of 159.80 μg/mL, both considered significant results.

## Recent Advances in Phytotherapy

Herbal medicines are products derived solely from plants, which represent an accessible and low-cost treatment [108]. Within this segment, Myrtaceae family has a large record of species with pharmacological and therapeutic potential, and many of these species are promising sources of bioactive compounds [109]. Within the genus *Syzygium,* there are several species that have high antioxidant capacities, which makes them candidates for the development of new herbal medicines for the treatment of autoimmune diseases [110].

According to Kauffmann and Ethur [111], the popular use of plants for medicinal purposes has aroused great interest for the development of new drugs, mainly from the Brazilian flora. The Myrtaceae family is included within this great interest, because the essential oils of *Eugenia uniflora* and *E. jambolana* have shown excellent antileishmanial activity. In addition, this property of *E. jambolana* may be related to its main constituent α-pinene.

In the endodontics field, the essential oil of *Psidium cattleianum* can be highlighted. Its major constituents viridiflorol (17.9%), β-caryophyllene (11.8%), 1.8-cineole (10.8%), and β-selinene (8.6%), may be related to the action of such oil against endodontic pathogenic bacteria. Thus, the essential oil of *Psidium cattleianum* can be a source of bioactive compounds for therapeutic solutions in this field [32]. *Psidium guajava* essential oil, which is characterized by β-caryophyllene (16.1%), α-humulene (11.9%), aromadendrene oxide (14.7%), δ-selinene (13.6%), and selin-11-en-4α-ol (12.5%), was also considered a strong candidate for the treatment of oral pathogens [112].

Eugenol is a monoterpene present as the main substance in the essential oil of *Syzygium aromaticum*, which has several biological actions such as antibacterial, antifungal, and anti-inflammatory activities. This monoterpene is indicated for use in dental treatments due to its anti-inflammatory, analgesic, and antiseptic potentials [113]. Furthermore, eugenol, eugenol acetate, and β-caryophyllene are strategic compounds for the treatment of Alzheimer's disease [114]. Similarly, the compounds α-pinene, β-limonene, β-phellandrene, *p*-cymene, and linalool are also explored for the treatment of Alzheimer's [115].

## CONCLUDING REMARKS

The essential oils of species of the Myrtaceae family present a diversified chemical composition that includes mainly terpenic compounds, which are also one of the main ones responsible for their biological properties. However, the explanation of the mechanisms of these biological activities is considered very complex, due to the synergism that exists among their chemical constituents.

Myrtaceae essential oils present antioxidant, neuroprotective, anti-inflammatory, cytotoxic, and antidiabetic properties, however the quantity of reports on antidiabetic activities is smaller (but still attractive) in comparison to the other effects. The essential oils of this family are promising for the development of new natural herbal medicines.

## CONSENT FOR PUBLICATION

Not applicable.

## CONFLICT OF INTEREST

The author declares no conflict of interest, financial or otherwise.

## ACKNOWLEDGEMENTS

We would like to thank the authors for their contributions and commitment to writing this scientific work, and the editors of this book for the invitation and vote of confidence.

## REFERENCES

[1]    Ríos, J-L. Essential Oils.*Essential Oils in Food Preservation, Flavor and Safety*; Elsevier, **2016**, pp. 3-10.
[http://dx.doi.org/10.1016/B978-0-12-416641-7.00001-8]

[2]    Aziz, Z.A.A.; Ahmad, A.; Setapar, S.H.M.; Karakucuk, A.; Azim, M.M.; Lokhat, D.; Rafatullah, M.; Ganash, M.; Kamal, M.A.; Ashraf, G.M. Essential Oils: Extraction Techniques, Pharmaceutical And Therapeutic Potential - A Review. *Curr. Drug Metab.,* **2018**, *19*(13), 1100-1110.
[http://dx.doi.org/10.2174/1389200219666180723144850] [PMID: 30039757]

[3]    Kandi, S.; Godishala, V.; Rao, P.; Ramana, K.V. Biomedical Significance of Terpenes: An Insight. *Biomed. Biotechnol.,* **2015**, *3*(1), 8-10.

[4]    Mewalal, R.; Rai, D.K.; Kainer, D.; Chen, F.; Külheim, C.; Peter, G.F.; Tuskan, G.A. Plant-Derived Terpenes: A Feedstock for Specialty Biofuels. *Trends Biotechnol.,* **2017**, *35*(3), 227-240.
[http://dx.doi.org/10.1016/j.tibtech.2016.08.003] [PMID: 27622303]

[5]    Petrović, J.; Stojković, D.; Soković, M. *Terpene core in selected aromatic and edible plants: Natural health improving agents,* 1st ed; Elsevier Inc., **2019**, Vol. 90, .

[6]    *Zuzarte; Salgueiro, Bioactive Essential Oils and Cancer*; Springer International Publishing: Cham, **2015**.

[7]    A, Ludwiczuk; K., Skalicka-Woźniak; M.I., Georgiev Terpenoids. In: *Pharmacognosy*; S, Badal; R., Delgoda, Eds.; Elsevier: Amisterdan, **2017**; pp. 233-266.
[http://dx.doi.org/10.1016/B978-0-12-802104-0.00011-1]

[8]    Cunha, O.H. L. L. T.; Lucena, E. M. P. de; Bonilla, "Exigências térmicas da floração à frutificação de quatro espécies de Myrtaceae em ambiente de Restinga," *Rev. Bras. Geogr. Física,* **2016**, *06*, 1275-1291.

[9]    Santos, M.P. *Almeida; Adonias, et al., "Prospecção tecnológica de eugenia uniflora l. (myrtaceae) technological prospection of eugenia uniflora l. (myrtaceae)*; Rev. GEINTEC, **2016**, Vol. 6, pp. 3109-3120.

[10]   de Araújo, F.F.; Neri-Numa, I.A.; de Paulo Farias, D.; da Cunha, G.R.M.C.; Pastore, G.M. Wild Brazilian species of Eugenia genera (Myrtaceae) as an innovation hotspot for food and pharmacological purposes. *Food Res. Int.,* **2019**, *121*, 57-72.
[http://dx.doi.org/10.1016/j.foodres.2019.03.018] [PMID: 31108783]

[11]   Lima, W.G.; Guedes-Bruni, R.R. Myrceugenia (Myrtaceae) ocorrentes no Parque Nacional do Itatiaia, Rio de Janeiro. *Rodriguésia,* **2004**, *55*(85), 73-94.
[http://dx.doi.org/10.1590/2175-78602004558505]

[12]   Bustos-Segura, C.; Dillon, S.; Keszei, A.; Foley, W.J.; Külheim, C. Intraspecific diversity of terpenes of Eucalyptus camaldulensis (Myrtaceae) at a continental scale. *Aust. J. Bot.,* **2017**, *65*(3), 257.
[http://dx.doi.org/10.1071/BT16183]

[13] Costa, A.V.; Pinheiro, P.F.; de Queiroz, V.T.; Rondelli, V.M.; Marins, A.K.; Valbon, W.R.; Pratissoli, D. Chemical Composition of Essential Oil from *Eucalyptus citriodora* Leaves and Insecticidal Activity Against *Myzus persicae* and *Frankliniella schultzei. J. Essent. Oil-Bear. Plants,* **2015**, *18*(2), 374-381.
[http://dx.doi.org/10.1080/0972060X.2014.1001200]

[14] Petrachaianan, T.; Chaiyasirisuwan, S.; Athikomkulchai, S.; Sareedenchai, V. Screening of acetylcholinesterase inhibitory activity in essential oil from Myrtaceae. *Thaiphesatchasan,* **2019**, *43*(1), 63-68.

[15] da Silva, A.G.; Alves, R.C.C.; Filho, C.M.B.; Bezerra-Silva, P.C.; Santos, L.M.M.; Foglio, M.A.; Navarro, D.M.A.F.; Silva, M.V.; Correia, M.T.S. Chemical Composition and Larvicidal Activity of the Essential Oil from Leaves of *Eugenia brejoensis* Mazine (Myrtaceae). *J. Essent. Oil-Bear. Plants,* **2015**, *18*(6), 1441-1447.
[http://dx.doi.org/10.1080/0972060X.2014.1000390]

[16] Neves, I.D.A.; Da F Rezende, S.R.; Kirk, J.M.; Pontes, E.G.; De Carvalho, M.G. Composition and larvicidal activity of essential oil of eugenia candolleana DC. (MYRTACEAE) against Aedes aegypti. *Rev. Virtual Quim.,* **2017**, *9*(6), 2305-2315.
[http://dx.doi.org/10.21577/1984-6835.20170138]

[17] Ramos, M.F.S.; Monteiro, S.S.; da Silva, V.P.; Nakamura, M.J.; Siani, A.C. Essential Oils From Myrtaceae Species of the Brazilian Southeastern Maritime Forest (Restinga). *J. Essent. Oil Res.,* **2010**, *22*(2), 109-113.
[http://dx.doi.org/10.1080/10412905.2010.9700275]

[18] Scalvenzi, L.; Grandini, A.; Spagnoletti, A.; Tacchini, M.; Neill, D.; Ballesteros, J.; Sacchetti, G.; Guerrini, A. Myrcia splendens (Sw.) DC. (syn. M. fallax (Rich.) DC.) (myrtaceae) essential oil from amazonian Ecuador: A chemical characterization and bioactivity profile. *Molecules,* **2017**, *22*(7), 1163.
[http://dx.doi.org/10.3390/molecules22071163] [PMID: 28704964]

[19] Silva, A.; Bomfim, H.; Magalhães, A.; Rocha, M.; Lucchese, A. COMPOSIÇÃO QUÍMICA E ATIVIDADE ANTINOCICEPTIVA EM MODELO ANIMAL DO ÓLEO ESSENCIAL DE Myrcia rostrata DC. (MYRTACEAE). *Quim. Nova,* **2018**, *41*(9), 982-988.
[http://dx.doi.org/10.21577/0100-4042.20170274]

[20] Medeiros, F.C.M.; Del Menezzi, C.H.S.; Vieira, R.F.; Fernandes, Y.F.M.; Santos, M.C.S.; Bizzo, H.R. Scents from Brazilian Cerrado: chemical composition of the essential oil from *Psidium laruotteanum* Cambess (Myrtaceae). *J. Essent. Oil Res.,* **2018**, *30*(4), 253-257.
[http://dx.doi.org/10.1080/10412905.2018.1462740]

[21] Fernandes, C.C.; Rezende, J.L.; Silva, E.A.J.; Silva, F.G.; Stenico, L.; Crotti, A.E.M.; Esperandim, V.R.; Santiago, M.B.; Martins, C.H.G.; Miranda, M.L.D. Chemical composition and biological activities of essential oil from flowers of Psidium guajava (Myrtaceae). *Braz. J. Biol.,* **2021**, *81*(3), 728-736.
[http://dx.doi.org/10.1590/1519-6984.230533] [PMID: 32876175]

[22] de Veras, B.O.; dos Santos, Y.Q.; da S. Oliveira, F.G Algrizea Minor Sobral, Faria & Proença (Myrteae, Myrtaceae): chemical composition, antinociceptive, antimicrobial and antioxidant activity of essential oil. In: *Nat. Prod. Res;* , **2020**; 34, pp. (20)3013-3017.

[23] An, N.T.G.; Huong, L.T.; Satyal, P.; Tai, T.A.; Dai, D.N.; Hung, N.H.; Ngoc, N.T.B.; Setzer, W.N. Mosquito larvicidal activity, antimicrobial activity, and chemical compositions of essential oils from four species of myrtaceae from central Vietnam. *Plants,* **2020**, *9*(4), 544.
[http://dx.doi.org/10.3390/plants9040544] [PMID: 32331486]

[24] Santos, L.S.; Fernandes Alves, C.C.; Borges Estevam, E.B.; Gomes Martins, C.H.; de Souza Silva, T.; Rodrigues Esperandim, V.; Dantas Miranda, M.L. Chemical Composition, *in vitro* Trypanocidal and Antibacterial Activities of the Essential Oil from the Dried Leaves of *Eugenia dysenterica* DC from

Brazil. *J. Essent. Oil-Bear. Plants,* **2019**, *22*(2), 347-355.
[http://dx.doi.org/10.1080/0972060X.2019.1626293]

[25]   Galheigo, M.R.U.; Prado, L.C.S.; Mundin, A.M.M.; Gomes, D.O.; Chang, R.; Lima, A.M.C.; Canabrava, H.A.N.; Bispo-da-Silva, L.B. Antidiarrhoeic effect of *Eugenia dysenterica* DC (Myrtaceae) leaf essential oil. *Nat. Prod. Res.,* **2016**, *30*(10), 1182-1185.
[http://dx.doi.org/10.1080/14786419.2015.1043633] [PMID: 26150261]

[26]   Monteiro, S.S.; Siani, A.C.; Nakamura, M.J.; Souza, M.C.; Ramos, M.F.S. Leaf Essential Oil from *Eugenia luschnathiana* and *Myrciaria tenella* (Myrtaceae) from Two Different Accesses in Southeastern Brazil. *J. Essent. Oil-Bear. Plants,* **2016**, *19*(7), 1675-1683.
[http://dx.doi.org/10.1080/0972060X.2016.1141074]

[27]   Durazzini, A.M.S.; Machado, C.H.M.; Fernandes, C.C.; Willrich, G.B.; Crotti, A.E.M.; Candido, A.C.B.B.; Magalhães, L.G.; Squarisi, I.S.; Ribeiro, A.B.; Tavares, D.C.; Martins, C.H.G.; Miranda, M.L.D. *Eugenia pyriformis* Cambess: a species of the Myrtaceae family with bioactive essential oil. *Nat. Prod. Res.,* **2019**, *0*(0), 1-5.
[http://dx.doi.org/10.1080/14786419.2019.1669031] [PMID: 31549535]

[28]   da Silva, V.P.; Alves, C.C.F.; Miranda, M.L.D.; Bretanha, L.C.; Balleste, M.P.; Micke, G.A.; Silveira, E.V.; Martins, C.H.G.; Ambrosio, M.A.L.V.; de Souza Silva, T.; Tavares, D.C.; Magalhães, L.G.; Silva, F.G.; Egea, M.B. Chemical composition and *in vitro* leishmanicidal, antibacterial and cytotoxic activities of essential oils of the Myrtaceae family occurring in the Cerrado biome. *Ind. Crops Prod.,* **2018**, *123*(July), 638-645.
[http://dx.doi.org/10.1016/j.indcrop.2018.07.033]

[29]   Ferreira, O.O.; da Cruz, J.N.; Franco, C.J.P.; Silva, S.G.; da Costa, W.A.; de Oliveira, M.S.; Andrade, E.H.A. First report on yield and chemical composition of essential oil extracted from myrcia eximia DC (Myrtaceae) from the Brazilian Amazon. *Molecules,* **2020**, *25*(4), 783.
[http://dx.doi.org/10.3390/molecules25040783] [PMID: 32059439]

[30]   Tietbohl, L.A.C.; Mello, C.B.; Silva, L.R.; Dolabella, I.B.; Franco, T.C.; Enríquez, J.J.S.; Santos, M.G.; Fernandes, C.P.; Machado, F.P.; Mexas, R.; Azambuja, P.; Araújo, H.P.; Moura, W.; Ratcliffe, N.A.; Feder, D.; Rocha, L.; Gonzalez, M.S. Green insecticide against Chagas disease: effects of essential oil from *Myrciaria floribunda* (Myrtaceae) on the development of *Rhodnius prolixus* nymphs. *J. Essent. Oil Res.,* **2020**, *32*(1), 1-11.
[http://dx.doi.org/10.1080/10412905.2019.1631894]

[31]   A. D. C, D. D. Abeywardhane K. W., and F. N. S., "Leaf Essential Oil Composition, Antioxidant Activity, Total Phenolic Content and Total Flavonoid Content of Pimenta Dioica (L.)Merr (Myrtaceae): A Superior Quality Spice Grown in Sri Lanka,". *Univers. J. Agric. Res.,* **2015**, *3*(2), 49-52.
[http://dx.doi.org/10.13189/ujar.2015.030203]

[32]   Chrystal, P.; Pereira, A.C.; Fernandes, C.C. Essential oil from Psidium cattleianum sabine (myrtaceae) fresh leaves: Chemical characterization and *in vitro* antibacterial activity against endodontic pathogens In: *Brazilian Arch. Biol. Technol*; , **2020**; 63, .

[33]   Dias, A.L.B.; Batista, H.R.F.; Estevam, E.B.B.; Alves, C.C.F.; Forim, M.R.; Nicolella, H.D.; Furtado, R.A.; Tavares, D.C.; Silva, T.S.; Martins, C.H.G.; Miranda, M.L.D. Chemical composition and *in vitro* antibacterial and antiproliferative activities of the essential oil from the leaves of *Psidium myrtoides* O. Berg (Myrtaceae). *Nat. Prod. Res.,* **2019**, *33*(17), 2566-2570.
[http://dx.doi.org/10.1080/14786419.2018.1457664] [PMID: 29611435]

[34]   Medeiros, F.C.M.; Del Menezzi, C.H.S.; Bizzo, H.R.; Vieira, R.F. Scents from Brazilian Cerrado: *Psidium myrsinites* DC. (Myrtaceae) leaves and inflorescences essential oil. *J. Essent. Oil Res.,* **2015**, *27*(4), 289-292.
[http://dx.doi.org/10.1080/10412905.2015.1037020]

[35]   Benelli, G.; Rajeswary, M.; Govindarajan, M. Towards green oviposition deterrents? Effectiveness of Syzygium lanceolatum (Myrtaceae) essential oil against six mosquito vectors and impact on four

aquatic biological control agents. *Environ. Sci. Pollut. Res. Int.,* **2018**, *25*(11), 10218-10227.
[http://dx.doi.org/10.1007/s11356-016-8146-3] [PMID: 27921244]

[36]   Oliveira, G.L.S. Determination *in vitro* of the antioxidant capacity of natural products by the DPPH•method: review study. *Rev. Bras. Plantas Med.,* **2015**, *17*(1), 36-44.
[http://dx.doi.org/10.1590/1983-084X/12_165]

[37]   Brewer, M.S. Natural Antioxidants: Sources, Compounds, Mechanisms of Action, and Potential Applications. *Compr. Rev. Food Sci. Food Saf.,* **2011**, *10*(4), 221-247.
[http://dx.doi.org/10.1111/j.1541-4337.2011.00156.x]

[38]   Diniz Do Nascimento, L.; Antônio Barbosa De Moraes, A.; Santana Da Costa, K. *Bioactive Natural Compounds and Antioxidant Activity of Essential Oils from Spice Plants: New Findings and Potential Applications,* **2020**.
[http://dx.doi.org/10.3390/biom10070988]

[39]   Larayetan, R.A.; Okoh, O.O.; Sadimenko, A.; Okoh, A.I. Terpene constituents of the aerial parts, phenolic content, antibacterial potential, free radical scavenging and antioxidant activity of Callistemon citrinus (Curtis) Skeels (Myrtaceae) from Eastern Cape Province of South Africa. *BMC Complement. Altern. Med.,* **2017**, *17*(1), 292.
[http://dx.doi.org/10.1186/s12906-017-1804-2] [PMID: 28583128]

[40]   Limam, H.; Ben Jemaa, M.; Tammar, S.; Ksibi, N.; Khammassi, S.; Jallouli, S.; Del Re, G.; Msaada, K. Variation in chemical profile of leaves essential oils from thirteen Tunisian Eucalyptus species and evaluation of their antioxidant and antibacterial properties. *Ind. Crops Prod.,* **2020**, *158*, 112964.
[http://dx.doi.org/10.1016/j.indcrop.2020.112964]

[41]   Hadidi, M.; Pouramin, S.; Adinepour, F.; Haghani, S.; Jafari, S.M. Chitosan nanoparticles loaded with clove essential oil: Characterization, antioxidant and antibacterial activities. *Carbohydr. Polym.,* **2020**, *236*, 116075.
[http://dx.doi.org/10.1016/j.carbpol.2020.116075] [PMID: 32172888]

[42]   da Silva, J.; Andrade, E.; Barreto, L.; da Silva, N.; Ribeiro, A.; Montenegro, R.; Maia, J. Chemical Composition of Four Essential Oils of Eugenia from the Brazilian Amazon and Their Cytotoxic and Antioxidant Activity. *Medicines (Basel),* **2017**, *4*(3), 51.
[http://dx.doi.org/10.3390/medicines4030051] [PMID: 28930266]

[43]   Toledo, A.G.; Souza, J.G.L.; Silva, J.P.B.; Favreto, W.A.J.; Costa, W.F.; Pinto, F.G.S. Chemical composition, antimicrobial and antioxidant activity of the essential oil of leaves of Eugenia involucrata DC. *Biosci. J.,* **2020**, *36*(2), 568-577.
[http://dx.doi.org/10.14393/BJ-v36n2a2020-48096]

[44]   Carneiro, N.S.; Alves, C.C.F.; Alves, J.M.; Egea, M.B.; Martins, C.H.G.; Silva, T.S.; Bretanha, L.C.; Balleste, M.P.; Micke, G.A.; Silveira, E.; Miranda, M.L.D. Chemical composition, antioxidant and antibacterial activities of essential oils from leaves and flowers of Eugenia klotzschiana Berg (Myrtaceae). *An. Acad. Bras. Cienc.,* **2017**, *89*(3), 1907-1915.
[http://dx.doi.org/10.1590/0001-3765201720160652] [PMID: 28767890]

[45]   da Costa, J.S.; Barroso, A.S.; Mourão, R.H.V.; da Silva, J.K.R.; Maia, J.G.S.; Figueiredo, P.L.B. Seasonal and Antioxidant Evaluation of Essential Oil from *Eugenia uniflora* L., Curzerene-Rich, Thermally Produced *in Situ. Biomolecules,* **2020**, *10*(2), 328.
[http://dx.doi.org/10.3390/biom10020328] [PMID: 32092893]

[46]   Imane, N.I.; Fouzia, H.; Azzahra, L.F.; Ahmed, E.; Ismail, G.; Idrissa, D.; Mohamed, K-H.; Sirine, F.; L'Houcine, O.; Noureddine, B. Chemical composition, antibacterial and antioxidant activities of some essential oils against multidrug resistant bacteria. *Eur. J. Integr. Med.,* **2020**, *35*, 101074.
[http://dx.doi.org/10.1016/j.eujim.2020.101074]

[47]   Santana, C.B. *J. G. de L. Souza, M. D. A. Coracini, et al., "Chemical composition of essential oil from Myrcia oblongata DC and potencial antimicrobial, antioxidant and acaricidal activity against Dermanyssus gallinae (Degeer, 1778)*; Biosci. J., **2018**, pp. 996-1009.

[48]   Okhale, S.E.; Ismail Buba, C.; Oladosu, P. Chemical Characterization, Antioxidant and Antimicrobial Activities of the Leaf Essential Oil of Syzgium guineense (Willd.) DC. var. Guineense (Myrtaceae) from Nigeria. *Int. J. Pharmacogn. Phytochem. Res.,* **2018,** *10*(11), 341-349.

[49]   José Serrano Vega, R.; Campos Xolalpa, N.; Josabad Alonso Castro, A. Terpenes from Natural Products with Potential Anti-Inflammatory Activity, **2018**.
[http://dx.doi.org/10.5772/intechopen.73215]

[50]   Swamy, M.K.; Akhtar, M.S.; Sinniah, U.R. *Antimicrobial properties of plant essential oils against human pathogens and their mode of action: An updated review*; Evidence-based Complement. Altern. Med, **2016,** Vol. 2016, .

[51]   Oliveira de Souza, L.I.; Bezzera-Silva, P.C.; do Amaral Ferraz Navarro, D.M.; da Silva, A.G.; dos Santos Correia, M.T.; da Silva, M.V.; de Figueiredo, R.C.B.Q. The chemical composition and trypanocidal activity of volatile oils from Brazilian Caatinga plants. *Biomed. Pharmacother.,* **2017,** *96*(November), 1055-1064.
[http://dx.doi.org/10.1016/j.biopha.2017.11.121] [PMID: 29217159]

[52]   Becker, N.A.; Volcão, L.M.; Camargo, T.M.; Freitag, R.A.; Ribeiro, G.A. "VITTALLE : Revista de Ciências da Saúde," *VITTALLE - Rev. Ciênc. Saúde (Porto Alegre),* **2007,** *29*(1), 22-30.

[53]   Costa, W.K.; Oliveira, J.R.S.; Oliveira, A.M.; Santos, I.B.S.; Cunha, R.X.; Freitas, A.F.S.; Silva, J.W.L.M.; Silva, V.B.G.; Aguiar, J.C.R.O.F.; Silva, A.G.; Navarro, D.M.A.F.; Lima, V.L.M.; Silva, M.V. Essential oil from Eugenia stipitata McVaugh leaves has antinociceptive, anti-inflammatory and antipyretic activities without showing toxicity in mice. *Ind. Crops Prod.,* **2020,** *144*, 112059.
[http://dx.doi.org/10.1016/j.indcrop.2019.112059]

[54]   do Nascimento, K.F.; Moreira, F.M.F.; Alencar Santos, J.; Kassuya, C.A.L.; Croda, J.H.R.; Cardoso, C.A.L.; Vieira, M.C.; Góis Ruiz, A.L.T.; Ann Foglio, M.; de Carvalho, J.E.; Formagio, A.S.N. Antioxidant, anti-inflammatory, antiproliferative and antimycobacterial activities of the essential oil of Psidium guineense Sw. and spathulenol. *J. Ethnopharmacol.,* **2018,** *210*, 351-358.
[http://dx.doi.org/10.1016/j.jep.2017.08.030] [PMID: 28844678]

[55]   Pina LTS, de Jesus AM, de Melo MS, Bispo RM, Alves PB, de Lima PCN, de Souza VRM, Silva GH, Júnior LJQ and M. S. and G. A. SM, Martins LRR, Ferreira AG, de Souza AR, "A New B-Triketone and Antinociceptive Effect from the Essential Oil of the Leaves of Calyptranthes restingae Sobral (Myrtaceae)," *Med. Aromat. Plants,* **2016,** *05*(03).

[56]   Gad, H.A.; Ayoub, I.M.; Wink, M. Phytochemical profiling and seasonal variation of essential oils of three Callistemon species cultivated in Egypt. *PLoS One,* **2019,** *14*(7), e0219571.
[http://dx.doi.org/10.1371/journal.pone.0219571] [PMID: 31295290]

[57]   Lorenzo-Leal, A.C.; Palou, E.; López-Malo, A.; Bach, H. Antimicrobial, Cytotoxic, and Anti-Inflammatory Activities of *Pimenta dioica* and *Rosmarinus officinalis* Essential Oils. *BioMed Res. Int.,* **2019,** *2019*, 1-8.
[http://dx.doi.org/10.1155/2019/1639726] [PMID: 31205934]

[58]   Lorençoni, M. F.; Figueira, M. M.; Toledo e Silva, M. V. Chemical composition and anti-inflammatory activity of essential oil and ethanolic extract of Campomanesia phaea (O. Berg.) Landrum leaves. *J. Ethnopharmacol,* **2020,** *252*, 112562.
[http://dx.doi.org/10.1016/j.jep.2020.112562]

[59]   Han, X.; Parker, T.L. Anti-inflammatory activity of clove ( *Eugenia caryophyllata* ) essential oil in human dermal fibroblasts. *Pharm. Biol.,* **2017,** *55*(1), 1619-1622.
[http://dx.doi.org/10.1080/13880209.2017.1314513] [PMID: 28407719]

[60]   González-Cofrade, L.; de las Heras, B.; Apaza Ticona, L.; Palomino, O.M. Molecular Targets Involved in the Neuroprotection Mediated by Terpenoids. *Planta Med.,* **2019,** *85*(17), 1304-1315.
[http://dx.doi.org/10.1055/a-0953-6738] [PMID: 31234214]

[61]   Chistiane, M.F.; Alisson, R.B.; Cassio, H.S.M.; Rivelilson, M.F.; Jose, E.N.F.; Emmanoel, V.C.;

Khaled, N.Z.R.; Joaquim, S.C.J. Antioxidant and anticholinesterase activities of the essential oil of Eugenia dysenterica DC. *Afr. J. Pharm. Pharmacol.,* **2017**, *11*(19), 241-249.
[http://dx.doi.org/10.5897/AJPP2015.4438]

[62]    da Costa, J.S.; de N. S., da Cruz; Setzer, W. N Essentials oils from Brazilian eugenia and syzygium species and their biological activities. In: *Biomolecules*; , **2020**; 10, pp. (8)1-36.

[63]    Barbosa, D. C.; Holandada Silva, V. N.; de Assis, C. R. D. Chemical composition and acetylcholinesterase inhibitory potential, in silico, of Myrciaria floribunda (H. West ex Willd.) O. Berg fruit peel essential oil. *Ind. Crops Prod,* **2020**, *151*, 112372.

[64]    Gomes da Rocha Voris, D.; dos Santos Dias, L.; Alencar Lima, J.; dos Santos Cople Lima, K.; Pereira Lima, J.B.; dos Santos Lima, A.L. Evaluation of larvicidal, adulticidal, and anticholinesterase activities of essential oils of Illicium verum Hook. f., Pimenta dioica (L.) Merr., and Myristica fragrans Houtt. against Zika virus vectors. *Environ. Sci. Pollut. Res. Int.,* **2018**, *25*(23), 22541-22551.
[http://dx.doi.org/10.1007/s11356-018-2362-y] [PMID: 29808407]

[65]    Maggio, A.; Loizzo, M.R.; Riccobono, L.; Bruno, M.; Tenuta, M.C.; Leporini, M.; Falco, T.; Leto, C.; Tuttolomondo, T.; Cammalleri, I.; La Bella, S.; Tundis, R. Comparative chemical composition and bioactivity of leaves essential oils from nine Sicilian accessions of *Myrtus communis* L. *J. Essent. Oil Res.,* **2019**, *31*(6), 546-555.
[http://dx.doi.org/10.1080/10412905.2019.1610089]

[66]    Siebert, D.A.; Tenfen, A.; Yamanaka, C.N.; de Cordova, C.M.M.; Scharf, D.R.; Simionatto, E.L.; Alberton, M.D. Evaluation of seasonal chemical composition, antibacterial, antioxidant and anticholinesterase activity of essential oil from *Eugenia brasiliensis* Lam. *Nat. Prod. Res.,* **2015**, *29*(3), 289-292.
[http://dx.doi.org/10.1080/14786419.2014.958736] [PMID: 25219800]

[67]    Zatelli, G.A.; Zimath, P.; Tenfen, A.; Mendes de Cordova, C.M.; Scharf, D.R.; Simionatto, E.L.; Alberton, M.D.; Falkenberg, M. Antimycoplasmic activity and seasonal variation of essential oil of *Eugenia hiemalis* Cambess. (Myrtaceae). *Nat. Prod. Res.,* **2016**, *30*(17), 1961-1964.
[http://dx.doi.org/10.1080/14786419.2015.1091455] [PMID: 26428391]

[68]    Yadav, M.; Jindal, D.K.; Parle, M.; Kumar, A.; Dhingra, S. Targeting oxidative stress, acetylcholinesterase, proinflammatory cytokine, dopamine and GABA by eucalyptus oil (Eucalyptus globulus) to alleviate ketamine-induced psychosis in rats. *Inflammopharmacology,* **2019**, *27*(2), 301-311.
[http://dx.doi.org/10.1007/s10787-018-0455-3] [PMID: 29464495]

[69]    Montalván, M.; Peñafiel, M.A.; Ramírez, J.; Cumbicus, N.; Bec, N.; Larroque, C.; Bicchi, C.; Gilardoni, G. Chemical composition, enantiomeric distribution, and sensory evaluation of the essential oils distilled from the ecuadorian species myrcianthes myrsinoides (Kunth) grifo and myrcia mollis (Kunth) dc. (Myrtaceae). *Plants,* **2019**, *8*(11), 511.
[http://dx.doi.org/10.3390/plants8110511] [PMID: 31731807]

[70]    Fagodia, S.K.; Singh, H.P.; Batish, D.R.; Kohli, R.K. Phytotoxicity and cytotoxicity of Citrus aurantiifolia essential oil and its major constituents: Limonene and citral. *Ind. Crops Prod.,* **2017**, *108*, 708-715.
[http://dx.doi.org/10.1016/j.indcrop.2017.07.005]

[71]    Costa, J.A.V.; Freitas, B.C.B.; Cruz, C.G.; Silveira, J.; Morais, M.G. Potential of microalgae as biopesticides to contribute to sustainable agriculture and environmental development. *J. Environ. Sci. Health B,* **2019**, *54*(5), 366-375.
[http://dx.doi.org/10.1080/03601234.2019.1571366] [PMID: 30729858]

[72]    Sparks, T.C.; Wessels, F.J.; Lorsbach, B.A.; Nugent, B.M.; Watson, G.B. *The new age of insecticide discovery-the crop protection industry and the impact of natural products*; Elsevier Inc, **2019**, Vol. 161, .

[73]    Guo, Y.; Liu, Z.; Hou, E.; Ma, N.; Fan, J.; Jin, C-Y.; Yang, R. Non-food bioactive natural forest

products as insecticide candidates: Preparation, biological evaluation and molecular docking studies of novel N-(1,3-thiazol-2- yl)carboxamides fused (+)-nootkatone from Chamaecyparis nootkatensis [D. Don] Spach. *Ind. Crops Prod.,* **2020**, *156*(August), 112864.
[http://dx.doi.org/10.1016/j.indcrop.2020.112864]

[74]    Tak, J. H.; Jovel, E.; Isman, M. B. *Contact, fumigant, and cytotoxic activities of thyme and lemongrass essential oils against larvae and an ovarian cell line of the cabbage looper, Trichoplusia ni,* **2016**.
[http://dx.doi.org/10.1007/s10340-015-0655-1]

[75]    Jesser, E.; Yeguerman, C.; Gili, V.; Santillan, G.; Murray, A.P.; Domini, C.; Werdin-González, J.O. Optimization and Characterization of Essential Oil Nanoemulsions Using Ultrasound for New Ecofriendly Insecticides. *ACS Sustain. Chem.& Eng.,* **2020**, *8*(21), 7981-7992.
[http://dx.doi.org/10.1021/acssuschemeng.0c02224]

[76]    Ahl, H.A.H.S.; Hikal, W.M.; Tkachenko, K.G. Essential Oils with Potential as Insecticidal Agents. *RE:view,* **2017**, *3*(4), 23-33.

[77]    Zengin, G.; Menghini, L.; Di Sotto, A.; Mancinelli, R.; Sisto, F.; Carradori, S.; Cesa, S.; Fraschetti, C.; Filippi, A.; Angiolella, L.; Locatelli, M.; Mannina, L.; Ingallina, C.; Puca, V.; D'Antonio, M.; Grande, R. Chromatographic analyses, *in vitro* biological activities, and cytotoxicity of cannabis sativa l. Essential oil: A multidisciplinary study. *Molecules,* **2018**, *23*(12), 3266.
[http://dx.doi.org/10.3390/molecules23123266] [PMID: 30544765]

[78]    Tilaoui, M.; Ait Mouse, H.; Jaafari, A.; Zyad, A. Comparative phytochemical analysis of essential oils from different biological parts of artemisia herba alba and their cytotoxic effect on cancer cells. *PLoS One,* **2015**, *10*(7), e0131799.
[http://dx.doi.org/10.1371/journal.pone.0131799] [PMID: 26196123]

[79]    Asif, M.; Yehya, A.H.S.; Al-Mansoub, M.A.; Revadigar, V.; Ezzat, M.O.; Khadeer Ahamed, M.B.; Oon, C.E.; Murugaiyah, V.; Abdul Majid, A.S.; Abdul Majid, A.M.S. Anticancer attributes of Illicium verum essential oils against colon cancer. *S. Afr. J. Bot.,* **2016**, *103*, 156-161.
[http://dx.doi.org/10.1016/j.sajb.2015.08.017]

[80]    Chua, L.K.; Lim, C.L.; Ling, A.P.K.; Chye, S.M.; Koh, R.Y. Anticancer Potential of Syzygium Species: a Review. *Plant Foods Hum. Nutr.,* **2019**, *74*(1), 18-27.
[http://dx.doi.org/10.1007/s11130-018-0704-z] [PMID: 30535971]

[81]    de S., Carneiro; V. C., de Lucena; L. B., Figueiró; R., Victório, C. P Larvicidal activity of plants from myrtaceae against aedes aegypti l. And simulium pertinax kollar (diptera). In: *Rev. Soc. Bras. Med. Trop;* , **2021**; 54, pp. 1-8.

[82]    Ashraf, A.; Sarfraz, R.A.; Rashid, M.A.; Mahmood, A.; Shahid, M.; Noor, N. Chemical composition, antioxidant, antitumor, anticancer and cytotoxic effects of *Psidium guajava* leaf extracts. *Pharm. Biol.,* **2016**, *54*(10), 1971-1981.
[http://dx.doi.org/10.3109/13880209.2015.1137604] [PMID: 26841303]

[83]    Younoussa, L.; Kenmoe, F.; Oumarou, M.K.; Batti, A.C.S.; Tamesse, J.L.; Nukenine, E.N. Combined Effect of Methanol Extracts and Essential Oils of *Callistemon rigidus* (Myrtaceae) and *Eucalyptus camaldulensis* (Myrtaceae) against *Anopheles gambiae* Giles larvae (Diptera: Culicidae). *Int. J. Zool.,* **2020**, *2020*(li), 1-9.
[http://dx.doi.org/10.1155/2020/4952041]

[84]    Batiha, G.E.S.; Alkazmi, L.M.; Wasef, L.G.; Beshbishy, A.M.; Nadwa, E.H.; Rashwan, E.K. Syzygium aromaticum l. (myrtaceae): Traditional uses, bioactive chemical constituents, pharmacological and toxicological activities. *Biomolecules,* **2020**, *10*(2), 1-17.
[PMID: 32019140]

[85]    Vuong, Q.V.; Chalmers, A.C.; Jyoti Bhuyan, D.; Bowyer, M.C.; Scarlett, C.J. Botanical, phytochemical, and anticancer properties of the eucalyptus species. *Chem. Biodivers.,* **2015**, *12*(6), 907-924.
[http://dx.doi.org/10.1002/cbdv.201400327] [PMID: 26080737]

[86]    Benelli, G.; Pavela, R. Beyond mosquitoes—Essential oil toxicity and repellency against bloodsucking insects. *Ind. Crops Prod.,* **2018**, *117*(February), 382-392.
[http://dx.doi.org/10.1016/j.indcrop.2018.02.072]

[87]    Filomeno, C.A.; Almeida Barbosa, L.C.; Teixeira, R.R.; Pinheiro, A.L.; de Sá Farias, E.; Ferreira, J.S.; Picanço, M.C. Chemical diversity of essential oils of Myrtaceae species and their insecticidal activity against Rhyzopertha dominica. *Crop Prot.,* **2020**, *137*, 105309.
[http://dx.doi.org/10.1016/j.cropro.2020.105309]

[88]    Jerônimo, L.B.; da Costa, J.S.; Pinto, L.C.; Montenegro, R.C.; Setzer, W.N.; Mourão, R.H.V.; da Silva, J.K.R.; Maia, J.G.S.; Figueiredo, P.L.B. Antioxidant and Cytotoxic Activities of Myrtaceae Essential Oils Rich in Terpenoids From Brazil. *Nat. Prod. Commun.,* **2021**, *16*(2), 1934578X2199615.
[http://dx.doi.org/10.1177/1934578X21996156]

[89]    da Silva, V.P.; Alves, C.C.F.; Miranda, M.L.D.; Bretanha, L.C.; Balleste, M.P.; Micke, G.A.; Silveira, E.V.; Martins, C.H.G.; Ambrosio, M.A.L.V.; de Souza Silva, T.; Tavares, D.C.; Magalhães, L.G.; Silva, F.G.; Egea, M.B. Chemical composition and *in vitro* leishmanicidal, antibacterial and cytotoxic activities of essential oils of the Myrtaceae family occurring in the Cerrado biome. *Ind. Crops Prod.,* **2018**, *123*, 638-645.
[http://dx.doi.org/10.1016/j.indcrop.2018.07.033]

[90]    Aranha, E.S.P.; de Azevedo, S.G.; dos Reis, G.G.; Silva Lima, E.; Machado, M.B.; de Vasconcellos, M.C. Essential oils from Eugenia spp.: *In vitro* antiproliferative potential with inhibitory action of metalloproteinases. *Ind. Crops Prod.,* **2019**, *141*(August), 111736.
[http://dx.doi.org/10.1016/j.indcrop.2019.111736]

[91]    Mazutti da Silva, S.; Rezende Costa, C.; Martins Gelfuso, G.; Silva Guerra, E.; de Medeiros Nóbrega, Y.; Gomes, S.; Pic-Taylor, A.; Fonseca-Bazzo, Y.; Silveira, D.; Magalhães, P. Wound healing effect of essential oil extracted from eugenia dysenterica DC (Myrtaceae) leaves. *Molecules,* **2018**, *24*(1), 2.
[http://dx.doi.org/10.3390/molecules24010002] [PMID: 30577426]

[92]    Sousa, R.M.F.; de Morais, S.A.L.; Vieira, R.B.K.; Napolitano, D.R.; Guzman, V.B.; Moraes, T.S.; Cunha, L.C.S.; Martins, C.H.G.; Chang, R.; de Aquino, F.J.T.; do Nascimento, E.A.; de Oliveira, A. Chemical composition, cytotoxic, and antibacterial activity of the essential oil from Eugenia calycina Cambess. leaves against oral bacteria. *Ind. Crops Prod.,* **2015**, *65*, 71-78.
[http://dx.doi.org/10.1016/j.indcrop.2014.11.050]

[93]    Aragão, F.B.; Palmieri, M.J.; Ferreira, A. Phytotoxic and cytotoxic effects of eucalyptus essential oil on lettuce (Lactuca sativa L.). *Allelopathy J.,* **2015**, *35*(2), 259-272.

[94]    Adnan, M. Bioactive potential of essential oil extracted from the leaves of Eucalyptus globulus (Myrtaceae). *J. Pharmacogn. Phytochem.,* **2019**, *8*(1), 213-216.

[95]    Park, C.; Jang, M.; Shin, E.; Kim, J. Myrtaceae plant essential oils and their β-triketone components as insecticides against drosophila suzukii. *Molecules,* **2017**, *22*(7), 1050.
[http://dx.doi.org/10.3390/molecules22071050] [PMID: 28672824]

[96]    Silva, M.V.S.G.; Silva, S.A.; Teixera, T.L.; De Oliveira, A.; Morais, S.A.L.; Da Silva, C.V.; Espindola, L.S.; Sousa, R.M.F. Essential oil from leaves of *Eugenia calycina* Cambes: Natural larvicidal against *AEDES AEGYPTI. J. Sci. Food Agric.,* **2021**, *101*(3), 1202-1208.
[http://dx.doi.org/10.1002/jsfa.10732] [PMID: 32789937]

[97]    Kauffmann, C.; Ethur, E.M.; Buhl, B.; Scheibel, T.; Machado, G.M.C.; Cavalheiro, M.M.C. Potential Antileishmanial Activity of Essential Oils of Native Species from Southern Brazil. *Environ. Nat. Resour. Res.,* **2016**, *6*(4), 18.
[http://dx.doi.org/10.5539/enrr.v6n4p18]

[98]    Rodrigues, K.A.F.; Amorim, L.V.; Dias, C.N.; Moraes, D.F.C.; Carneiro, S.M.P.; Carvalho, F.A.A. Syzygium cumini (L.) Skeels essential oil and its major constituent α-pinene exhibit anti-Leishmania activity through immunomodulation *in vitro. J. Ethnopharmacol.,* **2015**, *160*, 32-40.
[http://dx.doi.org/10.1016/j.jep.2014.11.024] [PMID: 25460590]

[99]   Kauffmann, C.; Giacomin, A.C.; Arossi, K.; Pacheco, L.A.; Hoehne, L.; Freitas, E.M.; Machado, G.M.C.; Cavalheiro, M.M.C.; Gnoatto, S.C.B.; Ethur, E.M. Antileishmanial *in vitro* activity of essential oil from Myrciaria plinioides, a native species from Southern Brazil. *Braz. J. Pharm. Sci.,* **2019**, *55*, e17584.
[http://dx.doi.org/10.1590/s2175-97902019000217584]

[100]  Kauffmann, C.; Ethur, E.M.; Arossi, K.; Hoehne, L.; de Freitas, E.M.; Machado, G.M.C.; Cavalheiro, M.M.C.; Flach, A.; da Costa, L.A.M.A.; Gnoatto, S.C.B. Chemical Composition and Evaluation Preliminary of Antileishmanial Activity *in vitro* of Essential Oil from Leaves of *Eugenia pitanga*, A Native Species of Southern of Brazil. *J. Essent. Oil-Bear. Plants,* **2017**, *20*(2), 559-569.
[http://dx.doi.org/10.1080/0972060X.2017.1281767]

[101]  Mahmoudvand, H.; Ezzatkhah, F.; Sharififar, F.; Sharifi, I.; Dezaki, E.S. Antileishmanial and cytotoxic effects of essential oil and methanolic extract of Myrtus communis L. *Korean J. Parasitol.,* **2015**, *53*(1), 21-27.
[http://dx.doi.org/10.3347/kjp.2015.53.1.21] [PMID: 25748705]

[102]  Albouchi, F.; Sifaoui, I.; Reyes-Batlle, M.; López-Arencibia, A.; Piñero, J.E.; Lorenzo-Morales, J.; Abderrabba, M. Chemical composition and anti- Acanthamoeba activity of Melaleuca styphelioides essential oil. *Exp. Parasitol.,* **2017**, *183*, 104-108.
[http://dx.doi.org/10.1016/j.exppara.2017.10.014] [PMID: 29103900]

[103]  Alam, F.; Shafique, Z.; Amjad, S.T.; Bin Asad, M.H.H. Enzymes inhibitors from natural sources with antidiabetic activity: A review. *Phytother. Res.,* **2019**, *33*(1), 41-54.
[http://dx.doi.org/10.1002/ptr.6211] [PMID: 30417583]

[104]  Cascaes, M.; Guilhon, G.; Andrade, E.; Zoghbi, M.; Santos, L. M. das Graças Bichara Zoghbi, and L. da Silva Santos, "Constituents and pharmacological activities of Myrcia (Myrtaceae): A review of an aromatic and medicinal group of plants,". *Int. J. Mol. Sci.,* **2015**, *16*(10), 23881-23904.
[http://dx.doi.org/10.3390/ijms161023881] [PMID: 26473832]

[105]  Tahir, H.U.; Sarfraz, R.A.; Ashraf, A.; Adil, S. Chemical Composition and Antidiabetic Activity of Essential Oils Obtained from Two Spices ( *Syzygium aromaticum* and *Cuminum cyminum* ). *Int. J. Food Prop.,* **2016**, *19*(10), 2156-2164.
[http://dx.doi.org/10.1080/10942912.2015.1110166]

[106]  Capetti, F.; Cagliero, C.; Marengo, A.; Bicchi, C.; Rubiolo, P.; Sgorbini, B. Bio-guided fractionation driven by *in vitro* α-amylase inhibition assays of essential oils bearing specialized metabolites with potential hypoglycemic activity. *Plants,* **2020**, *9*(9), 1242.
[http://dx.doi.org/10.3390/plants9091242] [PMID: 32967115]

[107]  Sen, A.; Kurkcuoglu, M.; Yildirim, A. Chemical and biological profiles of essential oil from different parts of myrtus communis l. Subsp. communis from turkey. *ACS Agric. Conspec. Sci.,* **2020**, *85*(1), 71-78.

[108]  Lucena, J.A.S.; Guedes, J.P.M. Revista Brasileira de Educação e Saúde ARTIGO DE REVISÃO Uso de fitoterápicos na prevenção e no tratamento da hipertensão arterial sistêmica Use of phytotherapics in the prevention and treatment of systemic arterial hypertension. *Rev. Bra. Edu. Saúde,* **2020**, *10*(1), 15-22.

[109]  Celaj, O.; Durán, A.G.; Cennamo, P. Phloroglucinols from Myrtaceae: attractive targets for structural characterization, biological properties and synthetic procedures. *Phytochemistry Reviews,* **2020**, *2*.

[110]  Cock, I.E.; Cheesman, M. *The Potential of Plants of the Genus Syzygium (Myrtaceae) for the Prevention and Treatment of Arthritic and Autoimmune Diseases,* 2nd ed; Elsevier Inc., **2019**.
[http://dx.doi.org/10.1016/B978-0-12-813820-5.00023-4]

[111]  Kauffmann, C.; Ethur, E.M. Myrtaceae Como Fonte Para Obtenção De Novos Candidatos a Fármacos Para O Potentiality of Species of the Family Myrtaceae As Source for Obtaining New Candidates for Drugs for the Treatment of. 2016,1983–0882. 56-74.

[112]   Silva, E.A.J.; Estevam, E.B.B.; Silva, T.S.; Nicolella, H.D.; Furtado, R.A.; Alves, C.C.F.; Souchie, E.L.; Martins, C.H.G.; Tavares, D.C.; Barbosa, L.C.A.; Miranda, M.L.D. Antibacterial and antiproliferative activities of the fresh leaf essential oil of Psidium guajava L. (Myrtaceae). *Braz. J. Biol.,* **2019**, *79*(4), 697-702.
[http://dx.doi.org/10.1590/1519-6984.189089] [PMID: 30462815]

[113]   Junior, Á.B.I.; Jonas, M. Plantas Medicinais E Fitoterápicos Úteis Na Odontologia Clínica : Uma Revisão. *Rev. Fac. Odontol. Univ. Fed. Bahia,* **2020**, *50*, 47-56.
[http://dx.doi.org/10.9771/revfo.v50i1.37116]

[114]   Benny, A.; Thomas, J. Essential Oils as Treatment Strategy for Alzheimer's Disease: Current and Future Perspectives. *Planta Med.,* **2019**, *85*(3), 239-248.
[http://dx.doi.org/10.1055/a-0758-0188] [PMID: 30360002]

[115]   Tetali, S.D. Terpenes and isoprenoids: a wealth of compounds for global use. *Planta,* **2019**, *249*(1), 1-8.
[http://dx.doi.org/10.1007/s00425-018-3056-x] [PMID: 30467631]

# CHAPTER 5

# Volatile Terpenoids of Annonaceae: Occurrence and Reported Activities

**Márcia M. Cascaes**[1,2,*], **Giselle M. S. P. Guilhon**[1], **Lidiane D. Nascimento**[2,3], **Angelo A. B. de Moraes**[2], **Sebastião G. Silva**[1], **Jorddy Neves Cruz**[2], **Oberdan O. Ferreira**[4], **Mozaniel S. Oliveira**[2] and **Eloisa H. A. Andrade**[1,2]

[1] *Program of Post-Graduation in Chemistry, Federal University of Pará, Belém, Brazil*

[2] *Adolpho Ducke Laboratory, Paraense Emílio Goeldi Museum, Belém, Brazil*

[3] *Program of Post-Graduation in Engineering of Natural Resources of Amazon, Federal University of Pará, Belém, Brazil*

[4] *Program of Post-Graduation in Biodiversity and Biotecnology-Bionorte, Federal University of Pará Belém, Brazil*

**Abstract:** Annonaceae includes 2,106 species. Some species of this family have an economic interest in the international fresh fruit market and are often used as raw materials for cosmetics, perfumes and folk medicine. The most cited species are mainly those belonging to the genera *Annona*, *Guatteria* and *Xylopia*. Chemical investigations indicate that the characteristic constituents of the Annonaceae are terpenoids, including mono and sesquiterpenoids, such as α-pinene, β-pinene, limonene, (*E*)-caryophyllene, bicyclogermacrene, caryophyllene oxide, germacrene D, spathulenol and β-elemene. Antimicrobial, antioxidant, larvicidal, antiproliferative, trypanocidal, antimalarial and anti-inflammatory effects have been described in these terpenes. This work is an overview of the chemical properties and biological effects of the volatile terpenoids from Annonaceae species.

**Keywords:** Antioxidant Potential, Biological Effects, Essential Oils, Monoterpenes, Sesquiterpenes.

## INTRODUCTION

Annonaceae are flowering plants that consist of trees, shrubs and lianas, which have a combination of striking characters, being one of the most uniform botanical families from both an anatomical and structural point of view. It is one of the most primitive of Angiosperms, belongs to the Magnoliopsida class, subclass Magnoliidae and order Magnoliales [1].

---

\* **Corresponding author Márcia M. Cascaes:** Program of Post-Graduation in Chemistry, Federal University of Pará, Brazil; E-mail: cascaesmm@gmail.com

**Mozaniel Santana de Oliveira & Antônio Pedro da Silva Souza Filho (Eds.)**

Annonaceae consists of 2,106 species, and more than 130 genera, concentrated in the Tropics, about 900 species are Neotropical, 450 are Afrotropical, and the remaining species are Indomalayan [2]. Annonaceae plays an important ecological role in terms of species diversity, especially in tropical forest ecosystems [3].

Some Annonaceae species are important in the international fresh fruit market, such as *Annona cherimola* Mill. ("cherimólia") and *Annona squamosa* L. ("pinha") [4]. In Brazil, some *Annona* fruits are very popular, such as those of *Annona crassiflora* Mart. ("araticum"), *Annona squamosa* L. ("fruta do conde") and *Annona muricata* L. ("graviola") [5]. In addition, some Annonaceae are often used as raw materials for cosmetics, perfumes and folk medicinal plants [6]. The most cited species in folk medicine are mainly those belonging to the genera *Annona*, *Guatteria* and *Xylopia* [7].

Numerous species of Annonaceae are odoriferous, and these fragrances are due to the presence of essential oils (EOs) [8]. In nature, EOs have many important functions, such as attracting insects or allelopathic communication between plants [9], in addition, they can act as antibacterials, antivirals, anti-inflammatories, and antifungals [10]. About 1% of these volatile constituents are known to date and are mainly represented by terpenoids, phenylpropanoids/benzenoids, fatty acids and amino acid derivatives [11].

According to a review published by Fournier and coworkers (1999) [12], the main volatile constituents of the EOs of Annonaceae species are monoterpene hydrocarbons in fruit and seed, sesquiterpene hydrocarbons in leaf, and oxygenated sesquiterpenes in bark and roots. After this review (1999), several papers have been published evidencing the presence of terpenoids in EOs from Annonaceae and their biological activities. The present work provides an overview of the chemical composition and the biological effects of the volatile terpenoids from Annonaceae species. Original articles published from 2015 to 2021 were considered for composition.

## CHEMICAL DIVERSITY OF VOLATILE TERPENOIDS

Terpenoids are natural products with incredibly diverse structures and activities. So far, more than 40,000 phytoterpenoids have been identified [13]. The terpenoids compose the largest class of plant secondary metabolites with many volatile representatives. Terpenoids originated from the universal five carbon precursors, isopentenyl diphosphate (IPP) and its allylic isomer dimethylallyl diphosphate (DMAPP) [11].

So far, more than 90 volatile terpenoids (>5%) have been obtained from different parts of Annonaceae species. Among these compounds, α-pinene, β-pinene, limonene, (E)-caryophyllene, bicyclogermacrene, caryophyllene oxide, germacrene D, spathulenol and β-elemene are the most dominant terpenes reported. The terpenes, the corresponding plant sources and references from which they are derived, are summarized in Table **1**.

**Table 1.** *Mono and Sesquiterpenoids Identified in Essential Oils of* Annonaceae *Species.*

| Annonaceae Species [Refs.] | Part of Plant | Number of Identified Compounds | Total of Identified Compounds (%) | Main Monoterpenoids (>5%) | Main Sesquiterpenoids (>5%) | Monoterpenoids (%) | Sesquiterpenoids (%) |
|---|---|---|---|---|---|---|---|
| *Alphonsea tonkinensis* A. DC [14] | Leaf | 40 | 98.7 | - | β-Elemene, β-caryophyllene, germacrene D, bicyclogermacrene and caryophyllene oxide | 5.6 | 92.9 |
| *A. tonkinensis* [14] | Stem | 40 | 99.9 | α-Pinene, β-pinene, and limonene | β-Caryophyllene, β-elemene, germacrene D and farnesol | 28 | 71.8 |
| *Anaxagorea Brevipes* Benth [15] | Leaf | 31 | 75.6 | - | Guaiol, γ-eudesmol, β-eudesmol and α-eudesmol | 3.3 | 72.3 |
| *Annona exsucca* DC. [16] | Leaf | 50 | 99.3 | Linalool | β-Elemene, (E)-caryophyllene, α-humulene, germacrene D, bicyclogermacrene | 13.9 | 84.5 |
| *A. exsucca* [16] | Leaf | 58 | 99.1 | *p*-Cymene, sylvestrene, terpinolene and linalool | Germacrene D and bicyclogermacrene | 62.7 | 36.4 |
| *A. Squamosa* L. [17] | Fruit | 33 | 86.0 | α-pinene, limonene, and β-cubebene | β-caryophyllene, spathulenol, caryophyllene oxide and α-cadinol | N.I | N.I |
| *A. leptopetala* (R.E.Fr.) H. Rainer [18] | Leaf | 37 | 98.1 | α-Limonene, linalool and α-terpineol | (E)-Caryophylene, bicyclogermacrene, spathulenol and guaiol | 44.1 | 55.9 |
| *A. muricata* L. [19] | Fruit | 31 | 99.98 | - | α-Muurolene, β-caryophyllene, δ-cadinene and α-cadinol | N.I | N.I |
| *A. sylvatica* A. St.-Hil Anelise [20] | Leaf | 36 | 98.97 | - | β-Selinene, (Z)-caryophyllene, γ-gurjunene and hinesol | NI | NI |
| *A. vepretorum* Mart. [21] | Leaf | 26 | 97.6 | α-Phellandrene, *o*-cymene and (E)-β-ocimene | Bicyclogermacrene and spathulenol | 30.1 | 67.4 |
| *A. vepretorum* [3] | Leaf | 19 | 93.9 | α-Pinene and limonene | Spathulenol and caryophyllene oxide | NI | NI |
| *A. vepretorum* [22] | Leaf | 16 | 100.0 | Limonene and (E)-β-ocimene | Germacrene D and bicyclogermacre | NI | NI |

*(Table 1) cont.....*

| Annonaceae Species [Refs.] | Part of Plant | Number of Identified Compounds | Total of Identified Compounds (%) | Main Monoterpenoids (>5%) | Main Sesquiterpenoids (>5%) | Monoterpenoids (%) | Sesquiterpenoids (%) |
|---|---|---|---|---|---|---|---|
| *Bocageopsis multiflora* (Mart.) R.E. Fr. [23] | Leaf | 61 | 95.0 | - | α-*trans*-Bergamotene, β-bisabolene, spathulenol and β-copaen-4-α-ol | 1.3 | 83.8 |
| *B. multiflora* [24] | Aerial parts | 23 | 87.6 | *cis*-Linalool oxide (furanoid) | 1-*epi*-Cubenol | 36.5 | 46.6 |
| *B. pleiosperma* Maas [25] | Leaf | 24 | 87.6 | - | (*E*)-α-Bergamotene, (*E*)-β-farnesene and β-bisabolene | NI | NI |
| *B. pleiosperma* [25] | Bark | 29 | 97.1 | - | β-Selinene, α-selinene, β-bisabolene and δ-cadinene | NI | NI |
| *B. pleiosperma* [25] | Twig | 23 | 72.6 | - | β-Bisabolene, (2*Z*,6*Z*)-farnesol and cryptomerione | NI | NI |
| *Cardiopetalum calophyllum* (Schltdl.) [26] | Flower | 25 | NI | - | Caryophyllene, germacrene D and germacrene B | 0.45 | 70.1% |
| *C. calophyllum* [26] | Fruit | 15 | NI | - | Germacrene D, germacrene B and spathulenol | 0.55 | 73.3 |
| *C. calophyllum* [26] | Leaf | 23 | NI | - | spathulenol, viridiflorol, (–)-isolongifolol acetate and (*Z*,*E*)-farnesol | 0.43 | 66.0 |
| *Cyathocalyx pruniferus* (Maingay ex Hook.f. & Thomson) J. Sinclair [27] | Leaf | 30 | 95.53 | α-Pinene | *E*-Caryophyllene, δ-cadinene and germacrene D | 78.2 | 17.4 |
| *Duguetia furfuracea* (A. St. -Hil.) Saff. [28] | Stem bark | 19 | 99.5 | - | Cyperene, α-gurjunene and bicyclogermacrene | - | 55.7 |
| *D. lanceolata* St. Hil. [29] | Branche | 37 | 92.9 | - | β-Elemene, β-caryophyllene, β-selinene, δ-cadinene, caryophyllene oxide, humulene epoxide II, β-eudesmol and cadina-1,4-dien-3-ol | 7.8 | 84.9 |
| *D. lanceolata* [30] | Leaf | 5 | 99.9 | - | *trans*-Muurola-4(14)-5-diene and β-bisabolene | - | 72.2 |
| *D. quitarensis* Benth. [24] | Aerial parts | 20 | 97.3 | α-Thujene | α-Copaene, (*E*)-caryophyllene and germacrene D | 23.7 | 39.8 |
| *D. gardneriana* Mart. [31] | Leaf | 4 | 96.0 | - | β-Bisabolene and elemicin | - | 96.0 |

*(Table 1) cont.....*

| Annonaceae Species [Refs.] | Part of Plant | Number of Identified Compounds | Total of Identified Compounds (%) | Main Monoterpenoids (>5%) | Main Sesquiterpenoids (>5%) | Monoterpenoids (%) | Sesquiterpenoids (%) |
|---|---|---|---|---|---|---|---|
| *Ephedranthus amazonicus* R.E. Fr [23] | Leaf | 63 | 98.0 | - | Cyclosativene, α-muurolene, spathulenol, caryophyllene oxide and humulene epoxide II | 0.6 | 95.0 |
| *Fusaea longifolia* Saff [24] | Aerial parts | 21 | 88.5 | - | (*E*)-Caryophyllene, β-selinene, *cis*-β-guaiene and (*Z*)-α-bisabolene | 0.1 | 87.6 |
| *Guatteria australis* A. ST.-HIL. [32] | Aerial parts | 24 | 94.26 | β-Pinene, *trans*-pinocarveol, *trans*-verbenol and myrtenol | Spathulenol and caryophyllene oxide | 41.92 | 52.65 |
| *G. australis* [7] | Leaf | 23 | 96.6 | - | (*E*)-Caryophyllene, germacrene D and germacrene B | 17.2 | 79.4 |
| *G. blepharophylla* Mart [23]. | Leaf | 24 | 99.0 | - | Palustrol, spathulenol and caryophyllene oxide | - | 94.4 |
| *Psidium guajava* [33] | Leaf | 13 | 100 | - | α-cadinol, β-caryophyllene, nerolidol, α-selinene, β-selinene and caryophyllene oxide | - | 100 |
| *G. elliptica* [9] | Leaf | 15 | 100 | - | Spathulenol, caryophyllene oxide and β-copaen-α-ol | - | 100 |
| *G. elliptica* [9] | Leaf | 34 | 100 | - | Spathulenol and caryophyllene oxide | - | 100 |
| *G. friesiana* (W.A.Rodrigues) Erkens & Maas [34] | Leaf | 8 | 93 | - | γ-Eudesmol, β-eudesmol and α-eudesmol, | - | 93 |
| *G. latifolia* (Mart.) R.E.Fr. [32] | Aerial parts | 25 | 73.24 | - | Spathulenol and caryophyllene oxide | 6.94 | 67.81 |
| *G. megalophylla* Diels [35] | Leaf | 34 | 88.7 | - | δ-Elemene, β-elemene, γ-muurolene, bicyclogermacrene and spathulenol | 1.4 | 87.3 |
| *G. pogonopus* Mart. [34] | Leaf | 19 | 88.4 | - | Germacrene D, γ-amorphene and spathulenol | - | 88.4 |
| *G. pogonopus* [36] | Leaf | 29 | 86.19 | α-Pinene and β-pinene | (*E*)-Caryophyllene, germacrene D, bicyclogermacrene and γ-patchoulene | 23.13 | 60.44 |
| *G. punctata* (Aubl.) R. A. Howard. [24] | Aerial parts | 23 | 79.3 | - | (*E*)-Caryophyllene, germacrene D, *cis*-β-guaiene and (*E*)-nerolidol | 0.6 | 75.9 |

(Table 1) cont.....

| Annonaceae Species [Refs.] | Part of Plant | Number of Identified Compounds | Total of Identified Compounds (%) | Main Monoterpenoids (>5%) | Main Sesquiterpenoids (>5%) | Monoterpenoids (%) | Sesquiterpenoids (%) |
|---|---|---|---|---|---|---|---|
| *G. sellowiana* Schltdl [32] | Aerial parts | 19 | 89.99 | - | (Z)-β-Farnesene, β-bisabolene, cis-α-bisabolene, spathulenol and caryophyllene oxide | 5.16 | 84.83 |
| *G. ferruginea* A. St.-Hil. [32] | Aerial parts | 22 | 88.33 | trans-Pinocarveol and myrtenol | (E,E)-α-Farnesene, spathulenol and caryophyllene oxide | 26.01 | 62.32 |
| *Isolona dewevrei* (De Wild. & T. Durand) Engl. & Diels [37] | Leaf | 68 | 96.1 | (E)-β-Ocimene | (E)-Caryophyllene, germacrene D, γ-elemene and germacrene B | N.I. | N.I. |
| *I. dewevrei* [38] | Leaf | 57 | 97.1 | - | (E)-Caryophyllene and germacrene D | 12.0 | 85.1 |
| *Onychopetalum amazonicum* R.E.Fr. [39] | Leaf | 25 | 87.8 | - | α-Copaene, (E)-caryophyllene, bicyclogermacrene, δ-cadinene, spathulenol and caryophyllene oxide | - | 87.8 |
| *O. amazonicum* [39] | Trunk bark | 19 | 92.2 | - | α-Gurjunene, allo-aromadendrene and α-epi-cadinol | - | 92.2 |
| *O. amazonicum* [39] | Twig | 25 | 75.0 | - | α-Gurjunene, α-ep-cadinol and cyperotundone | - | 75.0 |
| *O. periquino* (Rusby) D.M. Johnson & N.A. Murray [40] | Leaf | 13 | 91.3 | - | β-Elemene, β-selinene and spathulenol | - | 91.3 |
| *Polyalthia korintii* (DUNAL) BENTH. & HOOK.F. [41] | Leaf | 32 | 100 | α-Pinene and β-pinene | - | 75.55 | 22.6 |
| *Porcelia macrocarpa* R.E. Fries [42] | Leaf | 9 | 80.2 | - | Germacrene D and bicyclogermacrene | 0.52 | 81.5 |
| *P. macrocarpa* [42] | Ripe fruit | 65 | 99.6 | Neryl and geranyl formats | γ-Muurolene, δ-cadinene and dendrolasin | 44.8 | 37.1 |
| *Pseuduvaria macrophylla* (Oliv.) Merr. [43] | Leaf | 34 | 87.7 | - | α-Cadinol, δ-cadinene, germacrene D, α-copaene and bicyclogermacrene | - | 87.7 |

*(Table 1) cont.....*

| Annonaceae Species [Refs.] | Part of Plant | Number of Identified Compounds | Total of Identified Compounds (%) | Main Monoterpenoids (>5%) | Main Sesquiterpenoids (>5%) | Monoterpenoids (%) | Sesquiterpenoids (%) |
|---|---|---|---|---|---|---|---|
| *Unonopsis guatterioides* (A.DC.) R.E.Fr [44] | Leaf | 16 | 99.5 | - | α-Copaene, β-elemene, E-caryophyllene, α-humulene, *allo-*aromadendrene, germacrene D, bicyclogermacrene and spathulenol | NI | NI |
| *Uvaria grandiflora* Roxb. ex Hornem. [45] | Leaf | 29 | 96.5 | α-Phellandrene, limonene and 1,8-cineole | - | 56.9 | 14.3 |
| *U. microcarpa* Champ. ex Benth. [45] | Leaf | 14 | 99.9 | - | - | 4.6 | 4.1 |
| *U. chamae* p. Beauv. [46] | Leaf | 56 | 99.8 | Linalool | Germacrene D | N.I | N.I |
| *Xylopia aethiopica* (Dunal) A. Rich [47] | Fruit | 70 | 97.14 | α-Pinene, β-pinene and α-phellandrene | (*Z*)-γ-Bisabolene | 77.83 | 19.31 |
| *X. aethiopica* [48] | Fruit | 30 | 100 | Terpinen-4-ol and eugenol | β-Caryophyllene and Germacrene D | N.I | N.I |
| *X. aethiopica* [49] | Fresh fruit | 54 | 99.3 | α-Pinene, sabinene, β-pinene and 1,8-cineole | - | 80 | 19.3 |
| *X. aethiopica* [49] | Shade dried fruit | 54 | 99.6 | α-Pinene, sabinene, β-pinene and 1,8-cineole | - | 83.5 | 16.1 |
| *X. aethiopica* [50] | Seed | 52 | 100 | Terpin-4-ol and α-terpineol | β-Copaene | 38.07 | 44.28 |
| *X. aromatica* (Lam.) Mart. [23] | Leaf | 41 | 98.6 | *trans-*Pinocarveol, α-campholenal, camphor, dihydrocarveol and verbenone | Spathulenol | 54.5 | 44.1 |
| *X. aromatica* [51] | Leaf | 48 | 97.9 | - | Bicyclogermacrene, spathulenol, globulol, *cis*-guaia-3,9-die--11-ol and khusinol | 2.74 | 80.9 |
| *X. aromatica* [51] | Flowers | 29 | 99.9 | - | Bicyclogermacrene, 7-epi-α-eudesmol, khusinol and pentadecan-2-one | 3.44 | 68.9 |
| *X. frutescens* Aubl. [5] | Leaves | 43 | 90.2 | - | Caryophyllene, γ-cadinene and cadin-4-en-10-ol | 18.0 | 67.5 |

(Table 1) cont.....

| Annonaceae Species [Refs.] | Part of Plant | Number of Identified Compounds | Total of Identified Compounds (%) | Main Monoterpenoids (>5%) | Main Sesquiterpenoids (>5%) | Monoterpenoids (%) | Sesquiterpenoids (%) |
|---|---|---|---|---|---|---|---|
| *X. frutescens* [52] | Leaves | 23 | 91.2 | (*E*)-β-Ocimene | β-Elemene, (*E*)-caryophyllene, germacrene D and bicyclogermacrene | NI | NI |
| *X. hypolampra* Mildbr. [53] | stem bark | 28 | 90.3 | Eucalyptol, borneol and verbenone | - | 79.1 | 7.3 |
| *X. laevigata* (Mart.) R. E. Fries [54] | Leaves | 27 | 98.7 | - | Germacrene D, bicyclogermacrene, (*E*)-caryophyllene and germacrene B | NI | NI |
| *X. laevigata* [55] | Fresh fruits | 10 | 99.6 | α-Pinene, β-pinene and limonene | - | 95.0 | 4.6 |
| *X. laevigata* [52] | Leaves | 32 | 96.7 | - | (*E*)-Caryophyllene, γ-muurolene, germacrene D, bicyclogermacre, δ-cadinene and germacrene B | NI | NI |
| *X. langsdorffiana* St.-Hil. & Tul. [56] | Fruits | 9 | 100 | α-Pinene, camphene and *D*-limonene | Caryophyllene oxide | NI | NI |
| *X. sericea* A. St.-Hil. [57] | Fruits | 84 | 99.0 | - | Germacrene D, spathulenol and guaiol | 9.6 | 81.5 |

NI: Not informed.

## BIOACTIVITY OF THE ESSENTIAL OILS FROM ANNONACEAE SPECIES

It is generally accepted that the EOs chemistry determines its bioactivities. The EOs from Annonaceae species have been evaluated for their anti-inflammatory, antitumor, antibacterial, antioxidant and other pharmacological effects [23, 29, 58], which provide potential explanations for the use of some species in the treatment of various diseases in folk medicine. The terpenoids can play a central role in the recorded activities and are described as the main active constituents of the abovementioned effects [15, 39]. A detailed summary of pharmacological studies is given below.

### Acetylcholinesterase Inhibition

Alzheimer's disease (AD) is one of the most well-known neurodegenerative diseases, and it explains 50-60% of patients with dementia; most of the AD therapeutics available are acetylcholinesterase inhibitors [59]. The essential oil (EO) of *C. pruniferus*, rich in α-pinene (25.4%), germacrene D (20.2%), β-caryophyllene (10.8%) and δ-cadinene (6.4%), showed high inhibition against acetylcholinesterase, (I%: 75.5%, at 1,000 mg.mL$^{-1}$ concentration) [27].

*Xylopia aethiopica* EO from Nigeria, rich in eugenol (35.0%), terpinen-4-ol (7.2%) and germacrene D (5.4%), inhibited acetylcholinesterase (IC$_{50}$ = 18.5 µl.L$^{-1}$) and butyrylcholinesterase (IC$_{50}$ = 26.4 µl.L$^{-1}$) activities in a concentration-dependent manner (7.7-23.1 µl.L$^{-1}$) [48]. The EO from the seeds of *X. aethiopica*, also collected in Nigeria, with a high content terpinen-4-ol (11.8%) and α-terpineol (5.9%), showed significant anticholinesterase activity (IC$_{50}$ = 1.21 mg.mL$^{-1}$) [50]. The EO of *Pseuduvaria macrophylla*, which shows a high content by germacrene D (21.1%), bicyclogermacrene (10.5%), δ-cadinene (5.6%), α-copaene (5.1%) and α-cadinol (5.0%), showed weak inhibitory activity against acetylcholinesterase (I%: 32.5%) and butyrylcholinesterase (I%: 35.4%) assays [43].

## Antimicrobial Activities

Annonaceae species are an important source of new antimicrobial agents for combating resistant microorganisms, since that several EOs of this family had their antimicrobial properties evaluated and showed potentially relevant results. The EOs effect of *Xylopia aromatica*, rich in spathulenol (21.5%), dihydrocarveol (11.6%) and *trans*-pinocarveol (10.2%), and *Guatteria blepharophylla*, rich in caryophyllene oxide (55.7%), spathulenol (8.9%) and palustrol (6.5%), showed strong activity against Gram-positive bacteria *Streptococcus sanguinis* (MIC = 0.02 mg.mL$^{-1}$) [23].

The EOs from *Bocageopsis multiflora, Duguetia quitarensis, Fusaea longifolia and Guatteria punctata* were evaluated for their antibacterial activity [24]. Among the tested samples, the EO of *B. multiflora*, rich in *cis*-linalool oxide (furanoid) (33.1%) and 1-*epi*-cubenol (16.6%), showed high antibacterial activity against Gram-negative and Gram-positive strains with MIC values of 4.68 µg.mL$^{-1}$. The EO of *Duguetia quitarensis*, composed primarily of 4-heptanol (33.8%), α-thujene (18.4%), (*E*)-caryophyllene (14.4%), germacerne D (6.3%) and α-copaene (5.3%), was active against micro-organisms Gram-positive *Streptococcus mutans* and *Streptococcus pyogenes* with a MIC value of 37.5 µg.mL$^{-1}$. Essential oil of *F. longifolia*, rich in β-selinene (19.3%), *cis*-β-guaiene (18.3%), (*Z*)-α-bisabolene (12.0%) and (*E*)-caryophyllene (7.1%), was active against the Gram-negative *Pseudomonas aeruginosa* and Gram-positive *Streptococcus mutans* and methicillin-resistant *Staphylococcus aureus*, with a MIC value of 37.5 µg.mL$^{-1}$. The EO of *Guatteria punctata*, with a high content germacrene D (19.8%), (*E*)-caryophyllene (8.4%), (*E*)-nerolidol (9.9%) and *cis*-β-guaiene (5.5.%), was active against *S. mutans* and *S. pyogenes* with a MIC value of 4.68 µg.mL$^{-1}$ [24].

The EO from the leaves of *Anaxagorea brevipes*, composed primarily of β-eudesmol (13.16%), α-eudesmol (13.05%), γ-eudesmol (7.54%) and guaiol (5.12%), showed antibacterial and antifungal inhibitory effects against *Kocuria rhizophila, Staphylococcus aureus* penicillinase-negative, *Candida albicans* and *Candida parapsilosis* with MIC values varying 25.0 and 100.0 µg.mL$^{-1}$ [15].

The EOs from leaves, twigs and trunk bark of *Onychopetalum amazonicum* were evaluated for antimicrobial activities against four bacteria strains and five pathogenic fungi. The EO of the trunk bark exhibited activity against *Staphylococcus epidermidis, E.coli* and *Kocuria rhizophila*, with a MIC value of 62.5 µg.mL$^{-1}$. This important activity may be associated with the presence of sesquiterpene *allo*-aromadendrene (21.2%) [39].

The antibacterial activity of the EO of *Xylopia sericea* fruits was investigated and the results showed that the EO, with a high content of sesquiterpenes spathulenol (16.42%), guaiol (13.93%) and germacrene D (8.11%), has a high bacteriostatic effect against *S. aureus* (MIC = 7.8 µg.mL$^{-1}$), *Enterobacter cloacae* (MIC = 7.8 µg.mL$^{-1}$), *Bacillus cereus* (MIC = 15.6 µg.mL$^{-1}$) and *Klebsiella pneumonia* (MIC = 62.5 µg.mL$^{-1}$) [57].

The antimicrobial activity of the EOs from flowers and leaves of *Xylopia aromatica* was tested against Gram-positive and Gram-negative bacterial strains and fungi. The flower EO, rich in pentadecan-2-one (16.38%), bicyclogermacrene (9.74%), 7-*epi*-α-eudesmol (7.76%), khusinol (7.23%), *n*-tricosane (6.17%) and heptadecan-2-one (5.83%), and the leaf EO, rich in spathulenol (27.11%), khusinol (13.04%), bicyclogermacrene (8.52%), globulol (6.47%) and *cis*-guai--3,9-dien-11-ol (5.98%), exhibited the lowest MIC against *S. pyogenes* (200 and 100 µg.mL$^{-1}$, respectively) [51].

The EOs two *Guatteria elliptica* specimens collected at Paranapiacaba and Caraguatatuba (São Paulo, Brazil), with spathulenol (53.9%) and caryophyllene oxide (40.9%) as the major compounds, for Paranapiacaba and Caraguatatuba specimens, respectively, showed growth inhibition lower than 100% at the highest tested concentration (3 mg.mL$^{-1}$); and MIC values >3 mg.mL$^{-1}$ against all tested micro-organisms. The tests with EOs were considered inactive [9].

The EOs from four *Guatteria* species (*G. australis, G. ferruginea, G. latifolia* and *G. sellowiana*), rich in spathulenol (11.04 – 40.29%) and caryophyllene oxide (7.74 – 40.13%), showed strong antibacterial activity (MIC = 0.062-0.25 mg.ml$^{-1}$) against strains of *Rhodococcus equi* [32]. Essential oil from the leaves of *G. australis*, rich in germacrene B (50.6%), germacrene D (22.2%) and (*E*)-caryophyllene (8.9%), exhibited a small effect against *S. aureus* and *E. coli* (MIC = 250 µg/ml$^{-1}$) [7].

The antimicrobial activities from the EOs of leaves, twigs and bark of *Bocageopsis pleiosperma* were evaluated. The EO obtained from the bark exhibited a moderate effect against *Staphylococcus epidermidis* (MIC = 250 µg.mL$^{-1}$), while the other EOs did not exhibit antimicrobial activity. The sesquiterpene β-bisabolene was the main component in all aerial parts of the plant, with a higher concentration in the leaves (55.7%), followed by bark (38.5%) and twig (34.3%) [25].

The EO of stem bark from *Xylopia hypolampra*, rich in verbenone (20.2%), borneol (7.8%), eucalyptol (5.9%) and nopinone (5.5%), showed a weak inhibition against *Streptococcus pyogenes, Staphylococcus aureus* and *Escherichiacoli* [53].

The EO from the leaves of *Polyalthia korintii* showed no activity against *S. aureus*, *E. coli* and in a clinically isolated strain (*Klebsiella pneumonia*). The major constituents identified in the leaf EO were α-pinene (43.1%) and β-pinene (25.5%) [41]. The EO of dried fruits from *Xylopia aethiopica*, is rich in β-pinene (32.1%), β-phellandrene (10.7%), (*Z*)-γ-bisabolene (10.0%) and α-pinene (7.3%), showed high antifungal activity against *Aspergillus niger* and *Fusarium oxysporium* (MIC = 3000 ppm, both, and MFC = 3000 and 4000 ppm, respectively) [47].

The fruits of Atemoya, a hybrid of *Annona squamosa* and *Annona cherimola*, were subjected to different drying methods before obtaining the EOs. Different drying methods did not affect the EOs composition, as well as their antimicrobial activities against *S. aureus*, *E. coli*, *Listeria monocytogenes*, *Vibrio parahaemolyticus*, *Aspergillus niger*, *Penicillium italicum* and *Trichoderma reesei*. The MMC values varied from 0.188 to 6 mg.mL$^{-1}$. The main compound of EOs was spathulenol (ca 35%), therefore, according to the authors, spathulenol could be responsible for the observed antimicrobial activities [60].

**Anti-Inflammatory Activity**

Many Annonaceae species have been used for the treatment of inflammatory diseases in folk medicine [61]. Pharmacological studies showed that some terpenoids and EOs from this family have significant anti-inflammatory effects, such as caryophyllene oxide and the EO of *Duguetia lanceolata*; the EO from *D. lanceolata* branches, rich in β-elemene (8.3%), β-caryophyllene (6.2%), caryophyllene oxide (7.7%), β-eudesmol (7.2%), β-selinene (7.1%) and δ-cadinene (5.5%), played a crucial role as a protective factor against the carrageenan-induced acute inflammation [29].

The EO of atemoya fruits had an anti-inflammatory effect against J774A.1 macrophages. The results demonstrated that the EO can exhibit anti-inflammatory activity in the cell [17]. The anti-inflammatory activity of the EO from the underground stem bark of *Duguetia furfuracea*, rich in (*E*)-asarone (21.9%), bicyclogermacrene (16.7%), 2,4,5-trimethoxystyrene (16.1%), α-gurjunene (15%) and cyperene (7.8%), was investigated. The results indicated that this EO produces anti-inflammatory effect and supports the medicinal use of this species to treat inflammation and painful conditions [28].

## Antiproliferative and Cytotoxic Activities

The search for new drugs that display activity against several types of cancer has become one of the most interesting subjects in the field of natural products. Several EOs from Annonaceae species and their bioactive constituents have been evaluated for their antiproliferative and cytotoxic properties.

The antiproliferative activity of EO from the leaves of *Cardiopetalum calophyllum*, rich in spathulenol (28.78%), viridiflorol (9.99%) and (*Z,E*)-pharnesol (6.51%), was evaluated against tumor in different human tumor cell lines (breast adenocarcinoma, cervical adenocarcinoma and glioblastoma). The $IC_{50}$ values ranged from 216.8 to 353.5 µg.mL$^{-1}$, and selectivity was not observed. The authors justified the antiproliferative activity to the presence of the main constituents [58].

The EO fruit from the *Annona squamosa*, with a high level of spathulenol (32.5%), showed antineoplastic activity against hepatoma cell line with $IC_{50}$ lower than 55 µg/mL [62].

The cytotoxic, mutagenic and genotoxic profiles of the EO from leaves of *Xylopia laevigata* were investigated. The results demonstrated that the EO, rich in germacrene D (43.6%), bicyclogermacrene (14.6%), (*E*)-caryophyllene (7.9%) and germacrene B (7.3%), has mutagenic and anti-proliferative activities, which may be related to the cytotoxic effect of the EO major components [54].

The *in vitro* cytotoxicity of the EO of *Annona vepretorum* (alone, complexed with β-cyclodextrin and some of its major constituents) in tumor cell lines from different histotypes was evaluated. Besides that, the *in vivo* efficacy of these EO in mice inoculated with mouse melanoma (B16-F10) was evaluated. The results showed that sesquiterpene spathulenol and the EO, with a high concentration of bicyclogermacrene (35.71%), spathulenol (18.89%), (*E*)-β-ocimene (12.46%), α-phellandrene (8.08%) and *o*-cymene (6.24%), exhibited promising cytotoxicity. The *in vivo* tumour growth was inhibited by the treatment with the EO (inhibition of 34.46%), and the microencapsulation of the EO increased tumour growth inhibition (inhibition of 62.66%) [21].

Antitumor activity and toxicity of the volatile oil from *Annona leptopetala* leaves, rich in spathulenol (12.5%) and α-limonene (9.0%), were evaluated and showed *in vitro* and *in vivo* antitumor activity, mainly in the leukemia cell line, without major changes in toxicity parameters evaluated [18].

The *in vitro* cytotoxic activity of the EO from the fresh fruits of *Xylopia laevigata* and its major constituents, limonene, α-pinene and β-pinene, was evaluated against four tumor cell lines (mouse melanoma, human hepatocellular carcinoma, human promyelocytic leukemia and human chronic myelocytic leukemia) and one non-tumor cell (human peripheral blood mononuclear cells). Neither EO nor its major constituents, presented cytotoxic activity ($IC_{50} > 25.0$ µg.mL$^{-1}$) [55].

The *in vitro* and *in vivo* anti-leukemia potential of *Guatteria megalophylla* leaf EO were investigated. The *in vitro* cytotoxic potential of this EO was evaluated in human cancer cell lines (HL-60, MCF-7 CAL27, HSC-3, HepG2 and HCT116) and in human non-cancer cell line (MRC-5). The *in vivo* efficacy of this EO was evaluated in C.B17 SCID mice with HL-60 cell xenografts. The results demonstrated that this EO has anti-leukemia potential (with an $IC_{50}$ value of 12.51 µg.mL$^{-1}$ for HL-60 cells), and the major constituents spathulenol (27.7%), γ-muurolene (14.3%), bicyclogermacrene (10.4%), β-elemene (7.4%) and δ-elemene (5.1%), may play a central role for the recorded activities [35].

The antiproliferative activity of the EO from the leaves of *Anaxagorea brevipes* was investigated in a series of cancer cell lines and the bioactivity was described against MCF-7 (breast, TGI = 12.8 µg.mL$^{-1}$), NCI-H460 (lung, TGI = 13.0 µg.mL$^{-1}$) and PC-3 (prostate, TGI = 9.6 µg.mL$^{-1}$). The antiproliferative activity found was also attributed to the sesquiterpenes of the EO, β-eudesmol (13.16%), α-eudesmol (13.05%), γ-eudesmol (7.54%) and guaiol (5.12%) [15].

Antitumour activity and toxicity of *Xylopia langsdorffiana* EO, rich in α-pinene (34.5%) and limonene (31.7%), were evaluated. The oil caused *in vitro* and *in vivo* growth inhibition of tumor cells, without major changes in the toxicity parameters evaluated [56].

The EOs from two *Guatteria elliptica* specimens from the Paranapiacaba and Caraguatatuba showed an important antitumor activity against breast and prostate cancer cells ($IC_{50}$ = 7.0 and 5.35 µg.mL$^{-1}$, respectively) and a low cytotoxicity against normal fibroblast cells ($IC_{50} > 22.2$ and $IC_{10}$ = 18.5 µg.mL$^{-1}$, respectively) [9].

The EO of *Duguetia gardneriana*, with a high level of β-bisabolene (80.9%), exhibited cytotoxic effect. The $IC_{50}$ values were obtained for mouse melanoma, human hepatocellular carcinoma, human promyelocytic leukemia, and human chronic myelocytic leukemiacell lines (16.8, 19.1, 13.0, and 19.3 µg.mL$^{-1}$, respectively). The *in vivo* antitumor activity was evaluated using C57BL/6 mice subcutaneously inoculated with B16-F10 melanoma cells and revealed tumor growth inhibition rates of 5.37–37.52% at doses of 40 and 80 mg/kg/day, respectively [31].

The antiproliferative activity of EOs from four *Guatteria* species (*G. australis, G. ferruginea, G. latifolia* and *G. sellowiana*) was investigated. These EOs contained the oxygenated sesquiterpenes spathulenol (11.04 – 40.29%) and caryophyllene oxide (7.74 – 40.13%) as major constituents. The evaluation of antiproliferative activity showed strong selectivity (1.1-4.1 µg.ml$^{-1}$) against the tumor cell line ovarian cancer, even more active than the positive control doxorubicin (11.7 µg.ml$^{-1}$) [32].

The EO from the leaves of *Guatteria australis*, with a high concentration of germacrene B (50.6%), germacrene D (22.2%) and (*E*)-caryophyllene (8.9%), showed strong antiproliferative effect against NCI-ADR/RES (ovarian-resistant) and HT-29 (colon). The TGI values were equal to 31.0 and 32.8 µg.ml$^{-1}$, respectively [7].

**Larvicidal Activity**

The larvicidal effect of EOs from Annonaceae species has been tested against several disease vectors. The EOs of two *Duguetia* species were evaluated against *Artemia salina* and *Culex quinquefasciatus* larvaes. Essential oils from leaf, underground heart wood and underground stem bark from *D. furfuracea* showed potent activity against *A. salina* larvae (LC$_{50}$ 6.01, 7.79 and 9.98 µg.mL$^{-1}$, respectively). The major constituents were spathulenol (47.2%), bicyclogermacrene (26.4%) and caryophyllene oxide (5.2%) in leaf EO, (*E*)-asarone (21.9%), bicyclogermacrene (16.7%), 2,4,5-Trimethoxystyrene (16.1%), α-gurjunene (15.0%) and cyperene (7.8%) in underground stem bark EO and (*E*)-asarone (16.6%), cyperene (15.7%), spathulenol (14.2%), 2,4,5-trimethoxystyrene (13.2%), bicyclogermacrene (8.6%) and α-gurjunene (8.1%) in underground heart wood EO. The leaf EO from *D. lanceolata*, rich in α-selinene (11.0%), aristolochene (5.8%) (*E*)-caryophyllene (5.3%) and (*E*)-calamenene (5.2%), also showed potent activity against *A. salina* larvae (LC$_{50}$ 0.89 µg.mL$^{-1}$). The EOs from both species showed moderately active against *C. quinquefasciatus* since they exhibited LC$_{50}$ ranging from 57.8 to 121.7 µg.mL$^{-1}$ [63].

The EO of *Onychopetalum periquino*, with a high concentration of sesquiterpenes β-elemene (53.16%), spathulenol (11.94%) and β-selinene (9.25%), showed high larvicidal activity against *Aedes aegypti* larvae ($LC_{50}$ of 63.75 µg.mL$^{-1}$ with 100% mortality at 200 µg.mL$^{-1}$) [40]. The EOs from *Xylopia laevigata*, rich in germacrene D (27.0%), bicyclogermacrene (12.8%), (*E*)-caryophyllene (8.6%), γ-muurolene (8.6%) and δ-cadinene (6.8%), and *Xylopia frutescens,* with a high levels of bicyclogermacrene (23.2%), germacrene D (21.2%), (*E*)-caryophyllene (17.4%), β-elemene (6.3%) and (*E*)-β-ocimene (5.2%), showed no larvicidal activity [52].

## Trypanocidal and Antimalarial Activities

Chagas Disease, also known as American trypanosomiasis, is caused by the protozoan parasite *Trypanosoma cruzi.* With complex pathophysiology and a dynamic epidemiological profile, this disease remains an important public health concern and is an emerging disease in non-endemic countries; for the etiologic treatment, both in the acute and chronic phases, two main drugs are available to treat the disease: benznidazole and nifurtimox [64].

The EOs from *Bocageopsis multiflora, Duguetia quitarensis, Fusaea longifolia,* and *Guatteria punctata* were evaluated for their trypanocidal activity [24]. The results showed that these EOs were active at the tested concentrations. *Guatteria punctata* EO was the most active, with an $IC_{50}$ = 0.029 µg.mL$^{-1}$, being 34 times more active than the reference drug benznidazole. The authors related that the strong activity of this species can be attributed to the presence of germacrene D (19.8%) and (*E*)-caryophyllene (8.4%) in the composition of EO [24].

The EOs extracted from leaves of *Guatteria friesiana*, with a high level of β-eudesmol (51.9%), γ-eudesmol (18.9%) and α-eudesmol (12.6%), and of *Guatteria pogonopus*, rich in spathulenol (24.8%), γ-amorphene (14.7%) and germacrene D (11.8%), demonstrated potent trypanocidal and antimalarial activities with values of $IC_{50}$ lower than 41.3 µg.mL$^{-1}$ [34].

## Other Activities

The EOs of *Xylopia laevigata* and *Xylopia frutescens* showed a low degree of protection from *Aedes aegypti* landings, and, therefore, low repellent activity. The EO of *X. laevigata* has a high concentration of germacrene D (27.0%), bicyclogermacrene (12.8%), (*E*)-caryophyllene (8.6%), γ-muurolene (8.6%) and δ-cadinene (6.8%), while the *X. frutescens* EO was rich in bicyclogermacrene (23.2%), germacrene D (21.2%), (*E*)-caryophyllene (17.4%), β-elemene (6.3%) and (*E*)-β-ocimene (5.2%) [52].

The possible tocolytic effect of EO from *Annona leptopetala*, composed primarily of bicyclogermacrene (22.47%), *cis*-4-thujanol (17.37%) and germacrene (7.72%), on isolated rat uterus was investigated. The results showed that the EO exerts tocolytic activity [6].

The anticonvulsant, sedative, anxiolytic and antidepressant activities of the EO from *Annona vepretorum* leaves, rich in (*E*)-β-ocimene (42.59%), bicyclogermacrene (18.81%), germacrene D (12.19%) and limonene (10.02%), were investigated in mice. The results showed that the acute treatment with EO has anxiolytic, sedative, antiepileptic and antidepressant effects [22].

Gontijo and coworkers investigated the antiplasmodial activity of EO from *Xylopia sericea* leaves, composed primarily of α-pinene, β-pinene, *o*-cymene and *D*-limonene. Low growth inhibition (24.0 at 50 µg.mL$^{-1}$) of *Plasmodium falciparum* and low cytotoxicity to HepG2 cells (CC$_{50}$ 275.9 µg.mL$^{-1}$) were observed [65].

Sousa and coworkers demonstrated that the sesquiterpene caryophyllene oxide and the EO of *Duguetia lanceolata* branches, rich in β-elemene (8.3%), β-caryophyllene (6.2%), caryophyllene oxide (7.7%), β-eudesmol (7.2%), β-selinene (7.1%) and δ-cadinene (5.5%), have antinociceptive effect since they reduced the abdominal writhes in mice [29].

Insecticidal, antifungal and antiaflatoxigenic activities of EO of *Duguetia lanceolata* were evaluated on stored-grain deterioration agents. The main constituents of this EO were β-bisabolene (56.2%) and 2,4,5-trimethoxystyrene (19.1%). The results suggested that the EO has promising grain-protective properties against *Sitophilus zeamais* and *Zabrotes subfasciatus* showing activity comparable to an insecticide based on deltamethrin (positive control) [30].

The antinociceptvive effect of the EO of underground stem bark from *Duguetia furfuracea*, composed primarily by (*E*)-asarone (21.9%), bicyclogermacrene (16.7%), 2,4,5-trimethoxystyrene (16.1%), α-gurjunene (15%) and cyperene (7.8%), was investigated. The results showed that the antinociceptive activity of this EO is possibly mediated by adenosinergic and opioidergic pathways and its properties do not induce effects on motor coordination [28].

The EO from the leaves of *Guatteria australis*, with a high concentration of germacrene B (50.6%), germacrene D (22.2%) and (*E*)-caryophyllene (8.9%), showed antileishmanial activity against *Leishmania infantum* (IC$_{50}$ = 30.7 µg.ml$^{-1}$) [7].

The EO from *Unonopsis guatterioides* fresh leaves, rich in α-copaene, bicyclogermacrene and *trans*-caryophyllene (15.7% each), α-humullene (9.0%), *allo*-aromadendrene (8.4%) and spathulenol (7.3%), has shown phytotoxic effect on the germination, growth and development of monocotyledon (*Allium cepa*) and dicotyledon (*Lactuca sativa*) model plants [44].

## ANTIOXIDANT POTENTIAL

Antioxidants are widely used in the food industry for various reasons, including prevention of oxidation, neutralization of free radicals, preserving the food, and enhancing its flavor, aroma or color. As some synthetic antioxidants exhibit carcinogenic effects and could be toxic to nature, researchers encourage looking for natural antioxidants [66]. In several studies with EOs, antioxidant activity is related to compounds such as thymol, carvacrol, α-terpinene, β-terpinene, β-terpinolene, 1,8-cineole, eugenol and linalool, which present similar antioxidant activity to α-tocopherol [67].

*Xylopia sericea* EO was investigated for its antioxidant potential using different methods. The fruits EO is rich in spathulenol (16.42%), guaiol (13.93%) and germacrene D (8.11%) and it displayed a significant antioxidant activity in DPPH ($IC_{50}$ 49.1 μg.mL$^{-1}$), β-carotene/linoleic acid bleaching ($IC_{50}$ 6.9 μg.mL$^{-1}$), TAC ($IC_{50}$ 78.2 μg.mL$^{-1}$) and TBARS (80.0 μg.mL$^{-1}$) [57].

The EO of *Duguetia lanceolata* branches shows a high content on β-elemene (8.3%), β-caryophyllene (6.2%), caryophyllene oxide (7.7%), β-eudesmol (7.2%), β-selinene (7.1%) and δ-cadinene (5.5%). The antioxidant effects using DPPH assay ($EC_{50}$ 159.4 μg.mL$^{-1}$), reducing $Fe^{+3}$ power ($EC_{50}$ 187.8 μg.mL$^{-1}$) and lipid peroxidation inhibition (41.5%) were considered significant [29].

Leaf EOs from two *Guatteria elliptica* specimens, collected in Paranapiacaba and Caraguatatuba, showed low antioxidant potential ($EC_{50}$ = 7.24 and 28.68 mg.mL$^{-1}$ using DPPH assays, for Paranapiacaba and Caraguatatuba OEs, respectively). The difference in the $EC_{50}$ values could be at least in part attributed to the different content of major compounds [9].

The antioxidant potential of EO from the leaves of *Guatteria australis*, rich in germacrene B (50.6%), germacrene D (22.2%) and (*E*)-caryophyllene (8.9%), was evaluated using two methods. The antioxidant capacity was considered medium (TLC/DPPH, clear yellow spot) and small ($ORAC_{FL}$ assay, 457 μmolTE.g$^{1}$) [7].

The EOs obtained from the fruit pulp and leaves of *Annona muricata* showed a high total antioxidant capacity (49.03 gAAE/100 g and 50.88 gAAE/100 g for fruit pulp and leaf EOs, respectively). The $IC_{50}$ values from the DPPH assay were

244.8 $\mu$g.mL$^{-1}$ (leaf EO) and 512 $\mu$g.mL$^{-1}$ (fruit pulp EO). The EOs major compounds were $\delta$-cadinene (22.5%) and $\alpha$-muurolene (10.6%) in the leaf EO and Ç-sitosterol (19.8%) and 2-hydroxy-1-(hydroxymethyl) ethyl ester (13.4%) in the fruit pulp EO [19].

The EO of *Xylopia aethiopica* seeds is rich in terpinen-4-ol (11.8%) and $\alpha$-terpineol (5.9%) and it showed significant antioxidant potential (IC$_{50}$ value of DPPH = 2.19 mg.mL$^{-1}$) [50], while the dried fruits EO of this species, rich in $\beta$-pinene (32.1%), $\beta$-phellandrene (10.7%), $Z$-$\gamma$-bisabolene (10.0%) and $\alpha$-pinene (7.3%), showed a low antiradical activity (SC$_{50}$ = 594.5 $\mu$g/mL) when compared to that of BHT (SC$_{50}$ = 65.0 $\mu$g/mL) [47].

*Xylopia aethiopica* EO scavenged DPPH, NO, ABTS radicals and chelated Fe$^{2+}$. This EO strongly inhibited the linoleic acid peroxidation and the quinolinic acid-induced lipid peroxidation in rat brain homogenate *in vitro* [48].

The EOs of Atemoya fruits had the antioxidant activity measured by DPPH, nitric oxide, and reducing power methods; the results showed IC$_{50}$ values from 0.49 to 3.36 mg.mL$^{-1}$. Spathulenol (17.6-36.0%) was the main constituent of the EO [60].

## CONCLUDING REMARKS

This work summarizes the research progress on volatiles terpenoids of Annonaceae species and their activities. The large chemical variability that exists among EO from several species of this family was demonstrated. Terpenes, including mono and sesquiterpenoids, were identified as the main chemical constituents of the EOs. The compounds $\alpha$-pinene, $\beta$-pinene, limonene, (*E*)-caryophyllene, bicyclogermacrene, caryophyllene oxide, germacrene D, spathulenol and $\beta$-elemene were the most abundant. In addition, antimicrobial, antioxidant, larvicidal, trypanocidal, antimalarial, antiproliferative, anti-inflammatory and other activities of the EOs were described.

## CONSENT FOR PUBLICATION

Not applicable.

## CONFLICT OF INTEREST

The author declares no conflict of interest, financial or otherwise.

## ACKNOWLEDGEMENTS

We would like to thank the authors for their contributions and commitment to writing this scientific work, and the editors of this book for the invitation and vote of confidence.

## REFERENCES

[1]    Fechine, I.M.; Lima, M.A.; Navarro, V.R.; Cunha, E.V.L.; Silva, M.S.; Barbosa-Filho, J.M.; Maia, J.G.S. Alcalóides de Duguetia trunciflora Maas (Annonaceae). *Rev. Bras. Farmacogn.*, **2002**, *12*, 17-19.
[http://dx.doi.org/10.1590/S0102-695X2002000300009]

[2]    Tamokou, J.D.; Mbaveng, A.T.; Kuete, V. Antimicrobial Activities of African Medicinal Spices and Vegetables. In: *Medicinal Spices and Vegetables from Africa: Therapeutic Potential Against Metabolic, Inflammatory, Infectious and Systemic Diseases*; Kuete, V., Ed.; Elsevier Inc., **2017**; pp. 207-237.
[http://dx.doi.org/10.1016/B978-0-12-809286-6.00008-X]

[3]    da Silva Almeida, J.R.G.; Araújo, C.S.; de Oliveira, A.P.; Lima, R.N.; Alves, P.B.; Diniz, T.C. Chemical constituents and antioxidant activity of the essential oil from leaves of Annona vepretorum Mart. (Annonaceae). *Pharmacogn. Mag.*, **2015**, *11*(43), 615-618.
[http://dx.doi.org/10.4103/0973-1296.160462] [PMID: 26246740]

[4]    Lemos, E.E.P. A produção de anonáceas no Brasil. *Rev. Bras. Frutic.*, **2014**, *36*(spe1), 77-85.
[http://dx.doi.org/10.1590/S0100-29452014000500009]

[5]    Souza, I.L.L.; Correia, A.C.C.; Araujo, L.C.C.; Vasconcelos, L.H.C.; Silva, M.C.C.; Costa, V.C.O.; Tavares, J.F.; Paredes-Gamero, E.J.; Cavalcante, F.A.; Silva, B.A. Essential oil from Xylopia frutescens Aubl. reduces cytosolic calcium levels on guinea pig ileum: mechanism underlying its spasmolytic potential. *BMC Complement. Altern. Med.*, **2015**, *15*(1), 327.
[http://dx.doi.org/10.1186/s12906-015-0849-3]

[6]    Ferreira, P.; Martins, I.; Pereira, J.; Correia, A.; Sampaio, R.; Silva, M.; Costa, V.; Silva, M.; Cavalcante, F.; Silva, B. Tocolytic action of essential oil from Annona leptopetala R. E. Fries is mediated by oxytocin receptors and potassium channels. **2019**, 5.

[7]    Siqueira, C.A.T.; Serain, A.F.; Pascoal, A.C.R.F.; Andreazza, N.L.; de Lourenço, C.C.; Góis Ruiz, A.L.T.; de Carvalho, J.E.; de Souza, A.C.O.; Tonini Mesquita, J.; Tempone, A.G.; Salvador, M.J. Bioactivity and chemical composition of the essential oil from the leaves of *Guatteria australis* A.St.-Hil. *Nat. Prod. Res.*, **2015**, *29*(20), 1966-1969.
[http://dx.doi.org/10.1080/14786419.2015.1015017] [PMID: 25710362]

[8]    Rabelo, S.V.; Quintans, J. de S. S.; Costa, E. V.; Almeida, J. R. G. S.; Quintans, L. J. (2015). Annona species (Annonaceae) oils. In Essential Oils in Food Preservation, Flavor and Safety. Elsevier Inc **2015**.

[9]    Rajca Ferreira, A.K.; Lourenço, F.R.; Young, M.C.M.; Lima, M.E.L.; Cordeiro, I.; Suffredini, I.B.; Lopes, P.S.; Moreno, P.R.H. Chemical composition and biological activities of *Guatteria elliptica* R. E. Fries (Annonaceae) essential oils. *J. Essent. Oil Res.*, **2018**, *30*(1), 69-76.
[http://dx.doi.org/10.1080/10412905.2017.1371086]

[10]   Shaaban, H.A.E.; El-Ghorab, A.H.; Shibamoto, T. Bioactivity of essential oils and their volatile aroma components: Review. *J. Essent. Oil Res.*, **2012**, *24*(2), 203-212. [Review].
[http://dx.doi.org/10.1080/10412905.2012.659528]

[11]   Dudareva, N.; Negre, F.; Nagegowda, D.A.; Orlova, I. Plant Volatiles: Recent Advances and Future Perspectives. *Crit. Rev. Plant Sci.*, **2006**, *25*(5), 417-440.
[http://dx.doi.org/10.1080/07352680600899973]

[12]   Fournier, G.; Leboeuf, M.; Cavé, A. Annonaceae essential oils: A review. *J. Essent. Oil Res.,* **1999**, *11*(2), 131-142.
[http://dx.doi.org/10.1080/10412905.1999.9701092]

[13]   Mrudulakumari Vasudevan, U.; Lee, E.Y. Flavonoids, terpenoids, and polyketide antibiotics: Role of glycosylation and biocatalytic tactics in engineering glycosylation. *Biotechnol. Adv.,* **2020**, *41*107550
[http://dx.doi.org/10.1016/j.biotechadv.2020.107550] [PMID: 32360984]

[14]   Hung, N.V.; Dai, D.N.; Thai, T.H.; Thang, T.D.; Ogunwande, I.A.; Ogundajo, A.L. Essential Oil of *Alphonsea tonkinensis. Chem. Nat. Compd.,* **2018**, *54*(6), 1170-1171.
[http://dx.doi.org/10.1007/s10600-018-2584-8]

[15]   de Alencar, D.C.; Pinheiro, M.L.B.; Pereira, J.L.S.; de Carvalho, J.E.; Campos, F.R.; Serain, A.F.; Tirico, R.B.; Hernández-Tasco, A.J.; Costa, E.V.; Salvador, M.J. Chemical composition of the essential oil from the leaves of *Anaxagorea brevipes* (Annonaceae) and evaluation of its bioactivity. *Nat. Prod. Res.,* **2016**, *30*(9), 1088-1092.
[http://dx.doi.org/10.1080/14786419.2015.1101103] [PMID: 26586465]

[16]   Cascaes, M.M.; Silva, S.G.; Cruz, J.N.; Oliveira, M.S.; Oliveira, J.; Moraes, A.A.B.; De, B.; Da Costa, F.A.M.; Da Costa, K.S.; Nascimento, L.D.; Andrade, E.H.A.A. First report on the *Annona exsucca* DC. Essential oil and *in silico* identification of potential biological targets of its major compounds. *Nat. Prod. Res.,* **2021**, *7*, 1-4.
[PMID: 33678086]

[17]   Thang, T.D.; Dai, D.N.; Hoi, T.M.; Ogunwande, I.A. Study on the volatile oil contents of *Annona glabra* L., *Annona squamosa* L., *Annona muricata* L. and *Annona reticulata* L., from Vietnam. *Nat. Prod. Res.,* **2013**, *27*(13), 1232-1236.
[http://dx.doi.org/10.1080/14786419.2012.724413] [PMID: 22989376]

[18]   Brito, M.T.; Ferreira, R.C.; Beltrão, D.M.; Moura, A.P.G.; Xavier, A.L.; Pita, J.C.L.R.; Batista, T.M.; Longato, G.B.; Ruiz, A.L.T.G.; Carvalho, J.E.; Medeiros, K.C.P.; Santos, S.G.; Costa, V.C.O.; Tavares, J.F.; Diniz, M.F.F.M.; Sobral, M.V. Antitumor activity and toxicity of volatile oil from the leaves of *Annona leptopetala. Rev. Bras. Farmacogn.,* **2018**, *28*(5), 602-609.
[http://dx.doi.org/10.1016/j.bjp.2018.06.009]

[19]   Gyesi, J.N.; Opoku, R.; Borquaye, L.S. Chemical Composition, Total Phenolic Content, and Antioxidant Activities of the Essential Oils of the Leaves and Fruit Pulp of *Annona muricata* L. (Soursop) from Ghana. *Biochem. Res. Int.,* **2019**, *2019*, 1-9.
[http://dx.doi.org/10.1155/2019/4164576] [PMID: 31565436]

[20]   Formagio, A.S.N.; Vieira, M.C.; dos Santos, L.A.C.; Cardoso, C.A.L.; Foglio, M.A.; de Carvalho, J.E.; Andrade-Silva, M.; Kassuya, C.A.L. Composition and evaluation of the anti-inflammatory and anticancer activities of the essential oil from *Annona sylvatica* A. St.-Hil. *J. Med. Food,* **2013**, *16*(1), 20-25.
[http://dx.doi.org/10.1089/jmf.2011.0303] [PMID: 23297712]

[21]   Bomfim, L.M.; Menezes, L.R.A.; Rodrigues, A.C.B.C.; Dias, R.B.; Gurgel Rocha, C.A.; Soares, M.B.P.; Neto, A.F.S.; Nascimento, M.P.; Campos, A.F.; Silva, L.C.R.C.; Costa, E.V.; Bezerra, D.P. Antitumour Activity of the Microencapsulation of *Annona vepretorum* Essential Oil. *Basic Clin. Pharmacol. Toxicol.,* **2016**, *118*(3), 208-213.
[http://dx.doi.org/10.1111/bcpt.12488] [PMID: 26348780]

[22]   Diniz, T.C.; de Oliveira Júnior, R.G.; Miranda Bezerra Medeiros, M.A.; Gama e Silva, M.; de Andrade Teles, R.B.; dos Passos Menezes, P.; de Sousa, B.M.H.; Abrahão Frank, L.; de Souza Araújo, A.A.; Russo Serafini, M.; Stanisçuaski Guterres, S.; Pereira Nunes, C.E.; Salvador, M.J.; da Silva Almeida, J.R.G. Anticonvulsant, sedative, anxiolytic and antidepressant activities of the essential oil of *Annona vepretorum* in mice: Involvement of GABAergic and serotonergic systems. *Biomed. Pharmacother.,* **2019**, *111*, 1074-1087.
[http://dx.doi.org/10.1016/j.biopha.2018.12.114] [PMID: 30841421]

[23]    Alcântara, J.M.; de Lucena, J.M.V.M.; Facanali, R.; Marques, M.O.M.; da Paz Lima, M. Chemical composition and bactericidal activity of the essential oils of four species of Annonaceae growing in brazilian amazon. *Nat. Prod. Commun.,* **2017**, *12*(4)1934578X1701200
[http://dx.doi.org/10.1177/1934578X1701200437] [PMID: 30520609]

[24]    Bay, M.; Souza de Oliveira, J.V.; Sales Junior, P.A.; Fonseca Murta, S.M.; Rogério dos Santos, A.; Santos Bastos, I.; Puccinelli Orlandi, P.; Teixeira de Sousa Junior, P. *In vitro* Trypanocidal and Antibacterial Activities of Essential Oils from Four Species of the Family Annonaceae. *Chem. Biodivers.,* **2019**, *16*(11)e1900359
[http://dx.doi.org/10.1002/cbdv.201900359] [PMID: 31544347]

[25]    Soares, E.R.; da Silva, F.M.A.; de Almeida, R.A.; de Lima, B.R.; Koolen, H.H.F.; Lourenço, C.C.; Salvador, M.J.; Flach, A.; da Costa, L.A.M.A.; de Souza, A.Q.L.; Pinheiro, M.L.B.; de Souza, A.D.L. Chemical composition and antimicrobial evaluation of the essential oils of *Bocageopsis pleiosperma* Maas. *Nat. Prod. Res.,* **2015**, *29*(13), 1285-1288.
[http://dx.doi.org/10.1080/14786419.2014.996148] [PMID: 25562370]

[26]    Xavier, M.N.; Alves, C.C.F.; Cazal, C.M.; Santos, N.H. Chemical composition of the volatile oil of *Cardiopetalum calophyllum* collected in the Cerrado area. *Cienc. Rural,* **2016**, *46*(5), 937-942.
[http://dx.doi.org/10.1590/0103-8478cr20150371]

[27]    Salleh, W.M.N.H.W.; Khamis, S.; Nadri, M.H. Chemical composition and acetylcholinesterase inhibition of the essential oil of *Cyathocalyx pruniferus* (Maingay ex Hook.f. & Thomson). *J. Sinclair. Nat.Vol. Essent.Oils,* **2020**, *7*, 8-13.

[28]    Saldanha, A.A.; Vieira, L.; Ribeiro, R.I.M.A.; Thomé, R.G.; Santos, H.B.; Silva, D.B.; Carollo, C.A.; Oliveira, F.M.; Lopes, D.O.; Siqueira, J.M.; Soares, A.C. Chemical composition and evaluation of the anti-inflammatory and antinociceptive activities of Duguetia furfuracea essential oil: Effect on edema, leukocyte recruitment, tumor necrosis factor alpha production, iNOS expression, and adenosinergic and opioidergic systems. *J. Ethnopharmacol.,* **2019**, *231*, 325-336.
[http://dx.doi.org/10.1016/j.jep.2018.11.017] [PMID: 30445104]

[29]    Orlando, V.S.; Glauciemar, D.V.V.; Bruna, C.S.S.; C eacute lia, H.Y.; Ana, S.A.; Ailson, L.A.A.; Miriam, A.O.P.; Mirian, P.R.; Maria, S.A. *In- vivo* and *vitro* bioactivities of the essential oil of Duguetia lanceolata branches. *Afr. J. Pharm. Pharmacol.,* **2016**, *10*(14), 298-310.
[http://dx.doi.org/10.5897/AJPP2015.4497]

[30]    Ribeiro, L.P.; Domingues, V.C.; Gonçalves, G.L.P.; Fernandes, J.B.; Glória, E.M.; Vendramim, J.D. Essential oil from *Duguetia lanceolata* St.-Hil. (Annonaceae): Suppression of spoilers of stored-grain. *Food Biosci.,* **2020**, *36*100653
[http://dx.doi.org/10.1016/j.fbio.2020.100653]

[31]    Rodrigues, A.; Bomfim, L.; Neves, S.; Menezes, L.; Dias, R.; Soares, M.; Prata, A.; Rocha, C.; Costa, E.; Bezerra, D. Antitumor Properties of the Essential Oil From the Leaves of *Duguetia gardneriana*. *Planta Med.,* **2015**, *81*(10), 798-803.
[http://dx.doi.org/10.1055/s-0035-1546130] [PMID: 26125546]

[32]    Santos, A.R.; Benghi, T.G.S.; Nepel, A.; Marques, F.A.; Lobão, A.Q.; Duarte, M.C.T.; Ruiz, A.L.T.G.; Carvalho, J.E.; Maia, B.H.L.N.S. *In vitro* Antiproliferative and Antibacterial Activities of Essential Oils from Four Species of *Guatteria*. *Chem. Biodivers.,* **2017**, *14*(10)e1700097
[http://dx.doi.org/10.1002/cbdv.201700097] [PMID: 28719026]

[33]    Fernandes, C.C.; Rezende, J.L.; Silva, E.A.J.; Silva, F.G.; Stenico, L.; Crotti, A.E.M.; Esperandim, V.R.; Santiago, M.B.; Martins, C.H.G.; Miranda, M.L.D. Chemical composition and biological activities of essential oil from flowers of Psidium guajava (Myrtaceae). *Braz. J. Biol.,* **2021**, *81*(3), 728-736.
[http://dx.doi.org/10.1590/1519-6984.230533] [PMID: 32876175]

[34]    Meira, C.S.; Menezes, L.R.A.; dos Santos, T.B.; Macedo, T.S.; Fontes, J.E.N.; Costa, E.V.; Pinheiro, M.L.B.; da Silva, T.B.; Teixeira Guimarães, E.; Soares, M.B.P. Chemical composition and

antiparasitic activity of essential oils from leaves of *Guatteria friesiana* and *Guatteria pogonopus* (Annonaceae). *J. Essent. Oil Res.,* **2017**, *29*(2), 156-162.
[http://dx.doi.org/10.1080/10412905.2016.1210041]

[35]    Costa, R.G.A.; Anunciação, T.A.; Araujo, M.S.; Souza, C.A.; Dias, R.B.; Sales, C.B.S.; Rocha, C.A.G.; Soares, M.B.P.; Silva, F.M.A.; Koolen, H.H.F.; Costa, E.V.; Bezerra, D.P. *In vitro* and *in vivo* growth inhibition of human acute promyelocytic leukemia HL-60 cells by *Guatteria megalophylla* Diels (Annonaceae) leaf essential oil. *Biomed. Pharmacother.,* **2020**, *122*109713
[http://dx.doi.org/10.1016/j.biopha.2019.109713] [PMID: 31918282]

[36]    do N Fontes, J.E.; Ferraz, R.P.; Britto, A.C.S.; Carvalho, A.A.; Moraes, M.O.; Pessoa, C.; Costa, E.V.; Bezerra, D.P. Antitumor effect of the essential oil from leaves of *Guatteria pogonopus* (Annonaceae). *Chem. Biodivers.,* **2013**, *10*(4), 722-729.
[http://dx.doi.org/10.1002/cbdv.201200304] [PMID: 23576358]

[37]    Kambiré, D.A.; Boti, J.B.; Ouattara, Z.A.; Yapi, T.A.; Bighelli, A.; Tomi, F.; Casanova, J. Leaf essential oil from Ivorian *Isolona dewevrei* (Annonaceae): Chemical composition and structure elucidation of four new natural sesquiterpenes. *Flavour Fragrance J.,* **2021**, *36*(1), 22-33.
[http://dx.doi.org/10.1002/ffj.3612]

[38]    Kambiré, D.A.; Boti, J.B.; Yapi, T.A.; Ouattara, Z.A.; Bighelli, A.; Casanova, J.; Tomi, F. New Natural Oxygenated Sesquiterpenes and Chemical Composition of Leaf Essential Oil from Ivoirian *Isolona dewevrei* (De Wild. & T. Durand) Engl. & Diels. *Molecules,* **2020**, *25*(23), 5613.
[http://dx.doi.org/10.3390/molecules25235613] [PMID: 33260296]

[39]    de Lima, B.R.; da Silva, F.M.A.; Soares, E.R.; de Almeida, R.A.; da Silva Filho, F.A.; Pereira Junior, R.C.; Hernandez Tasco, Á.J.; Salvador, M.J.; Koolen, H.H.F.; de Souza, A.D.L.; Pinheiro, M.L.B. Chemical composition and antimicrobial activity of the essential oils of *Onychopetalum amazonicum* R.E.Fr. *Nat. Prod. Res.,* **2016**, *30*(20), 2356-2359.
[http://dx.doi.org/10.1080/14786419.2016.1163691] [PMID: 27033169]

[40]    De Lima, B.R.; Da Silva, F.M.A.; Soares, E.R.; De Almeida, R.A.; Maciel, J.B.; Fernandes, C.C.; De Oliveira, A.C.; Tadei, W.P.; Koolen, H.H.F.; De Souza, A.D.L.; Pinheiro, M.L.B. Chemical composition and larvicidal activity of the essential oil from the leaves of *Onychopetalum periquino* (Rusby). *D.M. Johnson & N.A. Murray. Nat. Prod. Res.,* **2019**, *0*, 1-4.
[PMID: 31135221]

[41]    Sherin, A.R.; Kukku, A.K.; Leela, N.K. Monoterpenes rich essential oils from the leaves of *Polyalthia korintii* (DUNAL) BENTH.& HOOK.F. (Annonaceae) from Kerala. *Asian J. Pharm. Health Sci.,* **2019**, *8*, 2019-2023.

[42]    da Silva, E.; Soares, M.; Mariane, B.; Vallim, M.; Pascon, R.; Sartorelli, P.; Lago, J. The seasonal variation of the chemical composition of essential oils from Porcelia macrocarpa R.E. Fries (Annonaceae) and their antimicrobial activity. *Molecules,* **2013**, *18*(11), 13574-13587.
[http://dx.doi.org/10.3390/molecules181113574] [PMID: 24189296]

[43]    Salleh, W.M.N.H.W.; Khamis, S.; Nafiah, M.A.; Abed, S.A. Chemical composition and anticholinesterase inhibitory activity of the essential oil of *Pseuduvaria macrophylla* (Oliv.) Merr. from Malaysia. *Nat. Prod. Res.,* **2019**, •••, 1-6.
[PMID: 31293176]

[44]    Yoshida, N.C.; Saffran, F.P.; Lima, W.G.; Freire, T.V.; de Siqueira, J.M.; Garcez, W.S. Chemical characterization and bioherbicidal potential of the essential oil from the leaves of *Unonopsis guatterioides* (A.DC.) R.E.Fr. (Annonaceae). *Nat. Prod. Res.,* **2019**, *33*(22), 3312-3316.
[http://dx.doi.org/10.1080/14786419.2018.1472595] [PMID: 29741113]

[45]    Thang, T.; Hoang, L.; Tuan, N.; Dai, D.; Ogunwande, I.; Hung, N. Analysis of the Leaf Essential Oils of *Uvaria grandiflora* Roxb. ex Hornem. and *Uvaria microcarpa* Champ. ex Benth. (Annonaceae) from Vietnam. *J. Essent. Oil-Bear. Plants,* **2017**, *20*(2), 496-501.
[http://dx.doi.org/10.1080/0972060X.2017.1321503]

[46]    Abu, T.; Rex-Ogbuku, E.; Idibiye, K. A review: secondary metabolites of *Uvaria chamae* p. Beauv. (Annonaceae) and their biological activities. *International Journal of Agriculture, Environment and Food Sciences,* **2018**, *2*(4), 177-185.
[http://dx.doi.org/10.31015/jaefs.18031]

[47]    Sokamte Tegang, A.; Ntsamo Beumo, T.M.; Jazet Dongmo, P.M.; Tatsadjieu Ngoune, L. Essential oil of *Xylopia aethiopica* from Cameroon: Chemical composition, antiradical and in *vitro* antifungal activity against some mycotoxigenic fungi. *J. King Saud Univ. Sci.,* **2018**, *30*(4), 466-471.
[http://dx.doi.org/10.1016/j.jksus.2017.09.011]

[48]    Adefegha, S.A.; Oboh, G.; Odubanjo, T.; Ogunsuyi, O.B. A comparative study on the antioxidative activities, anticholinesterase properties and essential oil composition of Clove (*Syzygium aromaticum*) bud and Ethiopian pepper (*Xylopia aethiopica*). *Riv. Ital. Sostanze Grasse,* **2015**, *92*, 257-268.

[49]    Thiam, A.; Guèye, M.T.; Ndiyae, I.; Diop, S.M.; Ndiaye, E.H.B.; Fauconnier, M-L.; Lognay, G. Effect of drying methods on the chemical composition of essential oils of *Xylopia aethiopicafruits* (Dunal) A. Richard (Annonaceae) from southern Senegal. *Am. J. Essent. Oils Nat. Prod.,* **2018**, *6*, 25-30.

[50]    Sulaimon, L.; Adisa, R.; Obuotor, E.; Lawal, M.; Moshood, A.; Muhammad, N. Chemical composition, antioxidant, and anticholine esterase activities of essential oil of *xylopia aethiopica* seeds. *Pharmacognosy Res.,* **2020**, *12*(2), 112-118.
[http://dx.doi.org/10.4103/pr.pr_47_19]

[51]    Nascimento, M.N.G.; Junqueira, J.G.M.; Terezan, A.P.; Severino, R.P.; Silva, T.S.; Martins, C.H.G.; Severino, V.G.P. Chemical composition and antimicrobial activity of essential oils from *Xylopia aromatica* (Annonaceae) flowers and leaves. Rev. *Virtual de Química,* **2018**, *10*, 1578-1590.
[http://dx.doi.org/10.21577/1984-6835.20180107]

[52]    Nascimento, A.M.D.; Maia, T.D.S.; Soares, T.E.S.; Menezes, L.R.A.; Scher, R.; Costa, E.V.; Cavalcanti, S.C.H.; La Corte, R. Repellency and Larvicidal Activity of Essential oils from *Xylopia laevigata, Xylopia frutescens, Lippia pedunculosa*, and Their Individual Compounds against *Aedes aegypti* Linnaeus. *Neotrop. Entomol.,* **2017**, *46*(2), 223-230.
[http://dx.doi.org/10.1007/s13744-016-0457-z] [PMID: 27844468]

[53]    Pedrali, A.; Robustelli della Cuna, F.S.; Grisoli, P.; Corti, M.; Brusotti, G. Chemical Composition and Antimicrobial Activity of the Essential Oil From the Bark of *Xylopia hypolampra. Nat. Prod. Commun.,* **2019**, *14*(6)1934578X1985702
[http://dx.doi.org/10.1177/1934578X19857022]

[54]    Pereira, T.S.; Machado Esquissato, G.N.; Costa, E.V.; Nogueira, P.C. de L.; Castro-Prado, M. A. A. Mutagenic and cytostatic activities of the *Xylopia laevigata* essential oil in human lymphocytes. *Nat. Prod. Res.,* **2019**, *14*, 1-4.

[55]    Costa, E.V.; Da Silva, T.B.; Costa, C.O.D.S.; Soares, M.B.P.; Bezerra, D.P. Chemical composition of the essential oil from the fresh fruits of *Xylopia laevigata* and its cytotoxic evaluation. *Nat. Prod. Commun.,* **2016**, *11*(3)1934578X1601100
[http://dx.doi.org/10.1177/1934578X1601100324] [PMID: 27169195]

[56]    Moura, A.P.G.; Beltrão, D.M.; Pita, J.C.L.R.; Xavier, A.L.; Brito, M.T.; Sousa, T.K.G.; Batista, L.M.; Carvalho, J.E.; Ruiz, A.L.T.G.; Della Torre, A.; Duarte, M.C.; Tavares, J.F.; da Silva, M.S.; Sobral, M.V. Essential oil from fruit of *Xylopia langsdorffiana* : antitumour activity and toxicity. *Pharm. Biol.,* **2016**, *54*(12), 3093-3102.
[http://dx.doi.org/10.1080/13880209.2016.1211154] [PMID: 27558915]

[57]    Mendes, R.F.; Pinto, N.C.C.; da Silva, J.M.; da Silva, J.B.; Hermisdorf, R.C.S.; Fabri, R.L.; Chedier, L.M.; Scio, E. The essential oil from the fruits of the Brazilian spice *Xylopia sericea* A. St.-Hil. presents expressive in-*vitro* antibacterial and antioxidant activity. *J. Pharm. Pharmacol.,* **2017**, *69*(3), 341-348.
[http://dx.doi.org/10.1111/jphp.12698] [PMID: 28134988]

[58]    Alves, C.C.F.; Oliveira, J.D.; Estevam, E.B.B.; Xavier, M.N.; Nicolella, H.D.; Furtado, R.A.; Tavares,

D.C.; Miranda, M.L.D. Antiproliferative activity of essential oils from three plants of the Brazilian Cerrado: Campomanesia adamantium (Myrtaceae), Protium ovatum (Burseraceae) and Cardiopetalum calophyllum (Annonaceae). *Braz. J. Biol.,* **2020**, *80*(2), 290-294.
[http://dx.doi.org/10.1590/1519-6984.192643] [PMID: 31017239]

[59]     Yoo, K.Y.; Park, S.Y. Terpenoids as potential anti-Alzheimer's disease therapeutics. *Molecules,* **2012**, *17*(3), 3524-3538.
[http://dx.doi.org/10.3390/molecules17033524] [PMID: 22430119]

[60]     Liu, T.T.; Chao, L.K.P.; Peng, C.W.; Yang, T.S. Effects of processing methods on composition and functionality of volatile components isolated from immature fruits of atemoya. *Food Chem.,* **2016**, *202*, 176-183.
[http://dx.doi.org/10.1016/j.foodchem.2016.01.111] [PMID: 26920282]

[61]     Siqueira, C.A.T.; Oliani, J.; Sartoratto, A.; Queiroga, C.L.; Moreno, P.R.H.; Reimão, J.Q.; Tempone, A.G.; Fischer, D.C.H. Chemical constituents of the volatile oil from leaves of *Annona coriacea* and *in vitro* antiprotozoal activity. *Rev. Bras. Farmacogn.,* **2011**, *21*(1), 0.
[http://dx.doi.org/10.1590/S0102-695X2011005000004]

[62]     Chen, Y.Y.; Peng, C.X.; Hu, Y.; Bu, C.; Guo, S.C.; Li, X.; Chen, Y.; Chen, J.W. Studies on chemical constituents and anti-hepatoma effects of essential oil from *Annona squamosa* L. pericarps. *Nat. Prod. Res.,* **2017**, *31*(11), 1305-1308.
[http://dx.doi.org/10.1080/14786419.2016.1233411] [PMID: 27687754]

[63]     Maia, D.S.; Lopes, C.F.; Saldanha, A.A.; Silva, N.L.; Sartori, Â.L.B.; Carollo, C.A.; Sobral, M.G.; Alves, S.N.; Silva, D.B.; de Siqueira, J.M. Larvicidal effect from different Annonaceae species on *Culex quinquefasciatus. Environ. Sci. Pollut. Res. Int.,* **2020**, *27*(29), 36983-36993.
[http://dx.doi.org/10.1007/s11356-020-08997-6] [PMID: 32577964]

[64]     Lidani, K.C.F.; Andrade, F.A.; Bavia, L.; Damasceno, F.S.; Beltrame, M.H.; Messias-Reason, I.J.; Sandri, T.L. Chagas Disease: From Discovery to a Worldwide Health Problem. *Front. Public Health,* **2019**, *7*, 166.
[http://dx.doi.org/10.3389/fpubh.2019.00166] [PMID: 31312626]

[65]     Gontijo, D.C.; Nascimento, M.F.A.; Brandão, G.C.; Oliveira, A.B. Phytochemistry and antiplasmodial activity of *Xylopia sericea* leaves. *Nat. Prod. Res.,* **2020**, *34*(24), 3526-3530.
[http://dx.doi.org/10.1080/14786419.2019.1577838] [PMID: 30810362]

[66]     Chandra, P.; Sharma, R.K.; Arora, D.S. Antioxidant compounds from microbial sources: A review. *Food Res. Int.,* **2019**, *129*.
[PMID: 32036890]

[67]     Nascimento, L. D.; Moraes, A. A. B.; Costa, K. S.; Galúcio, J. M. P.; Taube, P. S.; Costa, C. M. L.; Cruz, J. N. Andrade. E. H. A.; Faria, L. J. G. Bioactive Natural Compounds and Antioxidant Activity of Essential Oils from Spice Plants: New Findings and Potential Applications.

# CHAPTER 6

# Repellent Potential of Terpenoids Against Ticks

**Tássia L. Vale[1], Isabella C. Sousa[1], Caio P. Tavares[1], Matheus N. Gomes[1], Geovane F. Silva[1], Jhone R. S. Costa[1], Aldilene da Silva Lima, Claudia Q. Rocha** and **Livio Martins Costa-Júnior[1]**

[1] *Departamento de Patologia, Universidade Federal do Maranhão (UFMA), Brazil*

[2] *Departamento de Química, Universidade Federal do Maranhão (UFMA), Brazil*

**Abstract:** Substances used as repellents to avoid contact with ticks and tickborne disease are essential to control. Several compounds have been developed throughout human history to promote repellent activity, and in the last decades, synthetic repellents have been widely used. However, several humans, animal, and environmental health problems have been related to synthetic compounds. The use of natural molecules with low toxicity becomes an alternative to replace these compounds. The natural terpenoids from secondary plant metabolites are an essential group with repellency activity on different arthropods. This chapter addresses the primary terpenes with repellency activity, briefly identifying the effectiveness of tick repellents, test methodology, primary terpenes tested, and activity. The evaluated compound showed good repellent activity on different tick species and stages. However, through this chapter, we show the variations in the techniques used to evaluate the bioprospection of terpenes with possible repellent activity and a lack of *in vivo* repellency studies with terpenes. Finally, we emphasize the repellent activity of terpenes to encourage the use of natural compounds as a strategy to control ticks.

**Keywords:** Animals, Control, Natural Product, Repellent, Tick.

## INTRODUCTION

Repellent compounds are volatile chemicals that cause the arthropod to disorient its movements, removing it and thus preventing infestation or attack on the host (Fig. **1**) [1, 2]. Chemical repellents like DEET, IR3535, DEPA, Icaridin (picaridina) and Permethrin (synthetic pyrethroid) have been the most widely used repellents for repelling arthropods, such as insects and ticks [3, 4], with vehicle formulations in the form of a spray, lotion, and gel and can be applied to clothing or skin [5].

---

[*] **Corresponding author Livio Martins Costa-Júnior:** Departamento de Patologia, Universidade Federal do Maranhão, Brazil; E-mail: livioslz@yahoo.com

**Mozaniel Santana de Oliveira & Antônio Pedro da Silva Souza Filho (Eds.)**

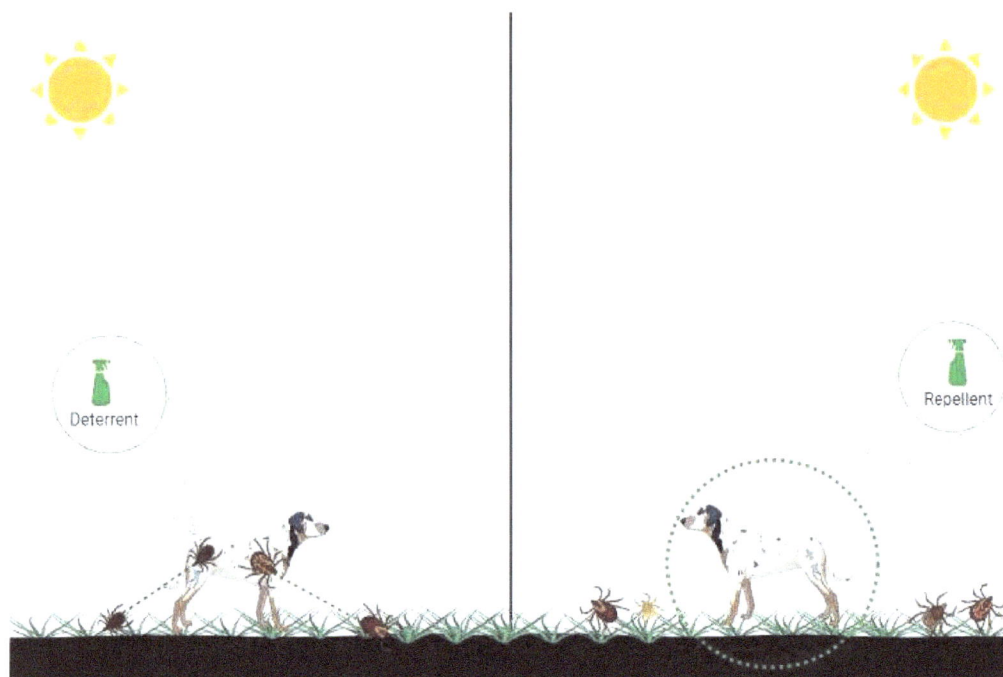

**Fig. (1).** Deterrent and repellent effect on dogs' ticks.

The repellent products were developed to promote the personal protection of humans against diseases transmitted by insects, such as malaria, dengue, zika, yellow fever, and chikungunya [5]. For many years DEET has been the most used and effective synthetic repellent for this activity, besides being an active compound for many commercial repellents. However, some reported toxicity cases can affect adults and children and may cause environmental and animal health risks [6 - 8]. In this context, growing research for safer, natural, available, and more effective methods for the control and repellency of arthropods parasites [9]. The use of natural products (NP) with repellent effects against arthropods has been promising. Repellent formulations containing citronella, lemon, and eucalyptus essential oil has registered as an insect repellent by US Environmental Protection Agency (US EPA) [10].

Plants have for centuries provided a variety of molecules that have a repellent effect against arthropods, with a large number of descriptions of natural repellents in the literature. NP that has a repellent effect are chemicals produced by the secondary metabolism of plants as a defense mechanism against predatory insects. This repellent action is mainly based on the production of terpenoids (isoprenoids), such as monoterpenes, sesquiterpenes, and phenols (Fig. **2**) [11]. However, classes such as alkaloids, quinones, nitrile, furanes, and lactones have also been described to perform this action [12]. Terpenes represent a diverse chemical group that is part of the secondary metabolism of plants, derivatives of hemiterpene units ($C_5$) and classified according to the number of carbons found in their chemical skeleton, which can range from monoterpene ($C_{10}$), sesquiterpenes ($C_{15}$) to polyterpenes ($>C_{40}$) units [13]. The acyclic and bicyclic isomeric skeletons and functional groups give terpenes the ability to form a wide variety of molecules [14].

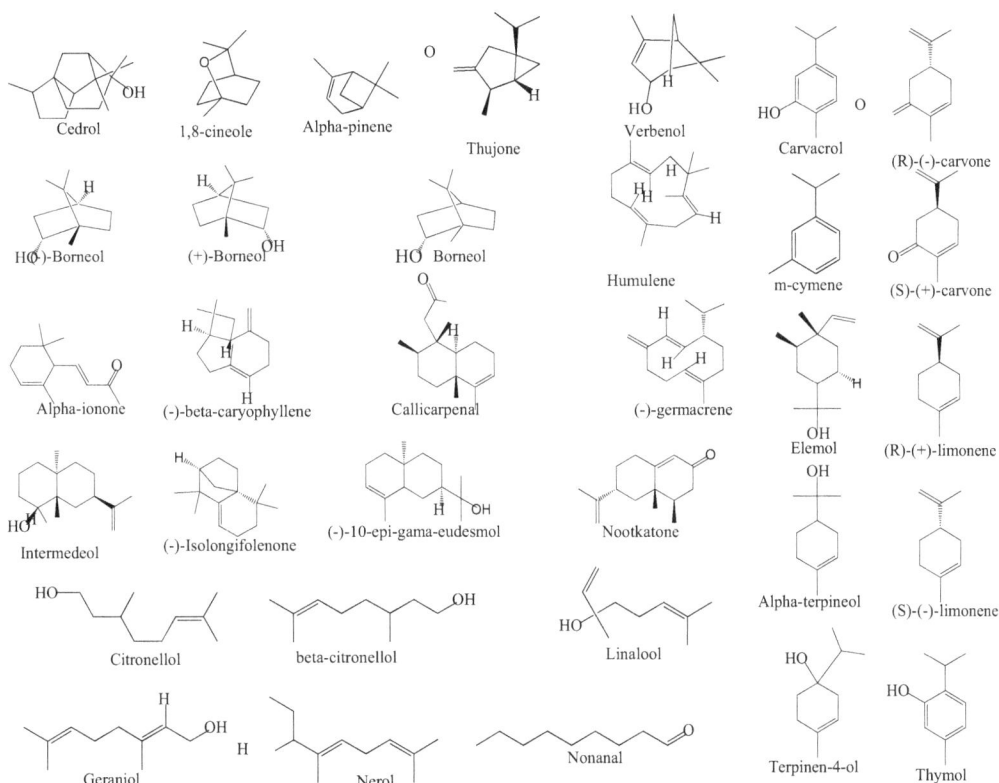

**Fig. (2).** Terpenoids tested as tick's repellent.

A variety of terpenes have high volatility and lipophilic characteristics, giving them the ability to penetrate the membrane. They are generally colorless and have aromatic odors [15, 16]. Many terpenes are used extensively in the perfumery, cosmetics, and food industries. These compounds show different biological activities; among them, acaricide and repellent against several species of arthropods and have as other functions pollination attractants, herbivore deterrents, antibacterial, anti-inflammatory, allelopathic toxic, antioxidants, thermotolerance, and photoprotection [17 - 20].

The repellent effect is induced by different isoprenoid metabolites, mainly acyclic, monocyclic, bicyclic monoterpenoid, diterpenoid, and sesquiterpenoid. Published data suggests that the repellent effect against arthropods is correlated with oxygenated components, with the occurrence of a hydroxyl group (OH) linked to a primary, secondary or aromatic carbon. However, depending on which carbon the hydroxyl group binds to, the repellent activity can be modulated [21]. These classes can be highly efficient in spatial repellency. However, the monoterpenoid is more volatile when compared to the sesquiterpenoids. Sesquiterpenoids have a higher molecular weight, with a 15-carbon structure, and thus their volatilization is lower, thus promoting a longer duration of their repellent efficiency [12]. This difference in repellency between the terpenoid classes is directly related to the terpenes' ability to interfere with the vectors' odorous receptors, terpene volatility, molecular weight, polarity, and the intermolecular forces among the molecules of the repellent compounds [22].

## Terpenoids Repellent Against Ticks

Arthropods, such as ticks, are responsible for vector-borne diseases to human and veterinary health. These diseases are limiting factors for animal production in tropical and subtropical regions of the world. In humans, they can cause severe toxic conditions and are a neglected public health problem in many parts of the world. Ticks are essential vectors, and tick bites can cause paralyzes, toxicoses, skin irritation, allergy, and secondary infections [23 - 25]. Tick control and disease prevention are mainly dependent on the use of chemical acaricides. Synthetic or natural repellents represent a viable form of protection against tick attacks. Several phytochemical defensives have been heavily exploited in the last decades for repellency against ticks.

Different terpenoids have been described to have repellent effects on tick species, about 33 terpenes (Fig. 2) have been reported to have this activity (Table 1). The studies already conducted show that terpenes have been tested on different ticks, such as *Hyalomma* sp., *Ixodes ricinus*, *Ixodes scapularis*, *Amblyomma americanum*, *Amblyomma sculptum*, *Riphicephalus annulatus*, *Riphicephalus, appendiculatus*, *Riphicephalus sanguineus*, and *Riphicephalus microplus*, and showed a repellent effect on the larval, nymph, and adult (Table 1).

Table 1. Terpenoids that show repellency against ticks, concentrations used, percentage of repellency, ticks species and stages repelled.

| Compound | Concentration/Repellency (%) | Species | Stage | References |
|---|---|---|---|---|
| β-Cyclocitral | 0.1 μL=90.0 | *R. appendiculatus* | * | Lwande *et al.*, 1999 |

*(Table 1) cont.....*

| Compound | Concentration/Repellency (%) | Species | Stage | References |
|---|---|---|---|---|
| β-Ocimene | 0.1 μL=77.8 | *R. appendiculatus* | * | Lwande *et al.*, 1999 |
| (+)-Borneol | 1.3 μg.cm$^{-2}$ = 33.3 | *I. ricinus* | N | Schubert *et al.*, 2017 |
| (-)-Borneol | 1.3 μg.cm$^{-2}$ = 26.6 | *I. ricinus* | N | Schubert *et al.*, 2017 |
| Borneol | 0.5%= 64.3 | *I. ricinus* | N | Pålsson *et al.*, 2008 |
| Carvacrol | 0.078% = 50.0 | *I. scapularis* | N | Dietrich *et al.*, 2006 |
| | 1.0 mg (AI). cm$^{-2}$ = 53.8 | *A. americanum* | A | Jordan *et al.*, 2012 |
| | 0.13 mg.cm$^{-2}$ = 50.0 | *R. microplus* | L | Lima *et al.*, 2019 |
| | 5%=100.0 | *I. ricinus* | L | Tabari *et al.*, 2017 |
| | 1.0 mg (AI).cm$^{-2}$ = 100.0 | *I. scapularis* | A | Jordan *et al.*, 2012 |
| beta-citronellol | 3.6% = 100.0 | *A. sculptum* | Un A | Ferreira *et al.*, 2017 |
| | 3.6% = 90.0 | *R. sanguineus* | Un A | Ferreira *et al.*, 2017 |
| (-)-beta-caryophyllene | 10%= 65.5 | *Ixodes ricinus* | N | Ashitani *et al.*, 2015 |
| Citronellol | 0.206 mg.cm$^{-2}$ =100.0 | *A. americanum* | N | Tabanca *et al.*, 2013 |
| Callicarpenal | 155 nmoles.cm$^{-2}$ = 98.0 | *I. scapularis* | N | Carroll *et al.*, 2007 |
| β-Caryophyllene | 0.1 μL= 27.8 | *R. appendiculatus* | * | Lwande *et al.*, 1999 |
| 1,8-Cineol | 0.5%= 70.4 | *I. ricinus* | N | Pålsson et al., 2008 |
| *m*-cymene | 0.1μl = 90.0 | *R. appendiculatus* | * | Lwande *et al.*, 1999 |
| Cedrol | 0.1mL=29.7 | *R. appendiculatus* | * | Ndungu *et al.*, 1999 |
| Cedrene | 0.1 μL= 86.7 | *R. appendiculatus* | * | Lwande *et al.*, 1999 |
| (S)-(+)-carvone | 4.14 mg.cm$^{-2}$ = 50.0 | *R. microplus* | L | Lima *et al.*, 2015 |
| (R)-(−)-carvone | 4.35 mg.cm$^{-2}$ = 50.0 | *R. microplus* | L | Lima *et al.*, 2015 |

*(Table 1) cont.....*

| Compound | Concentration/Repellency (%) | Species | Stage | References |
|---|---|---|---|---|
| *d*-Limonene | 0.1 µL= 27.2 | *R. appendiculatus* | * | Lwande *et al.*, 1999 |
| α-Ionone | 0.1 µL= 90.0 | *R. appendiculatus* | * | Lwande *et al.*, 1999 |
| β-Ionone | 0.1 µL= 48.7 | *R. appendiculatus* | * | Lwande *et al.*, 1999 |
| (−)-10-epi-gama-eudesmol Elemol | 0.103 mg.cm$^{-2}$ = 90.0 | *A. americanum* | N | Tabanca *et al.*, 2013 |
| | 5.157 nmole.cm$^{-2}$ = 50.0 | *I. scapularis* | N | Carroll *et al.*, 2010 |
| | 14783 nmole.cm$^{-2}$ = 50.0 | *A. americanum* | N | Carroll *et al.*, 2010 |
| Geraniol | 1% = 98.4 | *Hyalomma sp.* | * | Khallaayoune *et al.*, 2009 |
| | 0.206 mg.cm$^{-2}$ =90.0 | *A. americanum* | N | Tabanca *et al.*, 2013 |
| | 5% = 100.0 | *A. americanum* | Un, A | Bissinger *et al.*, 2014 |
| | 5% = 100.0. | *A. americanum* | Un, A | Bissinger *et al.*, 2014 |
| | 5% = 100.0 | *I. scapularis* | Un, A, | Bissinger *et al.*, 2014 |
| | 5% = 100.0 | *R. sanguineus* | Un, A | Bissinger *et al.*, 2014 |
| *trans*-Geranyl-acetone | 0.1 µL= 90.0 | *R. appendiculatus* | * | Lwande *et al.*, 1999 |
| *trans*-2-Methyl Cyclopentanol | 0.1 µL= 90.0 | *R. appendiculatus* | * | Lwande *et al.*, 1999 |
| (-)-Germacrene | 10%=70.2 | *I. ricinus* | N | Ashitani *et al.*, 2015 |
| Humulene | 10%= 96.8 | *Ixodes ricinus* | N | Ashitani *et al.*, 2015 |
| Intermedeol | 155mol. cm$^{-2}$ = 96.0 | *I.scapularis* | N | Carroll *et al.*, 2007 |
| (-)-Isolongifolenone | 78 nmol.cm$^{-2}$ =100.0 | *I. scapularis* | N | Zhang *et al.*, 2009 |
| Alpha-Ionone | 0.1µl = 90.0 | *R. appendiculatus* | * | Lwande *et al.*, 1999 |
| Linalool | 5%=50.24 | *I.ricinus* | L | Tabari *et al.*, 2017 |

*(Table 1) cont.....*

| Compound | Concentration/Repellency (%) | Species | Stage | References |
|---|---|---|---|---|
| (S)-(−)-limonene | 6.56 mg.cm$^{-2}$ = 50.0 | *R. microplus* | L | Lima *et al.*, 2015 |
| (R)-(+)-limonene | 7.81 mg.cm$^{-2}$ = 50.0 | *R. microplus* | L | Lima *et al.*, 2015 |
| Nerolidol | 0.1µl = 100 | *R. appendiculatus* | * | Lwande *et al.*, 1999 |
| Nootkatone | 0.089% = 50.0 | *I. scapularis* | N | Dietrich *et al.*, 2006 |
| | 1.0 mg (AI).cm$^{-2}$ = 100.0 | *I. scapularis* | A | Jordan *et al.*, 2012 |
| | 1.0 mg (AI).cm$^{-2}$ = 100.0 | *A. americanum* | A | Jordan *et al.*, 2012 |
| Nonanal | 0.1µl = 90.0 | *R. appendiculatus* | * | Lwande *et al.*, 1999 |
| Nerol | 0.1µl = 90.0 | *R. appendiculatus* | * | Lwande *et al.*, 1999 |
| Alpha-Pinene | 10%= 83.4 | *I. ricinus* | N | Tunón *et al.*, 2006 |
| 1-α-Terpineol | 0.1 µL= 89.9 | *R. appendiculatus* | * | Lwande *et al.*, 1999 |
| Alpha-terpineol | 65 µg.cm$^{-2}$ = 50.0 | *I. ricinus* | N | Schubert *et al.*, 2017 |
| Phytol | 0.1 mL= 48.4 | *R. appendiculatus* | * | Lwande *et al.*, 1999 |
| *trans*-Phytol | 0.1 mL= 48.4 | *R. appendiculatus* | * | Lwande *et al.*, 1999 |
| Terpinen-4-ol | 65 µg.cm$^{-2}$ = 66.6 | *I. ricinus* | N | Schubert *et al.*, 2017 |
| | 0.909% = 50.0 | *I. scapularis* | N | Dietrich *et al.*, 2006 |
| Thymol | 0.075 mg.cm$^{-2}$ = 66.7 | *A. americanum* | * | Carroll *et al.*, 2017 |
| | 10%= 85.7 | *R. annulatus* | L | Arafa *et al.*, 2020 |
| | 5%=100.0 | *I. ricinus* | L | Tabari *et al.*, 2017 |
| Thujone | 0.5%= 70.0 | *I. ricinus* | N | Pålsson *et al.*, 2008 |
| Verbenol | 0.5%= 68.6 | *I. ricinus* | N | Pålsson *et al.*, 2008 |

L=larvae, N=nymphs, A=adult, Un= unfed, Un A=unfed adult.

Most of the compounds with repellent effect, for example, monoterpenes such as limonene, citronellol, and citronellal, are common secondary metabolites of the essential oils that present arthropod-repellency activity. The repellent effect of terpenoids proves that they can be used as promising alternatives for producing new repellent products based on plant molecules or to assist synthetic products in prolonging their effect [26].

The repellent effect may be correlated to the structures of the terpenes and their interactions with a series of biochemical and physiological events that have been little explained to understand the mechanism of action of natural repellent compounds. Many of the terpenes reported the repellent activity in their structure oxygenated components, linked to a hydroxyl group on carbon 1, 2, or aromatic carbons [27]. Another factor that may contribute to repellency is the volatility of the terpenes, which can interfere with the location of their hosts. Arthropods rely predominantly upon the olfactory perception of volatility during host location [9]. Consequently, the terpenes cause a disorder in the tick's behavior in locating its host, preventing it from attacking. Olfactory sensilla ticks can detect volatile molecules, where it can be suggested that olfaction also is involved in the repellency process [21].

## Bioassays to Evaluate Repellent Compounds Against Ticks

For many decades, repellent development has been studied using different methods considering the biology and the species' behavior. The tick's behavior, mainly the negative geotropism of the majority of species, is used to evaluate the repellency effect [23]. The repellents could be classified into contact and spatial repellents. For contact repellents, the targets must be in touch with the treated surface before being repelled. In contrast, the spatial repellents must function at a distance from the application site, and the targets do not need to be in physical touch with the treated surface.

Among the disadvantages of researches into new tick repellents is the lack of a standardized test method. These studies differ in 1) species, 2) life stages of the ticks used, 3) the duration in which repellency and toxicity are assessed, 4) the quantity and formulations of the active ingredients of crude extracts, fractions, or essential oils, 5) use of solvent and 6) variability of the ticks' behaviors in assays [3, 28]. These variations in test methodologies and test conditions interfere with comparing studies and selecting molecules with possible repellent potential. Therefore, it is understood that the efficiency of all the bioassays will depend on essential factors such as 1) type of material used, 2) physical-chemical characteristics of the molecule used, 3) time of exposure and 4) the number of ticks used in the chosen method.

## Tick Climbing Bioassay

This method is the most used to analyzed tick repellent [29]. The principle is based on the negative geotropism of ticks (Fig. **3**). The tick is added to the base and needs to pass into a treated area with an experimental compound. The substrate used can be filter paper, glass stick, leather, bamboo stick, or other inert material or add the host odor. The number of specimens used depends on the stage and species of the tick.

A variation of the climbing assay is the Fingertip Bioassay, used as an experimental compound to impregnate part of the finger of a human volunteer. The tick is disposable at the base and permits a climb to the finger. The ticks that move up to the treated area are considered non-repelled, while those that retreat or fall off the treated surface are repelled [3, 28].

**Fig. (3).** Negative geotropism in *Rhipicephalus microplus* ticks and fingertip bioassay with *Amblyomma sculptum*.

## Olfactometer Bioassay

Bioassay using olfactometers to test repellent compounds on ticks can be performed. The assay with the attractive stimulus is typically associated with the search behavior of the hosts. The assays that use anemotaxic stimulants such as four-way, arena, or Y-shaped olfactometers, rely on the ectoparasite stimulus to guide themselves according to the prevailing wind plume [30, 31]. The double-choice Y-shaped olfactometer is one of the most used because it allows the insertion of two different odors during the analysis. Although the arena olfactometer is also used for repellency tests, it has a limitation when using only one odor for the test [27, 30].

## Petri Dish Bioassay

The Petri dish bioassay is well-known as the "Choice Assay," where half of the dish is treated and the other controlled (without treatment). After treatments of one side of the evaluated compound and the other with the same amount of solvent, the papers need to dry for ten minutes in a closed chapel to evaporate the solvent. Then, the ticks are placed in the center of the dish in groups of six, three males or three females. The positions found after 5, 10, and 30 minutes of testing are assessed to perform the analysis. Similar procedures could be achieved using watch glass and impregnating a circular filter paper adding in the glass center.

Although the methodology describes a time for the volatilization of the experimental compounds, these methods are not indicated for compounds that present fumigation activity since they do not have air circulation and kill ticks.

## Bioassay of the Falcon Tissue Flask Repellency

Jaenson e Dietrich [32, 33] used this bioassay in non-fed nymphs of *I. ricinus* using a Falcon flask. To perform this bioassay, fifty microliters of each test solution, negative and positive control, are applied to separated cotton cloths and attached with a rubber band on the open upper end (660 m2) of the flask. It is worth mentioning that it is necessary to drill the pipe wall to avoid air saturation with the substances' odors. In each repetition, five nymphs are used per group for five minutes. To simulate the host stimulus, the observer must hold the palm on the outer surface of the cloth during the period. For a tick to be considered attracted to the tissue, it must be separated from the surface of the flask. Ticks that are adhered to the tissue five minutes after the beginning of the test are registered as attracted, while ticks that do not attach can be registered as repelled [34].

## Moving Object Bioassay

This method was developed to increase the motivation of ticks to move while requiring search behavior by the host [35]. For this assay, heat, and movement are used to nurture attractions associated with hosts, allowing ectoparasites to exhibit their natural behavior of holding on to a host in the lab. It is necessary to use a vertical drum slowly rotating heated. In the drum, there is a surface that will be used as a base for the effective attachment of the parasite. A glass stick is then placed horizontally, ending in front of the drum, where the ticks will approach.

The distance between the drum and the glass stick is adjusted so that the tick reaches only the desired attachment location. As the drum rotates, the attachment region passes periodically, and the parasite can attach itself to the moving object, simulating the animal that gives. For the test analysis, are accounted for: ticks that approach/ do not approach the drum (spatial repellency), ticks fixed to the drum, and ticks that remain on the treated surface or fall off (contact repellency).

## Repellent Compound on Ticks of Medical Importance

The species of ticks usually describe as disease vectors to humans are *A. sculptum*, *A. aureolatum*, *A. ovale*, *Dermacentor andersoni*, and *Dermacentor variabilis* [36, 37]. These species are described as vectors of Rickettsia, the etiologic agent of the Rocky Mountain spotted fever (RMSF). Due to being involved in the transmission of numerous pathogens, these ticks create essential social and economic impacts associated with medical costs, loss of productivity, and death [38]. Considering the many damages created by ticks and the hardships found in effectively instituting pest control, a viable alternative to prevent diseases stemming from parasitism from these species is to reduce interaction with its vectors through the use of repellents [5].

The use of human repellents dates thousands of years ago, but it was only after World War II that it reached its apex, where soldiers employed chemical compounds to prevent bites from insects that were responsible for the transmission of some diseases. To prevent more deaths, numerous researches regarding long-lasting repellent compounds were initiated. One of the first repellent compounds to be developed was Dimethyl phthalate (DMP), followed by Indalone and ethyl hexanediol [10]. The main synthetic repellents found in the market are DEET, IR3535, Icaridin (Picaridin), DEPA, and permethrin (synthetic pyrethroid). But knowledge about the effects related to intoxication caused by synthetic repellents, in addition to the dissatisfaction regarding produce and environmental contamination, leads to a search for more ecologically viable alternatives [39].

Facing the issues occasioned by synthetic repellents, NP, namely plant-based products with volatile constituents, can be an essential source of molecules with repellent action. It is already possible to find citronella-based repellents in the market whose active components are citronellal and Citronellol [5]. Currently, the employment of NP extracted from plants has provoked interest around the world and comprises a promising strategy to tick control, especially essential oils and their chemical constituents. Knowing the importance of the compounds, it is then possible to list some Terpenoids repellent action over ticks of medical importance regarding public health, as seen in Table **1**.

## Repellent Compounds on Dogs' Ticks

Several tick species are capable of infecting dogs, such as *R. sanguineus*, *A. americanum*, *Haemaphysalis longicornis*, *I. scapularis*, *D. variabilis*, and *I. ricinus* [40 - 42]. Besides causing blood loss, irritability, and hypersensibility, these ticks are pathogen vectors [11, 22, 43]. Dogs are exposed to tick infestation in various situations, *e.g.*, during walks in parks and forests [44]. To prevent these ticks from transmitting pathogens, it is necessary to institute protection measures against ectoparasites [45]. One action is to avoid contact with ticks by employing parasite-repellent products. Such products are practical tools to reduce the possibility of interaction with ticks and, subsequently, pathogen transmission due to compounds that inhibit the host's attraction [46]. However, the process of parasite-repellent compounds occurs mainly due to interaction with irritant agents, also named deterrents [47] (Fig. **1**).

The main tick-repellent for dogs found in the market are based on synthetic compounds such as Pyrethroids and Phenylpyrazoles in the form of leashes, spray, spot-on, and pour-on [48 - 51]. These compounds are effective and relatively safe. However, there are concerns about their possible adverse effects on human and animal health [52, 53]. Also, there are reports of ticks resistant to the drugs, consequently reducing the acaricidal efficacy [54 - 56] significantly. In light of this issue, the employment of terpenes as repellents has been fully studied (Table **1**). However, studies on the effectiveness of these compounds as repellents have been carried out mainly *in vitro*. Until this work, there is no scientific literature on terpene-based repellents applied directly to dogs [57].

Based on *in vitro* tests, some products using essential oils are already being commercialized as repellents. Since some essential oils, such as clove oil, thyme oil, and cinnamon oil, are listed as "Generally recognized as safe" (GRAS) by Food and Drug Administration (FDA) and exempt from toxicity data requirements from EPA [58, 59]. These products, commercialized as repellents, are composed of essential oil mixtures such as Ultrashield Green (geraniol, citronella, rosemary, lemongrass, cedar e thyme oils), Pyranha Zero-bite (geraniol, thyme, and mint oils), and Flea+Tick Spray repellent (lemongrass, cinnamon, sesame and castor oils) [60]. The list of commercialized products based on natural compounds that produce repellent effects is very extensive [61].

Natural repellents can have a substantial effect against parasites, which could be beneficial in the prevention of tick infestation. Its low toxicity and appeal to be a green product, as they do not act on pollinating insects due to its high volatility, make this type of product increasingly attractive to consumers. When related to pets, such as dogs, due to their proximity to humans [62, 63]. However, it is necessary to note that the low toxicity does not exempt said products from other adverse effects in dogs and humans altogether [64]. Finally, the efficacy of tick repellents is poorly studied, which further hampers the use of new repellents [65] (Fig. **4**).

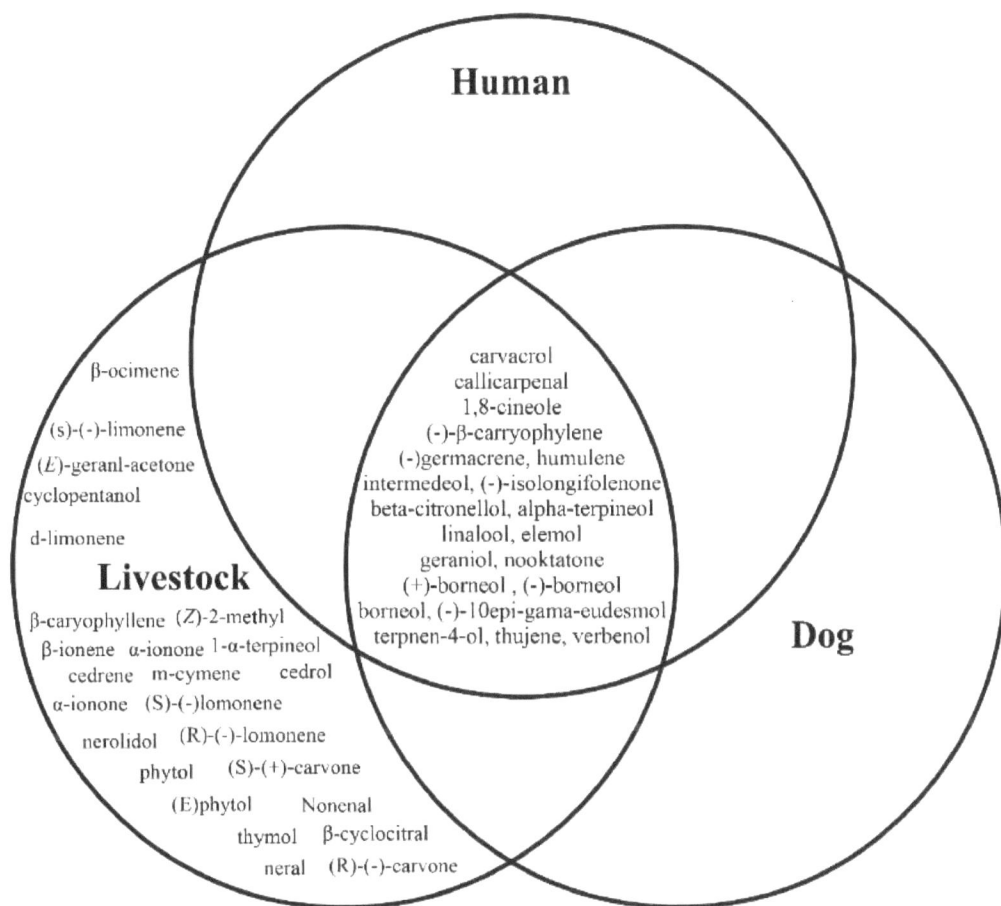

**Human**

β-ocimene

(s)-(-)-limonene

(E)-geranl-acetone

cyclopentanol

d-limonene

**Livestock**

β-caryophyllene (Z)-2-methyl

β-ionene  α-ionone  1-α-terpineol

cedrene  m-cymene  cedrol

α-ionone  (S)-(-)lomonene

nerolidol  (R)-(-)-lomonene

phytol  (S)-(+)-carvone

(E)phytol  Nonenal

thymol  β-cyclocitral

neral  (R)-(-)-carvone

carvacrol
callicarpenal
1,8-cineole
(-)-β-carryophylene
(-)germacrene, humulene
intermedeol, (-)-isolongifolenone
beta-citronellol, alpha-terpineol
linalool, elemol
geraniol, nooktatone
(+)-borneol , (-)-borneol
borneol, (-)-10epi-gama-eudesmol
terpnen-4-ol, thujene, verbenol

**Dog**

**Fig. (4).** Venn diagram of terpenes showed with repellent effect on ticks from human, livestock and dog.

## *Repellent Compounds on Livestock's Ticks*

The primary tick species that infest livestock animals are the southern cattle tick, *R. microplus* [54], One-host cattle fever tick, *R. annulatus* [66] and Brown ear tick *R. appendiculatus* [67] infesting cattle as primary hosts and, *A. sculptum* [68] and *A. cajennense* [69] generalist ticks common in horses. For this reason, these five species of ticks are among the most used in studies for the determination of the repellent activity of terpenoids.

A total of 27 terpenoids were tested as a repellent against ticks parasites of livestock animals (Table **1**). All these compounds are monoterpenes, sesquiterpenes, or diterpenes. The monoterpenes have the most compound with repellent activity (17 substances), followed by sesquiterpenes with eight compounds, and finally, the diterpenes with only two compounds. It is known that the constituents of essential oils are highly volatile. For this reason, researchers carried out the encapsulation of the monoterpene carvacrol with the yeast cell wall to increase the time of repellent activity of this molecule, delaying its volatilization [30].

Another species of tick widespread in horses is *Dermacentor nitens* (horse ear tick). However, there is a significant lack of studies to determine the repellent effect of natural products on this tick. The search in the scientific literature revealed only a single study using eugenol (phenylpropanoid) as a repellent for *D. nitens* larvae [70]. To date, no terpenoids have been evaluated for repellent activity on this tick species

## CONCLUDING REMARKS

Terpenoids have repellent activity against different species of ticks. This repellent activity is due to the various biochemical mechanisms involved, conferring degrees of repellency between the terpenoid compounds. Although there are several synthetic repellents for the control of ticks, terpenoids offer a safe alternative with low toxicity. This makes terpenoids attractive for human, animal, and environmental health. In addition, there is a range of molecules available, which can be essential tools as tick repellents. However, the lack of repellent *in vivo* studies with terpenoid compounds impedes a better understanding of how these compounds work in routine use. These problems directly reflect that few terpenes are commercially available, and their use against ticks is not widely used. Thus, the development of products with these molecules deserves to be further explored as a strategy to control ticks.

## CONSENT FOR PUBLICATION

Not applicable.

## CONFLICT OF INTEREST

The author declares no conflict of interest, financial or otherwise.

## ACKNOWLEDGEMENTS

The authors acknowledge the financial support received from CNPq (Brazilian National Council for Scientific and Technological Development), FAPEMA

(Maranhão State Research Foundation) and FINEP (Funding Authority for Studies and Projects) (PRONEM 01773/14 and IECT (Science and Technology Institute of Maranhão) Biotechnology). This study was financed in part by CAPES, Finance Code 001.

# REFERENCES

[1]     Dethier, V.G.; Browne, B.L.; Smith, C.N. The designation of chemicals in terms of the responses they elicit from insects. *J. Econ. Entomol.,* **1960**, *53*(1), 134-136.
        [http://dx.doi.org/10.1093/jee/53.1.134]

[2]     Ferreira, L.L.; Oliveira Filho, J.G.; Mascarin, G.M.; León, A.A.P.; Borges, L.M.F. *In vitro* repellency of DEET and β-citronellol against the ticks Rhipicephalus sanguineus sensu lato and Amblyomma sculptum. *Vet. Parasitol.,* **2017**, *239*, 42-45.
        [http://dx.doi.org/10.1016/j.vetpar.2017.04.021] [PMID: 28495195]

[3]     Bissinger, B.W.; Roe, R.M. Tick repellents: Past, present, and future. *Pestic. Biochem. Physiol.,* **2010**, *96*(2), 63-79.
        [http://dx.doi.org/10.1016/j.pestbp.2009.09.010]

[4]     Da Silva Lima, A.; De Carvalho, J.F.; Peixoto, M.G.; Blank, A.F.; Borges, L.M.F.; Costa Junior, L.M. Assessment of the repellent effect of *Lippia alba* essential oil and major monoterpenes on the cattle tick *Rhipicephalus microplus. Med. Vet. Entomol.,* **2016**, *30*(1), 73-77.
        [http://dx.doi.org/10.1111/mve.12140] [PMID: 26471008]

[5]     Tavares, M.; da Silva, M.R.M.; de Oliveira de Siqueira, L.B.; Rodrigues, R.A.S.; Bodjolle-d'Almeida, L.; dos Santos, E.P.; Ricci-Júnior, E. Trends in insect repellent formulations: A review. *Int. J. Pharm.,* **2018**, *539*(1-2), 190-209.
        [http://dx.doi.org/10.1016/j.ijpharm.2018.01.046] [PMID: 29410208]

[6]     Pitasawat, B.; Choochote, W.; Tuetun, B.; Tippawangkosol, P.; Kanjanapothi, D.; Jitpakdi, A.; Riyong, D. Repellency of aromatic turmeric Curcuma aromatica under laboratory and field conditions. *Journal of vector ecology: journal of the Society for Vector Ecology, 28*(2), 234-240.**2003,**

[7]     Elmhalli, F.H.; Pålsson, K.; Örberg, J.; Jaenson, T.G.T. Acaricidal effects of Corymbia citriodora oil containing para-menthane-3,8-diol against nymphs of Ixodes ricinus (Acari: Ixodidae). *Exp. Appl. Acarol.,* **2009**, *48*(3), 251-262.
        [http://dx.doi.org/10.1007/s10493-009-9236-4] [PMID: 19169833]

[8]     Pohlit, A. M.; Lopes, N. P.; Gama, R. A.; Tadei, W. P.; Andrade Neto, V. F. D. Patent literature on mosquito repellent inventions which contain plant essential oils-a review. *Volume 77*(6), 598-617.**2011,**

[9]     Frances, S.P.; Wirtz, R.A. Repellents: past, present, and future. *J. Am. Mosq. Control Assoc.,* **2005**, *21*(sp1) Suppl., 1-3.
        [http://dx.doi.org/10.2987/8756-971X(2005)21[1:RPPAF]2.0.CO;2] [PMID: 16921675]

[10]    Moore, S. J.; Debboun, M. History of insect repellents. (Insect repellents: principles, methods and uses), 3-29.**2007,**

[11]    Rehman, J.U.; Ali, A.; Khan, I.A. Plant based products: Use and development as repellents against mosquitoes: A review. *Fitoterapia,* **2014**, *95*, 65-74.
        [http://dx.doi.org/10.1016/j.fitote.2014.03.002] [PMID: 24631763]

[12]    da Silva, M.R.M.; Ricci-Júnior, E. An approach to natural insect repellent formulations: from basic research to technological development. *Acta Trop.,* **2020**, *212*105419
        [http://dx.doi.org/10.1016/j.actatropica.2020.105419] [PMID: 32119826]

[13]    Kiyama, R. Estrogenic terpenes and terpenoids: Pathways, functions and applications. *Eur. J. Pharmacol.,* **2017**, *815*, 405-415.

[http://dx.doi.org/10.1016/j.ejphar.2017.09.049] [PMID: 28970013]

[14] Marmulla, R.; Harder, J. Microbial monoterpene transformationsâ€"a review. *Front. Microbiol.*, **2014**, *5*, 346.
[http://dx.doi.org/10.3389/fmicb.2014.00346] [PMID: 25076942]

[15] Clarke, S., Ed. *Essential Chemistry for Aromatherapy E-Book*; Elsevier Health Sciences, **2009**.

[16] Oz, M.; Lozon, Y.; Sultan, A.; Yang, K.H.S.; Galadari, S. Effects of monoterpenes on ion channels of excitable cells. *Pharmacol. Ther.*, **2015**, *152*, 83-97.
[http://dx.doi.org/10.1016/j.pharmthera.2015.05.006] [PMID: 25956464]

[17] Kozioł, A.; Stryjewska, A.; Librowski, T.; Sałat, K.; Gaweł, M.; Moniczewski, A.; Lochyński, S. An overview of the pharmacological properties and potential applications of natural monoterpenes. *Mini Rev. Med. Chem.*, **2015**, *14*(14), 1156-1168.
[http://dx.doi.org/10.2174/1389557514666141127145820] [PMID: 25429661]

[18] Pateraki, I.; Kanellis, A.K. Stress and developmental responses of terpenoid biosynthetic genes in Cistus creticus subsp. creticus. *Plant Cell Rep.*, **2010**, *29*(6), 629-641.
[http://dx.doi.org/10.1007/s00299-010-0849-1] [PMID: 20364257]

[19] Tholl, D. Biosynthesis and biological functions of terpenoids in plants. *Biotechnology of isoprenoids*, , 63-106.**2015**

[20] Tetali, S.D. Terpenes and isoprenoids: a wealth of compounds for global use. *Planta*, **2019**, *249*(1), 1-8.
[http://dx.doi.org/10.1007/s00425-018-3056-x] [PMID: 30467631]

[21] Nerio, L.S.; Olivero-Verbel, J.; Stashenko, E. Repellent activity of essential oils: A review. *Bioresour. Technol.*, **2010**, *101*(1), 372-378.
[http://dx.doi.org/10.1016/j.biortech.2009.07.048] [PMID: 19729299]

[22] Norris, E.; Coats, J. Current and future repellent technologies: the potential of spatial repellents and their place in mosquito-borne disease control. *Int. J. Environ. Res. Public Health*, **2017**, *14*(2), 124.
[http://dx.doi.org/10.3390/ijerph14020124] [PMID: 28146066]

[23] Soares, S.F. Repelência de extratos de plantas e do deet (n, ndiethyl-m-toluamide) em amblyomma cajennense (acari: ixodidae). *Dissertação de Mestrado. Veterinária da Universidade Federal de Goiás*, **2008**.

[24] Ghosh, S.; Nagar, G. Problem of ticks and tick-borne diseases in India with special emphasis on progress in tick control research: a review. *J. Vector Borne Dis.*, **2014**, *51*(4), 259-270.
[PMID: 25540956]

[25] Solano-Gallego, L.; Sainz, Á.; Roura, X.; Estrada-Peña, A.; Miró, G. A review of canine babesiosis: the European perspective. *Parasit. Vectors*, **2016**, *9*(1), 336.
[http://dx.doi.org/10.1186/s13071-016-1596-0] [PMID: 27289223]

[26] Elmhalli, F. H.; Pålsson, K.; Örberg, J. Jaenson, T. G. Acaricidal effects **2009**.

[27] Pohlit, A. M.; Lopes, N. P.; Gama, R. A.; Tadei, W. P. Andrade Neto, V. F. D.Patent literature on mosquito repellent inventions which contain plant essential oils-a review. **2011**, *77*(6), 598-617.

[28] Adenubi, O.T.; Ahmed, A.S.; Fasina, F.O.; McGaw, L.J.; Eloff, J.N.; Naidoo, V. Pesticidal plants as a possible alternative to synthetic acaricides in tick control: A systematic review and meta-analysis. *Ind. Crops Prod.*, **2018**, *123*, 779-806.
[http://dx.doi.org/10.1016/j.indcrop.2018.06.075]

[29] de Oliveira Filho, J.G.; Ferreira, L.L.; Sarria, A.L.F.; Pickett, J.A.; Birkett, M.A.; Mascarin, G.M.; de León, A.A.P.; Borges, L.M.F. Brown dog tick, Rhipicephalus sanguineus sensu lato, infestation of susceptible dog hosts is reduced by slow release of semiochemicals from a less susceptible host. *Ticks Tick Borne Dis.*, **2017**, *8*(1), 139-145.
[http://dx.doi.org/10.1016/j.ttbdis.2016.10.010] [PMID: 28340941]

[30]    Lima, A.S.; Landulfo, G.A.; Costa-Junior, L.M. Repellent effects ofencapsulated carvacrol on the Rhipicephalus (Boophilus) microplus (Acari: Ixodidae). *J. Med. Entomol.,* **2019**, *56*(3), 881-885.
        [http://dx.doi.org/10.1093/jme/tjy240] [PMID: 30805609]

[31]    Nogueira, J.A.P.; Figueiredo, A.; Duarte, J.L.; de Almeida, F.B.; Santos, M.G.; Nascimento, L.M.; Fernandes, C.P.; Mourão, S.C.; Toscano, J.H.B.; Rocha, L.M.; Chagas, A.C.S. Repellency effect of Pilocarpus spicatus A. St.-Hil essential oil and nanoemulsion against Rhipicephalus microplus larvae. *Exp. Parasitol.,* **2020**, *215*107919
        [http://dx.doi.org/10.1016/j.exppara.2020.107919] [PMID: 32442440]

[32]    Jaenson, T.G.T.; Pålsson, K.; Borg-Karlson, A.K. Evaluation of extracts and oils of tick-repellent plants from Sweden. *Med. Vet. Entomol.,* **2005**, *19*(4), 345-352.
        [http://dx.doi.org/10.1111/j.1365-2915.2005.00578.x] [PMID: 16336298]

[33]    Dietrich, G.; Dolan, M. C.; Peralta-Cruz, J.; Schmidt, J.; Piesman, J.; Eisen, R. J. Repellent activity of fractioned compounds from Chamaecyparis nootkatensis essential oil against nymphal Ixodes scapularis (Acari: Ixodidae). *J. Med. Entomol.,* **2006**, *43*(5), 957-961.

[34]    Ashitani, T.; Garboui, S.S.; Schubert, F.; Vongsombath, C.; Liblikas, I.; Pålsson, K.; Borg-Karlson, A.K. Activity studies of sesquiterpene oxides and sulfides from the plant Hyptis suaveolens (Lamiaceae) and its repellency on Ixodes ricinus (Acari: Ixodidae). *Exp. Appl. Acarol.,* **2015**, *67*(4), 595-606.
        [http://dx.doi.org/10.1007/s10493-015-9965-5] [PMID: 26385208]

[35]    Nwanade, C.F.; Wang, M.; Wang, T.; Yu, Z.; Liu, J. Botanical acaricides and repellents in tick control: current status and future directions. *Exp. Appl. Acarol.,* **2020**, *81*(1), 1-35.
        [http://dx.doi.org/10.1007/s10493-020-00489-z] [PMID: 32291551]

[36]    Szabó, M.P.J.; Pinter, A.; Labruna, M.B. Ecology, biology and distribution of spotted-fever tick vectors in Brazil. *Front. Cell. Infect. Microbiol.,* **2013**, *3*, 27.
        [http://dx.doi.org/10.3389/fcimb.2013.00027] [PMID: 23875178]

[37]    Eisen, R.J.; Kugeler, K.J.; Eisen, L.; Beard, C.B.; Paddock, C.D. Tick-borne zoonoses in the United States: persistent and emerging threats to human health. *ILAR J.,* **2017**, *58*(3), 319-335.
        [http://dx.doi.org/10.1093/ilar/ilx005] [PMID: 28369515]

[38]    Drexler, N. A.; Traeger, M. S.; McQuiston, J. H.; Williams, V.; Hamilton, C. Medical and indirect costs associated with a Rocky Mountain spotted fever epidemic in Arizona, 2002–2011. *Am. J. Trop. Med. Hyg.,* **2015**, *93*(3), 549-551.

[39]    de Siqueira, S.L.; Kruse, M.H. [Agrochemicals and human health: contributions of healthcare professionals]. *Rev. Esc. Enferm. USP,* **2008**, *42*(3), 584-590.
        [PMID: 18856129]

[40]    Dantas-Torres, F. Biology and ecology of the brown dog tick, Rhipicephalus sanguineus. Parasites & vectors, *3*(1), 1-11.**2010**,

[41]    Minigan, J.N.; Hager, H.A.; Peregrine, A.S.; Newman, J.A. Current and potential future distribution of the American dog tick (Dermacentor variabilis, Say) in North America. *Ticks Tick Borne Dis.,* **2018**, *9*(2), 354-362.
        [http://dx.doi.org/10.1016/j.ttbdis.2017.11.012] [PMID: 29275873]

[42]    Porretta, D.; Mastrantonio, V.; Amendolia, S.; Gaiarsa, S.; Epis, S.; Genchi, C. Effects of global changes on the climatic niche of the tick Ixodes ricinus inferred by species distribution modelling. Parasites & vectors, **2013**, *6*(1), 1-8.

[43]    Ferreira Vilela, E. Castro Della Lucia, T. M.Feromónios de insetos (Biología, química e emprego no manejo de pragas. **1987**.

[44]    Jennett, A. L.; Smith, F. D. Tick infestation risk for dogs in a peri-urban park. Parasites &amp. *Vectors,* **2013**, *6*(1), 1-10.

[45]     Dantas-Torres, F.; Ketzis, J.; Mihalca, A.D.; Baneth, G.; Otranto, D.; Tort, G.P.; Watanabe, M.; Linh, B.K.; Inpankaew, T.; Jimenez Castro, P.D.; Borrás, P.; Arumugam, S.; Penzhorn, B.L.; Ybañez, A.P.; Irwin, P.; Traub, R.J. TroCCAP recommendations for the diagnosis, prevention and treatment of parasitic infections in dogs and cats in the tropics. *Vet. Parasitol.,* **2020**, *283*109167
[http://dx.doi.org/10.1016/j.vetpar.2020.109167] [PMID: 32580071]

[46]     Piesman, J.; Eisen, L. Prevention of tick-borne diseases. *Annu. Rev. Entomol.,* **2008**, *53*(1), 323-343.
[http://dx.doi.org/10.1146/annurev.ento.53.103106.093429] [PMID: 17877457]

[47]     Mcmahon, C.; Kröber, T. Ensaios *in vitro* para repelentes e dissuasores para carrapatos: diferentes efeitos dos produtos quando testados com estímulos atrativos ou de detenção. *Med. Vet. Entomol.,* **2003**, *17*(4), 370-378.
[http://dx.doi.org/10.1111/j.1365-2915.2003.00453.x] [PMID: 14651650]

[48]     Blair, J.; Fourie, J. J.; Varloud, M. Efficacy and speed of kill of a topically applied formulation of dinotefuran-permethrin-pyriproxyfen against weekly tick infestations with Rhipicephalus sanguineus (sensu lato) on dogs. *Parasites & vectors,* **2016**, *6*(1), 1-8.

[49]     Cutolo, A. A.; Galvis-Ovallos, F.; de Souza Neves, E.; Silva, F. O.; Chester, S. T. Repellent efficacy of a new combination of fipronil and permethrin against Lutzomyia longipalpis. Parasites & vectors, **2018**, *11*(1), 247.

[50]     de Burgh, S.; Hunter, K.; Jackson, C.; Chambers, M.; Klupiec, C.; Smith, V. Repellency Effect of an Imidacloprid / Flumethrin (Seresto®) Controlled Release Polymer Matrix Collar against the Australian Paralysis Tick (Ixodes holocyclus) in Dogs. *Parasitol. Res.,* **2017**, *116*(S1) Suppl. 1, 145-156.
[http://dx.doi.org/10.1007/s00436-017-5500-4] [PMID: 28717957]

[51]     Dumont, P.; Chester, T. S.; Gale, B.; Soll, M.; Fourie, J. J. Acaricidal efficacy of a new combination of fipronil and permethrin against Ixodes ricinus and Rhipicephalus sanguineus ticks. Parasites & vectors, **2015**, *8*(1), 1-5.

[52]     Abdel-Rahman, A.; Shetty, A.K.; Abou-Donia, M.B. Subchronic dermal application of N,N-diethyl m-toluamide (DEET) and permethrin to adult rats, alone or in combination, causes diffuse neuronal cell death and cytoskeletal abnormalities in the cerebral cortex and the hippocampus, and Purkinje neuron loss in the cerebellum. *Exp. Neurol.,* **2001**, *172*(1), 153-171.
[http://dx.doi.org/10.1006/exnr.2001.7807] [PMID: 11681848]

[53]     Corbel, V.; Stankiewicz, M.; Pennetier, C.; Fournier, D.; Stojan, J.; Girard, E. Evidência para inibição de colinesterases em sistemas nervosos de insetos e mamíferos pelo deet repelente de insetos. *BMC Biol.,* **2009**, *7*(1), 1-11.

[54]     Becker, S.; Webster, A.; Doyle, R.L.; Martins, J.R.; Reck, J.; Klafke, G.M. Resistance to deltamethrin, fipronil and ivermectin in the brown dog tick, Rhipicephalus sanguineus sensu stricto, Latreille (Acari: Ixodidae). *Ticks Tick Borne Dis.,* **2019**, *10*(5), 1046-1050.
[http://dx.doi.org/10.1016/j.ttbdis.2019.05.015] [PMID: 31175029]

[55]     Eiden, A.L.; Kaufman, P.E.; Oi, F.M.; Allan, S.A.; Miller, R.J. Detection of permethrin resistance and fipronil tolerance in Rhipicephalus sanguineus (Acari: Ixodidae) in the United States. *J. Med. Entomol.,* **2015**, *52*(3), 429-436.
[http://dx.doi.org/10.1093/jme/tjv005] [PMID: 26334817]

[56]     Eiden, A.L.; Kaufman, P.E.; Oi, F.M.; Dark, M.J.; Bloomquist, J.R.; Miller, R.J. Determination of metabolic resistance mechanisms in pyrethroid-resistant and fipronil-tolerant brown dog ticks. *Med. Vet. Entomol.,* **2017**, *31*(3), 243-251.
[http://dx.doi.org/10.1111/mve.12240] [PMID: 28639697]

[57]     Soutar, O.; Cohen, F.; Wall, R. Essential oils as tick repellents on clothing. *Exp. Appl. Acarol.,* **2019**, *79*(2), 209-219.
[http://dx.doi.org/10.1007/s10493-019-00422-z] [PMID: 31578646]

[58]     EPA US. http://www.epa.gov/oppbppd1/biopesticides/regtools/25b_list.htm**2004**.

[59]  Jiang, Z.L.; Akhtar, Y.; Zhang, X.; Bradbury, R.; Isman, M.B. Insecticidal and feeding deterrent activities of essential oils in the cabbage looper, Trichoplusia ni (Lepidoptera: Noctuidae). *J. Appl. Entomol.,* **2012**, *136*(3), 191-202.
[http://dx.doi.org/10.1111/j.1439-0418.2010.01587.x]

[60]  Eastep, J. *Efficacy of Natural Repellents in Ticks.,* **2018**.

[61]  George, D.R.; Finn, R.D.; Graham, K.M.; Sparagano, O.A.E. Present and future potential of plant-derived products to control arthropods of veterinary and medical significance. *Parasit. Vectors,* **2014**, *7*(1), 28.
[http://dx.doi.org/10.1186/1756-3305-7-28] [PMID: 24428899]

[62]  Isman, M.B. Botanical insecticides, deterrents, and repellents in modern agriculture and an increasingly regulated world. *Annu. Rev. Entomol.,* **2006**, *51*(1), 45-66.
[http://dx.doi.org/10.1146/annurev.ento.51.110104.151146] [PMID: 16332203]

[63]  Kiss, T.; Cadar, D.; Spînu, M. Tick prevention at a crossroad: New and renewed solutions. *Vet. Parasitol.,* **2012**, *187*(3-4), 357-366.
[http://dx.doi.org/10.1016/j.vetpar.2012.02.010] [PMID: 22424918]

[64]  Trumble, JT Caveat emptor: considerações de segurança para produtos naturais usados no controle de artrópodes. *Entomologista americano, 48*(1), 7-13.**2002**,

[65]  Bissinger, B.W.; Schmidt, J.P.; Owens, J.J.; Mitchell, S.M.; Kennedy, M.K. Activity of the plant-based repellent, TT-4302 against the ticks Amblyomma americanum, Dermacentor variabilis, Ixodes scapularis and Rhipicephalus sanguineus (Acari: Ixodidae). *Exp. Appl. Acarol.,* **2014**, *62*(1), 105-113.
[http://dx.doi.org/10.1007/s10493-013-9719-1] [PMID: 23907554]

[66]  Esteve-Gasent, M.D.; Rodríguez-Vivas, R.I.; Medina, R.F.; Ellis, D.; Schwartz, A.; Cortés Garcia, B.; Hunt, C.; Tietjen, M.; Bonilla, D.; Thomas, D.; Logan, L.L.; Hasel, H.; Alvarez Martínez, J.A.; Hernández-Escareño, J.J.; Mosqueda Gualito, J.; Alonso Díaz, M.A.; Rosario-Cruz, R.; Soberanes Céspedes, N.; Merino Charrez, O.; Howard, T.; Chávez Niño, V.M.; Pérez de León, A.A. Research on Integrated Management for Cattle Fever Ticks and Bovine Babesiosis in the United States and Mexico: Current Status and Opportunities for Binational Coordination. *Pathogens,* **2020**, *9*(11), 871.
[http://dx.doi.org/10.3390/pathogens9110871] [PMID: 33114005]

[67]  De Meneghi, D.; Stachurski, F.; Adakal, H. Experiências no controle do carrapato por acaricida no setor tradicional de gado na Zâmbia e Burkina Faso: possíveis implicações ambientais e de saúde pública. *Fronteiras em saúde pública,* **2016**, *4*, 239.

[68]  Muraro, L.S.; Nogueira, M.F.; Borges, A.M.C.M.; Souza, A.O.; Vieira, T.S.W.J.; de Aguiar, D.M. Detection of Ehrlichia sp. in Amblyomma sculptum parasitizing horses from Brazilian Pantanal wetland. *Ticks Tick Borne Dis.,* **2021**, *12*(3)101658
[http://dx.doi.org/10.1016/j.ttbdis.2021.101658] [PMID: 33556777]

[69]  Pires, M.S.; Santos, T.M.; Santos, H.A.; Vilela, J.A.R.; Peixoto, M.P.; Roier, E.C.R.; Silva, C.B.; Barreira, J.D.; Lemos, E.R.S.; Massard, C.L. Amblyomma cajennense infestation on horses in two microregions of the state of Rio de Janeiro, Brazil. *Rev. Bras. Parasitol. Vet.,* **2013**, *22*(2), 235-242.
[http://dx.doi.org/10.1590/S1984-29612013005000017] [PMID: 23778827]

[70]  Zeringóta, V.; Senra, T.O.S.; Calmon, F.; Maturano, R.; Faza, A.P.; Catunda-Junior, F.E.A.; Monteiro, C.M.O.; de Carvalho, M.G.; Daemon, E. Repellent activity of eugenol on larvae of Rhipicephalus microplus and Dermacentor nitens (Acari: Ixodidae). *Parasitol. Res.,* **2013**, *112*(7), 2675-2679.
[http://dx.doi.org/10.1007/s00436-013-3434-z] [PMID: 23636308]

# CHAPTER 7

# Use of Terpenoids to Control Helminths in Small Ruminants

**Dauana Mesquita-Sousa**[1], **Victoria Miro**[2], **Carolina R. Silva**[1], **Juliana R. F. Pereira**[1], **Livio M. Costa-Júnior**[1], **Guillermo Virkel**[2] and **Adrian Lifschitz**[2,*]

[1] *Laboratorio de controle de Parasitos, Departamento de Patologia, Universidade Federal do Maranhão (UFMA), São Luis, Maranhão, Brasil*

[2] *Laboratorio de Farmacología, Centro de Investigación Veterinaria de Tandil (CIVETAN) (CONICET-CICPBA-UNCPBA), Facultad de Ciencias Veterinarias, Universidad Nacional del Centro, Tandil, Argentina*

**Abstract:** Gastrointestinal nematodes affect the animal's health and cause economic losses in meat, milk, and wool production. Essential oils and their terpenoids have been shown to effectively control gastrointestinal nematodes and may be an alternative to control gastrointestinal nematodes. The great advantage of terpenoids is the possibility of acting on the parasite in a multidirectional way on the neuromuscular system and body structures of nematodes. The current chapter describes the pharmacological basis of the combination of terpenes and synthetic anthelmintics as an alternative for increasing antiparasitic efficacy. It is necessary to evaluate if these combinations show antagonist, additive or synergic effects at the pharmacokinetic and pharmacodynamic levels. The physicochemical properties, pharmacokinetic features and potential drug-drug interactions at the metabolism or transport level of monoterpenes may be relevant for obtaining effective concentrations against different nematodes. In this context, the prediction of absorption, distribution, metabolism and excretion (ADME) is essential to optimize the anthelmintic action of these compounds. The rapid absorption and elimination of monoterpenes after their oral administration may directly influence the drug concentration level attained at the target parasites and the resultant pharmacological effect. Therefore, investigations on the dose schedule, administration route and type of pharmaceutical formulation are necessary. The integration of *in vitro* assays, *in silico* analysis, and *in vivo* pharmaco-parasitological studies are relevant to corroborate the kinetic/metabolic interactions and the efficacy of bioactive natural products combined with synthetic anthelmintics

**Keywords:** Goat, Natural Product, Nematode, Sheep, Small Ruminant, Synthetic Anthelmintics, Terpenes.

---

\* **Corresponding Author Adrian Lifschitz:** Laboratorio de Farmacología, Centro de Investigación Veterinaria de Tandil (CIVETAN)(CONICET-CICPBA-UNCPBA), Facultad de Ciencias Veterinarias, Universidad Nacional del Centro, Tandil, Argentina; E-mail: adrianl@vet.unicen.edu.ar

# INTRODUCTION

Gastrointestinal nematodes are especially relevant for small ruminant production. These parasites affect the animal's health and cause economic losses in meat, milk, and wool production [1]. Essential oils and their terpenoids have been shown to effectively control gastrointestinal nematodes [2 - 4]. The terpenoids, compounds from plants, are alternatives to control the gastrointestinal nematodes [5]. However, the mechanism of action of these composts is not quite clear yet.

Since 1950, studies have been performed to better understand anthelmintic compounds' mechanism to control human and animal parasites [6]. The anthelmintic action is associated with the interference of the product in the biochemical process of the parasites. This interference may be related to energy production, muscular coordination, microtubule dynamic, and procedures that can take the parasite's death [7]. Thus, the mechanism of action of the anthelmintic may be invalidated, with alterations that happen in nematode strains, as to the development of parasite defense and are known as resistance [8]. The great advantage of terpenoids is the possibility of acting on the parasite in a multidirectional way [6]. Although the use of control strategies must be well elaborated and planned, it has the most significant effect. For this, a broad knowledge of the mechanism of action is required.

## Mechanism of Action of the Anthelmintic Compound

### Neuromuscular System and Motility Control

Cys-loop receptors are ligand-gated ion channels activated by several neurotransmitters, like acetylcholine, serotonin, glycine, and GABA [9]. The nervous system of nematodes includes an exclusive and diverse family of cys-loop receptors linked in rapid synaptic transmission, fundamental for worm sensory and locomotor functions [10]. The Cys-loop receptors target widely used anthelmintics, such as levamisole, piperazine, and ivermectin [11 - 13]. The Levamisol-sensitive nicotinic receptors (L-AChR) and GABA (A) (UNC-49) are two target muscular receptors of terpenoids that cause paralyzed effects. Thymol, carvacrol, and eugenol act as inhibitors of L-AChR and UNC-49 receptors from *Caenorhabditis elegans* muscle cells. This result is probably due to the double effects caused by terpenoids on muscle receptors that support antagonistic actions since L-ACHRs are involved in muscle contraction and UNC-49 receptors in muscle relaxation [11].

Terpenoids also act on other different transient receptors. There are 29 nAChR subunits present in *C. elegans*, demonstrating the importance of further studies to explore the selectivity of terpenoids in the nicotinic family [11]. Other terpenoids like carvone, pulegone and eugenol, were also identified as inhibitors of the nAChRs.

## Terpenoids with Action in GABA

γ-Aminobutyric acid (GABA) is a family of receptors widely distributed. Nematodes are responsible for regulating motility, feeding, and reproduction [14]. There are distinct forms of GABA receptors: GABAA and GABAB. GABAA is GABA-gated chloride channels located in post-synaptic membranes, while GABAB is G-protein coupled receptors located both in pre-and post-synaptic membranes [15, 16]. Some monoterpenes, such as thymol, thymoquinone, and borneol, are known as positive modulators for GABAA receptors [17]. Recently, with the use of *C. elegans,* it was identified that thymol and carvacrol might be causing paralyzing effects on the worm, linked to the critical receptors in its locomotion. Know well that the activation of neuronal GABA receptors generally results in hyperpolarization and muscle paralysis [11].

The blocking $Ca^{2+}$ channels and positive allosteric activation of the GABAA receptor were attributed to menthol. Menthol, which is well-known for producing a cooling effect, is a TRPM8 agonist. The GABAB receptor activity was found to inhibit TRPV1 sensitization, and TRPV1 activation triggers GABA release [18]. 1.8-cineole, menthol (both (-)- and (+)-), carvone (both (-)- and (+)-), pulegone, linalyl acetate, linalool, carvacrol, estragole, bisabolol, carvone (both (-)- and (+)-), terpinene-4-ol, are known to have analgesic properties targeting $Na^+$ and TRP channels [19]. TRPV1-4 are temperature-sensitive channels activated by heat stimuli, whereas TRPM8 and TRPA1 are temperature-sensitive channels activated by cold stimuli [17]. The study suggested the role of glutamatergic neurotransmission and transient receptor potential cation channels (TRP channels) in these actions. Also, monoterpenes with chemical similarity, *e.g.,* geraniol, limonene, α-phellandrene, and carvone, may similarly have anti-nociceptive action. These compounds may be ligands of the same receptors and have similar effects [20].

## The Action of the Terpenoids on Tubulin

Microtubules are involved in the regulation of various cellular functions, such as cell division, cell motility, intracellular trafficking, and maintenance of cell shape [21]. Commercial anthelmintics can interfere with microtubules. Benzimidazoles block the dimerization of the a and b-tubulin, thus inhibiting microtubules formation, mitosis, and resulting in worm mortality [22]. Some terpenoids have microtubules as the target of action. Citral was lethal to *Arabidopsis* seedlings, interfered with cell division, and in microtubules disrupted, without acting on actin filamentous [23].

## *Structural Alterations*

Nematodes cuticle is made up of concentric layers and plays a critical role in performing protective and selective absorption functions [24]. Molecules that induce significant morphological changes, such as nodulation along the body, degeneration of the body wall, and intense cytoplasmic vacuolization, are alternatives to new anthelmintics [25].

*D*-limonene, a lipophilic terpene with strong solvent capability, has already demonstrated efficacy against *H. contortus* infection in gerbils and sheep [26]. On *C. elegans*, *D*-limonene broke the waterproof protective layer of the worm's body surface and had strong penetrating properties so that worms die by suffocation [27]. *Mentha pulegium* essential oil, which has its main component a monoterpene pulegone, was also associated with the loss of the normal aspect of the parasite and the occurrence of structural alterations manifested by aggregates bubbles around the body surface [28]. Such damage could be attributed to lipophilic components, which have an excellent affinity for the cell membrane and could induce structural and chemical alterations in the cell membrane [29].

The genus *Eucalyptus* is well known for its essential oils and used in the fragrance and pharmaceutical industries [30]. The *in vitro* effects of *Eucalyptus citriodora* essential oil and its main constituent, citronellal, presented effects, completely inhibiting the motility of *H. contortus* [31]. Through Transmission Electron Microscopy, it was observed that when *H. contortus* was exposed to citronellal, it demonstrated ultrastructural alterations, such as the mitochondrial profile [31]. The loss of homeostasis of the parasites exposed to the treatments can explain motility failure, such as the destruction of the muscular layer after exposure to citronellal [32].

The effect of carvacryl acetate and carvacrol on sheep gastrointestinal nematodes may also be related to changes in the cuticle. Analysis through the Scanning Electron Microscopy shows the wrinkling of the cuticle as the primary change after *in vitro* exposure to these compounds. The cuticular modifications may interfere with the permeability of the cuticle and motility, hindering the maintenance of homeostasis within these parasites [33].

## Combination of Synthetic Anthelmintics and Terpenoids

Since the decade of 1960, when thiabendazole was reported, many synthetic anthelmintics have been developed to control nematodes in small ruminants [34]. These synthetic compounds have been a solution for many years, but resistant strains were selected due to overuse. The effect was a lowering efficacy of anthelmintics used are shown to be mainly in small ruminants production around the world [35 - 37].

The synthetics anthelmintic combination is an alternative to control resistant and susceptible parasites. The combination should be performed with different classes of anthelmintics to achieve good efficacy. Also, these combinations could show additive or synergic effects among the compounds [38, 39]. Although combining synthetic compounds increases the efficacy, the perspective is to select resistant strains to these mixtures over time [39]. Therefore, the research about natural products increased yearly, with many studies reporting terpenes as one alternative to control for small ruminant nematodes [2, 40]. The mixture of two or more terpenes demonstrated that more efficient combinations showed additive or synergic effects [3].

The strategy of combining terpenes and synthetic compound were used with commercial antibacterial and antifungal against resistant microorganisms. A combination of citral and norfloxacin and thymol and rifampicin was synergic against *Staphylococcus aureus* [41, 42]. The synergism was also shown in the combination of citronellol and amphotericin against a different strain of *Candida* spp. in the combination of terbinafine and the monoterpenes dihydrojasmone and terpinolene on dermatophytosis [43, 44]. Combining a low concentration of terpenes and synthetic anthelmintic is an alternative for increasing the efficacy of commercial synthetic compounds [38, 45].

The combination of albendazole or levamisole with r-carvone and s-carvone decreases the the hatchability of eggs and larval migration of *H. contortus*, respectively [45]. These authors showed damage to the egg wall resulting in an overflow of internal egg content when tested using albendazole and r-carvone. However, the synergism effect of the combination afore mentioned the antagonism was shown in some combinations [45]. The combination of synthetics, anthelmintics, and terpenes is a promising alternative to control of gastrointestinal nematode of small ruminants. However, pharmacological understanding and *in vivo* studies are necessary.

## Influence of Pharmacological Properties of Monoterpenes on their Anthelmintic Effect

Helminth parasites affecting domestic animals are located in several tissues in the body, such as the gastrointestinal tract (GIT), liver, bile ducts and lung parenchyma. Chemical compounds, including bioactive phytochemicals, require effective and sustained concentrations at the sites of parasite location to achieve their anthelmintic effect. In this context, the physicochemical properties and pharmacokinetic disposition of these molecules have a direct influence on their anthelmintic activity [46]. Lipophilicity determines the ability of the compounds to pass through the cell membranes of parasites, a critical step for their accumulation at the target parasites. The higher the amount of a drug gaining access to a cell, the greater the effect, a classic axiom in pharmaco-toxicology. Drug diffusion across parasite tegument is crucial to reach the specific receptor and to produce its effect. Monoterpenes are non-polar and lipophilic compounds that easily penetrate membranes [47].

There is scarce information on the pharmacokinetics of these phytochemicals in ruminants. Most of the published information on this issue comes from studies on humans and rats. In general, after oral administration, monoterpenes are rapidly absorbed and highly metabolized. The fast absorption of monoterpenes might be favoured by the small size and the lipophilic characteristics of these molecules [48]. Metabolites of monoterpenes resulted from both liver phase I and phase II metabolism. Oxydized metabolites of thymol and carvacrol were found in the systemic circulation after the oral administration of these monoterpenes to rats. Phase-II metabolites, *i.e.*, glucuronides or sulfates, were also detected in rats, rabbits, and humans [49, 50]. Due to their high clearance and short elimination half-lives, the accumulation of monoterpenes in body tissues seems to be unlikely [51]. Overall, the rapid absorption and elimination of monoterpenes after their oral administration may directly influence the drug concentration level attained at the target parasites and the resultant pharmacological effect. Therefore, investigations on the dose schedule, administration route and type of pharmaceutical formulation are necessary for the improvement of the anthelmintic efficacy of these phytochemical compounds.

The complex anatomy-physiology of ruminant's gastrointestinal tract resides in the evolutionary development of a series of three chambers, the rumen, reticulum, and omasum, anterior to the true stomach, the abomasum. Frequently, this accounts for the often-dramatic differences observed between ruminants and monogastric in the oral absorption of drugs [52]. Although the liver plays a pivotal role in the metabolism of foreign compounds, biotransformation may also occur in the gastrointestinal tract, particularly in the rumen, due to the fermentative environment. In comparison with the liver, where oxidative metabolism predominates, the ruminal microflora is very active in reductive reactions of xenobiotics [53]. Ruminal metabolism may play a role in reducing the systemic availability of orally administered compounds [54], thus being particularly important in the therapeutic outcome. In this context, bioactive natural products administered by the oral route should be stable in the ruminal environment to allow the active molecules to be in contact with the target gastrointestinal nematodes.

The chemical stability of carvacrol and thymol was evaluated in sheep ruminal content *in vitro*. Both monoterpenes were stable in the ruminal environment; the concentrations of unchanged (unmetabolized) carvacrol and thymol recovered after the incubation of both compounds were between 84-91% and 90-95%, respectively [55]. The lack of metabolic degradation of these phytochemicals was similar to that observed for synthetic anthelmintics such as monepantel [56, 57]. As carvacrol and thymol were found metabolically stable in the rumen, the oral administration of both compounds may assure their antiparasitic efficacy against GI nematodes. Among the different factors affecting the systemic availability of orally administered compounds in ruminants, the degree of adsorption to the particulate material of gastrointestinal contents plays a relevant role in their kinetic disposition. Thymol and carvacrol showed lower concentrations associated with the particulate material compared to those dissolved in the fluid phase of the ruminal content. The percentage of association to the particulate phase was higher for thymol (43-49%) than that observed for carvacrol (31-34%). The fact that most of the monoterpenes are kept in the fluid phase of ruminal content may imply a shorter residence time of these compounds in the rumen and a faster flow rate to the abomasum and small intestine [55].

The concurrent use of natural compounds and synthetic drugs for the control of nematode parasites may give rise to pharmacokinetic interactions that should be carefully evaluated. Exposure to plant-derived products may cause the induction or the inhibition of drug-metabolizing enzymes and/or transport proteins involved in the elimination of anthelmintic molecules. Certain anthelmintic drugs, such as albendazole and fenbendazole, are metabolized by different drug metabolizing enzymes from both hepatic and extrahepatic tissues. Metabolic conversions usually alter the polarity of the anthelmintic parent molecule and, consequently, the way in which the drug is distributed and excreted from the body. *In vivo* interference with the activity of certain drug metabolizing enzymes may give rise to pronounced modifications to both the pharmacokinetic behavior and the therapeutic outcome of active anthelmintic molecules. It has been shown that the flavin-monooxygenase (FMO) pathway is involved in roughly 65% of albendazole S-oxidation in sheep liver, whereas the rest of the production of the albendazole S-oxidized metabolite (*i.e.*, albendazole sulfoxide) depends on the cytochrome P450 (CYP) dependent metabolism [58]. In the presence of thymol, the metabolism of albendazole (ABZ) in sheep liver was decreased, particularly marked for its FMO-dependent production (54%). In contrast, the CYP-dependent production of the metabolite was less affected (25%) [59]. Sheep liver FMO and CYP1A1 enzyme activities were also significantly inhibited by thymol *in vitro* [59]. Overall, these *in vitro* investigations suggested that, in addition to its own anthelmintic effect, monoterpenes may potentiate ABZ anthelmintic activity by preventing its metabolic conversion into a less active metabolite. These observations are in agreement with other studies on the *in vitro* inhibition of the hepatic CYP-dependent metabolism [60]. For instance, carveol decreased the catalytic activities of CYP2B and CYP2C in rats [61]. Inhibition of CYP1A1- and CYP1B1-dependent enzyme activities by eugenol was also observed in rats, whereas geraniol inhibited the catalytic activity of CYP2B6 in human liver microsomes [62, 63]. In order to confirm the clinical relevance of these interactions, further *in vivo* studies in ruminant species are necessary.

Efflux membrane transporters are energy-dependent protein complexes responsible for removing several compounds, including endogenous molecules and xenobiotics, from cells through active transport. They contribute to the preservation of cell homeostasis by removing potentially toxic compounds from the intracellular space; however, they may also limit the penetration of certain compounds. Therefore, the activity of these transport proteins may affect the bioavailability and elimination of numerous drugs and other xenobiotics in animals and humans. *P*-glycoprotein (*P*-gp) is one of the best-known efflux membrane transporters [64]. Modulation of *P*-gp accounts for the enhancement of the systemic exposure of antiparasitic drugs in ruminants. Besides, transport-related drug-drug interactions in parasite location tissues may contribute to enhance drug accumulation and efficacy against resistant worms [65].

Several monoterpenes were found to inhibit *P*-gp or to down-regulate its gene/protein expression. For instance, menthol led to a higher accumulation of rhodamine 123, a *P*-glycoprotein-specific substrate, in Caco-2 cells in a dose-dependent manner [66]. Citronellal caused down-regulation of *P*-gp mediated digoxin transport. However, these authors did not find a relationship between the monoterpenes' molecular structure and their inhibitory potential [67]. Among terpenoids, diterpenes have shown greater *P*-gp modulation properties than monoterpenes [68]. Recently, the presence of carvone significantly increased Rho123 accumulation in cattle intestinal explants. The presence of carvone also increased the intestinal concentration of the synthetic anthelmintic abamectin, a macrocyclic lactone [69]. Molecular docking studies may help to understand the interaction between monoterpenes and *P*-gp. Whereas abamectin showed specific binding to this transport protein *in silico*, this fact was not observed for carvone [69]. Therefore, the potential interaction of carvone with *P*-gp seemed to be unlikely. Thus, the influence of carvone on drug absorption and accumulation may be explained by other mechanisms. Different bioactive phytochemicals may increase intestinal absorption by enhancing enterocyte membrane permeability or by opening paracellular tight junctions [70].

The prediction of absorption, distribution, metabolism and excretion (ADME) helps to reduce the number of animal experiments, contributing to the so-called

3Rs principle (replacement, reduction and refinement). These methods use statistical and learning approaches, molecular descriptors and experimental data [71]. Table **1** shows ADME prediction for different monoterpenes. The enzymes CYP2C9 and CYP2C19 are well-known in the metabolism of carvone to carveol. However, it is not known if these enzymes are either induced or inhibited [72]. In ADME prediction, the carveol and carvone have shown inhibition activity on CYP2C9 and CYP 2C19 and were substrates for CYP 3A4. *In silico* analysis corroborated that carveol and carvone have a moderate capacity for binding to plasma proteins, dismissed a possible plasma transport mechanism [71]. The prediction also shows that geraniol and limonene will have a wide distribution across the blood-brain barrier (BBB). The molecules that pass-through BBB produce a quick, coordinated and effectively response [73]. Carvone's skin permeability is more effective than other compounds. This is an essential parameter for the transdermal delivery of drugs [74]. The ADME predictions showed that citronellal and citral would behave as *P*-gp inhibitors. Caco-2 and MDCK prediction are useful tools for measuring the absorption of the molecules [74]. Caco-2 permeability is more effective for nerol when compared to other compounds. Eugenol has the worst intestinal absorption (HIA) ~ 96.8% but has shown higher effective permeability across MDCK cells.

**Table 1. Predicted ADME properties of compounds terpenoids.**

| ID | Carveol | Carvone | Citral | Citronellal | Eugenol | Geraniol | Limonene | Menthol | Nerol | *p*-Cimene |
|---|---|---|---|---|---|---|---|---|---|---|
| BBB | 5.079 | 1.060 | 2.008 | 1.732 | 2.255 | 6.741 | 8.278 | 6.255 | 6.741 | 4.969 |
| Caco-2 | 37.936 | 47.742 | 13.967 | 13.725 | 46.886 | 8.756 | 23.631 | 39.490 | 87.568 | 23.433 |
| CYP 2C19 inhibition | Inhibitor | Inhibitor | Inhibitor | Inhibitor | Inhibitor | Inhibitor | Inhibitor | Inhibitor | Inhibitor | Inhibitor |
| CYP 2C9 inhibition | Inhibitor | Inhibitor | Inhibitor | Inhibitor | Inhibitor | Inhibitor | Inhibitor | Inhibitor | Inhibitor | Inhibitor |
| CYP 2D6 inhibition | No | No | No | No | No | No | No | No | No | No |
| CYP 2D6 substrate | No | No | No | No | Weakly | No | No | No | No | No |
| CYP 3A4 inhibition | No | No | No | No | No | No | No | Inhibitor | No | Inhibitor |
| CYP 3A4 substrate | Substrate | Substrate | Substrate | Weakly | No | Substrate | Substrate | Non | Substrate | Weakly |
| HIA | 100.000 | 100.000 | 100.000 | 100.000 | 96.774 | 100.000 | 100.000 | 100.000 | 100.000 | 100.000 |
| MDCK | 97.813 | 97.545 | 252.255 | 265.594 | 342.148 | 271.032 | 238.434 | 96.054 | 271.032 | 23.750 |

*(Table 1) cont.....*

| ID | Carveol | Carvone | Citral | Citronellal | Eugenol | Geraniol | Limonene | Menthol | Nerol | *p-*Cimene |
|---|---|---|---|---|---|---|---|---|---|---|
| PGP inhibition | No | No | Inhibitor | Inhibitor | No | No | Inhibitor | No | No | No |
| PPB | 57.945 | 58.045 | 100.000 | 100.000 | 100.000 | 100.000 | 100.000 | 100.000 | 100.000 | 100.000 |
| Skin Permeability | -1.401 | -1.403 | -0.961 | -1.019 | -1.310 | -1.059 | -0.834 | -1.607 | -1.059 | -0.805 |

BBB: blood–brain barrier (C.brain/C.blood); Caco-2: Caco2-cell model; CYP: cytochrome P450; HIA: human intestinal absorption model (HIA, %); MDCK: Madin–Darby canine kidney (nm/s); PGP inhibition: Inhibitor *P*-glycoprotein; PPB: plasma protein binding (%); Skin permeability: skin permeability (cm/h).

The physicochemical properties, pharmacokinetic features and potential drug-drug interactions at the metabolism or transport level of monoterpenes may be relevant for obtaining effective concentrations against different nematodes. In this context, the prediction of ADME is essential to optimize the anthelmintic use of these phytochemicals in ruminants.

### *In Vivo* Anthelmintic Effect of Monoterpenes

Abomasal and intestinal nematodes are among the most pathogenic gastrointestinal parasites in sheep, cattle, and goats. An efficient deworming program is essential for obtaining rational control and sustained productivity. However, after years of intensive use, resistance to synthetic drugs has been spread worldwide [75]. There is an urgent need to find novel pharmacological tools in order to ensure efficient control of gastrointestinal nematodes. A wide variety of naturally sourced terpenes were shown to possess anthelmintic activity *in vitro* [45, 76, 77], but it is necessary to demonstrate whether these effects can be observed *in vivo*. Although *in vitro* assays supply useful information, it is necessary to know the *in vivo* fate of terpenes after their administration [59]. This issue is crucial, as active compounds need to attain effective concentrations for some time at the sites of parasite location [78].

The anthelmintic activity of thymol was observed *in vitro*, particularly against *H. contortus* eggs, larvae and adult parasites [2, 3, 40, 45]. To evaluate whether effective concentrations of thymol can be obtained after the oral administration to sheep, the kinetic disposition of this monoterpene was evaluated in lambs. Thymol was chemically stable in metabolically active ruminal content of lambs and can be detected in the bloodstream after the oral administration of two high doses (150 mg/kg). Absorption of thymol after its oral administration was markedly fast in lambs, with a peak plasma concentration achieved between 1- and 2-hours post-administration. The highest thymol concentration detected in plasma was 1.9 µg/mL [59]. However, *in vitro* tests on *Haemonchus* spp. eggs and larvae have corroborated that thymol concentrations necessary to obtain an efficacy above 90% were between 4 and 6 mg/mL [2, 3, 40]. Similarly, the concentrations necessary to affect the motility of adults of *Haemonchus* spp. were above 400 µg/mL [40]. It is evident that kinetics disposition features of thymol in sheep do not ensure a high exposure of parasites to the monoterpene as the measured concentrations *in vivo* were several times below the minimal effective concentrations. In agreement with these observations, administering one dose of 300 mg/kg of thymol was ineffective in reducing the egg count in faeces of infected sheep [2]. A lack of reduction of faecal egg counts was also observed after the administration of two oral doses of 150 mg/kg of this monoterpene to infected sheep [59]. On the contrary, André *et al.* [40] obtained an efficacy of 59% after the administration of thymol at 250 mg/kg to naturally infected sheep. Altogether these *in vivo* tests of efficacy suggest a low *in vivo* exposure of parasites after the oral administration of thymol, leading to zero or low efficacy. Further research on its potential anthelmintic use may include the study of pharmaceutical strategies for providing sustained and effective concentrations. For instance, the encapsulation technique appears as a strategy to improve solubility, stability and the pharmacological response of thymol and other monoterpenes.

A semisynthetic derivative of carvacrol, carvacryl acetate, was synthesized in an attempt to reduce collateral effects. In fact, carvacryl acetate was shown to be safe; in an acute toxicity trial in mice, the LD50 of carvacryl acetate was 1544 (mg/kg), whereas for carvacrol was 919 mg/kg [33]. Further, carvacryl acetate and carvacrol showed similar *in vitro* activity in a larval development test; the former compound showed a higher effect on the motility of adult specimens of *H. contortus*. After 24 h of exposure at 200 mg/mL, carvacryl acetate inhibited worm motility by 100%, while the inhibition of carvacrol was 41.8%. In addition, both carvacrol and carvacryl acetate displayed morphological alterations in the cuticle and the vulvar flap of adult specimens of treated *H. contortus*, suggesting that these compounds have the same mechanism of action [79]. The *in vivo* trial addressed to evaluate the efficacy of carvacryl acetate showed a reduction faecal egg counts of 66% after its oral administration to infected sheep at 250 mg/kg. A similar improved *in vitro* and *in vivo* performance was shown after the acetylation of thymol [40]. Clearly, the aforementioned investigations show that chemical-pharmaceutical modifications are relevant to increase the exposure and effectiveness of bioactive phytochemicals.

Carvone is another terpene whose anthelmintic activity was tested *in vitro* and *in vivo*. The *in vitro* activity of R-carvone has been shown, particularly against *H. contortus*. The 99% lethal concentration of carvone that could inhibit hatching of

*H. contortus* was 366 µg/mL, and this effect was increased after the combination with different phytochemicals [3]. Therefore, carvone showed a higher potency compared to other monoterpenes such as carvacrol and thymol. Taking advantage of these properties, the evaluation of the long-term administration of encapsulated carvone plus anethole was recently evaluated in sheep [80]. Two oral doses (20 mg/kg and 50 mg/kg) of the bioactive phytochemical combination were administered to sheep for 45 days. A marked reduction of faecal egg counts was the main effect observed after the administration of the highest dose rate; total adult worm counts remained similar compared to the control group. Although further studies are necessary in order to find the ideal dose of carvone and the appropriate pharmaceutical formulation, the reduction faecal egg counts when animals are highly infected with gastrointestinal nematodes, can lead to a reduction in the infection of pasture. This fact may help to improve worm control programs.

Considering the *in vitro* and *in vivo* low pharmacological potency shown by terpenes as anthelmintic compounds, the combination with synthetic drugs may be an interesting approach to increase the clinical efficacy of phytochemicals. The rationale behind the use of combinations is the lower probability of parasites to be resistant to multiple active compounds compared with that observed after the administration of a formulation with a single active molecule [38, 82]. Despite the diverse chemical groups available to control parasitic diseases in ruminants, macrocyclic lactones (MLs) and benzimidazoles (BZD) have been the most widely used drugs. Considering the high level of resistance to both families [82] there is an urgent need to search for novel strategies to extend the life span of antiparasitic agents. The administration of different anthelmintic compounds in combination may lead to unpredictable pharmacokinetic (PK) and/or pharmacodynamic (PD) drug-drug interactions (DDI). PK interactions occurring at the absorption and metabolism/transport/excretion processes may affect the anthelmintic response [38]. PD interactions may take place when one drug may alter the intensity of the pharmacological effects of another drug given concurrently, resulting in additive, synergistic or antagonistic pharmacological effects [81, 83].

There is scarce information about the combinations involving terpenes and synthetic anthelmintics. Albendazole showed a pharmacodynamic synergism in combination with carvone after *in vitro* evaluation on eggs hatch assay [45]. As this synthetic anthelmintic is extensively metabolized in the liver by mixed function oxidases, an *in vitro* metabolic interaction was corroborated between albendazole and thymol [55]. In fact, thymol inhibited the hepatic sulphoxidation and sulphonation of the synthetic compound [55, 59]. Based on these *in vitro* findings, it was initially suggested that, in addition to its own anthelmintic effect, thymol might improve albendazole anthelmintic activity by preventing its metabolic conversion into a less active metabolite [55]. In a further *in vivo* investigation, the pharmacokinetic and the efficacy of the combination of albendazole and thymol were evaluated in lambs naturally infected with resistant gastrointestinal nematodes [59].

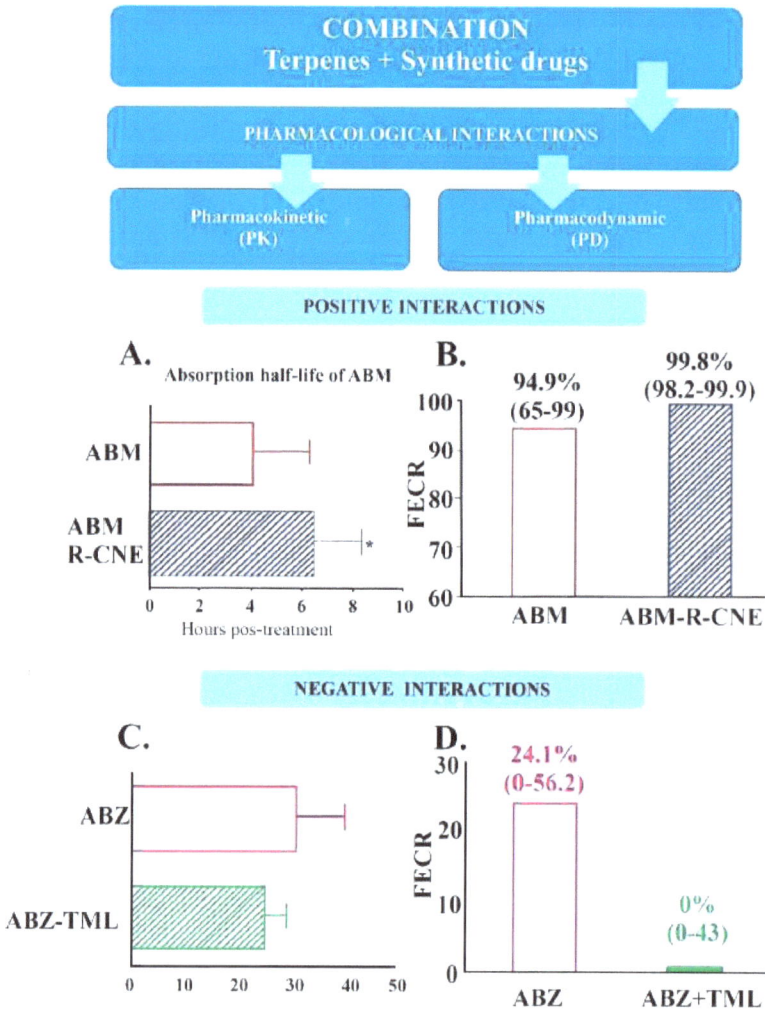

**Fig. (1).** Drug-drug interactions at pharmacokinetic (PK) and pharmacodynamics (PD) level after the combined administrations of terpenes and synthetic compounds **(A)** Positive PK interaction. The co-admiration of abamectin (AMB) and carvone (CNE) prolonged the absorption hael0life of ABM by 57% in lambs (Miró *et al.*, 2020c). **(b)** positive PD interaction. The *in vivo* efficacy of AMB against resistant gastrointestinal nematodes increased from 94.9 to 99.8% in the presence of CNE (Miró *et al.*, 2020c). **(C)** negative PK interaction. Albendazole suphoxide (ABZSO) systemic exposure measured as the area under concentration *vs.* time curve were lower after the combined treatment with albendazole and thymol (ABZ+TML) (Miró *et al.*, 2020b). **(D)** Negative PD interaction. The presence of TML decreases the *in vivo* efficacy of ABZ from 24.1% to 0% (Miró *et al.*, 2020b).

However, although the hepatic metabolism of albendazole was inhibited by thymol, the combined *in vivo* administration of the monoterpene and albendazole did not improve the kinetic plasma disposition of albendazole metabolites. Additional *in vitro* studies showed that thymol also inhibited the ruminal sulphoreduction of albendazole sulphoxide into albendazole, a metabolic reaction that renders a more active anthelmintic product. This metabolic step is considered of vital importance for the antiparasitic efficacy of benzimidazole thioethers. Therefore, the combined treatment of parasitized sheep with thymol and albendazole led to a negative pharmacokinetic interaction. In addition, the anthelmintic treatment of thymol with the synthetic drug caused a decline in the faecal egg count reduction test from 24% when albendazole was administrated alone to 0% when it was administrated in combination with the monoterpene [59].

Phytochemicals may also modulate the absorption and excretion processes of synthetic anthelmintics. The combination of carvone and the macrocyclic lactone abamectin was evaluated to test the potential modulation of the *P*-glycoprotein-dependent excretion of the synthetic drug. *In vitro* assays with intestinal explants showed that carvone improves the tissue accumulation of abamectin. Moreover, a prolonged absorption plasma half-life of the macrocyclic lactone was observed in naturally infected sheep receiving a combination of abamectin and carvone. An increased antiparasitic efficacy, from 94.9 (65-99%) to 99.8% (98-99.9%), was observed after the combined treatment [69]. However, a lack of interaction of carvone with the transport protein *P*-glycoprotein was revealed *in silico* [69]. Therefore, the beneficial effect of carvone on the pharmacokinetics of abamectin was attributed to the effect of the natural product on the enterocyte membrane permeability [70]. The potential drug-drug interactions at the pharmacokinetic and pharmacodynamic level after the combined administration of terpenes and synthetic anthelmintic compounds are shown in Fig. (1).

## CONCLUDING REMARKS

Altogether, these observations clearly remark that the integration of *in vitro* assays, *in silico* analysis, and *in vivo* pharmaco-parasitological studies are relevant to corroborate the kinetic/metabolic interactions and the efficacy of bioactive natural products combined with synthetic anthelmintic. The development of pharmacology-based information is critical for the design of successful strategies for parasite control using pharmaceutical formulations of bioactive phytochemicals that ensure a high efficacy with low toxicity.

## CONSENT FOR PUBLICATION

Not applicable.

## CONFLICT OF INTEREST

The author declares no conflict of interest, financial or otherwise.

## ACKNOWLEDGEMENT

Declared none.

## REFERENCES

[1]    Doyle, SR; Illingworth, CJ; Laing, R; Bartley, DJ; Redman, E; Martinelli, A Population genomic and evolutionary modelling analyses reveal a single major QTL for ivermectin drug resistance in the pathogenic nematode, Haemonchus contortus. BMC genomic **2019**, *20*(1), 1-19.
       [http://dx.doi.org/10.1186/s12864-019-5592-6]

[2]    Ferreira, L.E.; Benincasa, B.I.; Fachin, A.L.; França, S.C.; Contini, S.S.H.T.; Chagas, A.C.S.; Beleboni, R.O. Thymus vulgaris L. essential oil and its main component thymol: Anthelmintic effects against Haemonchus contortus from sheep. *Vet. Parasitol.,* **2016**, *228*, 70-76.
       [http://dx.doi.org/10.1016/j.vetpar.2016.08.011] [PMID: 27692335]

[3]    Katiki, L.M.; Barbieri, A.M.E.; Araujo, R.C.; Veríssimo, C.J.; Louvandini, H.; Ferreira, J.F.S. Synergistic interaction of ten essential oils against Haemonchus contortus *in vitro. Vet. Parasitol.,* **2017**, *243*(June), 47-51.
       [http://dx.doi.org/10.1016/j.vetpar.2017.06.008] [PMID: 28807309]

[4]    Sousa, A.I.P.; Silva, C.R.; Costa-Júnior, H.N.; Silva, N.C.S.; Pinto, J.A.O.; Blank, A.F.; Soares, A.M.S.; Costa-Júnior, L.M. Essential oils from *Ocimum basilicum* cultivars: analysis of their composition and determination of the effect of the major compounds on *Haemonchus contortus* eggs. *J. Helminthol.,* **2021**, *95*, e17.
       [http://dx.doi.org/10.1017/S0022149X21000080] [PMID: 33745470]

[5]    Aderibigbe, S.A.; Idowu, S.O. Anthelmintic activity of <i>Ocimum gratissimum</i> and <i>Cymbopogon citratus</i> leaf extracts against <i>Haemonchus placei</i> adult worm. *J. Pharm. Bioresour.,* **2020**, *17*(1), 8-12.
       [http://dx.doi.org/10.4314/jpb.v17i1.2]

[6]    Mukherjee, N.; Mukherjee, S.; Saini, P.; Roy, P.; Babu, S.P. Phenolics and Terpenoids; the Promising New Search for Anthelmintics: A Critical Review. *Mini Rev. Med. Chem.,* **2016**, *16*(17), 1415-1441.
       [http://dx.doi.org/10.2174/1389557516666151120121036] [PMID: 26586122]

[7]    Abongwa, M.; Martin, R.J.; Robertson, A.P. A brief review on the mode of action of antinematodal drugs. *Physiol. Behav.,* **2016**, *176*(12), 139-148.https://www.sciendo.com/article/10.1515/acve-201--0013 [Internet].

[8]    Doyle, S.R.; Cotton, J.A. Genome-wide Approaches to Investigate Anthelmintic Resistance. *Trends Parasitol.,* **2019**, *35*(4), 289-301.
       [http://dx.doi.org/10.1016/j.pt.2019.01.004] [PMID: 30733094]

[9]    Lynagh, T.; Pless, S.A. Principles of agonist recognition in Cys-loop receptors. *Front. Physiol.,* **2014**, *5*, 160.
       [http://dx.doi.org/10.3389/fphys.2014.00160] [PMID: 24795655]

[10]   Accardi, M.V.; Beech, R.N.; Forrester, S.G. Nematode cys-loop GABA receptors: biological function, pharmacology and sites of action for anthelmintics. *Invert. Neurosci.,* **2012**, *12*(1), 3-12.
       [http://dx.doi.org/10.1007/s10158-012-0129-6] [PMID: 22430311]

[11]   Hernando, G.; Turani, O.; Bouzat, C. *Caenorhabditis elegans* muscle Cys-loop receptors as novel targets of terpenoids with potential anthelmintic activity. *PLoS Negl. Trop. Dis.,* **2019**, *13*(11), e0007895.

[http://dx.doi.org/10.1371/journal.pntd.0007895] [PMID: 31765374]

[12]　Kwaka, A.; Hassan Khatami, M.; Foster, J.; Cochrane, E.; Habibi, S.A.; de Haan, H.W.; Forrester, S.G. Molecular characterization of binding loop E in the nematode cys-loop GABA receptor. *Mol. Pharmacol.,* **2018**, *94*(5), 1289-1297.
[http://dx.doi.org/10.1124/mol.118.112821] [PMID: 30194106]

[13]　Rayes, D.; Flamini, M.; Hernando, G.; Bouzat, C. Activation of single nicotinic receptor channels from *Caenorhabditis elegans* muscle. *Mol. Pharmacol.,* **2007**, *71*(5), 1407-1415.
[http://dx.doi.org/10.1124/mol.106.033514] [PMID: 17314321]

[14]　Wolstenholme, A.J.; Fairweather, I.; Prichard, R.; von Samson-Himmelstjerna, G.; Sangster, N.C. Drug resistance in veterinary helminths. *Trends Parasitol.,* **2004**, *20*(10), 469-476.
[http://dx.doi.org/10.1016/j.pt.2004.07.010] [PMID: 15363440]

[15]　Enna, S.J. *The GABA receptors. In the GABA receptors*; Humana Press, **2007**, pp. 1-21.
[http://dx.doi.org/10.1007/978-1-59745-465-0]

[16]　Bormann, J. The 'ABC' of GABA receptors. *Trends Pharmacol. Sci.,* **2000**, *21*(1), 16-19.
[http://dx.doi.org/10.1016/S0165-6147(99)01413-3] [PMID: 10637650]

[17]　Koyama, S.; Heinbockel, T. The effects of essential oils and terpenes in relation to their routes of intake and application. *Int. J. Mol. Sci.,* **2020**, *21*(5), 1558.
[http://dx.doi.org/10.3390/ijms21051558] [PMID: 32106479]

[18]　Bergman, M.E.; Davis, B.; Phillips, M.A. Medically useful plant terpenoids: biosynthesis, occurrence, and mechanism of action. *Molecules,* **2019**, *24*(21), 3961.
[http://dx.doi.org/10.3390/molecules24213961] [PMID: 31683764]

[19]　Kobayashi, K.; Fukuoka, T.; Obata, K.; Yamanaka, H.; Dai, Y.; Tokunaga, A.; Noguchi, K. Distinct expression of TRPM8, TRPA1, and TRPV1 mRNAs in rat primary afferent neurons with aδ/c-fibers and colocalization with trk receptors. *J. Comp. Neurol.,* **2005**, *493*(4), 596-606.
[http://dx.doi.org/10.1002/cne.20794] [PMID: 16304633]

[20]　Chirumbolo, S.; Bjørklund, G. The Antinociceptive Activity of Geraniol. *Basic Clin. Pharmacol. Toxicol.,* **2017**, *120*(2), 105-107.
[http://dx.doi.org/10.1111/bcpt.12683] [PMID: 28000392]

[21]　Redeker, V. Mass spectrometry analysis of C-terminal posttranslational modifications of tubulins. *Methods Cell Biol.,* **2010**, *95*, 77-103.
[http://dx.doi.org/10.1016/S0091-679X(10)95006-1] [PMID: 20466131]

[22]　Velan, A.; Hoda, M. *In-silico* comparison of inhibition of wild and drug-resistant *Haemonchus contortus* β-tubulin isotype-1 by glycyrrhetinic acid, thymol and albendazole interactions. *J. Parasit. Dis.,* **2021**, *45*(1), 24-34.
[http://dx.doi.org/10.1007/s12639-020-01274-w] [PMID: 33746383]

[23]　Chaimovitsh, D.; Abu-Abied, M.; Belausov, E.; Rubin, B.; Dudai, N.; Sadot, E. Microtubules are an intracellular target of the plant terpene citral. *Plant J.,* **2010**, *61*(3), 399-408.
[http://dx.doi.org/10.1111/j.1365-313X.2009.04063.x] [PMID: 19891702]

[24]　Balqis, U.; Hambal, M.; Rinidar, F.A.; Athaillah, F.; Ismail, ; Azhar, ; Vanda, H.; Darmawi, Cuticular surface damage of *Ascaridia galli* adult worms treated with *Veitchia merrillii* betel nuts extract *in vitro. Vet. World,* **2017**, *10*(7), 732-737.
[http://dx.doi.org/10.14202/vetworld.2017.732-737] [PMID: 28831213]

[25]　Sant'Anna, V.; Railbolt, M.; Oliveira-Menezes, A.; Calogeropoulou, T.; Pinheiro, J.; de Souza, W. Ultraestructural study of effects of alkylphospholipid analogs against nematodes. *Exp. Parasitol.,* **2018**, *187*, 49-58.
[http://dx.doi.org/10.1016/j.exppara.2018.02.004] [PMID: 29496523]

[26]　Squires, J.M.; Foster, J.G.; Lindsay, D.S.; Caudell, D.L.; Zajac, A.M. Efficacy of an orange oil emulsion as an anthelmintic against *Haemonchus contortus* in gerbils (Meriones unguiculatus) and in

sheep. *Vet. Parasitol.,* **2010**, *172*(1-2), 95-99.
[http://dx.doi.org/10.1016/j.vetpar.2010.04.017] [PMID: 20452126]

[27]   Piao, X.; Sun, M.; Yi, F. Evaluation of Nematocidal Action against *Caenorhabditis elegans* of Essential Oil of Flesh Fingered Citron and Its Mechanism. *J. Chem.,* **2020**, *2020*, 1-9. [Internet].
[http://dx.doi.org/10.1155/2020/1740938]

[28]   Sebai, E.; Abidi, A.; Serairi, R.; Marzouki, M.; Saratsi, K.; Darghouth, M.A.; Sotiraki, S.; Akkari, H. Essential oil of *Mentha pulegium* induces anthelmintic effects and reduces parasite-associated oxidative stress in rodent model. *Exp. Parasitol.,* **2021**, *225*, 108105.
[http://dx.doi.org/10.1016/j.exppara.2021.108105] [PMID: 33812980]

[29]   Ben Arfa, A.; Combes, S.; Preziosi-Belloy, L.; Gontard, N.; Chalier, P. Antimicrobial activity of carvacrol related to its chemical structure. *Lett. Appl. Microbiol.,* **2006**, *43*(2), 149-154.
[http://dx.doi.org/10.1111/j.1472-765X.2006.01938.x] [PMID: 16869897]

[30]   Batish, D.R.; Singh, H.P.; Kohli, R.K.; Kaur, S. Eucalyptus essential oil as a natural pesticide. *For. Ecol. Manage.,* **2008**, *256*(12), 2166-2174. [Internet].
[http://dx.doi.org/10.1016/j.foreco.2008.08.008]

[31]   Araújo-Filho, J.V.; Ribeiro, W.L.C.; André, W.P.P.; Cavalcante, G.S.; Rios, T.T.; Schwinden, G.M.; Rocha, L.O.; Macedo, I.T.F.; Morais, S.M.; Bevilaqua, C.M.L.; Oliveira, L.M.B. Anthelmintic activity of *Eucalyptus citriodora* essential oil and its major component, citronellal, on sheep gastrointestinal nematodes. *Rev. Bras. Parasitol. Vet.,* **2019**, *28*(4), 644-651.
[http://dx.doi.org/10.1590/s1984-29612019090] [PMID: 31800886]

[32]   Brunet, S.; Fourquaux, I.; Hoste, H. Ultrastructural changes in the third-stage, infective larvae of ruminant nematodes treated with sainfoin (*Onobrychis viciifolia*) extract. *Parasitol. Int.,* **2011**, *60*(4), 419-424.
[http://dx.doi.org/10.1016/j.parint.2010.09.011] [PMID: 21787880]

[33]   Andre, W.P.P.; Ribeiro, W.L.C.; Cavalcante, G.S.; Santos, J.M.L.; Macedo, I.T.F.; Paula, H.C.B.; de Freitas, R.M.; de Morais, S.M.; Melo, J.V.; Bevilaqua, C.M.L. Comparative efficacy and toxic effects of carvacryl acetate and carvacrol on sheep gastrointestinal nematodes and mice. *Vet. Parasitol.,* **2016**, *218*, 52-58.
[http://dx.doi.org/10.1016/j.vetpar.2016.01.001] [PMID: 26872928]

[34]   Nixon, SA; Dela, C; Woods, DJ; Costa-Junior, LM; Zamanian, M; Martin, RJ Where are all the anthelmintics? Challenges and opportunities on the path to new anthelmintics. Inter Jour Parasitol, [Internet]. **2020**, *18*, 8-16.
[http://dx.doi.org/10.1016/j.ijpddr.2020.07.001]

[35]   Chaparro, J.J.; Villar, D.; Zapata, J.D.; López, S.; Howell, S.B.; López, A.; Storey, B.E. Multi-drug resistant Haemonchus contortus in a sheep flock in Antioquia, Colombia. *Vet. Parasitol. Reg. Stud. Rep.,* **2017**, *10*, 29-34.
[http://dx.doi.org/10.1016/j.vprsr.2017.07.005] [PMID: 31014594]

[36]   Höglund, J.; Enweji, N.; Gustafsson, K. First case of monepantel resistant nematodes of sheep in Sweden. *Vet. Parasitol. Reg. Stud. Rep.,* **2020**, *22*(September), 100479.
[http://dx.doi.org/10.1016/j.vprsr.2020.100479] [PMID: 33308757]

[37]   Mphahlele, M.; Tsotetsi-Khambule, A.M.; Moerane, R.; Komape, D.M.; Thekisoe, O.M.M. Anthelmintic resistance and prevalence of gastrointestinal nematodes infecting sheep in Limpopo Province, South Africa. *Vet. World,* **2021**, *14*(2), 302-313.
[http://dx.doi.org/10.14202/vetworld.2021.302-313] [PMID: 33776295]

[38]   Lanusse, C.; Canton, C.; Virkel, G.; Alvarez, L.; Costa-Junior, L.; Lifschitz, A. Strategies to Optimize the Efficacy of Anthelmintic Drugs in Ruminants. *Trends Parasitol.,* **2018**, *34*(8), 664-682.
[http://dx.doi.org/10.1016/j.pt.2018.05.005] [PMID: 29960843]

[39]   Luque, S.; Lloberas, M.; Cardozo, P.; Virkel, G.; Farias, C.; Viviani, P.; Lanusse, C.; Alvarez, L.; Lifschitz, A. Combined moxidectin-levamisole treatment against multidrug-resistant gastrointestinal

nematodes: A four-year efficacy monitoring in lambs. *Vet. Parasitol.,* **2021**, *290*(290), 109362.
[http://dx.doi.org/10.1016/j.vetpar.2021.109362] [PMID: 33524780]

[40]   André, W.P.P.; Cavalcante, G.S.; Ribeiro, W.L.C.; Santos, J.M.L.; Macedo, I.T.F.; Paula, H.C.B.;
       Morais, S.M.; Melo, J.V.; Bevilaqua, C.M.L. Anthelmintic effect of thymol and thymol acetate on
       sheep gastrointestinal nematodes and their toxicity in mice. *Rev. Bras. Parasitol. Vet.,* **2017**, *26*(3),
       323-330.
       [http://dx.doi.org/10.1590/s1984-29612017056] [PMID: 28977246]

[41]   Gupta, P.; Patel, D.K.; Gupta, V.K.; Pal, A.; Tandon, S.; Darokar, M.P. Citral, a monoterpenoid
       aldehyde interacts synergistically with norfloxacin against methicillin resistant Staphylococcus aureus.
       *Phytomedicine,* **2017**, *34*(April), 85-96.
       [http://dx.doi.org/10.1016/j.phymed.2017.08.016] [PMID: 28899514]

[42]   Valliammai, A.; Selvaraj, A.; Yuvashree, U.; Aravindraja, C.; Karutha Pandian, S. *sarA*-Dependent
       Antibiofilm Activity of Thymol Enhances the Antibacterial Efficacy of Rifampicin Against
       *Staphylococcus aureus. Front. Microbiol.,* **2020**, *11*, 1744.
       [http://dx.doi.org/10.3389/fmicb.2020.01744] [PMID: 32849374]

[43]   Silva, D.; Diniz-Neto, H.; Cordeiro, L.; Silva-Neta, M.; Silva, S.; Andrade-Júnior, F.; Leite, M.;
       Nóbrega, J.; Morais, M.; Souza, J.; Rosa, L.; Melo, T.; Souza, H.; Sousa, A.; Rodrigues, G.; Oliveira-
       Filho, A.; Lima, E. (R)-(+)-β-Citronellol and (S)-(−)-β-Citronellol in Combination with Amphotericin
       B against Candida Spp. *Int. J. Mol. Sci.,* **2020**, *21*(5), 1785.
       [http://dx.doi.org/10.3390/ijms21051785] [PMID: 32150884]

[44]   Pinto, Â.V.; Oliveira, J.C.; Costa de Medeiros, C.A.; Silva, S.L.; Pereira, F.O. Potentiation of
       antifungal activity of terbinafine by dihydrojasmone and terpinolene against dermatophytes. *Lett. Appl.
       Microbiol.,* **2021**, *72*(3), 292-298.
       [http://dx.doi.org/10.1111/lam.13371] [PMID: 32790923]

[45]   Silva, CR; Lifschitz, AL; Macedo, SR; Campos, NR; Viana-Filho, M; Alcântara, AC Combination of
       synthetic anthelmintics and monoterpenes: Assessment of efficacy, and ultrastructural and biophysical
       properties of Haemonchus contortus using atomic force microscopy. Vet Parasitol, **2021**, *290*, 109345.
       [http://dx.doi.org/10.1016/j.vetpar.2021.109345]

[46]   Lanusse, C.E.; Prichard, R.K. Relationship between pharmacological properties and clinical efficacy
       of ruminant anthelmintics. *Vet. Parasitol.,* **1993**, *49*(2-4), 123-158.
       [http://dx.doi.org/10.1016/0304-4017(93)90115-4] [PMID: 8249240]

[47]   Aldred, E.M.; Buck, C.; Vall, K. Terpenes-Chapter 22. Pharmacology, Churchill Livingstone
       [Internet]. **2009**, 167-174.
       [http://dx.doi.org/10.1016/B978-0-443-06898-0.00022-0]

[48]   Kohlert, C.; Schindler, G.; März, R.W.; Abel, G.; Brinkhaus, B.; Derendorf, H.; Gräfe, E.U.; Veit, M.
       Systemic availability and pharmacokinetics of thymol in humans. *J. Clin. Pharmacol.,* **2002**, *42*(7),
       731-737.
       [http://dx.doi.org/10.1177/009127002401102678] [PMID: 12092740]

[49]   Austgulen, L; Solheim, E; Scheline, R Metabolism in rats of *p*-cymene derivatives: carvacrol and
       thymol. Pharmacol Toxicol. **1987**, *61*, 98-102.
       [http://dx.doi.org/10.1111/j.1600-0773.1987.tb01783.x]

[50]   Takada, M.; Agata, I.; Sakamoto, M.; Yagi, N.; Hayashi, N. On the metabolic detoxication of thymol
       in rabbit and man. *J. Toxicol. Sci.,* **1979**, *4*(4), 341-349.
       [http://dx.doi.org/10.2131/jts.4.341] [PMID: 548583]

[51]   Kohlert, C.; van Rensen, I.; März, R.; Schindler, G.; Graefe, E.U.; Veit, M. Bioavailability and
       pharmacokinetics of natural volatile terpenes in animals and humans. *Planta Med.,* **2000**, *66*(6), 495-
       505.
       [http://dx.doi.org/10.1055/s-2000-8616] [PMID: 10985073]

[52]   Koritz, GD Influence of ruminant gastrointestinal physiology on the pharmacokinetics of drugs in

dosage forms administered orally. J Vet Pharmacol. Ther. **1983**, 151-163.
[http://dx.doi.org/10.1007/978-94-009-6604-8]

[53]    Virkel, G.; Lifschitz, A.; Pis, A.; Lanusse, C. *In vitro* ruminal biotransformation of benzimidazole sulphoxide anthelmintics: enantioselective sulphoreduction in sheep and cattle. *J. Vet. Pharmacol. Ther.,* **2002**, *25*(1), 15-23.
[http://dx.doi.org/10.1046/j.1365-2885.2002.00373.x] [PMID: 11874522]

[54]    Irazoqui, I.; Rodriguez, A.; Birriel, E.; Gabay, M.; Lavaggi, M.; Repetto, J.; Cajarville, C.; Gonzalez, M.; Cerecetto, H. Anaerobic metabolism of the agro-pesticide nitroxinil by bovine ruminal fluid. *Drug Metab. Lett.,* **2015**, *8*(2), 101-108.
[http://dx.doi.org/10.2174/1872312809666141208150629] [PMID: 25496284]

[55]    Miró, V.; Lifschitz, A.; Viviani, P.; Rocha, C.; Lanusse, C.; Costa, L. *In vitro* inhibition of the hepatic S-oxygenation of the anthelmintic albendazole by the natural monoterpene thymol in sheep. *Xenobiotica,* **2019**, 1-7.
[http://dx.doi.org/10.1080/00498254.2019.1644390] [PMID: 31305200]

[56]    Ballent, M.; Virkel, G.; Maté, L.; Viviani, P.; Lanusse, C.; Lifschitz, A. Hepatic biotransformation pathways and ruminal metabolic stability of the novel anthelmintic monepantel in sheep and cattle. *J. Vet. Pharmacol. Ther.,* **2016**, *39*(5), 488-496.
[http://dx.doi.org/10.1111/jvp.12296] [PMID: 26923886]

[57]    Lifschitz, A.; Virkel, G.; Ballent, M.; Sallovitz, J.; Pis, A.; Lanusse, C. Moxidectin and ivermectin metabolic stability in sheep ruminal and abomasal contents. *J. Vet. Pharmacol. Ther.,* **2005**, *28*(5), 411-418.
[http://dx.doi.org/10.1111/j.1365-2885.2005.00674.x] [PMID: 16207302]

[58]    Virkel, G.; Lifschitz, A.; Sallovitz, J.; Pis, A.; Lanusse, C. Comparative hepatic and extrahepatic enantioselective sulfoxidation of albendazole and fenbendazole in sheep and cattle. *Drug Metab. Dispos.,* **2004**, *32*(5), 536-544.
[http://dx.doi.org/10.1124/dmd.32.5.536] [PMID: 15100176]

[59]    Miró, M.V.; e Silva, C.R.; Viviani, P.; Luque, S.; Lloberas, M.; Costa-Júnior, L.M.; Lanusse, C.; Virkel, G.; Lifschitz, A. Combination of bioactive phytochemicals and synthetic anthelmintics: *In vivo* and *in vitro* assessment of the albendazole-thymol association. *Vet. Parasitol.,* **2020**, *281*, 109121.
[http://dx.doi.org/10.1016/j.vetpar.2020.109121] [PMID: 32361524]

[60]    Wink, M.; Ashour, M.L.; Youssef, F.S.; Gad, H.A. Inhibition of cytochrome P450 (CYP3A4) activity by extracts from 57 plants used in traditional chinese medicine (TCM). *Pharmacogn. Mag.,* **2017**, *13*(50), 300-308.
[http://dx.doi.org/10.4103/0973-1296.204561] [PMID: 28539725]

[61]    Shimada, T.; Shindo, M.; Miyazawa, M. Species differences in the metabolism of (+)- and (-)-limonenes and their metabolites, carveols and carvones, by cytochrome P450 enzymes in liver microsomes of mice, rats, guinea pigs, rabbits, dogs, monkeys, and humans. *Drug Metab. Pharmacokinet.,* **2002**, *17*(6), 507-515.
[http://dx.doi.org/10.2133/dmpk.17.507] [PMID: 15618705]

[62]    Han, E.H.; Hwang, Y.P.; Jeong, T.C.; Lee, S.S.; Shin, J.G.; Jeong, H.G. Eugenol inhibit 7,12-dimethylbenz[a]anthracene-induced genotoxicity in MCF-7 cells: Bifunctional effects on CYP1 and NAD(P)H:quinone oxidoreductase. *FEBS Lett.,* **2007**, *581*(4), 749-756.
[http://dx.doi.org/10.1016/j.febslet.2007.01.044] [PMID: 17275817]

[63]    Seo, K.A.; Kim, H.; Ku, H.Y.; Ahn, H.J.; Park, S.J.; Bae, S.K.; Shin, J.G.; Liu, K.H. The monoterpenoids citral and geraniol are moderate inhibitors of CYP2B6 hydroxylase activity. *Chem. Biol. Interact.,* **2008**, *174*(3), 141-146.
[http://dx.doi.org/10.1016/j.cbi.2008.06.003] [PMID: 18611395]

[64]    Virkel, G.; Ballent, M.; Lanusse, C.; Lifschitz, A. Role of ABC Transporters in Veterinary Medicine: Pharmaco- Toxicological Implications. *Curr. Med. Chem.,* **2019**, *26*(7), 1251-1269.

[http://dx.doi.org/10.2174/0929867325666180201094730] [PMID: 29421996]

[65] Lifschitz, A.; Entrocasso, C.; Alvarez, L.; Lloberas, M.; Ballent, M.; Manazza, G.; Virkel, G.; Borda, B.; Lanusse, C. Interference with *P*-glycoprotein improves ivermectin activity against adult resistant nematodes in sheep. *Vet. Parasitol.,* **2010**, *172*(3-4), 291-298.
[http://dx.doi.org/10.1016/j.vetpar.2010.04.039] [PMID: 20605686]

[66] Eid, S.Y.; El-Readi, M.Z.; Eldin, E.E.M.N.; Fatani, S.H.; Wink, M. Influence of combinations of digitonin with selected phenolics, terpenoids, and alkaloids on the expression and activity of P-glycoprotein in leukaemia and colon cancer cells. *Phytomedicine,* **2013**, *21*(1), 47-61.
[http://dx.doi.org/10.1016/j.phymed.2013.07.019] [PMID: 23999162]

[67] Yoshida, N.; Takagi, A.; Kitazawa, H.; Kawakami, J.; Adachi, I. Inhibition of *P*-glycoprotein-mediated transport by extracts of and monoterpenoids contained in Zanthoxyli Fructus. *Toxicol. Appl. Pharmacol.,* **2005**, *209*(2), 167-173.
[http://dx.doi.org/10.1016/j.taap.2005.04.001] [PMID: 15890377]

[68] Yu, J.; Zhou, P.; Asenso, J.; Yang, X.D.; Wang, C.; Wei, W. Advances in plant-based inhibitors of P-glycoprotein. *J. Enzyme Inhib. Med. Chem.,* **2016**, *31*(6), 867-881.
[http://dx.doi.org/10.3109/14756366.2016.1149476] [PMID: 26932198]

[69] Miró, M.V.; Luque, S.; Cardozo, P.; Lloberas, M.; Sousa, D.M.; Soares, A.M.S.; Costa-Junior, L.M.; Virkel, G.L.; Lifschitz, A.L. Plant-Derived Compounds as a Tool for the Control of Gastrointestinal Nematodes: Modulation of Abamectin Pharmacological Action by Carvone. *Front. Vet. Sci.,* **2020**, *7*, 601750.
[http://dx.doi.org/10.3389/fvets.2020.601750] [PMID: 33392294]

[70] Zhao, Q.; Luan, X.; Zheng, M.; Tian, X.H.; Zhao, J.; Zhang, W.D.; Ma, B.L. Synergistic mechanisms of constituents in herbal extracts during intestinal absorption: focus on natural occurring nanoparticles. *Pharmaceutics,* **2020**, *12*(2), 128.
[http://dx.doi.org/10.3390/pharmaceutics12020128] [PMID: 32028739]

[71] Moroy, G.; Martiny, V.Y.; Vayer, P.; Villoutreix, B.O.; Miteva, M.A. Toward *in silico* structure-based ADMET prediction in drug discovery. *Drug Discov. Today,* **2012**, *17*(1-2), 44-55.
[http://dx.doi.org/10.1016/j.drudis.2011.10.023] [PMID: 22056716]

[72] Zehetner, P.; Höferl, M.; Buchbauer, G. Essential oil components and cytochrome P450 enzymes: a review. *Flavour Fragrance J.,* **2019**, *34*(4), 223-240.
[http://dx.doi.org/10.1002/ffj.3496]

[73] Zhang, Y.; Long, Y.; Yu, S.; Li, D.; Yang, M.; Guan, Y. Natural volatile oils derived from herbal medicines: A promising therapy way for treating depressive disorder. *Pharmacol. Res.,* **2020**, *105376*
[http://dx.doi.org/10.1016/j.phrs.2020.105376] [PMID: 33316383]

[74] Ghannay, S.; Kadri, A.; Aouadi, K. Synthesis, *in vitro* antimicrobial assessment, and computational investigation of pharmacokinetic and bioactivity properties of novel trifluoromethylated compounds using *in silico* ADME and toxicity prediction tools. *Monatsh. Chem.,* **2020**, *151*(2), 267-280.
[http://dx.doi.org/10.1007/s00706-020-02550-4]

[75] Kaplan, R.M.; Vidyashankar, A.N. An inconvenient truth: Global worming and anthelmintic resistance. *Vet. Parasitol.,* **2012**, *186*(1-2), 70-78.
[http://dx.doi.org/10.1016/j.vetpar.2011.11.048] [PMID: 22154968]

[76] Mengistu, G.; Hoste, H.; Karonen, M.; Salminen, J.P.; Hendriks, W.H.; Pellikaan, W.F. The *in vitro* anthelmintic properties of browse plant species against Haemonchus contortus is determined by the polyphenol content and composition. *Vet. Parasitol.,* **2017**, *237*, 110-116.
[http://dx.doi.org/10.1016/j.vetpar.2016.12.020] [PMID: 28262394]

[77] Oliveira, A.F.; Costa Junior, L.M.; Lima, A.S.; Silva, C.R.; Ribeiro, M.N.S.; Mesquista, J.W.C.; Rocha, C.Q.; Tangerina, M.M.P.; Vilegas, W. Anthelmintic activity of plant extracts from Brazilian savanna. *Vet. Parasitol.,* **2017**, *236*, 121-127.
[http://dx.doi.org/10.1016/j.vetpar.2017.02.005] [PMID: 28288755]

[78]   Lifschitz, A.; Imperiale, F.; Virkel, G.; Muñoz Cobeñas, M.; Scherling, N.; DeLay, R.; Lanusse, C. Depletion of moxidectin tissue residues in sheep. *J. Agric. Food Chem.,* **2000**, *48*(12), 6011-6015.https://pubs.acs.org/doi/abs/10.1021/jf0000880
[http://dx.doi.org/10.1021/jf0000880] [PMID: 11141269]

[79]   Garcia-Bustos, J.F.; Sleebs, B.E.; Gasser, R.B. An appraisal of natural products active against parasitic nematodes of animals. *Parasit. Vectors,* **2019**, *12*(1), 306.
[http://dx.doi.org/10.1186/s13071-019-3537-1] [PMID: 31208455]

[80]   Katiki, L.M.; Araujo, R.C.; Ziegelmeyer, L.; Gomes, A.C.P.; Gutmanis, G.; Rodrigues, L.; Bueno, M.S.; Veríssimo, C.J.; Louvandini, H.; Ferreira, J.F.S.; Amarante, A.F.T. Evaluation of encapsulated anethole and carvone in lambs artificially- and naturally-infected with Haemonchus contortus. *Exp. Parasitol.,* **2019**, *197*, 36-42.
[http://dx.doi.org/10.1016/j.exppara.2019.01.002] [PMID: 30633915]

[81]   Ballent, M.; Canton, C.; Dominguez, P.; Bernat, G.; Lanusse, C.; Virkel, G.; Lifschitz, A. Pharmacokinetic-pharmacodynamic assessment of the ivermectin and abamectin nematodicidal interaction in cattle. *Vet. Parasitol.,* **2020**, *279*, 109010.
[http://dx.doi.org/10.1016/j.vetpar.2019.109010] [PMID: 32035291]

[82]   Kotze, A.C.; Prichard, R.K. Anthelmintic Resistance in Haemonchus contortus. *Adv. Parasitol.,* **2016**, *93*, 397-428.
[http://dx.doi.org/10.1016/bs.apar.2016.02.012] [PMID: 27238009]

[83]   Foucquier, J.; Guedj, M. Analysis of drug combinations: current methodological landscape. *Pharmacol. Res. Perspect.,* **2015**, *3*(3), e00149.
[http://dx.doi.org/10.1002/prp2.149] [PMID: 26171228]

# CHAPTER 8

# Terpenes Behavior in Soil

**Marcia M. Mauli[1], Adriana M. Meneghetti[2] and Lúcia H. P. Nóbrega[3]**

[1] *Secretaria do Estado da Educação do Paraná (Seed), Paraná, Brasil*

[2] *Universidade Tecnológica Federal do Paraná, Brazil*

[3] *Universidade Estadual do Oeste do Paraná, Brazil*

**Abstract:** Soil is a complex and dynamic system in constant change due to its natural processes, as well as interaction among physical, chemical and biological characteristics that take part in it. However, the greatest transformation occurred due to the farm business and the adopted management system. Thus, man can manipulate some soil characteristics and make it more suitable for cropping development. Although anthropic action cannot fully control how soil characteristics interact, it is possible to track them. The action of chemical substances should not be disregarded, a product of the secondary metabolism of plants, since they interfere with plant's ability to compete and survive. Such substances can act out as protectors against herbivores and pathogens. They can be attractive or repellent agents in plant-plant competition and plant-microorganism symbiosis. They can also influence the interaction between plant matter and soil organisms. Among these substances, terpenoids are highlighted as the most structurally diverse chemical family in the class of secondary metabolites that are part of natural products. This knowledge allows a better understanding of nutrient decomposition and cycling processes, the influence of environmental factors on production and terpenoid variability in some plants with medicinal and economic importance.

**Keywords:** Allelochemicals in Soil, Ecological Interactions in Soil, Secondary Metabolism.

## INTRODUCTION

### Terpenoids

Secondary metabolites of plants can be classified into three chemically distinct groups: phenolic compounds, nitrogen compounds and terpenoids [1]. Terpenoids, also known as isoprenoids or terpenes, constitute the most chemically structurally diverse family of the secondary metabolite class, which are part of natural products. The term terpenoid should rather be more used than terpene, which

---

\* **Corresponding author Marcia M. Mauli:** Secretaria do Estado da Educação do Paraná (Seed), Paraná, Brasil; E-mail: marcia.m.mauli@gmail.com

**Mozaniel Santana de Oliveira & Antônio Pedro da Silva Souza Filho (Eds.)**

should be used for compounds that are alkenes. The first known terpene structures were α-pinene and camphor, isolated from turpentine [2].

According to Zhou, Picherskym [3], Christianson [4], and Priya *et al.* [5], there are more than 80,000 terpenoid compounds. The terpenome is responsible for almost a third of all compounds described in the Dictionary of Natural Products (http://dnp.chemnetbase.com). These compounds have broad physiological functions, including respiration, photosynthesis, growth, development, reproduction, defense and environmental sensing [3, 4, 6 - 8]. There are also terpenoids derived from animals (cholesterol, dolichol, ubiquinone), which take part in the formation of cell membranes, glycoprotein biosynthesis and intracellular electron transport [9], and others that come from plants (tocopherol, brassinolid and gibberellin). They are responsible for growth regulation and cellular defense [10], and several ecological functions (volatile monoterpenes attract pollinators and sesquiterpenes are present in floral aromas) [3]. These volatile compounds often play an essential role in a plant's defense system, both directly and indirectly, as volatile compounds that repel or attract other insects, respectively [2, 11].

In nature, they also play a significant role in plant-environment interactions, plant-plant communication, and plant-insect and plant-animal interactions [12, 13]. Isoprenoids or terpenoids not only serve as vital allelochemicals in plant defense, but also in several other secondary metabolic processes and plant communication. Some terpenoids are commercially useful, such as pharmaceuticals, flavorings, and biofuels. They are used in food, cosmetic and agricultural industries [14 - 16]. Terpenes can also serve as a source of new drugs or as prototypes for the development of effective pharmaco-therapeutic agents [17, 18].

Despite their structural diversity, all terpenoids are derived from the repetitive bonding of five branching carbons: isopentane (1) - these monomers are referred to as isoprene units (2) - Terpenoids begin with two isoprene-like building blocks, isopentenyl diphosphate (IPP) (3) and dimethylallyl diphosphate (DMAPP) (4) (Fig. **1**) [19, 20].

These isoprenic isomers are grouped into categories of natural products, based on their structures, in two pathways for their biosynthesis, and have evolved in different taxonomical organisms [21]. The plants usually use two metabolically separated pathways for IPP and DMAPP biosynthesis in different cell compartments, the mevalonate and non-mevalonate pathways [20, 22].

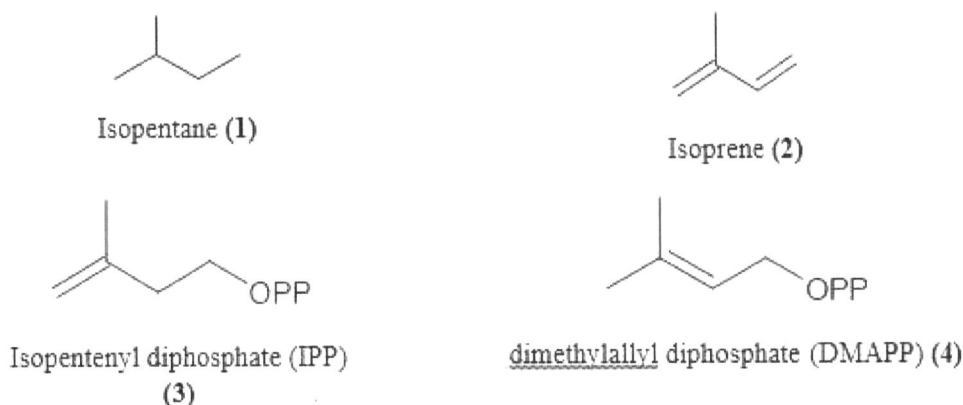

Isopentane (1)

Isoprene (2)

Isopentenyl diphosphate (IPP)
(3)

dimethylallyl diphosphate (DMAPP) (4)

**Fig. (1).** 2D isoprene unit formulas.

The non-mevalonate pathway, also known as 2-C-methyl-*D*-erythritol 4-phosphate (MEP) or 1-deoxy-*D*-xylulose 5-phosphate (DXP) pathway, simultaneously produces IPP and DMAPP from the condensation reaction between a pyruvate molecule and 3-phosphate gliceraldehyde, located in plastids, while mevalonic acid (MVA) pathway (Fig. **2**) synthesizes IPP from the reaction of three molecules of Acetyl-CoA to form mevalonic acid. This last acid, after undergoing pyro-phosphorylation, decarboxylation and dehydration reactions, results in IPP and is distributed among cytoplasm, endoplasmic reticulum and peroxisomes in eukaryotes [2, 4, 21, 22], and, despite this compartmentalization, there is some evidence of exchange limited number of common precursors among plastids and cytosol [13, 19, 20].

IPP is a phosphorus-activated compound that becomes its DMAPP isomer, which is biosynthesized by the MEP pathway that occurs in chloroplasts and has an oxygen-pyrophosphate group (OPP). After its oxygen protonation and allelic cation formation, dimerization occurs with geranyl diphosphate formation (GPP) [23 - 25]. Terpenoids are derived from the precursor compounds of IPP and DMAPP and can be classified according to the amount of isoprene residues. They exist as single-unit hemiterpenoid ($C_5$), monoterpenoid ($C_{10}$), sesquiterpenoid ($C_{15}$), diterpenoid ($C_{20}$), sesteterpenoid ($C_{25}$), triterpenoid ($C_{30}$), tetraterpenoid ($C_{40}$), and polyesterpenoids (> $C_{40}$) and are sub-classified in terms of the degree of cyclization into acyclic, monocyclic or bicyclic [2, 24].

Terpenoids formation occurs by the complete addition of their building blocks (IPP and DMAPP), which are biological equivalents of isoprene, first a head-to-tail condensation of IPP and DMAPP occurs, producing geranyl diphosphate (GPP), a monoterpenoid precursor. The IPP successive addition results in the formation of sesquiterpenoid and diterpenoid precursors, farnesyl diphosphate

(FPP) and geranylgeranyl diphosphate (GGPP), respectively. Thus, FPP and GGPP are condensed head to head to form squalene and phytene, respectively. Then, these linearized precursors are cyclized and modified by oxidation and acetylation to form several terpenoids [20, 11, 20, 22, 26, 27], which inform that the plants have an MVA pathway into the cytoplasm to form sesquiterpenoids and triterpenoids, while MEP pathway in plastids forms monoterpenoids, diterpenoids and tetraterpenoids, respectively.

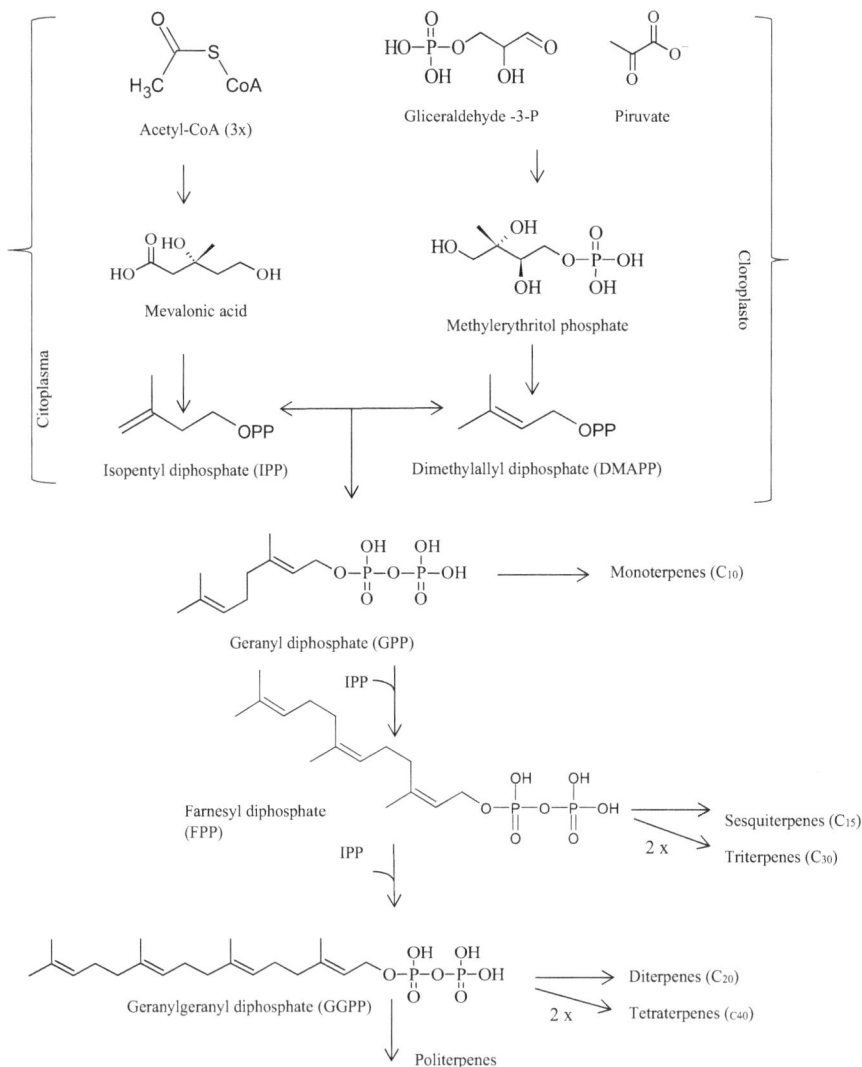

**Fig. (2).** Scheme of terpenoid synthesis of both metabolic pathways, the mevalonate pathway (MEV) and metyl-erythritol -phosphate pathway (MEP) and their classification according to isoprene units. Source: Adapted from [23].

Synthetic pathways have also been reported, in addition to the natural ones [25, 28, 29]. The central structures of terpenes are post-modified by P450s (P450s) cytochromes, which play a vital role in donating several bioactivities to the terpenoids.

## Biosynthesis of IPP (Isopentyl Diphosphate) and DMAPP (Dimethylallyl Diphosphate)

### *The Mevalonic Acid Pathway (MVA)*

MVA produces one IPP molecule, and consumes 3 acetyl-CoAs, 3 ATPs, plus a reduction of 2 NPDPH. This begins with the condensation of two molecules of acetyl-CoA to form acetoacetyl-CoA, a reaction catalyzed by acetyl-CoA acetyltransferase (AACT). Then, there is an additional condensation of a third molecule of acetyl-CoA to produce 3-hydroxy-3-methyl-glutaryl-CoA (HMG-Coa), a reaction catalyzed by hydroxymethylglutaryl-CoA synthase (HMGS). The limiting step of carbon flow rate through MVA pathway is HMG-CoA reduction, into the endoplasmic reticulum, to mevalonate by HMG-CoA reductase (HMGR), consuming two molecules of NADPH in the process [25, 29]. The resulting MVA is activated by two phosphorylation steps catalyzed by mevalonate kinase (MVK) and phosphomevalonate kinase (PMK). The final step of MVA pathway is ATP-dependent decarboxylation catalyzed by mevalonate pyrophosphate decarboxylase (MPD) to produce IPP as the product. Two structurally unrelated IPPs: DMAPP isomerases (IDI-1 and IDI-2), are then responsible for the interconversion of IPP and DMAPP [11, 30, 31, 34].

## Methylerithritol-Phosphate Pathway (MEP)

MEP was accepted exclusively as a source of biosynthesis for isoprenodes. However, inconsistencies were observed in experiments with isotopic marking, which showed a lack of specific incorporation to a presumed start of materials and intermediates of several terpenoids [16, 32]. While a potential acetolactate pathway was refuted, some proposals either resembled MVA-based biosynthesis, or could not be tested experimentally. However, it was repeatedly concluded that Acetyl-CoA worked as a universal precursor to IPP [32 - 34].

In 1996, Michel Rohmer discovered the first stage of the alternative MEP pathway that was found in prokaryote plastids (but was not present in humans). This pathway produces IPP and DMAPP for terpenoid synthesis, to which *D*-glyceraldehyde-3-phosphate (GAP) and pyruvate undergo condensation and reduction, at the expense of 3 ATPs and 3-equivalent NADPH. Condensation of pyruvate and *D*-glyceraldehyde 3-phosphate (GAP) to obtain 1-deoxy-*D*-xylulose-5-phosphate (DXP) is a reaction catalyzed by the thiamine-diphosphate,

1-deoxy-*D*-xylulose-dependent enzyme 5-phosphate synthase (DXS). After DXP formation, it is reductively isomerized by DXP reduct-isomerase (DXR) in 2-C-methyl-*D*-erythritol 4-phosphate (MEP), from which MEP pathway gets its name. MEP is then activated by the enzyme 2-C-methyl-*D*-erythritol-2,4 cyclodiphosphate (MEcPP) synthase (MDS) and then the subsequent cyclization of MEcPP occurs [11, 16].

Both final steps are catalyzed by two iron-sulfur-containing enzymes, and MEcPP is reduced to 1-hydroxy-2-methyl-2-butenyl-4-diphosphate (HMBPP) by ferric sulfur protein HMBPP synthase (HDS) and again reductase (HDR) is responsible for the ring opening and reductive dehydration of MEcPP to produce 4-hydroxy-3-methylbutenyl 1-diphosphate (HMBPP), and by reductive dehydration of HMBPP to produce IPP and DMAPP. Thus, IDI is not essential for the survival of many organisms that use the MEP pathway, although it can play a role in modulating the IPP / DMAPP ratio in a cell [11, 30 - 33, 35].

## Hemiterpenes $C_5$

Isoprene is the most abundant hemiterpene in plants, as well as a volatile compound synthesized from DMAPP, and widely distributed in the plant kingdom. There are mosses, ferns, gymnosperms and angiosperms, and some trees, such as poplar and plants from humid tropics, among the species that synthesize isoprene. Isopropenes are emitted to the atmosphere in order to protect leaves from overcoming short periods of high temperatures, and this increases the plant's tolerance to ozone and oxygen-reactive species [11]. Hemiterpenes also act out as signaling agents, and when a plant is damaged, they emit volatile compounds such as methacrolein (5), hexenal compounds, monoterpenes and methyljasmonate (6) (Fig. 3), that are perceived by plants in the neighborhood, allowing a faster reaction when there is an attack [36].

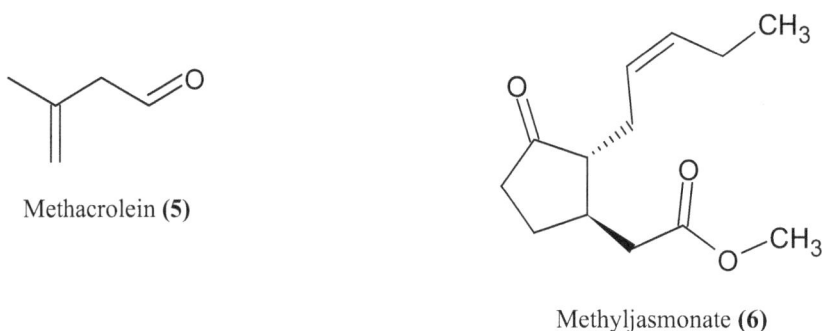

Methacrolein (5)

Methyljasmonate (6)

**Fig. (3).** 2D hemiterpene unit formulas.

## Monoterpenes C$_{10}$

Monoterpenes are isoprene dimers, originating from a DMAPP molecule and an IPP molecule, united, in most cases, by head and tail, forming transgeranyl diphosphate (GPP). This can be folded into mono, bi and tricyclic and undergoes transformations to produce several monoterpenes [37 - 40]. Monoterpenes are volatile lipophilic compounds that occur in coniferous resins, essential oils and floral aromas and contribute to the characteristic flavor and aroma of many plants [11, 38 - 40]. As an example (Fig. **4**), limonene, formed up from geranyl diphosphate, by carbocation cyclization, which, in the end, involves the loss of a proton by cation to form an alkene.

Geranyl diphosphate (GPP)                                               Limonene

**Fig. (4).** Formation of limonene from geranyl diphosphate. Source [38].

According to their structure showed in Fig. (**5**), monoterpenes are separated into two groups: the acyclic group has an open chain (mycrene (7), ocimene (8)) and the cyclic one can be mono - (limonene, Perillyl acid (9)), to several cyclic ones (camphor (10) pinenes (11), sabinene (12)) [40].

Myrcene (**7**)

Perillyl acid (**9**)

Camphor (**10**)

α-pinene (**11**)

Ocimene (**8**)

Sabinene (**12**)

Geraniol (**13**)

**Fig. (5).** Structural formulas of monoterpenoids.

Monoterpenes such as mircene and ocimene are components of essential oils of hops and laurel, while geraniol (13) (monoterpenic alcohol) is a component of essential oils of rose, geranium and other flower essences; limonene was found in the pine bark, in lemon and turpentine, as well as in cumin oil (Fig. 6) [40, 41]. Since they are volatile, a large amount can be grouped in specialized structures, as in *Lamiaceae* and *Asteraceae*, as they have large trichomes with secretory cells that produce terpenes and secrete them in a shared subcuticular storage cavity [42]. Likewise, conifers accumulate oleoresin, which is a complex mixture of mono-, sesqui- and diterpenoids, in bubbles or resin ducts, covered by a layer of epithelial cells that synthesize and secrete terpenes in the lumen.

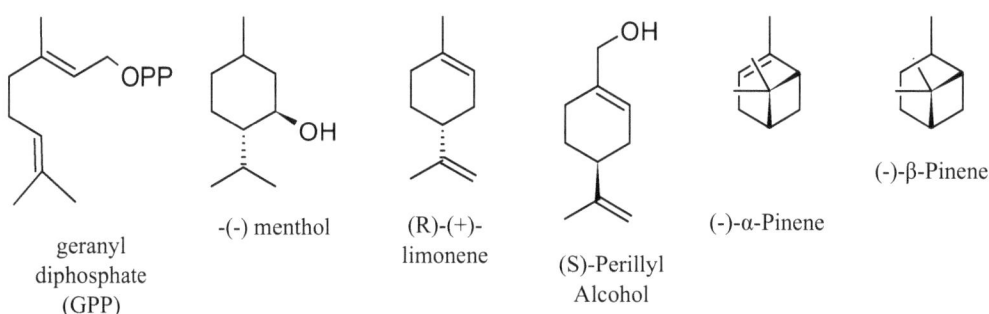

**Fig. (6).** Structures of some monoterpenes derived from geranyl diphosphate (Source [38]).

The physiological function of monoterpenes is a defense, used to attract pollinators and make some plant-plant communication and plant-insect interactions [43]. They were particularly studied, taking conifers and the bark beetle as examples, which damage the tissue; thereafter, the plant secrets an oleoresin, named turpentine, which consists of biologically active mono and sesquiterpenes, such as limonene and pinene. After turpentine evaporates, the remaining non-volatile resin solidifies, trapping predators and sealing the damage [44].

Despite their toxicity, oleoresin monoterpenes work as olfactory signals that help bark beetles to find their host. Monoterpenes ingested by beetles are converted into pheromones that attract more beetles or serve as anti-aggregating signals. In addition, conifer monoterpenes participate in tritrophic interactions and attract predatory insects that feed themselves on bark beetles of trees [45].

Low-molecular 10-carbon monoterpenoids and 15-carbon sesquiterpenoids are emitted by plants as volatile compounds, and play roles in direct defenses against pathogens and herbivores and indirectly in attracting parasites to plants [46 - 48]. Monoterpenoids such as linalool **(14)** act out against bacterial rice blight caused by *Xanthomonas oryzae,* and recently, rice (S)-limonene synthase (OsTPS19) acts out directly against the fungus *Magnaporthe oryzae* [49 - 51].

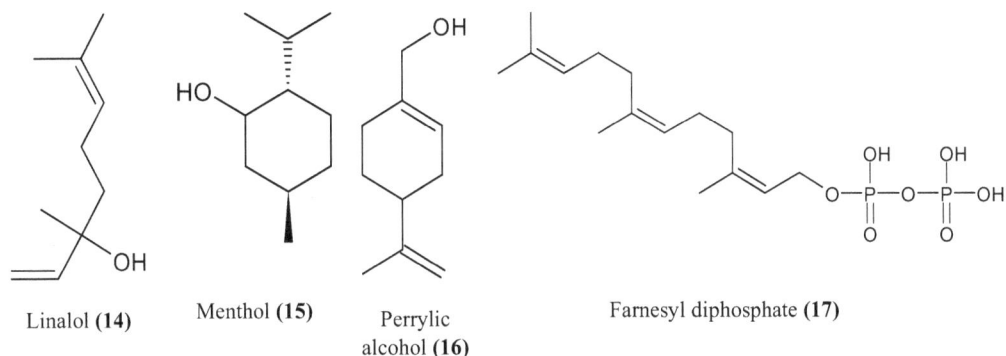

Linalol (**14**)      Menthol (**15**)      Perrylic alcohol (**16**)      Farnesyl diphosphate (**17**)

**Fig. (7).** Monoterpenes used in cosmetics.

Many monoterpenes are used in perfumery, aromatherapy, as cosmetics and insecticides. Menthol (**15**), a constituent of essential oils of *Mentha* species, is one of the most used monoterpenes in pharmaceutical products, oral health products, chewing gums and tobacco products [42].

Two monoterpenes with a promising anticancer effect are perrylic alcohol (**16**) and (+) - (R) - limonene (Fig. **7**) [52 - 66]. Both compounds induce apoptosis and suppress the translation of 3-hydroxy-3-methylglutaryl-CoA reductase (HMG-CoA), an enzyme of MVA pathway. This enzyme is a promising target for anti-tumor compounds, because many proteins involved in cell growth are prenylated (addition of hydrophobic molecules to a protein), and tumor cells have high levels of HMG-CoA reductase. HMG-CoA reductase suppression is enough to decrease terpene biosynthesis in humans, since animals do not have the alternative DXP pathway [52].

## Sesquiterpenes C$_{15}$

Sesquiterpenes are C$_{15}$ terpenoids terpenoids shown in the Fig. (**8**), gathered from three isoprene units and formed by DMAPP condensation with two IPP molecules. The central intermediate C$_{15}$, farnesyl diphosphate (FPP), (**17**) can be folded in mono, bi- or tricyclic systems. They are usually oxygenated hydrocarbons, including lactones, alcohols, acids, aldehydes and ketones. They are found mainly in higher plants, in other marine organisms and fungi [40, 50].

Sesquiterpenes that present pentacyclic lactone, named as sesquiterpene lactones, and occur in families such as *Asteraceae*, have a bitter taste and serve as a dietary deterrent to herbivores [50 - 54]. Sesquiterpenes are usually less volatile than monoterpenes [24]. Firstly, it was assumed that all sesquiterpenes are produced *via* cytosolic, and MVA. However, some sesquiterpenes are originally from isoprene units, supplied by DXP [11, 38, 39, 55, 56], or in both biosynthetic routes, by transporting isoprenoid precursors from plastids to the cytosol [57].

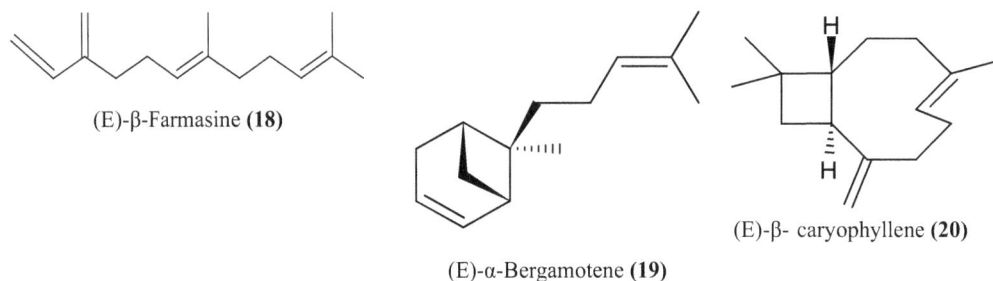

(E)-β-Farmasine **(18)**

(E)-α-Bergamotene **(19)**

(E)-β- caryophyllene **(20)**

**Fig. (8).** Structural formulas of sesquiterpenoids.

Among the important sesquiterpenes, abscisic acid is a phytohormone and stomatal closure and seed dormancy during drought are due to it. Other sesquiterpenes participate in tritrophic interactions among plants, herbivores or parasites. In maize plants (*Zea mays* L.), infested and damaged with lepidopteran larvae, sesquiterpenes (E)-β-farnesene **(18)** and (E)-α-bergamotene **(19)** attract the parasitic wasp *Cotesia marginiventris* [50, 58].

Sesquiterpene (E)-β-karyophylene **(20)**, released by leaves and roots of maize after the attack of *Diabrotica virgifera* beetle larvae, indirectly defends maize crops by attracting parasitic nematode *Heterorhabditis megidis* [46, 49, 51, 59, 60].

## Diterpenes $C_{20}$

Diterpenoids are originally from plastids in the DXP pathway and are synthesized from DMAPP and three IPP molecules, producing metabolic $C_{20}$ geranylgeranyl diphosphate (GGPP) shown in the Fig. **(9)**. GGPP is a phytyl precursor of chlorophyll and plastoquinone lipophilic side chain [61, 62]. As in smaller terpenoids, GGPP also undergoes very different cyclization and structural rearrangements. Diterpenoids such as abietic acid **(21)** and levopimaric acid **(22)** are constituents of oleoresin extracted from conifers and work as a defense against herbivores and pathogens.

After distillation removal of mono and sesquiterpenes (turpentine) from oleoresin, solid diterpene fraction is named colophonium and used in the arches  of  stringed instruments, while mono and sesquiterpenes are used as turpentine oil to remove paints and varnishes [52, 61].

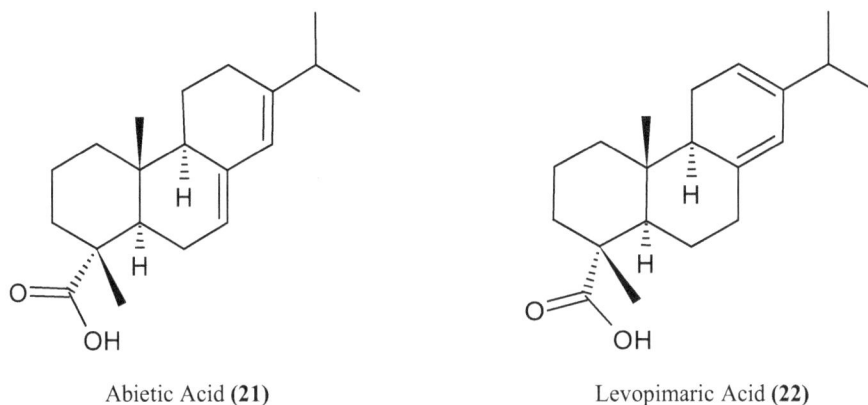

Abietic Acid (**21**)                              Levopimaric Acid (**22**)

**Fig. (9).** Structural formulas of diterpenoids.

Gibberellins are tetracyclic diterpenes that act out as phytohormones and promote shoots elongation, flowering and seed germination [63].

Some diterpenoids play important roles during development processes as intermediates in general metabolism, among them, there are biosynthesis of chlorophyll, plastoquinones and phytohormones gibberellin (GA) [62, 64 - 67]. However, most diterpenoids are specialized metabolites that often represent molecules in individual plant species or families, and the biosynthesis commonly specific to tissue or organ underlies strict regulation by environmental stimuli [10, 37, 52, 68].

## Triterpenoids $C_{30}$

Triterpenoids are synthesized *via* the MVA pathway showed in Fig. (**10**), from two FPP molecules, connected by tail-to-tail condensation of $C_{30}$ precursor, squalene, which consists of six units of $C_5$ isoprene with the following cyclization and generation of triterpenoid structures: steroids, sterols, saponins (glycosides) and others [38, 67]. The importance of triterpenoids lies in their physiological role in cells, which are essential to forming cell membranes and are the basis for indicators formation, such as sterol hormones and cognate receptors [68, 69].

Squalene is cyclized into lanosterol (a primary cholesterol and precursor to ergosterol) and into cycloartenol (B-sitosterol precursor) [40, 72, 73]. Squalene (**23**) is observed in many plants and animals, including humans, and is an intermediate during phytosterol or cholesterol biosynthesis.

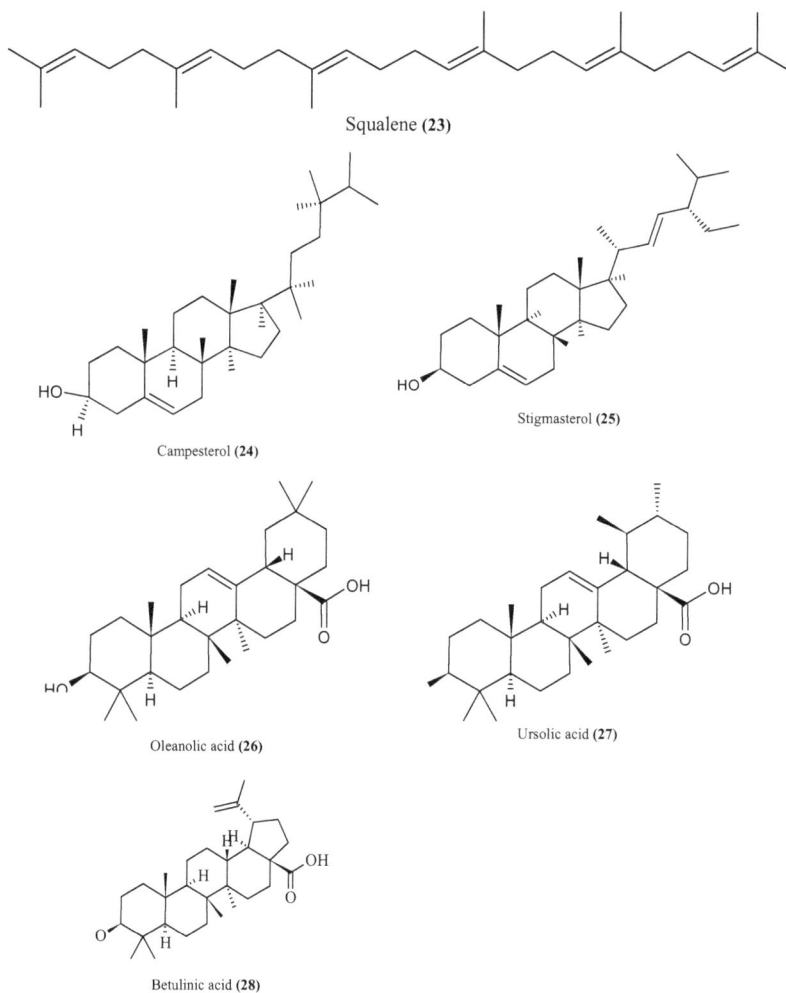

Squalene (**23**)

Campesterol (**24**)

Stigmasterol (**25**)

Oleanolic acid (**26**)

Ursolic acid (**27**)

Betulinic acid (**28**)

**Fig. (10).** Triterpenoids biosynthetic pathways of triterpenoids.

Phytosterols inhibit cholesterol absorption in animals, and in contrast to animals, whose cholesterol is the main sterol, many plant sterols are methylated or ethylated in $C_{24}$ of the side chain; campesterol (**24**) and stigmasterol (**25**) are constituents of biomembranes in plants and influence on their permeability. Phytosterol esters are used as cholesterol-lowering food additives [24, 38]. Triterpenes and triterpenoids with potential anti-aging activity are abundant in plants. For example, oleanolic acid (**26**), ursolic acid (**27**) and betulinic acid (**28**) are pentacyclic triterpene compounds detected in leaves, roots and fruits of many plant species (Fig. **10**) [11, 70, 73].

# Tetraterpenoids $C_{40}$

Tetraterpenoids are synthesized from two GGPP molecules by tail-to-tail addition and take part of only one group of compounds, the catotenoids. The addition of double bonds produces an extended conjugated system with *trans*-configuration, responsible for yellow, orange and red of carotenoids, or even, or when one or both ends of tetraterpenoid chain are cyclized in a six-membered ring [20, 24, 67].

When carotenoids have a hydroxy or epoxy function, they are classified as xanthophylls. Carotenoids present a physiological function in plants and are part of the light collection complex. They also act out as an accessory to chlorophyll pigment, attract pollinators and seed dispersers in flowers and fruits. And, some of them are essential for human health, such as α-carotene, β-carotene and β-cryptoxanthin, precursors of vitamin A (Fig. **11**) [2, 74].

## Secondary Metabolites

### *Allelochemicals Behavior in the Environment*

There is a significant interest in allelochemicals due to their extensive use in food, pharmaceuticals, cosmetics, agriculture and industries. Most terpenoids have been found in plants, where some play roles in general physiology, although most of them are restricted to a specific lineage and act out in a multitude of ecological functions [75].

Studies have shown that secondary metabolites affect the ability of plants to compete and survive, playing an important role in plants, such as protecting against herbivores and pathogenic micro-organisms, acting as attractive or even repellent, as well as agents in plant-plant competition and plant-microorganisms symbiosis. Secondary metabolites also act out in ecosystem processes, such as plant succession, litter decomposition, *etc.* Thus, they influence the interaction between plant matter and soil organisms, ergo, they are considered the most important soil and ecosystem functioning boosters. It is possible to predict the processes of decomposition and cycling of nutrients in ecosystems and future impacts of changes in their functioning when this relationship is understood [76 - 79].

Although secondary metabolites have been extensively studied on soil, their behavior in soil still needs to be understood in several aspects. They are released into the environment mainly by four routes: (i) volatilization and diffusion away from the plant tissues, (ii) leaching of plant material over soil, (iii) exudation by roots, and (iv) litter decomposition [80].

Except for volatilization, the other pathways release allelopathic compounds in soil. This plant-soil system, as well as the phytotoxic activity of allelochemicals, are influenced by several factors such as climatic (for example, solar radiation, temperature, precipitation), chemical nature of allelochemical, soil factors (for example, texture, pH, ion exchange capacity, organic matter content, nutrients dynamization, humidity and microbial ecology), and plant factors, both from the donor and the target plants (*e.g.*, species, botanical variety, growth stages, parts of the plant) [81].

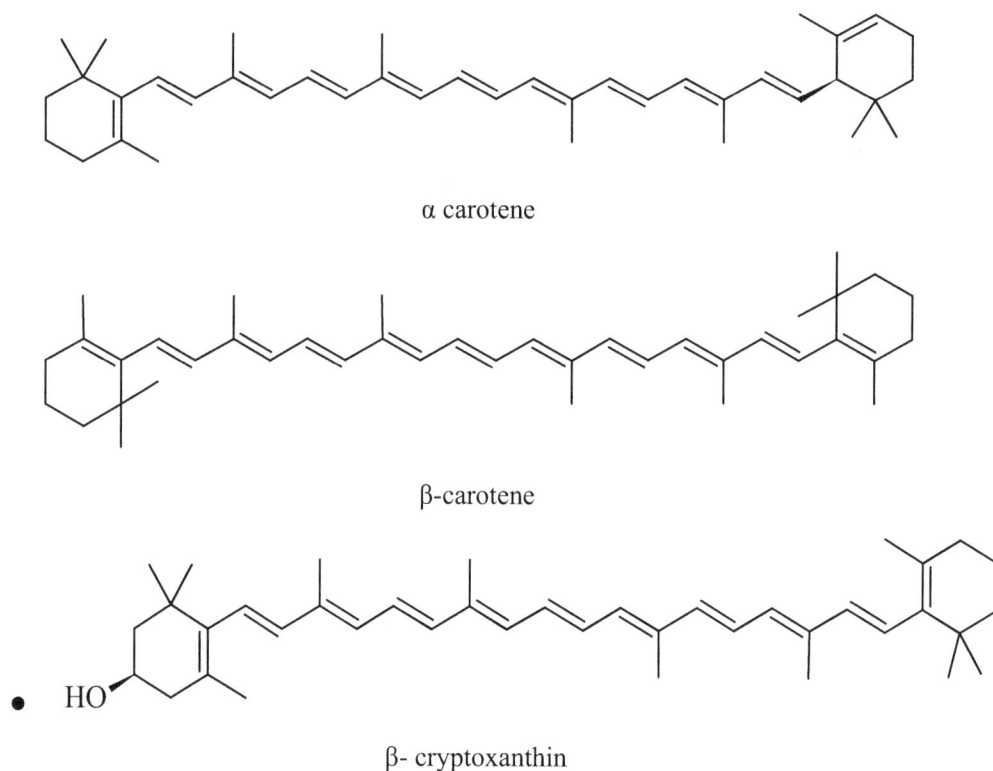

α carotene

β-carotene

β- cryptoxanthin

**Fig. (11).** Structural formula of the main carotenoids.

Lima, Kaplan and Cruz [82] highlighted the influence of environmental factors on the production and variability of terpenoids in some plants with medicinal and economic importance. According to them, the production and variability of special metabolites in plants can be evidenced under different conditions of light, temperature, levels of nutrition and water. Monoterpenoids and sesquiterpenoids produced by different groups of plants are metabolites often subject to these abiotic factors. Low light intensity generally decreases monoterpene production. Small daily variations in temperatures stimulate terpenoid production, while extreme values cause their decrease. Precipitation rates, however, do not follow a pattern and vary among species. While, nitrogen and phosphorus content increases in soil favor greater yield in essential oil content.

Root exudation is the most important pathway to release secondary metabolites in soil. However, in order to be adsorbed by the target plant's roots, allelochemicals must be present in the soil solution [83]. Therefore, a factor that can determine the phytotoxic activity of allelochemicals is their concentration in such a solution. However, once present in soil, these compounds are submitted to processes of retention, transformation and transport [84]. Retention is a physical, dynamic and reversible process, which consists of the interaction, also called adsorption or sorption, among allelochemicals and soil, water and air particles, which implies allelochemicals' mobility in soil [85].

Transformation is a biochemical process, which can be positive or negative, and occurs through micro-organisms present in the soil. They convert allelochemicals into more active, less active or totally inactive compounds. And in addition, it reduces the amount of the original allelochemical available for transportation [84]. Transportation is the movement of allelochemicals in soil and is directly related to retention and transformation processes. The interaction among them is influenced by the chemical nature of a compound, organisms involved, soil properties and environmental conditions [86].

Once allelochemicals are released into the soil system, several factors can influence retention, transportation and transformation processes and their availability, since the rhizosphere is the most active site of plant-microorganism interactions [86]. According to Weidenhamer [87], allelochemicals are continuously removed or immobilized from soil solution by leaching, microbial degradation, adsorption to soil particles and absorption by the plant. Besides an increase in nutrient solubilization, availability and efficiency of use, allelochemicals usually decrease or inhibit plant mineral absorption [88].

According to Vidal [89], allelochemical activity is also limited by time and space. The time restriction occurs because allelochemicals are not released all at once from decomposing residues. In addition, degradation and removal processes reduce the available concentration of these allelochemicals in soil solution. Space restrictions, on the other hand, occur because the allelochemical action would be spatially limited to an area near the donor plant or decomposing residues, due to some reduced concentration.

Besides the allelochemical activity, factors that influence their biosynthesis and accumulation in donor plant must be considered since they can show some genetic, ontogenic, morphogenetic and environmental character. Allelochemical content depends on several environmental factors, such as light, temperature, soil solution, soil fertility and salinity. And yet, for most plants, a change in an individual factor can cause some change in the amount of allelochemical produced and stored, even if other factors keep on constant [90].

It is also known that some of these factors do not act out by themselves and they show correlations with one another. They may also present a combined action on secondary metabolisms, such as rainfall index and seasonality, temperature and altitude, water deficiency and high temperatures, among others. So, both amount and sometimes the nature of active constituents are not constant throughout the year. Thus, it is important to understand the synergistic effects of several environmental factors on secondary plant metabolism [90, 91].

As it was already observed, water stress often has significant consequences on secondary metabolites concentration in plants, which can lead to an increase in their production. This will depend on the stress degree and period in which it occurs, with short-term effects that seem to increase production, but, at long-term, it is observed an opposite effect. On the other hand, continuous rain can result in loss of water-soluble substances from leaves and roots by leaching [92].

In agriculture, it is common management to add nutrients, mainly nitrogen, to increase biomass production. However, nutrients can affect not only primary metabolism, but also the production of different secondary metabolites. It is observed that nutritional stress usually results in an increase in secondary metabolites concentration [77, 93].

Over-fertilization induces the plants to store and release a wide variety of terpenoids. This diversity is considered as a result of selection pressures, and abiotic and biotic factors, which work concomitantly in plant species [94]. Variation in terpenoids' distribution induced by environmental or seasonal stimuli can vary depending on plant's chemotype, which suggests that plants that are genetically different may have different responses to environmental stress [95]. According to Copolovici and Niinemets [96], stress conditions can increase or inhibit terpenoid production, changing its emission pattern and amount.

The usual emission of terpenoids is higher during the day than at night. Emissions of monoterpenes and sesquiterpenes are generally higher during summer than winter. However, emissions can also vary depending on changes in temperature, radiation, water availability or physiological changes [97, 98]. Terpenoid emissions usually present stronger and faster short-term responses to environmental factors due to the absence of terpenoid reservoirs that reduce the direct influence of the environment [99].

On the other hand, some studies have shown that secondary metabolites of a wild plant species, sampled directly in its habitat, remained in constant concentrations during two years of study. Therefore, such observation shows that, in some cases, secondary metabolism may not change due to climatic, temporal or environmental factors [100].

## Allelochemicals Behavior in Soil

Allelopathy is a result of different kinds of interactions, including plant-plant, plant-insect and plant-microorganism, in which the allelopathic agent can be plant or micro-organism. However, allelochemicals presence in donor plants and their antimicrobial activities in the laboratory, using artificial media, such as agar, do not characterize allelopathic activity under natural conditions, since soil micro-organisms consume a large number of organic molecules and, thus, inhibitory compounds may not be gathered at enough levels to become toxic.

In addition, the chemical nature of the compound, texture, structure, aeration, temperature, organic matter and soil pH, as well as the involved microbial species, must be considered [86].

Allelochemical activity can be divided into direct and indirect in a simplified way. Some changes were observed in soil properties, its nutritional conditions and micro-organisms' activity in the indirect activity. During the direct activity, it is possible to observe allelochemicals binding to membranes of receiving plant, as well as penetrating cells, and directly interfering in its metabolism [101].

The study of these metabolites in soils is still far from being fully understood and is challenging for three main reasons: the great diversity and complexity of compounds, and their chemical change in soils and soil organisms [6].

There are diverse soils because of their composition and behavior, as well as many living organisms and plant organs. The chemical constitution of vegetal origin in the soil is dynamic once plant materials are constantly added and will result in new compounds according to quality and amount of residues. They can also be incorporated or remain on non-mobilized soil, influencing so in their decomposition.

In addition, the descriptions of plant material and soil, as well as climatic conditions, among other factors, will influence this process and will give rise to other compounds in soil, which can be temporarily immobilized by living beings, adsorbed to colloids, or they can leave the environment by leaching. These new products can be toxic, beneficial or harmless to plants depending on their concentration and plant susceptibility [102, 103]. Furthermore, it is essential to establish an allelopathic interference that depends on several factors, such as concentration, movement and persistence of compounds in the environment [86]. Almeida [103] highlights oxygen availability as a limiting factor in this process. Under aerobic conditions, organic compounds are fast metabolized by micro-organisms. The microbial activity ends under anaerobic conditions, and acetic, lactic, butyric acids and other organic acids, phenolic compounds, and amino acids are produced, among other intermediates, with phytotoxic potential evidenced in the laboratory. This anaerobic condition can be found in soaked or compacted soils, for example.

According to Kobayashi [83] and Vidal and Bauman [89], adsorption-desorption balance also influences allelochemicals concentration in soil solution and is closely influenced by physical soil (texture, structure, organic matter content, humidity and aeration), chemical (reaction, ion exchange capacity, nutrient dynamics, $O_2$ and $CO_2$ concentrations) and biological characteristics (soil micro-organisms). These factors are closely associated with one another and have multiple effects on the retention, transport and transformation processes of allelochemicals in soil [86].

Besides, since clay minerals differ greatly from one another, clay typology also influences allelochemicals availability. A high ion exchange capacity means greater allelochemicals retention and, therefore, less bioavailability, consequently, phytotoxicity. It also regulates soil pH, increases its temperature and favors chemical degradation that is carried out by micro-organisms. A well-structured soil has high porosity, balance among liquid and gaseous phases and high organic matter content. Thus, it increases the transformation process worked by micro-organisms and decreases allelochemicals leaching. In sandy soils, aerobic micro-organisms degrade allelochemicals very fast [86].

So, results on soil texture influence of allelochemicals phytotoxicity are still contradictory since some authors have found greater inhibitory activity in clayey soils and others in sandy substrates. However, according to Scavo *et al.* [86], the first hypothesis is the most realistic. In fact, clays, due to their high specific surface and negative surface charges, decrease water infiltration, increase cation exchange capacity, and reduce allelochemicals leaching. This is the most important factor that affects their phytotoxic behavior. Moreover, there is less aeration in clayey soils than in sandy soils and, so aerobic micro-organisms degrade allelochemicals in a very slow way.

Lima, Kaplan and Cruz [82] reported on the complex effect of soil on plant growth, development, dry matter production and, especially, special metabolites production. They researched results obtained by other authors and emphasized that a stimulating effect on essential oil production was detected in alkaline soil in *Valeriana officinalis* L. and *Matricaria chamomilla* L. (Berbec; El-Badry & Hilal *apud* Bernáth [104]). A study with *Mentha arvensis* L. showed an increase in oil yield of approximately 100% when a greater supply of nitrogen and phosphorus was provided to soil (Bains *et al. apud* Bernáth [104]). In *Ocimum basilicum* L., Wahab (*apud* Bernáth [104]), the interaction of nitrogen, phosphorus and potassium was observed with a percentage increase in oil content. Mártonfi *et al.* [105] studied the relationship between chemotype patterns variation of *Thymus pulegioides* and soil chemistry. A correlation analysis was suggested when carbonates increased on soil since there was a decrease in the chemotype diversity of populations. Linalool chemotype ratio has increased, and phenol chemotype has decreased. On the other hand, the increase in linalool chemotype caused a decrease in chemotypes diversity. The opposite was observed concerning this diversity with higher production of carvacrol chemotype. Studies that were carried out with maize have also shown that fertilization has strongly affected volatiles emission [106].

According to Scavo *et al.* [86], allelochemicals, when released in soil, can act directly on the target organism, and can be degraded or transformed by micro-organisms that are in the soil. They can also induce a third species to produce another compound that interferes with donor plants and also causes changes in soil abiotic factors that affect the target plants.

It should be said that these allelochemicals in soil are subject to degradation processes by microbial decomposition, photolysis, oxidation, and removal or transfer processes such as volatilization and adsorption [89]. Soil biota also affects the chemical nature of products, as well as the decomposition rate. Decomposer communities are often adapted to degrade the litter they usually find, which mostly comes from plant species in nearby regions [107 - 109]. Nevertheless, allelochemicals, when released into rhizosphere, influence solubilization, release and absorption of mineral nutrients, depending on their kinds and concentration, and can improve the efficiency of nutrient application, making fertilization techniques effective and sustainable [88, 86].

## Terpenes in Soil

The terpenoid group stands out for its volatile forms, showing its role in chemical communication among organisms and their association with resistance mechanisms against biotic and abiotic stresses [6, 60]. Terpenoids have several functions, from attracting pollinators to protecting plants from herbivores to acting as toxic insecticides and repellents [77]. They can act out by direct fungi toxic action and affect spore germination and colony development, and/or indirectly, by phytoalexins induction, they induce plants to be resistant [110]. In addition to interfering with seeds germination, highlighting the influence on the secondary metabolism of plants, they can be harmful in nutrient capture, protein synthesis, water assimilation, in biochemical processes of germination, among others [111].

They also have a complex impact on soil micro-organisms, as they can inhibit the activity and growth of certain microbial groups while stimulating others [112]. Studies by White [113] showed that monoterpenes inhibit nitrogen mineralization and liquid nitrification in soil. Their causes may be related to the direct action of monoterpenes in an enzyme that takes part in ammonium pathway oxidation [114], or growth inhibition of a bacterium that oxidizes ammonium to nitrite, *Nitrosomonas europaea* [115]. According to these authors, other terpenoids, such as betapinene, may have stimulating effects on the growth of the same bacterium.

Root exudation represents the most important pathway for allelochemicals release. Once released into soil, allelochemicals interact with their organic and inorganic phases, as well as with soil micro-organisms. The set of these interactions results in allelochemicals bioavailability and their phytotoxic level [86]. However, according to Lobon *et al.* [116], diterpenes are incorporated mainly by leaching leaves and decomposing litter. Terpenoid concentrations decrease relatively with soil depth and show a positive correlation with soil C/N ratio [114, 117]. The depth decrease in secondary metabolites concentrations may suggest that these compounds are not easily leached into lower layers of a soil or, more likely, that they are partially degraded in the upper layer [113].

Lima, Kaplan & Cruz [82] concluded in their research that some precautions must be taken when conducting fieldwork and in qualitative and quantitative analyses regarding special metabolites, such as terpenoids. Prior knowledge of crop conditions is essential to stimulate the increase or maintenance of active ingredient production. Gobbo-Neto & Lopes [91], who studied factors of influence on secondary metabolites content in medicinal plants, reported that little information is available on micronutrient impact on secondary metabolites production in plants and on the relationship between pH or micro-organisms' soil and secondary metabolism. The authors pointed out that, in agriculture, nutrients addition, mainly N, is used to increase biomass production. However, nutrients affect not only primary metabolism, but also influence secondary metabolites production.

Gobbo-Neto & Lopes [91] also pointed out that most studies about the influence of nutrients in the soil are correlated with the intensity of incident light and focused on discussions on resource allocation (hypotheses of carbon/nutrient balance). These studies aim at establishing a relationship among N, C availability and light, showing a positive correlation with carbon/nutrient ratio (C/N); that is, in soils that are poor in nutrients, in relation to the least growth rate, it is usually registered higher production of secondary metabolites. However, these hypotheses have not been fully proven, and there are controversial results and attempts to refine C/N balance hypotheses. On the other hand, nutritional stress usually results in an increase in secondary metabolites concentration, except in the case of N and S deficiency, in which secondary metabolites production, containing these elements is reduced. On the other hand, metabolites derived from mevalonate do not seem to show consistent correlations with changes in nitrogen, phosphorus or potassium availability.

Inderjit & Weiner [118] recorded support in the literature for the hypothesis that the most important effects of compounds released in soil, through plants on other plants, occur through indirect effects. And so, they confirm that most phenomena widely known as allelopathic interference are better conceptualized and investigated in terms of soil chemical ecology. Even when direct plant-plant allelochemical interference occurs, the levels of allelochemicals in the environment and their effects on plants are strongly influenced by abiotic and biotic components of the soil ecosystem.

Although reference to allelochemicals, in general, covers some release of toxic substances, allelochemical effects in the field may be due to (i) direct harmful effects of chemicals released from donor plants, (ii) degraded or transformed products from released chemicals, (iii) the effect of chemicals released on physical, chemical and biological factors in soil, and (iv) release induction of biologically active chemicals by a third species, as shown in Table **1**, proposed by those authors.

**Table 1. Classification of potential allelopathic effects defined by Inderjit & Weiner [118].**

| **I. Allelopathic plant-to-plant interference (allelopathy in the strict sense)** |
|---|
| Plant A produces compound X, which interferes on plant B |
| II. Indirect ecological interactions of soil (indirect allelopathic effects) |
| A. Indirect allelopathy |
| 1. Decomposition-mediated plant-to-plant allelopathy |
| Plant A produces compound X, which is degraded or otherwise transformed by micro-organism C in compound Y, which interferes on Plant B |
| 2. Induced allelopathy |
| Plant A produces compound X, which is released and induces organism D to produce compound Z that interferes with plant B |

(Table 1) cont.....

| I. Allelopathic plant-to-plant interference (allelopathy in the strict sense) |
|---|
| B. Indirect toxicity |
| Compound X interacts with soil ecosystem and causes generation of compound Z (which is not a product of decomposition of compound X), which interferes on plant B |
| C. Indirect environmental effects |
| Compound X causes some change on soil environment, which affects nutritional status of soil, thus reducing growth, survival or reproductive production of plant B, without toxic effects |

Inderjit & Weiner [118] also suggest that indirect effects of allelochemicals may be more important for plant communities than direct effects of allelochemicals released by plants on other plants. And that these products can influence abiotic components of the ecosystem, such as availability and inorganic ions accumulation, as well as allelochemicals activities are influenced by ecological factors, such as nutrient restriction, light regime and humidity deficiency. They also pointed out that terpenoids increased $N$-ammonium immobilization by soil organisms, instead of inhibiting nitrification. And they cited Rice [80], who suggested that during succession, the nitrification rate is reduced, maybe due to allelochemicals that, under poor amounts of nutrients, can increase their production and, thus, influence allelochemical interference. Another considered aspect are bioassays with plant and soil residues, which are frequently used to demonstrate phytotoxicity and allelopathy, generating reports of inhibitory effects of secondary compounds released by plant residues. Adding these residues to soil can influence nutrient mobilization and soil pH and further influence nutrient immobilization and microbial activity.

Kapoor *et al.* [119] pointed out that the role of arbuscular mycorrhiza (AM) symbiosis to improve secondary metabolites accumulation, especially terpenoids, has obtained some recognition over the last two decades. Also, there is a wide increase in terpenoid production on shoots of medicinal plants colonized by AM. These authors reported that AM effects on terpenoids production in different plants, for at least two decades, have been comprehensively summarized in recent analyses, with few studies related to AM in total terpenoids, and most of them concerned about essential oil production and AM effect on individual essential oil components or on a specific terpenoid. When the discussion is about the increase in plant growth and biomass, Kapoor *et al.* [119] stated that the quantitative increase in terpenoids in a plant is due to its increased concentration (biosynthesis) and its biomass increase in tissue in which they are synthesized/stored (for example, bud, leaves, flowers and fruits). It is highly recognized that AM symbiosis improves the biomass of the shoot part of a plant, resulting in greater photosynthetic capacity and, thus, greater production. The highest nutritional absorption of phosphorus promoted by AM positively influences terpenoid biosynthesis.

According to Adamczyyk *et al.* [120], knowledge about both concentration and role of terpenes is scarce in soil. Some of the few available results have shown the presence of superior terpenes and monoterpenes in boreal forest soil, playing an important role in controlling the decomposition process in this environment. As several sets of enzymes drive decomposition, the influence of compounds in litter seems to have a significant role in the enzymatic activity that they act. Thus, there is some evidence that terpenes can decrease enzymatic activities in soil. The inhibitory effect of terpenes on enzymes does not need to be perceived as a negative factor for decomposition processes. Reactions among terpenes and enzymes can lead to enzyme stabilization in soil.

It should be stated that under field conditions, the soil decomposition process is submitted to several factors, including a mixture of different secondary plant compounds. Its combined effect can be weakened or strengthened. In addition, abiotic factors such as humidity and temperature can also change the influence of terpenes in decomposition processes [121].

Soil ecological processes cause quantitative and qualitative variation in chemicals present in the soil environment. There must have a lot of research before a complex interaction of soil microbial ecology, and allelochemical phytotoxicity can be understood. So, the observed patterns of plant growth due to microbial-mediated allelochemical production or release can sometimes be incorrectly attributed to direct plant-plant allelopathy. It is very difficult to eliminate micro-organisms influence on allelochemicals and to demonstrate a direct influence of chemicals leached by a donor plant in individuals and populations nearby [118]. A better understanding of allelochemicals behavior on soil can be positively applied in agroecosystems to control weed plants and pests in agricultural management. In addition, it can be used to manage soil nutrients dynamic, improve the efficiency of nutrient use by plants and avoid the toxicity of heavy metals [86].

**CONSENT FOR PUBLICATION**

Not applicable.

**CONFLICT OF INTEREST**

The author declares no conflict of interest, financial or otherwise.

**ACKNOWLEDGEMENT**

Declared none.

# REFERENCES

[1] Taiz, L.; Zeiger, E.; Møller, I.M.; Murphy, A. Fisiologia e desenvolvimento vegetal. Artmed, 6ª ed. 888p **2017**.

[2] Tetali, S.D. Terpenes and isoprenoids: a wealth of compounds for global use. *Planta,* **2019**, *249*(1), 1-8.
[http://dx.doi.org/10.1007/s00425-018-3056-x] [PMID: 30467631]

[3] Zhou, F.; Pichersky, E. More is better: the diversity of terpene metabolism in plants. *Curr. Opin. Plant Biol.,* **2020**, *55*, 1-10.
[http://dx.doi.org/10.1016/j.pbi.2020.01.005] [PMID: 32088555]

[4] Christianson, D.W. Structural and chemical cytology of terpenoid cyclases. *Chem. Rev.,* **2017**, *117*(17), 11570-11648.
[http://dx.doi.org/10.1021/acs.chemrev.7b00287] [PMID: 28841019]

[5] Priya, P.; Yadav, A.; Chand, J.; Yadav, G. Terzyme: a tool for identification and analysis of the plant terpenome. *Plant Methods,* **2018**, *14*(1), 4.
[http://dx.doi.org/10.1186/s13007-017-0269-0] [PMID: 29339971]

[6] Chomel, M.; Guittonny-Larchevêque, M.; Fernandez, C.; Gallet, C.; DesRochers, A.; Paré, D.; Jackson, B.G.; Baldy, V. Plant secondary metabolites: a key driver of litter decomposition and soil nutrient cycling. *J. Ecol.,* **2016**, *104*(6), 1527-1541.
[http://dx.doi.org/10.1111/1365-2745.12644]

[7] Tiago, O.; Maicon, N.; Ivan, R.C.; Diego, N.F.; Vinícius, J.S.; Mauricio, F.; Alan, J.P.; Velci, Q.S. Plant secondary metabolites and its dynamical systems of induction in response to environmental factors: A review. *Afr. J. Agric. Res.,* **2017**, *12*(2), 71-84.
[http://dx.doi.org/10.5897/AJAR2016.11677]

[8] Priya, P.; Kumari, S.; Yadav, G. Quantification of the plant terpenome: predicted *versus* actual emission potentials. *Indian J. Plant. Physiol.,* **2016**, *21*(4), 569-575.
[http://dx.doi.org/10.1007/s40502-016-0256-x]

[9] Tarkowská, D.; Strnad, M. Isoprenoid-derived plant signaling molecules: biosynthesis and biological importance. *Planta,* **2018**, *247*(5), 1051-1066.
[http://dx.doi.org/10.1007/s00425-018-2878-x] [PMID: 29532163]

[10] Tholl, D. Biosynthesis and biological functions of terpenoids in plants. *Biotechnol. Isoprenoids,* **2015**, *148*, 63-106.
[http://dx.doi.org/10.1007/10_2014_295]

[11] Kumari, S.; Priya, P.; Misra, G.; Yadav, G. Structural and biochemical perspectives in plant isoprenoid biosynthesis. *Phytochem. Rev.,* **2013**, *12*(2), 255-291.
[http://dx.doi.org/10.1007/s11101-013-9284-6]

[12] Pichersky, E.; Gershenzon, J. The formation and function of plant volatiles: perfumes for pollinator attraction and defense. *Curr. Opin. Plant Biol.,* **2002**, *5*(3), 237-243.
[http://dx.doi.org/10.1016/S1369-5266(02)00251-0] [PMID: 11960742]

[13] Aharoni, A.; Jongsma, M.A.; Kim, T. R.I., M.; Giri, A.P.; Francel W. A. Verstappen, F. W.A.; Schwab, W.; Bouwmeester. H.J. Metabolic engeneering of terpenoid biosynthesis in plants. *Phytochem. Rev.,* **2006**, *5*, 49-58.
[http://dx.doi.org/10.1007/s11101-005-3747-3]

[14] Ajikumar, P.K.; Tyo, K.; Carlsen, S.; Mucha, O.; Phon, T.H.; Stephanopoulos, G. Terpenoids: opportunities for biosynthesis of natural product drugs using engineered microorganisms. *Mol. Pharm.,* **2008**, *5*(2), 167-190.
[http://dx.doi.org/10.1021/mp700151b] [PMID: 18355030]

[15] Immethun, C.M.; Hoynes-O'Connor, A.G.; Balassy, A.; Moon, T.S. Microbial production of isoprenoids enabled by synthetic biology. *Front. Microbiol.,* **2013**, *4*, 75.

[http://dx.doi.org/10.3389/fmicb.2013.00075] [PMID: 23577007]

[16]   Tippmann, S.; Chen, Y.; Siewers, V.; Nielsen, J. From flavors and pharmaceuticals to advanced biofuels: Production of isoprenoids in *Saccharomyces cerevisiae*. *Biotechnol. J.,* **2013**, *8*(12), 1435-1444.
       [http://dx.doi.org/10.1002/biot.201300028] [PMID: 24227704]

[17]   Zhang, C.; Hong, K. Production of terpenoids by synthetic biology approaches. *Front. Bioeng. Biotechnol.,* **2020**, *8*, 347.
       [http://dx.doi.org/10.3389/fbioe.2020.00347] [PMID: 32391346]

[18]   Diniz, L.R.L.; Perez-Castillo, Y.; Elshabrawy, H.A.; Filho, C.S.M.B.; de Sousa, D.P. Bioactive terpenes and their derivatives as potential SARS-CoV-2 proteases inhibitors from molecular modeling studies. *Biomolecules,* **2021**, *11*(1), 74.
       [http://dx.doi.org/10.3390/biom11010074] [PMID: 33430299]

[19]   Borges, L.P.; Amorim, V.A. Metabólitos secundários de plantas. *Rev. Agrotecnol.,* **2020**, *11*, 54-67.

[20]   Wang, Q.; Quan, S.; Xiao, H. Towards efficient terpenoid biosynthesis: manipulating IPP and DMAPP supply. *Bioresour. Bioprocess.,* **2019**, *6*(1), 6.
       [http://dx.doi.org/10.1186/s40643-019-0242-z]

[21]   Pulido, P.; Perello, C.; Rodriguez-Concepcion, M. New insights into plant isoprenoid metabolism. *Mol. Plant,* **2012**, *5*(5), 964-967.
       [http://dx.doi.org/10.1093/mp/sss088] [PMID: 22972017]

[22]   Liao, P.; Hemmerlin, A.; Bach, T.J.; Chye, M.L. The potential of the mevalonate pathway for enhanced isoprenoid production. *Biotechnol. Adv.,* **2016**, *34*(5), 697-713.
       [http://dx.doi.org/10.1016/j.biotechadv.2016.03.005] [PMID: 26995109]

[23]   García, A.A.; Carril, E.P. Metabolismo secundário de plantas. Reduca (Biología). *Serie Fisiologia Vegetal,* **2009**, *2*, 119-145.

[24]   Dewick, P.M. *Medical natural product: A biosynthetic approach,* 3rd ed; Jonh Wiley & Sons: Chichester, **2009**, p. 546.
       [http://dx.doi.org/10.1002/9780470742761]

[25]   Chatzivasileiou, A.O.; Ward, V.; Edgar, S.M.; Stephanopoulos, G. Two-step pathway for isoprenoid synthesis. *Proc. Natl. Acad. Sci. USA,* **2019**, *116*(2), 506-511.
       [http://dx.doi.org/10.1073/pnas.1812935116] [PMID: 30584096]

[26]   Berthelot, K.; Estevez, Y.; Deffieux, A.; Peruch, F. Isopentenyl diphosphate isomerase: A checkpoint to isoprenoid biosynthesis. *Biochimie,* **2012**, *94*(8), 1621-1634.
       [http://dx.doi.org/10.1016/j.biochi.2012.03.021] [PMID: 22503704]

[27]   Henry, L.K.; Thomas, S.T.; Widhalm, J.R.; Lynch, J.H.; Davis, T.C.; Kessler, S.A.; Bohlmann, J.; Noel, J.P.; Dudareva, N. Contribution of isopentenyl phosphate to plant terpenoid metabolism. *Nat. Plants,* **2018**, *4*(9), 721-729.
       [http://dx.doi.org/10.1038/s41477-018-0220-z] [PMID: 30127411]

[28]   Kang, A.; George, K.W.; Wang, G.; Baidoo, E.; Keasling, J.D.; Lee, T.S. Isopentenyl diphosphate (IPP)-bypass mevalonate pathways for isopentenol production. *Metab. Eng.,* **2016**, *34*, 25-35.
       [http://dx.doi.org/10.1016/j.ymben.2015.12.002] [PMID: 26708516]

[29]   Clomburg, J.M.; Qian, S.; Tan, Z.; Cheong, S.; Gonzalez, R. The isoprenoid alcohol pathway, a synthetic route for isoprenoid biosynthesis. *Proc. Natl. Acad. Sci. USA,* **2019**, *116*(26), 12810-12815.
       [http://dx.doi.org/10.1073/pnas.1821004116] [PMID: 31186357]

[30]   Zhao, L.; Chang, W.; Xiao, Y.; Liu, H.; Liu, P. Methylerythritol phosphate pathway of isoprenoid biosynthesis. *Annu. Rev. Biochem.,* **2013**, *82*(1), 497-530.
       [http://dx.doi.org/10.1146/annurev-biochem-052010-100934] [PMID: 23746261]

[31]   Bergman, M.E.; Davis, B.; Phillips, M.A. Medically useful plant terpenoids: biosyntheisis, ocurrence,

and mechanism of action. *Molecules,* **2019**, *24*(21), 3961.
[http://dx.doi.org/10.3390/molecules24213961] [PMID: 31683764]

[32]　Frank, A.; Groll, M. The methylerythritol phosphate pathway to isoprenoids. *Chem. Rev.,* **2017**, *117*(8), 5675-5703.
[http://dx.doi.org/10.1021/acs.chemrev.6b00537] [PMID: 27995802]

[33]　Rohmer, M. Diversity in isoprene unit biosynthesis: The methylerythritol phosphate pathway in bacteria and plastids. *Pure Appl. Chem.,* **2007**, *79*(4), 739-751.
[http://dx.doi.org/10.1351/pac200779040739]

[34]　Chang, W.; Song, H.; Liu, H.; Liu, P. Current development in isoprenoid precursor biosynthesis and regulation. *Curr. Opin. Chem. Biol.,* **2013**, *17*(4), 571-579.
[http://dx.doi.org/10.1016/j.cbpa.2013.06.020] [PMID: 23891475]

[35]　Berthelot, K.; Estevez, Y.; Deffieux, A.; Peruch, F. Isopentenyl diphosphate isomerase: A checkpoint to isoprenoid biosynthesis. *Biochimie,* **2012**, *94*(8), 1621-1634.
[http://dx.doi.org/10.1016/j.biochi.2012.03.021] [PMID: 22503704]

[36]　Baldwin, I.; Paschold, A.; Halitschke, R. Volatile signaling in plant-plant interactions: "Talking Trees" in the Genomics Era Science **2006**, *311*, 812-815.

[37]　Gershenzon, J.; Dudareva, N. The function of terpene natural products in the natural world. *Nat. Chem. Biol.,* **2007**, *3*(7), 408-414.
[http://dx.doi.org/10.1038/nchembio.2007.5] [PMID: 17576428]

[38]　Osbourn, A.E.; Lanzotti, V. Plant-derived natural products – synthesis, function and application. Ed. Springer, New York, 584 **2009**.

[39]　Barrero, A.F.; Herrador, M.M.; Arteaga, P.; Arteaga, J.F.; Arteaga, A.F. Communic acids: occurrence, properties and use as chirons for the synthesis of bioactive compounds. *Molecules,* **2012**, *17*(2), 1448-1467.
[http://dx.doi.org/10.3390/molecules17021448] [PMID: 22310167]

[40]　Proshkina, E.; Plyusnin, S.; Babak, T.; Lashmanova, E.; Maganova, F.; Koval, L.; Platonova, E.; Shaposhnikov, M.; Moskalev, A. Terpenoids as Potential Geroprotectors. *Antioxidants,* **2020**, *9*(6), 529.
[http://dx.doi.org/10.3390/antiox9060529]

[41]　Ghaffari, T.; Kafil, H.S.; Asnaashari, S.; Farajnia, S.; Delazar, A.; Baek, S.C.; Hamishehkar, H.; Kim, K.H. Chemical composition and antimicrobial activity of essential oils from the aerial parts of Pinus eldarica grown in northwestern Iran. *Molecules,* **2019**, *24*(17), 3203.
[http://dx.doi.org/10.3390/molecules24173203] [PMID: 31484421]

[42]　Croteau, R.B.; Davis, E.M.; Ringer, K.L.; Wildung, M.R. (−)-Menthol biosynthesis and molecular genetics. *Naturwissenschaften,* **2005**, *92*(12), 562-577.
[http://dx.doi.org/10.1007/s00114-005-0055-0] [PMID: 16292524]

[43]　Mahmoud, S.S.; Croteau, R.B. Strategies for transgenic manipulation of monoterpene biosynthesis in plants. *Trends Plant Sci.,* **2002**, *7*(8), 366-373.
[http://dx.doi.org/10.1016/S1360-1385(02)02303-8] [PMID: 12167332]

[44]　Phillips, M.A.; Croteau, R.B. Resin-based defenses in conifers. *Trends Plant Sci.,* **1999**, *4*(5), 184-190.
[http://dx.doi.org/10.1016/S1360-1385(99)01401-6] [PMID: 10322558]

[45]　Trapp, S.; Croteau, R. Defensive resin biosynthesis in conifers. *Annu. Rev. Plant Physiol. Plant Mol. Biol.,* **2001**, *52*(1), 689-724.
[http://dx.doi.org/10.1146/annurev.arplant.52.1.689] [PMID: 11337413]

[46]　Köllner, T.G.; Held, M.; Lenk, C.; Hiltpold, I.; Turlings, T.C.J.; Gershenzon, J.; Degenhardt, J. A maize (E)-β-caryophyllene synthase implicated in indirect defense responses against herbivores is not expressed in most American maize varieties. *Plant Cell,* **2008**, *20*(2), 482-494.
[http://dx.doi.org/10.1105/tpc.107.051672] [PMID: 18296628]

[47] Vaughan, M.M.; Wang, Q.; Webster, F.X.; Kiemle, D.; Hong, Y.J.; Tantillo, D.J.; Coates, R.M.; Wray, A.T.; Askew, W.; O'Donnell, C.; Tokuhisa, J.G.; Tholl, D.; Tholl, D. Formation of the unusual semivolatile diterpene rhizathalene by the Arabidopsis class I terpene synthase TPS08 in the root stele is involved in defense against belowground herbivory. *Plant Cell,* **2013,** *25*(3), 1108-1125.
[http://dx.doi.org/10.1105/tpc.112.100057] [PMID: 23512856]

[48] Erb, M.; Veyrat, N.; Robert, C.A.M.; Xu, H.; Frey, M.; Ton, J.; Turlings, T.C.J. Indole is an essential herbivore-induced volatile priming signal in maize. *Nat. Commun.,* **2015,** *6*(1), 6273.
[http://dx.doi.org/10.1038/ncomms7273] [PMID: 25683900]

[49] Taniguchi, S.; Hosokawa-Shinonaga, Y.; Tamaoki, D.; Yamada, S.; Akimitsu, K.; Gomi, K. Jasmonate induction of the monoterpene linalool confers resistance to rice bacterial blight and its biosynthesis is regulated by JAZ protein in rice. *Plant Cell Environ.,* **2014,** *37*(2), 451-461.
[http://dx.doi.org/10.1111/pce.12169] [PMID: 23889289]

[50] Cheng, A.X.; Xiang, C.Y.; Li, J.X.; Yang, C.Q.; Hu, W.L.; Wang, L.J.; Lou, Y.G.; Chen, X.Y. The rice (E)-β-caryophyllene synthase (OsTPS3) accounts for the major inducible volatile sesquiterpenes. *Phytochemistry,* **2007,** *68*(12), 1632-1641.
[http://dx.doi.org/10.1016/j.phytochem.2007.04.008] [PMID: 17524436]

[51] Muchlinski, A. Chen. X.; Lovell, J.T.; Köllner, T.G.; Pelot, K.A.; Zerbe, P.; Ruggiero, M.; Callaway Lll, L.; Laliberte, S.; Chen, F.; Tholl, D. Biosynthesis and emission of stress-induced volatile terpenes in roots and leaves of switchgrass (Panicum virgatum L.). *Front. Plant Sci.,* **2019,** 10.

[52] Schmelz, E.A.; Huffaker, A.; Sims, J.W.; Christensen, S.A.; Lu, X.; Okada, K.; Peters, R.J. Biosynthesis, elicitation and roles of monocot terpenoid phytoalexins. *Plant J.,* **2014,** *79*(4), 659-678.
[http://dx.doi.org/10.1111/tpj.12436] [PMID: 24450747]

[53] Chen, F.; Tholl, D.; Bohlmann, J.; Pichersky, E. The family of terpene synthases in plants: a mid-size family of genes for specialized metabolism that is highly diversified throughout the kingdom. *Plant J.,* **2011,** *66*(1), 212-229.
[http://dx.doi.org/10.1111/j.1365-313X.2011.04520.x] [PMID: 21443633]

[54] Heinrich, M.; Robles, M.; West, J.E.; Ortiz de Montellano, B.R.; Rodriguez, E. Ethnopharmacology of Mexican asteraceae (Compositae). *Annu. Rev. Pharmacol. Toxicol.,* **1998,** *38*(1), 539-565.
[http://dx.doi.org/10.1146/annurev.pharmtox.38.1.539] [PMID: 9597165]

[55] Dudareva, N.; Andersson, S.; Orlova, I.; Gatto, N.; Reichelt, M.; Rhodes, D.; Boland, W.; Gershenzon, J. The nonmevalonate pathway supports both monoterpene and sesquiterpene formation in snapdragon flowers. *Proc. Natl. Acad. Sci. USA,* **2005,** *102*(3), 933-938.
[http://dx.doi.org/10.1073/pnas.0407360102] [PMID: 15630092]

[56] Piel, J.; Donath, J.; Bandemer, K.; Boland, W. Mevalonate-independent biosynthesis of terpenoid volatiles in plants: induced and constitutive emission of volatiles. *Angew. Chem. Int. Ed.,* **1998,** *37*(18), 2478-2481.
[http://dx.doi.org/10.1002/(SICI)1521-3773(19981002)37:18<2478::AID-ANIE2478>3.0.CO;2-Q] [PMID: 29711361]

[57] Bick, J.A.; Lange, B.M. Metabolic cross talk between cytosolic and plastidial pathways of isoprenoid biosynthesis: unidirectional transport of intermediates across the chloroplast envelope membrane. *Arch. Biochem. Biophys.,* **2003,** *415*(2), 146-154.
[http://dx.doi.org/10.1016/S0003-9861(03)00233-9] [PMID: 12831836]

[58] Schnee, C.; Köllner, T.G.; Held, M.; Turlings, T.C.J.; Gershenzon, J.; Degenhardt, J. The products of a single maize sesquiterpene synthase form a volatile defense signal that attracts natural enemies of maize herbivores. *Proc. Natl. Acad. Sci. USA,* **2006,** *103*(4), 1129-1134.
[http://dx.doi.org/10.1073/pnas.0508027103] [PMID: 16418295]

[59] Rasmann, S.; Köllner, T.G.; Degenhardt, J.; Hiltpold, I.; Toepfer, S.; Kuhlmann, U.; Gershenzon, J.; Turlings, T.C.J. Recruitment of entomopathogenic nematodes by insect-damaged maize roots. *Nature,* **2005,** *434*(7034), 732-737.

[http://dx.doi.org/10.1038/nature03451] [PMID: 15815622]

[60]   Pinto-Zevallos, D.M.; Martins, C.B.C.; Pellegrino, A.C.; Zarbin, P.H.G. Compostos orgânicos voláteis na defesa induzida das plantas contra insetos herbívoros. *Quim. Nova,* **2013**, *36*(9), 1395-1405.
[http://dx.doi.org/10.1590/S0100-40422013000900021]

[61]   Karunanithi, P.S.; Zerbe, P. Terpene synthases as metabólic gatekeepers in the evolution of plant terpenoid chemical diversity. *Front. Plant Sci.,* **2019**, *10*, 1166.
[http://dx.doi.org/10.3389/fpls.2019.01166] [PMID: 31632418]

[62]   Murphy, K.M.; Zerbe, P. Specialized diterpenoid metabolism in monocot crops: Biosynthesis and chemical diversity. *Phytochemistry,* **2020**, *172*, 112289.
[http://dx.doi.org/10.1016/j.phytochem.2020.112289] [PMID: 32036187]

[63]   Bishopp, A.; Mähönen, A.P.; Helariutta, Y. Signs of change: hormone receptors that regulate plant development. *Development,* **2006**, *133*(10), 1857-1869.
[http://dx.doi.org/10.1242/dev.02359] [PMID: 16651539]

[64]   Liu, M.; Lu, S. Plastoquinone and ubiquinone in plants: biosynthesis, physiological function and metabolic engineering. *Front. Plant Sci.,* **2016**, *7*, 1898.
[http://dx.doi.org/10.3389/fpls.2016.01898] [PMID: 28018418]

[65]   Salazar-Cerezo, S.; Martínez-Montiel, N.; García-Sánchez, J.; Pérez-y-Terrón, R.; Martínez-Contreras, R.D. Gibberellin biosynthesis and metabolism: A convergent route for plants, fungi and bacteria. *Microbiol. Res.,* **2018**, *208*, 85-98.
[http://dx.doi.org/10.1016/j.micres.2018.01.010] [PMID: 29551215]

[66]   Mo, H.; Elson, C.E. Studies of the isoprenoid-mediated inhibition of mevalonate synthesis applied to cancer chemotherapy and chemoprevention. *Exp. Biol. Med. (Maywood),* **2004**, *229*(7), 567-585.
[http://dx.doi.org/10.1177/153537020422900701] [PMID: 15229351]

[67]   Maeda, H.A. Evolutionary diversification of primary metabolism and contribution to plant chemical diversity. *Front. Plant Sci.,* **2019**, *10*, 881.
[http://dx.doi.org/10.3389/fpls.2019.00881] [PMID: 31354760]

[68]   Celedon, J.M.; Bohlmann, J. Oleoresin defenses in conifers: chemical diversity, terpene synthases and limitations of oleoresin defense under climate change. *New Phytol.,* **2019**, *224*(4), 1444-1463.
[http://dx.doi.org/10.1111/nph.15984] [PMID: 31179548]

[69]   Cárdenas, P.D.; Almeida, A.; Bak, S. Evolution of structural diversity of triterpenoids. *Front. Plant Sci.,* **2019**, *10*, 1523.
[http://dx.doi.org/10.3389/fpls.2019.01523] [PMID: 31921225]

[70]   Pérez-Camino, M.C.; Cert, A. Quantitative determination of hydroxy pentacyclic triterpene acids in vegetable oils. *J. Agric. Food Chem.,* **1999**, *47*(4), 1558-1562.
[http://dx.doi.org/10.1021/jf980881h] [PMID: 10564016]

[71]   Jaeger, R.; Cuny, E. Terpenoids with special pharmacological significance: A review. *Nat. Prod. Commun.,* **2016**, *11*(9), 1934578X1601100.
[http://dx.doi.org/10.1177/1934578X1601100946] [PMID: 30807045]

[72]   Jiang, Z.; Kempinski, C.; Chappell, J. Extraction and analysis of terpenes/terpenoids. *Curr. Protoc. Plant Biol.,* **2016**, *1*(2), 345-358.
[http://dx.doi.org/10.1002/cppb.20024] [PMID: 27868090]

[73]   Jäger, S.; Trojan, H.; Kopp, T.; Laszczyk, M.; Scheffler, A. Pentacyclic triterpene distribution in various plants - rich sources for a new group of multi-potent plant extracts. *Molecules,* **2009**, *14*(6), 2016-2031.
[http://dx.doi.org/10.3390/molecules14062016] [PMID: 19513002]

[74]   Croteau, R.; Kutchan, T.M. Lewis. N.G.; Natural products (secondary metabolites). In: *Biochemistry and molecular biology of plants*; Buchanan, B.; Gruissem, W.; Jones, R., Eds.; American Society of Plant Physiologists: Rockville, **2000**; pp. 1250-1318.

[75]   Pichersky, E.; Raguso, R.A. Why do plants produce so many terpenoid compounds? *New Phytol.,* **2018**, *220*(3), 692-702.
       [http://dx.doi.org/10.1111/nph.14178] [PMID: 27604856]

[76]   Viegas Júnior, C. Terpenos com atividade inseticida: uma alternativa para o controle químico de insetos. *Quim. Nova,* **2003**, *26*(3), 390-400.
       [http://dx.doi.org/10.1590/S0100-40422003000300017]

[77]   Taiz, L.; Zeiger, E. Fisiologia vegetal. Artmed, 5ª ed, Porto Alegre, **2013**, 918.

[78]   Khan, T.; Abbasi, B.H.; Khan, M.A. The interplay between light, plant growth regulators and elicitors on growth and secondary metabolism in cell cultures of Fagonia indica. *Journal of Photochemistry & Photobiology,* **2018**, *185*, 153-160.

[79]   Kortbeek, R.W.; Van Der Gragt, M.; Bleeker, P.M. Endogenous plant metabolites against insects. *Eur. J. Plant Pathol.,* **2018**, *54*, 67-90.

[80]   Rice, E.L. *Allelopathy,* 2nd ed; Academic Press: Orlando, **1984**, 424, .

[81]   Scavo, A.; Restuccia, A.; Mauromicale, G. Allelopathy: Principles and basic aspects for agroecosystem control. In: *Sustainable agriculture reviews. Springer, 28*; Gaba, S.; Smith, B.; Lichtfouse, E., Eds.; Cham, **2018**; pp. 47-101.
       [http://dx.doi.org/10.1007/978-3-319-90309-5_2]

[82]   Lima, H.R.P.; Kaplan, M.A.C.; Cruz, A.V.M.C. Influência dos fatores abióticos na produção e variabilidade de terpenóides em plantas. *Floresta Ambient.,* **2003**, *10*, 71-77.

[83]   Kobayashi, K. Factors affecting phytotoxic activity of allelochemicals in soil. *Weed Biol. Manage.,* **2004**, *4*(1), 1-7.
       [http://dx.doi.org/10.1111/j.1445-6664.2003.00112.x]

[84]   Cheng, H.H. A conceptual framework for assessing allelochemicals in the soil environmental. **1992**.

[85]   Cheng, H.H.; Koskinen, W.C. Effects of "aging" on bio reactive chemical retention, transformation, and transport in soil. *Molecular environmental soil science at the interfaces in the earth's critical zone.,* **2010**, , 184-186.
       [http://dx.doi.org/10.1007/978-3-642-05297-2_55]

[86]   Scavo, A.; Abbate, C.; Mauromicale, G. Plant allelochemicals: agronomic, nutritional and ecological relevance in the soil system. *Plant Soil,* **2019**, *442*(1-2), 23-48.
       [http://dx.doi.org/10.1007/s11104-019-04190-y]

[87]   Weidenhamer, J.D. Distinguishing resource competition and chemical interference: overcoming the methodological impasse. *Agron. J.,* **1996**, *88*(6), 866-875.
       [http://dx.doi.org/10.2134/agronj1996.00021962003600060005x]

[88]   Jabran, K.; Farooq, M.; Aziz, T.; Siddique, K.H.M. Allelopathy and crop nutrition. In: *Allelopathy: current trends and future applications*; Cheema, ZA; Farooq, M.; Wahid, A., Eds.; Springer: Verlag Berlin Heidelberg, Germany, **2013**; pp. 337-348.
       [http://dx.doi.org/10.1007/978-3-642-30595-5_14]

[89]   Vidal, R.A.; Bauman, T.T. Destino de aleloquímicos no solo. *Cienc. Rural,* **1997**, *27*(2), 351-357.
       [http://dx.doi.org/10.1590/S0103-84781997000200032]

[90]   Liu, Y.; Fang, S.; Yang, W.; Shang, X.; Fu, X. Light quality affects flavonoid production and related gene expression in Cyclocarya paliurus. *J. Photochem. Photobiol. B,* **2018**, *179*, 66-73.
       [http://dx.doi.org/10.1016/j.jphotobiol.2018.01.002] [PMID: 29334625]

[91]   Gobbo-Neto, L.; Lopes, N.P. Plantas medicinais: fatores de influência no conteúdo de metabólitos secundários. *Quim. Nova,* **2007**, *30*(2), 374-381.
       [http://dx.doi.org/10.1590/S0100-40422007000200026]

[92]   Medina, E.; Olivares, E.; Diaz, M. Water stress and light intensity effects on growth and nocturnal acid

accumulation in a terrestrial CAM bromeliad (Bromelia humilis Jacq.) under natural conditions. *Oecologia,* **1986**, *70*(3), 441-446.
[http://dx.doi.org/10.1007/BF00379509] [PMID: 28311933]

[93]    Cipollini, D.F., Jr; Redman, A.M. Age-dependent effects of jasmonic acid treatment and wind exposure on foliar oxidase activity and insect resistance in tomato. *J. Chem. Ecol.,* **1999**, *25*(2), 271-281.
[http://dx.doi.org/10.1023/A:1020842712349]

[94]    Ormeño, E.; Fernandez, C. Effect of soil nutrient on production and diversity of volatile terpenoids from plants. *Curr. Bioact. Compd.,* **2012**, *8*(1), 71-79.
[http://dx.doi.org/10.2174/157340712799828188] [PMID: 23097639]

[95]    Arrabal, C.; García-Vallejo, M.C.; Cadahia, E.; Cortijo, M.; Fernández de Simón, B. Seasonal variations of lipophilic compounds in needles of two chemotypes of Pinus pinaster Ait. *Plant Syst. Evol.,* **2014**, *300*(2), 359-367.
[http://dx.doi.org/10.1007/s00606-013-0888-5]

[96]    Copolovici, L.; Niinemets, Ü. Environmental impacts on plant volatile emission. *Deciphering chemical language of plant communication.,* **2016**, , 35-59.
[http://dx.doi.org/10.1007/978-3-319-33498-1_2]

[97]    Mu, Z.; Llusià, J.; Liu, D.; Ogaya, R.; Asensio, D.; Zhang, C.; Peñuelas, J. Seasonal and diurnal variations of plant isoprenoid emissions from two dominant species in Mediterranean shrubland and forest submitted to experimental drought. *Atmos. Environ.,* **2018**, *191*, 105-115.
[http://dx.doi.org/10.1016/j.atmosenv.2018.08.010]

[98]    Bartwal, A.; Mall, R.; Lohani, P.; Guru, S.K.; Arora, S. Role of secondary metabolites and brassinosteroids in plant defense against environmental stresses. *J. Plant Growth Regul.,* **2013**, *32*(1), 216-232.
[http://dx.doi.org/10.1007/s00344-012-9272-x]

[99]    Ormeño, E.; Goldstein, A.; Niinemets, U. Extracting and trapping biogenic volatile organic compounds stored in plant species. TracTrend. *Anal. Chem.,* **2011**, *30*(7), 978-989.

[100]   Sakamoto, H.T.; Gobbo-Neto, L.; Cavalheiro, A.J.; Lopes, N.P.; Lopes, J.L.C. Quantitative HPLC analysis of sesquiterpene lactones and determination of chemotypes in Eremanthus seidelii MacLeish & Schumacher (Asteraceae). *J. Braz. Chem. Soc.,* **2005**, *16*(6b), 1396-1401.
[http://dx.doi.org/10.1590/S0103-50532005000800016]

[101]   Ferreira, A.; Aquila, M.E.A. Alelopatia: uma área emergente da ecofisiologia. **2000**, 299-340.

[102]   Blum, U. Allelopathy: a soil system perspective. In: *Gonzalez, L. Allelopathy: A physiological process with ecological implications*; Reigos, M.J.; Pedrol, N., Eds.; Springer: Dordrecht, The Netherlands, **2006**; pp. 299-340.

[103]   Almeida, F.S. A alelopatia e as plantas. *Fundação Instituto Agronômico do Paraná, Londrina (Circular Técnica 53),* **1988**, 60.

[104]   Bernáth, J. Production ecology of secondary plant products. *Herbs, spices and medicinal plants. Recent advances in Botany, Horticulture and Pharmacology.,* **1992**, *Vol. 1*, 185-234.

[105]   Mártonfi, P.; Grejtovský, A.; Repčák, M. Chemotype pattern differentiation of Thymus pulegioides on different substrates. *Biochem. Syst. Ecol.,* **1994**, *22*(8), 819-825.
[http://dx.doi.org/10.1016/0305-1978(94)90086-8]

[106]   Gouinguené, S.P.; Turlings, T.C.J. The effects of abiotic factors on induced volatile emissions in corn plants. *Plant Physiol.,* **2002**, *129*(3), 1296-1307.
[http://dx.doi.org/10.1104/pp.001941] [PMID: 12114583]

[107]   Ayres, E.; Steltzer, H.; Berg, S.; Wall, D.H. Soil biota accelerate decomposition in high-elevation forests by specializing in the breakdown of litter produced by the plant species above them. *J. Ecol.,* **2009**, *97*(5), 901-912.

[http://dx.doi.org/10.1111/j.1365-2745.2009.01539.x]

[108] Strickland, M.S.; Lauber, C.; Fierer, N.; Bradford, M.A. Testing the functional significance of microbial community composition. *Ecology,* **2009**, *90*(2), 441-451. a
[http://dx.doi.org/10.1890/08-0296.1] [PMID: 19323228]

[109] Strickland, M.S.; Osburn, E.; Lauber, C.; Fierer, N.; Bradford, M.A. Litter quality is in the eye of the beholder: initial decomposition rates as a function of inoculum characteristics. *Funct. Ecol.,* **2009**, *23*(3), 627-636. b
[http://dx.doi.org/10.1111/j.1365-2435.2008.01515.x]

[110] Hillen, T.; Schwan-Estrada, K.R.F.; Mesquini, R.M.; Cruz, M.E.S.; Stangarlin, J.R.; Nozaki, M. Atividade antimicrobiana de óleos essenciais no controle de alguns fitopatógenos fúngicos *in vitro* e no tratamento de sementes. *Rev. Bras. Plantas Med.,* **2012**, *14*(3), 439-445.
[http://dx.doi.org/10.1590/S1516-05722012000300003]

[111] Maraschin-Silva, F.; Aqüila, M.E.A. Potencial alelopático de espécies nativas na germinação e crescimento inicial de Lactuca sativa L. (Asteraceae). *Acta Bot. Bras.,* **2006**, *20*(1), 61-69.
[http://dx.doi.org/10.1590/S0102-33062006000100007]

[112] Amaral, J.A.; Knowles, R. Inhibition of methane consumption in forest soils by monoterpenes. *J. Chem. Ecol.,* **1998**, *24*(4), 723-734.
[http://dx.doi.org/10.1023/A:1022398404448]

[113] White, C. The role of monoterpenes in soil nitrogen cycling processes in ponderosa pine. *Biogeochemistry,* **1991**, *12*(1), 43-68.
[http://dx.doi.org/10.1007/BF00002625]

[114] White, C.S. Monoterpenes: Their effects on ecosystem nutrient cycling. *J. Chem. Ecol.,* **1994**, *20*(6), 1381-1406.
[http://dx.doi.org/10.1007/BF02059813] [PMID: 24242344]

[115] Ward, B.B.; Courtney, K.J.; Langenheim, J.H. Inhibition of Nitrosomonas europaea by monoterpenes from Coastal Redwood (Sequoia sempervirens) in whole-cell studies. *J. Chem. Ecol.,* **1997**, *23*(11), 2583-2598.
[http://dx.doi.org/10.1023/B:JOEC.0000006668.48855.b7]

[116] Chaves Lobón, N.; Ferrer de la Cruz, I.; Alías Gallego, J. Autotoxicity of diterpenes present in leaves of Cistus ladanifer L. *Plants,* **2019**, *8*(2), 27-38.
[http://dx.doi.org/10.3390/plants8020027] [PMID: 30678267]

[117] Kanerva, S.; Smolander, A. Microbial activities in forest floor layers under silver birch, Norway spruce and Scots pine. *Soil Biol. Biochem.,* **2007**, *39*(7), 1459-1467.
[http://dx.doi.org/10.1016/j.soilbio.2007.01.002]

[118] Inderjit, ; Weiner, J. Plant allelochemical interference or soil chemical ecology? *Perspect. Plant Ecol. Evol. Syst.,* **2001**, *4*(1), 3-12.
[http://dx.doi.org/10.1078/1433-8319-00011]

[119] Kapoor, R.; Anand, G.; Gupta, P.; Mandal, S. Insight into the mechanisms of enhanced production of valuable terpenoids by arbuscular mycorrhiza. *Phytochem. Rev.,* **2017**, *16*(4), 677-692.
[http://dx.doi.org/10.1007/s11101-016-9486-9]

[120] Adamczyk, S.; Adamczyk, B.; Kitunen, V.; Smolander, A. Monoterpenes and higher terpenes may inhibit enzyme activities in boreal forest soil. *Soil Biol. Biochem.,* **2015**, *87*, 59-66.
[http://dx.doi.org/10.1016/j.soilbio.2015.04.006]

[121] Asensio, D.; Yuste, J.C.; Mattana, S.; Ribas, À.; Llusià, J.; Peñuelas, J. Litter VOCs induce changes in soil microbial biomass C and N and largely increase soil $CO_2$ efflux. *Plant Soil,* **2012**, *360*(1-2), 163-174.

# Potential Use of Terpenoids in Weed Management

**Mozaniel Santana de Oliveira**[1,*]**, Jordd Nevez Cruz**[1]**, Eloisa Helena de Aguiar Andrade**[1] **and Antônio Pedro da Silva Souza Filho**[2]

[1] *Museu Paraense Emilio Goeldi, Av. Perimetral, 1901 - Terra Firme, Belém-PA, 66077-830, Brazil*

[2] *Embrapa Amazônia Oriental, Tv. Dr. Eneas Pinheiro, s/n - Marco, Belém-PA, 66095-903, Brazil*

**Abstract:** Invasive plants represent a source of economic damage to the agricultural system, and their management has become indispensable from an agronomic point of view, as such plants are known for their competitiveness for resources such as water, light, nutrients, and space. Their control is performed in some cases, such as in Brazil, through the use of pesticides, which can be harmful to human health and other animals. With the change of habits and the search for a better quality of life, the use of these chemicals in management areas is increasingly less encouraged. A possible ecological alternative would be the use of natural products, as secondary metabolites have been shown as potential promoters of phytotoxic activity. Among the allelochemicals produced naturally, terpenoids can be highlighted because their chemical variability can help in the sustainable management of invasive plants.

**Keywords:** Ecology, Essential oils, Natural Products, Terpenoids, Weed.

## INTRODUCTION

In tropical regions, where acidic and low fertility soils predominate, and environmental conditions are highly favorable to the development of biotic agents that are harmful to crops, the success of agricultural activities has always been linked to the use of growth stimulants and agricultural pesticides [1]. While these techniques have ensured satisfactory success, both in terms of productivity and meeting market needs, this scenario has undergone profound changes in recent decades, requiring new paradigms that consider both the values of modern society and the requirements that valorize responsible agriculture in relation to the preservation of natural resources, wildlife, and the humans themselves, who have been seeking food that is increasingly free of chemical residues [2].

---

[*] **Corresponding author Mozaniel de Oliveira:** Museu Paraense Emilio Goeldi, Av. Perimetral, 1901 - Terra Firme, Belém-PA, 66077-830, Brazil; E-mail: mozaniel.oliveira@yahoo.com.br

In this context, weeds are one of the most recurrent problems affecting agricultural production and, consequently, the returns on investments applied [3] [4, 5]. Among the species that infest agricultural areas are those with broad leaves, especially from the families Leguminosae, Malvaceae, Lamiaceae, Convolvulaceae, and Asteraceae [6], and those with narrow leaves, especially from the families Cyperaceae and Poaceae [7 - 9]. The species of these families are characterized by aggressiveness and a high capacity to compete with plants of economic interest, constituting the main component of crop maintenance costs [10, 11]. The control of these species is relevant for crop productivity, and the control methods employed by the producers generate dissatisfaction that promotes insecurity in the sector, especially in relation to chemical products [12, 13].

Many of the current herbicides in use in agriculture have resulted from various weed species being resistant to these products. In recent decades, the number of resistant plant breeds and species has increased significantly in different parts of the world [14 - 16]. In Brazil, the number of herbicide-resistant plants has also increased as a result of the systematic use of herbicides with the same site of action [17, 18]. The use of allelochemicals for the formulation of innovative products can face the challenge of controlling plants resistant to the current products in use, improving the agricultural system, and mitigating the social dissatisfaction arising from the use of herbicides [19, 20].

Allelochemicals can also offer new and innovative molecules with the potential for direct use in the management of weeds, or even make it possible to obtain products as efficient as commercial herbicides [21, 22], without posing any risk to the environment or even to humans, since they have a low permanence rate in the environment, and are quickly degraded by soil microorganisms [23]. Among the various possibilities for this purpose, the terpenoid class deserves to be highlighted due to the wide chemical diversity of its components, which can be classified as monoterpenes ($C_{10}$), sesquiterpenes ($C_{15}$), diterpenes ($C_{20}$), sesterterpenes ($C_{25}$), triterpenes ($C_{30}$), tetraterpenes ($C_{40}$), and polyterpenes ($>C_{40}$) [24].

These compounds have shown phytotoxic activity on invasive plants [25 - 28], which can constitute an advantageous tool to be considered in the strategies of the current agriculture model. Compounds with phytotoxic activity are referred to in the literature as allelochemicals [29 - 34], and in Fig. (**1**), it is possible to observe a form of interaction between plants called allelopathy, in which one of the species produces allelochemicals capable of inhibiting the development of the other one.

Therefore, this work seeks to gather recent information that expresses, in all possibilities, the real potential of using terpenoids in different strategies in weed management.

**Fig. (1).** Illustrative interactions between plants.

## *Volatile Terpenoids*

Terpenoids are an important group of highly diverse chemicals produced by plants that play a leading role in plant defense, and can provide chemical molecules with possibilities to be used in pest and disease management, as well as in other human demands, such as in hygiene and health [35 - 38]. Reports of terpenoid-producing plants are widely found in the literature. In Table **1**, some good examples are listed. In the terpenoid, class included the essential oils (monoterpenes, diterpenes, sesquiterpenes), and in higher molecular weight compounds, triterpenes and tetraterpenes can be found [39 - 41].

**Table 1. Main volatile compounds are present in essential oils of different species.**

| Species | Family | Plant Fraction | Compound | Refs. |
|---|---|---|---|---|
| *Origanum vulgare* L. | Lamiaceae | Leaves and branches | Sabinene, myrcene, *p*-cymene, 1,8-cineole, β-ocimene, γ-terpinene, sabinene hydrate, linalool, α-terpineol, carvacrol methyl ether, linalyl acetate, thymol and carvacrol. Among the sesquiterpenes β-caryophyllene, germacrene D, germacrene *D*-4-ol, spathulenol, caryophyllene oxide | [48] |
| *Thymus daenensis* and *Thymus vulgaris* | Lamiaceae | Aerial parts | Thymol, carvacrol, *p*-cymene, and terpinene | [49] |
| *Salvia sclarea* | Lamiaceae | Aerial parts | Linalyl acetate, linalool, and germacrene D | [50] |
| *Cinnamomum verum* | Lauraceae | Bark | Cinnamaldehyde | [51] |
| *Laurus nobilis* | Lauraceae | Leaves | 1,8-Cineole, sabinene, and linalool | [52] |
| *Rosemary officinalis* | Lamiaceae | Not informed | 1,8-cineole | [53] |
| *Pogostemon cablin* | Lamiaceae | Leaves | β-patchoulene, cariofilene, γ-patchoulene, α-patchoulene, β-guaiene, and α-selinene | [54] |
| *Coriandrum sativum L* | Apiaceae | Seeds, Flowers, and Leaves | Linalool, γ-terpinene, α-pinene, *p*-cymene, camphor, geranyl acetate, Benzofuran,2,3-dihydro, hexadecanoic acid, methyl ester, 2,4a-epiox--3,4,5,6,7,8,-hexahydro-2,5,5,8a-tetramethyl-2h-1-benzofuran, 2-methyoxy-4-vinylphenol, 2,3,5,6-tetrafluroanisole, 2,6-dimethyl-3-aminobenzoquinone, dodecanoic acid Decanal, *trans*-2-decenal, 2-decen-1-ol, cyclodecane, cis-2-dodecena, Dodecanal, and dodecan-1-ol | [55] |
| *Origanum vulgare* L. | Lamiaceae | Leaves | α-Pinene, Camphene, β-Pinene, Myrcene, α-Terpinene, ρ-Cymene, Cineole, γ-Terpinene, Linalool, Thymol, Carvacrol, Ethyl caprate, β-Bisabolene, (E)-β-Farnesene, and Ledol. | [56] |
| Piper aduncum L. | Piperaceae | Leaves | 1,8-cineole | [57] |
| Lavanda | Lamiaceae | Aerial parts | *D*-limolene, Eucalyptol, linalyl acetate, camphora, linalol, and caryophyllene | [58] |
| Zingiber officinale Rosc | Zingiberaceae | Aerial parts | 6-Gingerol | [59] |

(Table 1) cont.....

| Species | Family | Plant Fraction | Compound | Refs. |
|---------|--------|----------------|----------|-------|
| *Piper corcovadensis* (Miq.) C. DC | Piperaceae | Fresh leaves | 1-butyl-3,4-methylenedioxybenzene, terpinolene, *trans* -caryophyllene, α-pinene, δ-cadinene, and Limonene | [60] |
| *Piper cernuum* | Piperaceae | Aerial parts | Camphen. | [61] |

Essential oils are formed by complex mixtures of volatile substances composed mainly of terpenoids [42], which are responsible for the flavor of plants, and can be found in different families and parts, such as leaves, barks, roots, flowers, fruits, and seeds [43]. They are also among the main components exercising an important function in plant defense and as suppliers of molecules with great potential for the development of innovative products for direct use in agriculture or in the formulation of new defensive agents [44].

The composition of the oils is quite varied, as well as the concentration of each component. These aspects may also vary depending on the species, plant fraction, and time of the year [45, 46]. Plant species belonging to the same family and genus have differences in chemical composition and in the concentration of each component [47]. Table **1** lists the chemical composition of essential oils rich in terpenoids of different species, and are among the bioactive products with the greatest potential for use in controlling seed germination and weed development, representing an excellent opportunity for use in the agricultural model that is currently active.

**Monoterpenes with Phytotoxic Potential**

Monoterpenes are formed by two $C_5$ isoprene structures, resulting in a $C_{10}$ structure. Their 2D chemical structures can be seen in Fig. (**2**). When considering the phytotoxic activity of essential oils on the germination and development of weeds, the intensity of the effects is directly dependent on three factors: i) the chemical composition of the oils; ii) concentration of each component; and iii) bioactivity of the major compounds [62, 63] Oils with highly expressive phytotoxic activity are, in general, composed prevalently of monoterpenes and diterpenes [64].

Among monoterpenes, those oxygenated and non-oxygenated have shown phytotoxic activity [62, 65], however, in some cases, oxygenated monoterpenes have shown greater phytotoxic activity in relation to hydrocarbon monoterpenes [66]. Thus, identifying chemical components, and the concentration of each compound as major substances, is a good indicator to predict the phytotoxic potential of certain oils.

In a recent study with three species of Piperaceae, Jaramillo-Colorado *et al.* [67], analyzed the chemical composition and phytotoxic activity of four species: *P. dilatatum, P. divaricatum, P. hispidum,* and *P. sanctifelici.* The compounds identified in the highest concentration were the monoterpenes eugenol, methyl eugenol, γ-elemene, apiol, (*E*)-caryophyllene, δ-3-carene, limonene, *p*-cymene, β-pinene, nerolidol, limonene, δ-3-carene, and *p*-cymene; and the species that showed the highest phytotoxic activity were *P. dilatatum* and *P. divaricatum*. In addition, monoterpenes, both oxygenated and hydrocarbons, may have different specificities when tested on the same recipient plant species, suppressing specific parts, such as germination, and radicle and hypocotyl elongation [62]. In Table **2**, different monoterpenes with potential phytotoxic activity can be seen.

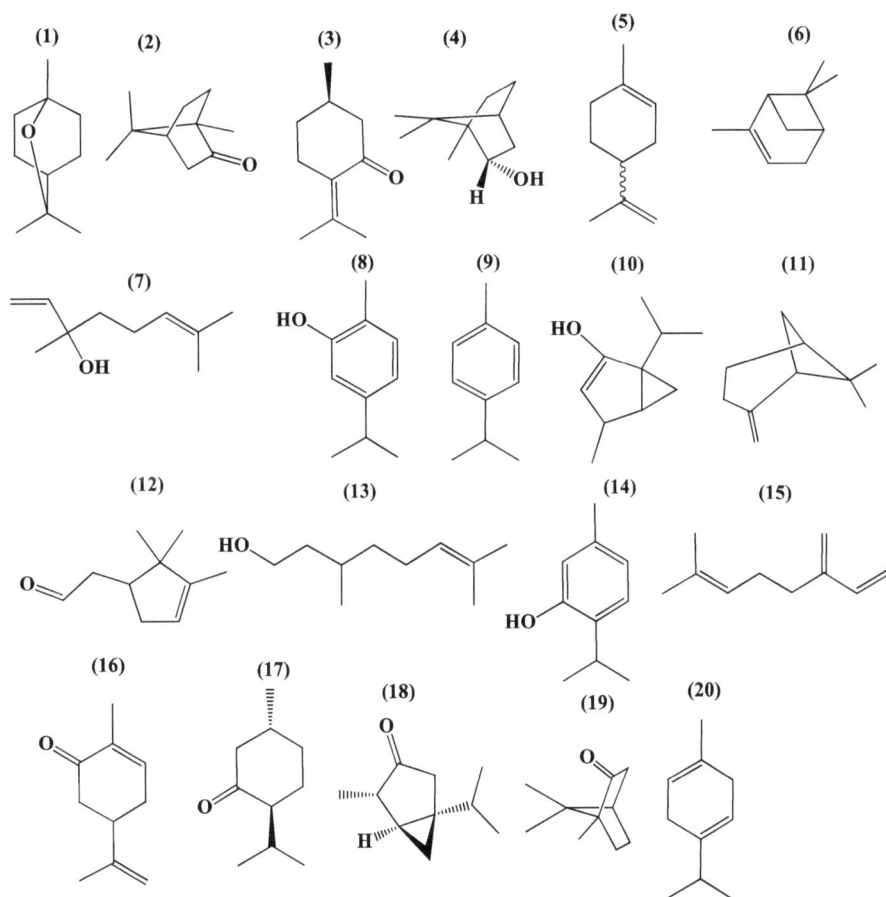

**Fig. (2).** Monoterpenes with potential phytotoxic activity. (1) = 1,8 cineole, (2) = camphor, (3) = pulegone, (4) = borneol, (5) = limonene, (6) = α-pinene, (7) = linalool, (8) = carvocrol, (9) = *p*-cymene, (10) = thujenol, (11) = β-pinene, (12) = α-campholenal, (13) = citronellol, (14) = thymol, (15) = β-myrcene, (16) = carvone, (17) = menthone, (18) = α-thujene, (19) = camphor, (20) γ-terpinene.

**Table 2. Oxygenated and non-oxygenated monoterpenes with potential phytotoxic activity.**

| Species with Phytotoxic Potential | Compounds | Recipient Species | Refs. |
|---|---|---|---|
| Comercial monoterpene | Citronellal, linalool, citronellol, and 1,8-cineole | *Cassia occidentalis* | [72] |
| Comercial monoterpene | Citronellol | *Ageratum conyzoides* L., *Chenopodium album* L., *Parthenium hysterophorus* L., *Malvastrum coromandelianum* (L.), Garcke, *Cassia occidentalis* L., and *Phalaris minor Retz.* | [73] |
| *Eucalyptus citriodora* | Citronelal and Citronelol | *Triticum aestivum* L., *Oryza sativa* L.) *Amaranthus viridis* L. *Echinochloa crus-galli* (L.) Beauv. | [74] |
| *Hyptis suaveolens* | α-phellandrene, α-pinene, allo-ocimene, limonene, β-thujene, γ-terpinene, and o-cymene. | *Oryza sativ.* and *Echinochloa crus-galli* | [75] |
| *A. rosaeodora, C. odorata, C. sativa inflorescences, C. citratus, C. limon, C. nobilis, C. sempervirens, C. longa rhizomes, H. perforatum, I. verum, M. alternifolia, Mentha × piperita, M. spicata, M. fistulosa, O. micranthum, O. basilicum, O. quixos, P. capitatum, P. nigra, P. cablin, S. aromaticum, T.s vulgari, V. zizanoidess, Z. officinale* | α-Pinene, Camphene, Sabinene, Myrcene, α-Phellandrene, Limonene, and γ-Terpinene. | *Solanum lycopersicum* L. | [76] |
| *Citrus aurantiifolia* | Limonene and citral | *Avena fatua, Echinochloa crus-galli,* and *Phalaris minor* | [66] |
| *Artemisia scoparia* Waldst. & Kit | β-myrcene, *p*-cymene, and dl-limonene. | *Avena sativa,* and *Triticum aestivum* | [77] |
| *Origanum acutidens* | carvacrol, thymol, and *p*-cymene | *Amaranthus retroflexus, Chenopodium album,* and *Rumex crispus* | [78] |
| *Achillea millefolium, Acorus calamus, Carum carvi, Chamomilla recutita, Foeniculum vulgare, Lavandula angustifolia, Melissa officinalis, Mentha × piperita, Salvia officinalis, Solidago canadensis, Tanacetum vulgare,* and *Thymus vulgaris* | carvone, limonene, thymol, menthone, menthol, α-thujone, and camphor | *Avena sativa, Brassica napus,* and *Zea mays* | [79] |

*(Table 2) cont.....*

| Species with Phytotoxic Potential | Compounds | Recipient Species | Refs. |
|---|---|---|---|
| Comercial monoterpene | (±)-β-citronellol, (±)-citronellal, (-)-α-pinene, (-)-β-pinene, α-terpinene, γ-terpinene, α-terpineol, 1,8-cineole, citral, thymol, carvacrol, α+β-thujone, camphene, (±)-camphor, (-)-borneol, *p*-cymene, myrcene, menthone, (±)-menthol, geraniol, geranyl acetate, linalool, linalyl acetate, (R)-(-)-α-phellandrene, estragole, (R)-(-)-carvon, limonene | *Raphanus sativus* L. (radish) and *Lepidium sativum* L. | [71] |
| *Hyssopus officinalis, Lavandula angustifolia, Majorana hortensis, Melissa officinalis, Ocimum basilicum, Origanum vulgare, Salvia officinalis and Thymus vulgaris (Lamiaceae), Verbena officinalis (Verbenaceae), Pimpinella anisum, Foeniculum vulgare, and Carum carvi (Apiaceae)* | β-Pinene, Carvone, Linalyl acetate, *iso*-Pinocamphone, Borneol, *o*-Cymene, Geraniol, (-)-Citronellal, Linalol, 1,8-Cineole, and cis-Anethole | Raphanus sativus, Lactuca sativa, and Lepidium sativum | [80] |
| *S. aromaticum* | Eugenol, Eugenol acetate | *M. pudica* and *S. obtusifolia* | [44] |

It is possible to observe that monoterpenes can be considered phytotoxins alone or in complex mixtures (Table **2**). Major compounds present in essential oils have been evaluated individually, as part of understanding the overall phytotoxicity effects of these oils on weeds [44]. In this sense, other components have already been isolated and tested for phytotoxic activity, as are the cases of camphor, 1,8-cineol, nerol, and neryl isovalerate [68], among others [69 - 71].

## Sesquiterpenes with Phytotoxic Potential

Sesquiterpenes are composed of three isoprene structures, forming a 15-carbon molecule [81]. These compounds have several industrial applications [82], and in agriculture, are widely known for being potential bioherbicides. For instance, species like *Copaifera duckei* (Dwyer), *Copaifera martii* (Hayne), and *Copaifera reticulata* (Ducke) have essential oils rich in sesquiterpenes like germacrene D, β-caryophyllene, α-humulene, δ-elemene, and δ-cadinene; and these essential oils have been shown to have phytotoxic potential against invasive Amazonian species present in management areas, such as *Mimosa pudica* L. and *Senna obtusifolia* (L.), at different intensity values depending on the recipient plant [83].

In the essential oils of *Vitex negundo* L, the sesquiterpene hydrocarbon β-Caryophyllene is the major component. Such oils demonstrated cytotoxic activity against seed germination of two plant species: *Avena fatua* L. and *Echinochloa crus-galli* (L.) [84]. In studies on sesquiterpene lactones, these substances were isolated from *Inula viscosa*: (4E,7R*,8R*,10S*)-3-oxo-germacra-4,11(13)-dien-8β-12-olide, its 11,13-dihydro analogue, (5R*,7R*,8R*,10R*)-1,1--methylene-5β-hydroxy-eudesm-1(15),11(13)-dien-8β-12-olide, and (7R*,8R*)-1,4-dimethyl-4-hydroxy-secoeudesm-5(10),11(13)-dien-8β-12-olide. The authors noted that the isolated molecules had effects of up to 100% inhibition of *C. campestris* and *O. crenata* germination [85].

To summarize the data, Table **1** was elaborated, in which different species of plants and their major volatile and non-volatile components (sesquiterpenoids) can be observed as promoters of phytotoxic activity. Their 2D chemical structures are shown in Fig. (**3**).

Table 3. Volatile and non-volatile sesquiterpenes with potential phytotoxic activity.

| Species with Phytotoxic Potential | Compounds | Recipient Species | Refs. |
|---|---|---|---|
| *Cynara cardunculus* L. | Aguerin B, Grosheimin, 11,13-dihydroxy-8-Deoxygrosheimin, Deacylcynaropicrin, and 11,13-dihydro-Deacylcynaropicrin | *Amaranthus retroflexus* L. and *Portulaca oleracea* L | [86] |
| *Coprophilous Fungus* Penicillium | 3R,6R-dihydroxy-9,7(11)-dien-8-oxoeremophilane, isopetasol, sporogen AO-1, and dihydrosporogen AO-1. | *Amaranthus hypochondriacus* | [87] |
| *Saussurea lappa* | Lappalone | *Lepidium sativum* L., *Alium cepa* L., *Lactuca sativa* L., and *Solanum lycopersicum* L. | [88] |
| *Bidens sulphurea* L. | Reynosin (2), santamarine (3), 11-epiartesin, 11α-13-dihydroreynosin, 1β,4β-dihydroxyarbusculin, 1β,6α-dihydroxyeudesm-4(5)-ene, clovane-2β-9α-diol, 10-hydrox--15-oxo-αcadinol, 10-oxoisodauc-3-en-15-al, 10-hydroxy-6,-0-epoxy4(14)-isodaucane, caryolane-1,9 β-diol, alloromadendrane4β,10α-diol, alloromadendrane-4β,10β-diol, α-dimorphecolic acid, (9Z,12S,13E,15Z)-12-hydroxyoctade-a-9,13,15-trienoic acid, (9Z,12S,13E,15Z)-1--hydroxyoctadeca-9,13,15-trienoic acid methyl ester, and stigmasterol. | *Amaranthus viridis,* and *Panicum maximum* | [89] |
| *Ferula pseudalliacea* | methyl galbanate, ethyl galbanate, fekrynol acetate, farnesiferol B, and kamonolol acetate. | *Nicotiana tabacum* L. cv. Burley 21 | [90] |

*(Table 3) cont.....*

| Species with Phytotoxic Potential | Compounds | Recipient Species | Refs. |
|---|---|---|---|
| *Stereum complicatum* | sterostrein X, sterostrein Y, hirsutenol G, sterpurol C, sterostrein H, sterostrein P., and sterostrein Q. | *Lactuca sativa, Agrostis stolonifera,* and *Lemna paucicostata* | [91] |
| *Ligularia cymbulifera* | Ligulacymirin A. and Ligulacymirin B | *A. thaliana* | [92] |
| *Eupatorium adenophorum Spreng* | γ-cadinene, γ-muurolene, 3-acetoxyamorpha-4,7(11)-die-e-8-one, and bornyl acetate | *Phalaris minor* | [93] |
| Plant Essential Oils | γ-cadinene, γ-muurolene, and 3-acetoxyamorpha-4,7(1-)-diene-8-one, β-caryophyllene, (Z)-caryophyllene, germacrene D, hexahydrofarnesyl acetone, β-caryophyllene, caryophyllene oxide, and β-bisabolene | Systematic review | [94] |
| *Drimys brasiliensis* Miers | polygodial, polygodial acetal, dendocarbin L, and (+)-fuegin | *Barbarea verna* (Mill.), *Echinochloa crus-galli* (L.) and *Ipomoea grandifolia* (Dammer) | [95] |

## Diterpenes with Phytotoxic Potential

Diterpenes are highly diverse molecules formed by four isoprene structures, resulting in a 20-carbon structure, and exhibit diverse functions in mediating antagonistic and/or beneficial interactions in and among organisms [96]. In the study performed by Berrueab and Russell [97], the authors described 602 new compounds. In addition, several potential biological activities are reported in the literature [98, 99], such as the phytotoxic potential of harzianelactone A, harzianelactone B, harzianone A, harzianone B, harzianone C, harzianone D, and harziane [100].

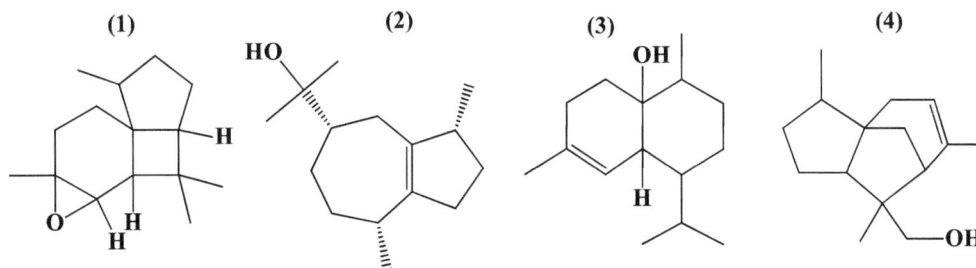

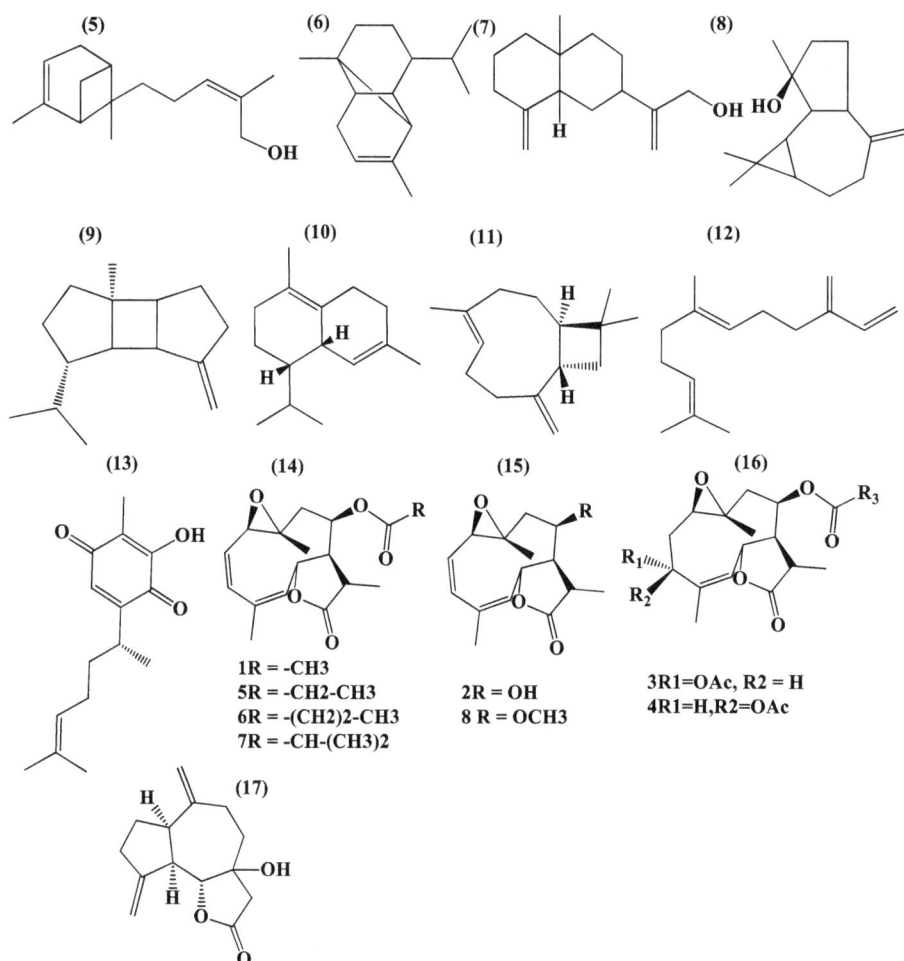

**Fig. (3).** Chemical structures of sesquiterpenes with potential phytotoxic activity, (1) = Italicene epoxide, (2) = Guaiol, (3) 1,10-di-epi-Cubenol, (4) = 8-Cedren-13-ol, (5) = (Z)-α-*trans*-Bergamotol, (6) = α-Copaene, (7) = β-Costol, (8) = Spathulenol, (9) = ß-Bourbonene, (10) = δ-Cadinene, (11) = ß-Caryophyllene, (12) = ß-Farnesene, (13) = Isoperezone, (14-17) = sesquiterpene lactones.

In the work of Cimmino *et al.* [101], the authors evaluated the phytotoxic effect of the diterpene 1S,2S,3S,4S,5S,9R,10S,12S,13S)-1,12-acetoxy-2,3-hydroxy- 6-oxopimara-7(8),15-dien-18-oic acid-2,18-lactone on weeds. The authors reported that at a concentration of 2 mg/mL, the toxin caused necrotic lesions on *Mercurialis annua*, *Cirsium arvense*, and *Setaria viride*. Other diterpenes such as 2-oxokovalenic acid and 12,19-hydroxyferruginol obtained from *Tectona grandis*, presented high phytotoxic activity against *Allium cepa* L. (onion), dicots *Lycopersicon esculentum* Will. (tomato), *Lepidium sativum* L. (cress), and *Lactuca sativa* L [102].

Several other articles report the possibility of using terpenoids isolated from the most different species as molecules with potential use in agriculture for the control of invasive plant species. They can be seen in Table **4**, and are summarized in different studies that demonstrated the effectiveness of this important class of phytochemicals. In addition, the chemical structures of some diterpenoids can be seen in Fig. (**4**).

**Table 4. Diterpenoids with potential phytotoxic activity.**

| Species with Phytotoxic Potential | Compounds | Recipient Species | Refs. |
|---|---|---|---|
| *Sphaeropsis sapinea* f. sp. Cupressi and *Diplodia mutila* | Sphaeropsidins B and C | *Cupressus sempervirens* L | [103] |
| *Salvia miniata* Fernald | Ten clerodane diterpenoids | *Papaver rhoeas* L. and *Avena sativa* L. | [104] |
| *Blakiella bartsiifolia* (S.F. Blake) | 15,16-epoxy-2-hydroxy-3,13(16),14-clerodatrien-20-oic acid (bartsiifolic acid) (2) and Z-15,1--dihydroxy-3,13-clerodien-20-oic acid (barthydrolic acid), junceic acid, 1,20-epox--1,3(20),6(E),10(E),14-phytapentaen-18-methyl-19-oic acid (blakielic acid) (4), 1,20-epox--1,3(20),10(E),14-phytapentaen-18-methyl-19-oic acid (blakifolic acid) and 1,20-epox--1,3(20),6,14-phytatetraen-19-methyl-18-oic acid (dihydrocentipedic acid) | *Allium cepa* and *Lactuca sativa* | [104] |
| *Eragrostis plana* | neocassa-12(17),15-dien-3-oneand, and neocassa-12(13),15-diene-3,14-dione | *Ipomoea grandifolia* and *Euphorbia heterophylla* | [105] |
| *Eragrostis plana* | neocassa-1,12(13),15-triene-3,14-dione, 19-norneocassa-1,12(13),15-triene-3,14-dione, and 14-hydroxyneocassa-1,12(17),15-triene-3-one, 3 | *Lemna paucicostata (L.)* Hegelm | [106] |
| *Leucas áspera* | (rel 5S,6R,8R,9R,10S,13S,15S,16R)-6-acetoxy-9,13;15,16-diepoxy-15-hyd-oxy-16-methoxylabdane, and (rel 5S,6R,8R,9R,10S,13S,15R,16R)-6-acetoxy-9,13;15-16-diepoxy-15-hydroxy-16-methoxylabdane | *Lepidum ativum* and *Echinochloa crus-galli* | [107] |
| *Euphorbia esula* L | ellagic acid, 3,3'-di-O-methylellagic acid and 3,3',4'-tri-O-methylellagic acid | *Arabidopsis thaliana* | [108] |
| *Cupressus sempervirens* L | Sphaeropsidins D and E | *Cupressus macrocarpa, C. sempervirens* and *C. arizonica* | [109] |

Fig. (4).   Chemical structures of some diterpenoids. (1)  =  19-norneocassa1,12(13),15-triene-3-14-dione,(2)=14-hydroxyneocassa 1,12(17),15-triene-3-one, (3) = rel 5S,6R,8R,9R,10S,13S,15R,16R)-6-acetoxy-9,13;15,16-diepoxy-15-hydroxy-16-methoxylabdane, (4) = Perezone; (5)= Isoperezone, (6)= Dihydroperezone; (7) = Dihydroisoperezone; (8) = Anilidoperezone, (9) = junceic acid, (10) = 5,16-epoxy-2-hydroxy-3,13(16),14-clerodatrien-20-oic acid, (11) = Z-15,16-dihydroxy-3,13-clerodien-20-oic acid, (11) = barthydrolic acid, (12) = 1,20-epoxy-1,3(20),6(E),10(E),14-phytapentaen-18-methyl-19-oic acid, (13) = blakielic acid, (14) = 1,20-epoxy-1,3(20),10(E),14-phytapentaen-18-methyl-19-oic, and (15) = 1,20-epoxy-1, 3(20),6,14-phytatetraen-19-methyl-18-oic acid.

## CONCLUDING REMARKS

The use of allelochemicals obtained from the secondary metabolism of plants can be an ecologically viable alternative for the control of invasive plants. For this purpose, essential oils have shown efficiency, which may be related to their diversified chemical composition. Compounds belonging to the class of monoterpenes, sesquiterpenes, and diterpenes are promising allelochemicals with phytotoxic potential. In relation to monoterpenes, they stand out in relation to others, such as linalool and 1.8 cineole; among sesquiterpenes, β-karyophylene; and among diterpenes, sphaeropsidins B and C stand out. It must be considered that the concentration used in the experiments can also be an important factor, in addition to the osmotic activity of water.

## CONSENT FOR PUBLICATION

Not applicable.

## CONFLICT OF INTEREST

The author declares no conflict of interest, financial or otherwise.

## ACKNOWLEDGMENTS

The author Mozaniel Santana de Oliveira thanks PCI-MCTIC/MPEG, as well as CNPq for the scholarship process number: [301194/2021-1].

## REFERENCES

[1]    Carvalho, M.M.X.; Nodari, E.S.; Nodari, R.O. "Defensivos" ou "agrotóxicos"? História do uso e da percepção dos agrotóxicos no estado de Santa Catarina, Brasil, 1950-2002. *Hist. Cienc. Saude Manguinhos,* **2017**, *24*(1), 75-91.http://www.scielo.br/scielo.php?script=sci_arttext&pid=S0104-59702017000100075&lng=pt&tlng=pt
[http://dx.doi.org/10.1590/s0104-59702017000100002] [PMID: 28380166]

[2]    Beck, J.J.; Alborn, H.T.; Block, A.K.; Christensen, S.A.; Hunter, C.T.; Rering, C.C.; Seidl-Adams, I.; Stuhl, C.J.; Torto, B.; Tumlinson, J.H. Interactions Among Plants, Insects, and Microbes: Elucidation of Inter-Organismal Chemical Communications in Agricultural Ecology. *J. Agric. Food Chem.,* **2018**, *66*(26), 6663-6674.https://pubs.acs.org/doi/10.1021/acs.jafc.8b01763
[http://dx.doi.org/10.1021/acs.jafc.8b01763] [PMID: 29895142]

[3]    Radhakrishnan, R.; Alqarawi, A.A.; Abd Allah, E.F. Bioherbicides: Current knowledge on weed control mechanism. *Ecotoxicol. Environ. Saf.,* **2018**, *158*, 131-138.https://linkinghub.elsevier.com/retrieve/pii/S0147651318303129
[http://dx.doi.org/10.1016/j.ecoenv.2018.04.018] [PMID: 29677595]

[4]    van Evert, F.K.; Fountas, S.; Jakovetic, D.; Crnojevic, V.; Travlos, I.; Kempenaar, C. Big Data for weed control and crop protection. *Weed Res.,* **2017**, *57*(4), 218-233.http://doi.wiley.com/10.1111/wre.12255 [Internet].
[http://dx.doi.org/10.1111/wre.12255]

[5]    Macías, F.A.; Mejías, F.J.R.; Molinillo, J.M.G. Recent advances in allelopathy for weed control: from knowledge to applications. *Pest Manag. Sci.,* **2019**, *75*(9), 2413-2436.http://doi.wiley.com/

10.1002/ps.5355
[http://dx.doi.org/10.1002/ps.5355] [PMID: 30684299]

[6]     Mallick, S; Plant, R; Centre, R. Weed flora of Rourkela and adjoining areas of Sundargarh district , Odisha , WEED FLORA OF ROURKELA AND ADJOINING AREAS OF. J Econ Taxon Bot. **2016**, *39*, 7.

[7]     de Vries, F.T. Chufa (cyperus esculentus, cyperaceae): A weedy cultivar or a cultivated weed? *Econ. Bot.,* **1991**, *45*(1), 27-37.http://link.springer.com/10.1007/BF02860047 [Internet].
[http://dx.doi.org/10.1007/BF02860047]

[8]     Arriola, P.E.; Ellstrand, N.C. Crop-to-weed gene flow in the genus *Sorghum* (Poaceae): Spontaneous interspecific hybridization between johnsongrass, *Sorghum halepense*, and crop sorghum, *S. bicolor. Am. J. Bot.,* **1996**, *83*(9), 1153-1159.http://doi.wiley.com/10.1002/j.1537-2197.1996.tb13895.x [Internet].
[http://dx.doi.org/10.1002/j.1537-2197.1996.tb13895.x]

[9]     Hjorth, M.; Mondolot, L.; Buatois, B.; Andary, C.; Rapior, S.; Kudsk, P.; Mathiassen, S.K.; Ravn, H.W. An easy and rapid method using microscopy to determine herbicide effects in Poaceae weed species. *Pest Manag. Sci.,* **2006**, *62*(6), 515-521.http://doi.wiley.com/10.1002/ps.1194 [http://dx.doi.org/10.1002/ps.1194] [PMID: 16628541]

[10]    Jayabarathi, T.; Yazdani, A.; Ramesh, V.; Raghunathan, T. Combined heat and power economic dispatch problem using the invasive weed optimization algorithm. *Front. Energy,* **2014**, *8*(1), 25-30.http://link.springer.com/10.1007/s11708-013-0276-4 [Internet].
[http://dx.doi.org/10.1007/s11708-013-0276-4]

[11]    Barisal, A.K.; Prusty, R.C. Large scale economic dispatch of power systems using oppositional invasive weed optimization. *Appl. Soft Comput.,* **2015**, *29*, 122-137.https://linkinghub.elsevier.com/retrieve/pii/S1568494614006462 [Internet].
[http://dx.doi.org/10.1016/j.asoc.2014.12.014]

[12]    Abouziena, H.F.; Haggag, W.M. Weed Control in Clean Agriculture: A Review1. *Planta Daninha,* **2016**, *34*(2), 377-392.http://www.scielo.br/scielo.php?script=sci_arttext&pid=S0100-83582016000-200377&lng=en&tlng=en
[http://dx.doi.org/10.1590/S0100-83582016340200019]

[13]    Jabran, K.; Mahajan, G.; Sardana, V.; Chauhan, B.S. Allelopathy for weed control in agricultural systems. *Crop Prot.,* **2015**, *72*, 57-65.https://linkinghub.elsevier.com/retrieve/pii/S0261219415000782 [Internet].
[http://dx.doi.org/10.1016/j.cropro.2015.03.004]

[14]    Forouzesh, A.; Zand, E.; Soufizadeh, S.; Samadi Foroushani, S. Classification of herbicides according to chemical family for weed resistance management strategies - an update. *Weed Res.,* **2015**, *55*(4), 334-358.http://doi.wiley.com/10.1111/wre.12153 [Internet].
[http://dx.doi.org/10.1111/wre.12153]

[15]    Nandula, V.K.; Riechers, D.E.; Ferhatoglu, Y.; Barrett, M.; Duke, S.O.; Dayan, F.E.; Goldberg-Cavalleri, A.; Tétard-Jones, C.; Wortley, D.J.; Onkokesung, N.; Brazier-Hicks, M.; Edwards, R.; Gaines, T.; Iwakami, S.; Jugulam, M.; Ma, R. Herbicide Metabolism: Crop Selectivity, Bioactivation, Weed Resistance, and Regulation. *Weed Sci.,* **2019**, *67*(2), 149-175.https://www.cambridge.org/core/product/identifier/S0043174518000887/type/journal_article [Internet].
[http://dx.doi.org/10.1017/wsc.2018.88]

[16]    Wu, C.; Varanasi, V.; Perez-Jones, A. A nondestructive leaf-disk assay for rapid diagnosis of weed resistance to multiple herbicides. *Weed Sci.,* **2021**, *69*(3), 274-283. https://www.cambridge.org/core/product/identifier/S0043174521000151/type/journal_article [Internet].
[http://dx.doi.org/10.1017/wsc.2021.15]

[17]    Lopez Ovejero, R.F.; Takano, H.K.; Nicolai, M.; Ferreira, A.; Melo, M.S.C.; Cavenaghi, A.L.;

Christoffoleti, P.J.; Oliveira, R.S., Jr Frequency and Dispersal of Glyphosate-Resistant Sourgrass ( *Digitaria insularis* ) Populations across Brazilian Agricultural Production Areas. *Weed Sci.,* **2017**, *65*(2), 285-294.https://www.cambridge.org/core/product/identifier/S004317451600031X/type/journal_article [Internet]. [http://dx.doi.org/10.1017/wsc.2016.31]

[18]   Alcántara-de la Cruz, R.; Moraes de Oliveira, G.; Bianco de Carvalho, L. Herbicide Resistance in Brazil: Status, Impacts, and Future Challenges. In: Kontogiannatos D, Kourti A, Mendes KF, editors. Pests, Weeds and Diseases in Agricultural Crop and Animal Husbandry Production [Internet]. 1st ed. Londo:   IntechOpen   **2020**,   30.https://www.intechopen.com/books/pests-weeds-and-disease-
-in-agricultural-crop-and-animal-husbandry-production/herbicide-resistance-in-brazil-st-
tus-impacts-and-future-challenges

[19]   Kong, C.H.; Xuan, T.D.; Khanh, T.D.; Tran, H.D.; Trung, N.T. Allelochemicals and Signaling Chemicals in Plants. *Molecules,* **2019**, *24*(15), 2737.https://www.mdpi.com/1420-3049/24/15/2737 [http://dx.doi.org/10.3390/molecules24152737] [PMID: 31357670]

[20]   Scavo, A.; Abbate, C.; Mauromicale, G. Plant allelochemicals: agronomic, nutritional and ecological relevance in the soil system. *Plant Soil,* **2019**, *442*(1-2), 23-48. http://link.springer.com/ 10.1007/s11104-019-04190-y [Internet]. [http://dx.doi.org/10.1007/s11104-019-04190-y]

[21]   Reigosa, M.J.; Gonzalez, L.; Sanches-Moreiras, A.; Duran, B.; Puime, D.; Fernadez, D.A. Comparison of physiological effects of allelochemicals and commercial herbicides. *Allelopathy J.,* **2001**, *8*(2), 211-220.

[22]   Durán-Serantes, B.; González, L.; Reigosa, M.J. Comparative physiological effects of three allelochemicals and two herbicides on Dactylis glomerata. *Acta Physiol. Plant.,* **2002**, *24*(4), 385-392.http://link.springer.com/10.1007/s11738-002-0034-4 [Internet]. [http://dx.doi.org/10.1007/s11738-002-0034-4]

[23]   Das, C.; Dey, A.; Bandyopadhyay, A. **2021**.http://link.springer.com/10.1007/978-981-15-8127-4_12

[24]   Davis, E.M.; Croteau, R. Cyclization Enzymes in the Biosynthesis of Monoterpenes, Sesquiterpenes, and Diterpenes. In: F.J. L, J.C. V, editors. Biosynthesis. Topics in Current Chemistry. Amisterdan; **2000**, 53-95.http://link.springer.com/10.1007/3-540-48146-X_2

[25]   He HB, Wang HB, Fang CX, Lin YY, Zeng CM, Wu LZ, *et al.* Herbicidal effect of a combination of oxygenic   terpenoids   on   Echinochloa   crus-galli.   Weed   Res.   **2009**,   *49*(2), 183-92.http://doi.wiley.com/10.1111/j.1365-3180.2008.00675.x

[26]   Araniti, F.; Sánchez-Moreiras, A.M.; Graña, E.; Reigosa, M.J.; Abenavoli, M.R. Terpenoid *trans* - caryophyllene inhibits weed germination and induces plant water status alteration and oxidative damage in adult *Arabidopsis. Plant Biol.,* **2017**, *19*(1), 79-89.http://doi.wiley.com/10.1111/plb.12471 [http://dx.doi.org/10.1111/plb.12471] [PMID: 27173056]

[27]   Li, J.Y.; Lin, S.X.; Zhang, Q.; Li, L.; Hu, W.W.; He, H.B. Phenolic acids and terpenoids in the soils of different weed-suppressive circles of allelopathic rice. *Arch. Agron. Soil Sci.,* **2020**, *66*(2), 266-278.https://www.tandfonline.com/doi/full/10.1080/03650340.2019.1610560 [Internet]. [http://dx.doi.org/10.1080/03650340.2019.1610560]

[28]   Umehara, M.; Hanada, A.; Yoshida, S.; Akiyama, K.; Arite, T.; Takeda-Kamiya, N.; Magome, H.; Kamiya, Y.; Shirasu, K.; Yoneyama, K.; Kyozuka, J.; Yamaguchi, S. Inhibition of shoot branching by new terpenoid plant hormones. *Nature,* **2008**, *455*(7210), 195-200. http://www.nature.com/ articles/nature07272 [http://dx.doi.org/10.1038/nature07272] [PMID: 18690207]

[29]   Chotsaeng, N.; Laosinwattana, C.; Charoenying, P. Herbicidal Activities of Some Allelochemicals and Their Synergistic Behaviors toward Amaranthus tricolor L. *Molecules,* **2017**, *22*(11), 1841.http://www.mdpi.com/1420-3049/22/11/1841

[http://dx.doi.org/10.3390/molecules22111841] [PMID: 29077029]

[30]  Yang, X.F.; Lei, K.; Kong, C.H.; Xu, X.H. Effect of allelochemical tricin and its related benzothiazine derivative on photosynthetic performance of herbicide-resistant barnyardgrass. *Pestic. Biochem. Physiol.,* **2017**, *143*, 224-230.https://linkinghub.elsevier.com/retrieve/pii/S0048357517302468 [http://dx.doi.org/10.1016/j.pestbp.2017.08.010] [PMID: 29183596]

[31]  Ghimire, B.K.; Hwang, M.H.; Sacks, E.J.; Yu, C.Y.; Kim, S.H.; Chung, I.M. Screening of Allelochemicals in *Miscanthus sacchariflorus* Extracts and Assessment of Their Effects on Germination and Seedling Growth of Common Weeds. *Plants,* **2020**, *9*(10), 1313.https://www.mdpi.com/2223-7747/9/10/1313 [http://dx.doi.org/10.3390/plants9101313] [PMID: 33028036]

[32]  Latif, S.; Chiapusio, G.; Weston, L.A. *Allelopathy and the Role of Allelochemicals in Plant Defence.,* **2017**.https://linkinghub.elsevier.com/retrieve/pii/S0065229616301203 [http://dx.doi.org/10.1016/bs.abr.2016.12.001]

[33]  Chaïb, S.; Pistevos, J.C.A.; Bertrand, C.; Bonnard, I. Allelopathy and allelochemicals from microalgae: An innovative source for bio-herbicidal compounds and biocontrol research. *Algal Res.,* **2021**, *54*, 102213.https://linkinghub.elsevier.com/retrieve/pii/S2211926421000321 [Internet]. [http://dx.doi.org/10.1016/j.algal.2021.102213]

[34]  Mushtaq, W.; Siddiqui, M.B.; Hakeem, K.R. Role of Allelochemicals in Agroecosystems. In: *Allelopathy-Potential for Green Agriculture,* 1st ed; Mushtaq, W.; Siddiqui, M.B.; Hakeem, K.R., Eds.; Springer: Cham, **2020**; pp. 45-52. http://link.springer.com/ 10.1007/978-3-030-40807-7_5 [Internet] [http://dx.doi.org/10.1007/978-3-030-40807-7_5]

[35]  Thoppil, R.J.; Bishayee, A. Terpenoids as potential chemopreventive and therapeutic agents in liver cancer. *World J. Hepatol.,* **2011**, *3*(9), 228-249.http://www.wjgnet.com/1948-5182/full/v3/i9/228.htm [http://dx.doi.org/10.4254/wjh.v3.i9.228] [PMID: 21969877]

[36]  Huang, M.; Lu, J.J.; Huang, M.Q.; Bao, J.L.; Chen, X.P.; Wang, Y.T. Terpenoids: natural products for cancer therapy. *Expert Opin. Investig. Drugs,* **2012**, *21*(12), 1801-1818. http://www.tandfonline.com/ doi/full/10.1517/13543784.2012.727395 [http://dx.doi.org/10.1517/13543784.2012.727395] [PMID: 23092199]

[37]  Caputi, L.; Aprea, E. Use of terpenoids as natural flavouring compounds in food industry. *Recent Pat. Food Nutr. Agric.,* **2011**, *3*(1), 9-16. http://www.eurekaselect.com/openurl/content.php?genre= article&issn=2212-7984&volume=3&issue=1&spage=9 [http://dx.doi.org/10.2174/2212798411103010009] [PMID: 21114471]

[38]  Santana de Oliveira, M.; Pereira da Silva, V.M.; Cantão Freitas, L.; Gomes Silva, S.; Nevez Cruz, J.; Aguiar Andrade, E.H. Extraction Yield, Chemical Composition, Preliminary Toxicity of *Bignonia nocturna* (Bignoniaceae) Essential Oil and *in Silico* Evaluation of the Interaction. *Chem. Biodivers.,* **2021**, *18*(4), e2000982. [http://dx.doi.org/10.1002/cbdv.202000982] [PMID: 33587821]

[39]  Ludwiczuk, A.; Skalicka-Woźniak, K.; Georgiev, M.I. Terpenoids. In: *Pharmacognosy*; Badal, S.; Delgoda, R., Eds.; Elsevier: Amisterdan, **2017**; pp. 233-66. https://linkinghub.elsevier.com/ retrieve/pii/B9780128021040000111

[40]  Tholl, D. *Biosynthesis and Biological Functions of Terpenoids in Plants.,* **2015**. http://link.springer.com/10.1007/10_2014_295 [http://dx.doi.org/10.1007/10_2014_295]

[41]  Lu, J.J.; Dang, Y.Y.; Huang, M.; Xu, W.S.; Chen, X.P.; Wang, Y.T. Anti-cancer properties of terpenoids isolated from Rhizoma Curcumae – A review. *J. Ethnopharmacol.,* **2012**, *143*(2), 406-411.https://linkinghub.elsevier.com/retrieve/pii/S0378874112004783 [http://dx.doi.org/10.1016/j.jep.2012.07.009] [PMID: 22820242]

[42]  de Oliveira, M.S.; Silva, S.G.; da Cruz, J.N.; Ortiz, E.; da Costa, W.A.; Bezerra, F.W.F. Supercritical

$CO_2$ Application in Essential Oil Extraction. In: *Industrial Applications of Green Solvents –,* 2nd ed; Inamuddin, R.M.; Asiri, A.M., Eds.; Materials Research Foundations: Millersville, PA, USA, **2019**; Vol. II, pp. 1-28.http://www.mrforum.com/product/9781644900314-1 [Internet]

[43]   Santana de Oliveira, M.; Almeida da Costa, W.; Gomes Silva, S. *Essential Oils - Bioactive Compounds, New Perspectives and Applications,* 1st ed; IntechOpen: London, **2020**. https://www.intechopen.com/books/essential-oils-bioactive-compounds-new-perspectives-and-applications
[http://dx.doi.org/10.5772/intechopen.87266]

[44]   de Oliveira, M.S.; da Costa, W.A.; Pereira, D.S.; Botelho, J.R.S.; de Alencar Menezes, T.O.; de Aguiar Andrade, E.H.; da Silva, S.H.M.; da Silva Sousa Filho, A.P.; de Carvalho, R.N. Chemical composition and phytotoxic activity of clove ( Syzygium aromaticum ) essential oil obtained with supercritical CO 2. *J. Supercrit. Fluids,* **2016**, *118*, 185-193. [Internet].
[http://dx.doi.org/10.1016/j.supflu.2016.08.010]

[45]   Silva, S.G.; Figueiredo, P.L.B.; Nascimento, L.D.; da Costa, W.A.; Maia, J.G.S.; Andrade, E.H.A. Planting and seasonal and circadian evaluation of a thymol-type oil from Lippia thymoides Mart. & Schauer. *Chem. Cent. J.,* **2018**, *12*(1), 113. https://bmcchem.biomedcentral.com/articles/10.1186/s13065-018-0484-4
[http://dx.doi.org/10.1186/s13065-018-0484-4] [PMID: 30421173]

[46]   Figueiredo, P.L.B.; Silva, S.G.; Nascimento, L.D.; Ramos, A.R.; Setzer, W.N.; da Silva, J.K.R.; Andrade, E.H.A. Seasonal Study of Methyleugenol Chemotype of *Ocimum campechianum* Essential Oil and Its Fungicidal and Antioxidant Activities. *Nat. Prod. Commun.,* **2018**, *13*(8), 1934578X1801300.
[http://dx.doi.org/10.1177/1934578X1801300833]

[47]   Rodrigues, T.L.M.; Castro, G.L.S.; Viana, R.G.; Gurgel, E.S.C.; Silva, S.G.; de Oliveira, M.S. Physiological performance and chemical compositions of the Eryngium foetidum L. (Apiaceae) essential oil cultivated with different fertilizer sources. *Nat. Prod. Res.,* **2020**, *34*(19), 1-5.https://www.tandfonline.com/doi/full/10.1080/14786419.2020.1795653 [Internet].
[PMID: 32691619]

[48]   Lukas, B.; Schmiderer, C.; Novak, J. Essential oil diversity of European Origanum vulgare L. (Lamiaceae). *Phytochemistry,* **2015**, *119*, 32-40. https://linkinghub.elsevier.com/retrieve/pii/S003194221530087X
[http://dx.doi.org/10.1016/j.phytochem.2015.09.008] [PMID: 26454793]

[49]   Askary, M.; Behdani, M.A.; Parsa, S.; Mahmoodi, S.; Jamialahmadi, M. Water stress and manure application affect the quantity and quality of essential oil of Thymus daenensis and Thymus vulgaris. *Ind. Crops Prod.,* **2018**, *111*, 336-344. https://linkinghub.elsevier.com/retrieve/pii/S09266690-17306647
[http://dx.doi.org/10.1016/j.indcrop.2017.09.056]

[50]   Cui, H.; Zhang, X.; Zhou, H.; Zhao, C.; Lin, L. Antimicrobial activity and mechanisms of Salvia sclarea essential oil. *Bot. Stud. (Taipei, Taiwan),* **2015**, *56*(1), 16. http://www.as-botanicalstudies.com/content/56/1/16
[http://dx.doi.org/10.1186/s40529-015-0096-4] [PMID: 28510825]

[51]   Zhang, Y.; Liu, X.; Wang, Y.; Jiang, P.; Quek, S. Antibacterial activity and mechanism of cinnamon essential oil against Escherichia coli and Staphylococcus aureus. *Food Control,* **2016**, *59*, 282-289.https://linkinghub.elsevier.com/retrieve/pii/S0956713515300219 [Internet].
[http://dx.doi.org/10.1016/j.foodcont.2015.05.032]

[52]   Caputo, L.; Nazzaro, F.; Souza, L.; Aliberti, L.; De Martino, L.; Fratianni, F.; Coppola, R.; De Feo, V. Laurus nobilis: Composition of essential oil and its biological activities. *Molecules,* **2017**, *22*(6), 930.
[http://dx.doi.org/10.3390/molecules22060930] [PMID: 28587201]

[53]   Turasan, H.; Sahin, S.; Sumnu, G. Encapsulation of rosemary essential oil. *Lebensm. Wiss. Technol.,* **2015**, *64*(1), 112-119.https://linkinghub.elsevier.com/retrieve/pii/S0023643815003990 [Internet].

[http://dx.doi.org/10.1016/j.lwt.2015.05.036]

[54]    Costa, G.A.; Carvalho Filho, J.L.S.; Deschamps, C. Yield and composition of the patchouli (Pogostemon cablin) essential oil according to the extraction time. *Rev. Bras. Plantas Med.,* **2013,** *15*(3), 319-324. http://www.scielo.br/scielo.php?script=sci_arttext&pid=S1516-05722013000300002 &lng=pt&tlng=pt [Internet].
[http://dx.doi.org/10.1590/S1516-05722013000300002]

[55]    Mandal, S.; Mandal, M. Coriander (Coriandrum sativum L.) essential oil: Chemistry and biological activity. *Asian Pac. J. Trop. Biomed.,*    **2015,** *5*(6), 421-428. http://linkinghub.elsevier.com/ retrieve/pii/S2221169115000647 [Internet].
[http://dx.doi.org/10.1016/j.apjtb.2015.04.001]

[56]    Cui, H.; Zhang, C.; Li, C.; Lin, L. Antibacterial mechanism of oregano essential oil. *Ind. Crops Prod.,* **2019,** *139*, 111498.https://linkinghub.elsevier.com/retrieve/pii/S0926669019305102 [Internet].
[http://dx.doi.org/10.1016/j.indcrop.2019.111498]

[57]    Oliveira, GL; Cardoso, SK.; Lara, Junior CR; Vieira, TM. Chemical study and larvicidal activity against Aedes aegypti of essential oil of Piper aduncum L. (Piperaceae). *An Acad Bras Cienc,* **2013,** *85*(4), 1227-34. Available from: http://www.scielo.br/scielo.php?script=sci_arttext& pid=S0001-37652013000401227&lng=en&tlng=en

[58]    Silva, GL DA; Luft, C.; Lunardelli, A.; Amaral, RH.; Melo, Dads; Donadio, MVF Antioxidant, analgesic and anti-inflammatory effects of lavender essential oil. *An Acad Bras Cienc,* **2015,** *87*(2 suppl), 1397-308. http://www.scielo.br/scielo.php?script=sci_arttext&pid=S0001-37652015000-301397&lng=en&tlng=en

[59]    Mahboubi, M. Zingiber officinale Rosc. essential oil, a review on its composition and bioactivity. *Clin Phytoscience,* **2019,** *5*(1), 6. Available from: https://clinphytoscience.springeropen.com/articles/ 10.1186/s40816-018-0097-4

[60]    da Silva, M.F.R.; Bezerra-Silva, P.C.; de Lira, C.S.; de Lima Albuquerque, B.N.; Agra Neto, A.C.; Pontual, E.V.; Maciel, J.R.; Paiva, P.M.G.; Navarro, D.M.A.F. Composition and biological activities of the essential oil of Piper corcovadensis (Miq.) C. DC (Piperaceae). *Exp. Parasitol.,* **2016,** *165*, 64-70.https://linkinghub.elsevier.com/retrieve/pii/S0014489416300431
[http://dx.doi.org/10.1016/j.exppara.2016.03.017] [PMID: 26993082]

[61]    Girola, N.; Figueiredo, C.R.; Farias, C.F.; Azevedo, R.A.; Ferreira, A.K.; Teixeira, S.F.; Capello, T.M.; Martins, E.G.A.; Matsuo, A.L.; Travassos, L.R.; Lago, J.H.G. Camphene isolated from essential oil of Piper cernuum (Piperaceae) induces intrinsic apoptosis in melanoma cells and displays antitumor activity   *in    vivo.    Biochem.    Biophys.    Res.    Commun.,*    **2015,**    *467*(4),    928-934. https://linkinghub.elsevier.com/retrieve/pii/S0006291X15307415
[http://dx.doi.org/10.1016/j.bbrc.2015.10.041] [PMID: 26471302]

[62]    Andrés, M.F.; Rossa, G.E.; Cassel, E.; Vargas, R.M.F.; Santana, O.; Díaz, C.E.; González-Coloma, A. Biocidal effects of Piper hispidinervum (Piperaceae) essential oil and synergism among its main components. *Food Chem. Toxicol.,* **2017,** *109*(Pt 2), 1086-1092. https://linkinghub.elsevier.com/ retrieve/pii/S0278691517301862
[http://dx.doi.org/10.1016/j.fct.2017.04.017] [PMID: 28416272]

[63]    Parra Amin, J.E.; Cuca, L.E.; González-Coloma, A. Antifungal and phytotoxic activity of benzoic acid derivatives from inflorescences of Piper cumanense. *Nat. Prod. Res.,* **2019,** 1-9. https:// www.tandfonline.com/ doi/full/10.1080/14786419.2019.1662010 [Internet].
[PMID: 31502484]

[64]    Valarezo, E.; Flores-Maza, P.; Cartuche, L.; Ojeda-Riascos, S.; Ramírez, J. Phytochemical profile, antimicrobial and antioxidant activities of essential oil extracted from Ecuadorian species Piper ecuadorense sodiro. *Nat. Prod. Res.,*   **2020,** 1-6. https://www.tandfonline.com/doi/full/10.1080/ 14786419.2020.1813138
[PMID: 32851854]

[65] Ibáñez, M.; Blázquez, M. Phytotoxicity of Essential Oils on Selected Weeds: Potential Hazard on Food Crops. *Plants,* **2018**, *7*(4), 79.http://www.mdpi.com/2223-7747/7/4/79 [http://dx.doi.org/10.3390/plants7040079] [PMID: 30248993]

[66] Fagodia, S.K.; Singh, H.P.; Batish, D.R.; Kohli, R.K. Phytotoxicity and cytotoxicity of Citrus aurantiifolia essential oil and its major constituents: Limonene and citral. *Ind. Crops Prod.,* **2017**, *108*, 708-715.https://linkinghub.elsevier.com/retrieve/pii/S0926669017304624 [Internet]. [http://dx.doi.org/10.1016/j.indcrop.2017.07.005]

[67] Jaramillo-Colorado, B.E.; Pino-Benitez, N.; González-Coloma, A. Volatile composition and biocidal (antifeedant and phytotoxic) activity of the essential oils of four Piperaceae species from Choco-Colombia. *Ind. Crops Prod.,* **2019**, *138*, 111463. https://linkinghub.elsevier.com/retrieve/pii/ S0926669019304728 [Internet]. [http://dx.doi.org/10.1016/j.indcrop.2019.06.026]

[68] Jassbi, A.R.; Zamanizadehnajari, S.; Baldwin, I.T. Phytotoxic volatiles in the roots and shoots of Artemisia tridentata as detected by headspace solid-phase microextraction and gas chromatographic-mass spectrometry analysis. *J. Chem. Ecol.,* **2010**, *36*(12), 1398-1407. http://link.springer.com/ 10.1007/s10886-010-9885-0 [http://dx.doi.org/10.1007/s10886-010-9885-0] [PMID: 21086024]

[69] Kaur, S.; Singh, H.P.; Mittal, S.; Batish, D.R.; Kohli, R.K. Phytotoxic effects of volatile oil from Artemisia scoparia against weeds and its possible use as a bioherbicide. *Ind. Crops Prod.,* **2010**, *32*(1), 54-61.https://linkinghub.elsevier.com/retrieve/pii/S0926669010000543 [Internet]. [http://dx.doi.org/10.1016/j.indcrop.2010.03.007]

[70] Chowhan, N.; Singh, H.P.; Batish, D.R.; Kohli, R.K. Phytotoxic effects of β-pinene on early growth and associated biochemical changes in rice. *Acta Physiol. Plant.,* **2011**, *33*(6), 2369-2376.http://link.springer.com/10.1007/s11738-011-0777-x [Internet]. [http://dx.doi.org/10.1007/s11738-011-0777-x]

[71] Martino, L.D.; Mancini, E.; Almeida, L.F.R.; Feo, V.D. The antigerminative activity of twenty-seven monoterpenes. *Molecules,* **2010**, *15*(9), 6630-6637.http://www.mdpi.com/1420-3049/15/9/6630 [http://dx.doi.org/10.3390/molecules15096630] [PMID: 20877249]

[72] Singh, H.P.; Batish, D.R.; Kaur, S.; Ramezani, H.; Kohli, R.K. Comparative phytotoxicity of four monoterpenes against Cassia occidentalis. *Ann. Appl. Biol.,* **2002**, *141*(2), 111-116.http://doi.wiley.com/10.1111/j.1744-7348.2002.tb00202.x [Internet]. [http://dx.doi.org/10.1111/j.1744-7348.2002.tb00202.x]

[73] Singh, H.P.; Batish, D.R.; Kaur, S.; Kohli, R.K.; Arora, K. Phytotoxicity of the volatile monoterpene citronellal against some weeds. *Z. Naturforsch. C J. Biosci.,* **2006**, *61*(5-6), 334-340.https://www.degruyter.com/document/doi/10.1515/znc-2006-5-606/html [http://dx.doi.org/10.1515/znc-2006-5-606] [PMID: 16869489]

[74] Batish, D.R.; Singh, H.P.; Setia, N.; Kaur, S.; Kohli, R.K. Chemical composition and phytotoxicity of volatile essential oil from intact and fallen leaves of Eucalyptus citriodora. *Z. Naturforsch. C J. Biosci.,* **2006**, *61*(7-8), 465-471. [http://dx.doi.org/10.1515/znc-2006-7-801] [PMID: 16989303]

[75] Sharma, A.; Singh, H.P.; Batish, D.R.; Kohli, R.K. Chemical profiling, cytotoxicity and phytotoxicity of foliar volatiles of Hyptis suaveolens. *Ecotoxicol. Environ. Saf.,* **2019**, *171*, 863-870.https://linkinghub.elsevier.com/retrieve/pii/S0147651318313939 [http://dx.doi.org/10.1016/j.ecoenv.2018.12.091] [PMID: 30665103]

[76] Rolli, E.; Marieschi, M.; Maietti, S.; Sacchetti, G.; Bruni, R. Comparative phytotoxicity of 25 essential oils on pre- and post-emergence development of Solanum lycopersicum L.: A multivariate approach. *Ind. Crops Prod.,* **2014**, *60*, 280-290.https://linkinghub.elsevier.com/retrieve/pii/S0926669014003549 [Internet]. [http://dx.doi.org/10.1016/j.indcrop.2014.06.021]

[77] Pal Singh, H.; Kaur, S.; Mittal, S.; Batish, D.R.; Kohli, R.K. Phytotoxicity of major constituents of the volatile oil from leaves of Artemisia scoparia Waldst. & Kit. *Z. Naturforsch. C J. Biosci.,* **2008**, *63*(9-10), 663-666.
[http://dx.doi.org/10.1515/znc-2008-9-1009] [PMID: 19040104]

[78] Kordali, S.; Cakir, A.; Ozer, H.; Cakmakci, R.; Kesdek, M.; Mete, E. Antifungal, phytotoxic and insecticidal properties of essential oil isolated from Turkish Origanum acutidens and its three components, carvacrol, thymol and *p*-cymene. *Bioresour. Technol.,* **2008**, *99*(18), 8788-8795.https://linkinghub.elsevier.com/retrieve/pii/S0960852408003696
[http://dx.doi.org/10.1016/j.biortech.2008.04.048] [PMID: 18513954]

[79] Synowiec, A.; Kalemba, D.; Drozdek, E.; Bocianowski, J. Phytotoxic potential of essential oils from temperate climate plants against the germination of selected weeds and crops. J Pest Sci (2004) [Internet]. **2017**, *90*(1), 407-19.http://link.springer.com/10.1007/s10340-016-0759-2

[80] De Almeida, L.F.R.; Frei, F.; Mancini, E.; De Martino, L.; De Feo, V. Phytotoxic activities of Mediterranean essential oils. *Molecules,* **2010**, *15*(6), 4309-4323. http://www.mdpi.com/ 1420-3049/15/6/4309
[http://dx.doi.org/10.3390/molecules15064309] [PMID: 20657443]

[81] Le Bideau, F.; Kousara, M.; Chen, L.; Wei, L.; Dumas, F. Tricyclic Sesquiterpenes from Marine Origin. *Chem. Rev.,* **2017**, *117*(9), 6110-6159.https://pubs.acs.org/doi/10.1021/acs.chemrev.6b00502
[http://dx.doi.org/10.1021/acs.chemrev.6b00502] [PMID: 28379015]

[82] Kramer, R.; Abraham, W.R. Volatile sesquiterpenes from fungi: what are they good for? *Phytochem. Rev.,* **2012**, *11*(1), 15-37.http://link.springer.com/10.1007/s11101-011-9216-2 [Internet].
[http://dx.doi.org/10.1007/s11101-011-9216-2]

[83] Gurgel, E.S.C.; de Oliveira, M.S.; Souza, M.C.; da Silva, S.G.; de Mendonça, M.S. Souza Filho AP da S. Chemical compositions and herbicidal (phytotoxic) activity of essential oils of three Copaifera species (Leguminosae-Caesalpinoideae) from Amazon-Brazil. *Ind. Crops Prod.,* **2019**, 142.

[84] Issa, M.; Chandel, S.; Pal Singh, H.; Rani Batish, D.; Kumar Kohli, R.; Singh Yadav, S.; Kumari, A. Appraisal of phytotoxic, cytotoxic and genotoxic potential of essential oil of a medicinal plant Vitex negundo. *Ind. Crops Prod.,* **2020**, *145*, 112083. https://linkinghub.elsevier.com/retrieve/pii/ S0926669019310933 [Internet].
[http://dx.doi.org/10.1016/j.indcrop.2019.112083]

[85] Andolfi, A.; Zermane, N.; Cimmino, A.; Avolio, F.; Boari, A.; Vurro, M.; Evidente, A. Inuloxins A–D, phytotoxic bi-and tri-cyclic sesquiterpene lactones produced by Inula viscosa: Potential for broomrapes and field dodder management. *Phytochemistry,* **2013**, *86*, 112-120. https://linkinghub. elsevier.com/retrieve/pii/S0031942212004335
[http://dx.doi.org/10.1016/j.phytochem.2012.10.003] [PMID: 23137725]

[86] Scavo, A.; Rial, C.; Molinillo, J.M.G.; Varela, R.M.; Mauromicale, G.; Macías, F.A. Effect of Shading on the Sesquiterpene Lactone Content and Phytotoxicity of Cultivated Cardoon Leaf Extracts. *J. Agric. Food Chem.,* **2020**, *68*(43), 11946-11953.https://pubs.acs.org/doi/10.1021/acs.jafc.0c03527
[http://dx.doi.org/10.1021/acs.jafc.0c03527] [PMID: 33052675]

[87] Del Valle, P.; Figueroa, M.; Mata, R. Phytotoxic eremophilane sesquiterpenes from the coprophilous fungus Penicillium sp. G1-a14. *J. Nat. Prod.,* **2015**, *78*(2), 339-342. https://pubs.acs.org/doi/ 10.1021/np5009224
[http://dx.doi.org/10.1021/np5009224] [PMID: 25603174]

[88] Cárdenas, D.M.; Cala, A.; Molinillo, J.M.G.; Macías, F.A. Preparation and phytotoxicity study of lappalone from dehydrocostuslactone. *Phytochem. Lett.,* **2017**, *20*, 66-72. https://linkinghub. elsevier.com/retrieve/pii/S1874390017301192 [Internet].
[http://dx.doi.org/10.1016/j.phytol.2017.04.017]

[89] da Silva, B.P.; Nepomuceno, M.P.; Varela, R.M.; Torres, A.; Molinillo, J.M.G.; Alves, P.L.C.A.; Macías, F.A. Phytotoxicity Study on *Bidens sulphurea* Sch. Bip. as a Preliminary Approach for Weed

Control. *J. Agric. Food Chem.*, **2017**, *65*(25), 5161-5172. https://pubs.acs.org/doi/10.1021/acs.jafc.7b01922
[http://dx.doi.org/10.1021/acs.jafc.7b01922] [PMID: 28605187]

[90]   Dastan, D.; Salehi, P.; Ghanati, F.; Gohari, A.R.; Maroofi, H.; Alnajar, N. Phytotoxicity and cytotoxicity of disesquiterpene and sesquiterpene coumarins from Ferula pseudalliacea. *Ind. Crops Prod.*, **2014**, *55*, 43-48.https://linkinghub.elsevier.com/retrieve/pii/S0926669014000727 [Internet].
[http://dx.doi.org/10.1016/j.indcrop.2014.01.051]

[91]   Perera, W.H.; Meepagala, K.M.; Wedge, D.E.; Duke, S.O. Sesquiterpenoids from culture of the fungus Stereum complicatum (Steraceae): structural diversity, antifungal and phytotoxic activities. *Phytochem. Lett.*, **2020**, *37*, 51-58.https://linkinghub.elsevier.com/retrieve/pii/S187439002030080X [Internet].
[http://dx.doi.org/10.1016/j.phytol.2020.03.012]

[92]   Chen, J.; Zheng, G.; Zhang, Y.; Aisa, H.A.; Hao, X.J. Phytotoxic Terpenoids from *Ligularia cymbulifera* Roots. *Front. Plant Sci.*, **2017**, *7*, 2033. http://journal.frontiersin.org/article/10.3389/fpls.2016.02033/full
[http://dx.doi.org/10.3389/fpls.2016.02033] [PMID: 28119715]

[93]   Ahluwalia, V.; Sisodia, R.; Walia, S.; Sati, O.P.; Kumar, J.; Kundu, A. Chemical analysis of essential oils of Eupatorium adenophorum and their antimicrobial, antioxidant and phytotoxic properties. J Pest Sci (2004) [Internet]. **2014**, *87*(2), 341-9.http://link.springer.com/10.1007/s10340-013-0542-6

[94]   Abd-ElGawad, A.M.; El Gendy, A.E.N.G.; Assaeed, A.M.; Al-Rowaily, S.L.; Alharthi, A.S.; Mohamed, T.A.; Nassar, M.I.; Dewir, Y.H.; Elshamy, A.I. Phytotoxic Effects of Plant Essential Oils: A Systematic Review and Structure-Activity Relationship Based on Chemometric Analyses. *Plants*, **2020**, *10*(1), 36.https://www.mdpi.com/2223-7747/10/1/36
[http://dx.doi.org/10.3390/plants10010036] [PMID: 33375618]

[95]   Anese, S.; Jatobá, L.J.; Grisi, P.U.; Gualtieri, S.C.J.; Santos, M.F.C.; Berlinck, R.G.S. Bioherbicidal activity of drimane sesquiterpenes from Drimys brasiliensis Miers roots. *Ind. Crops Prod.*, **2015**, *74*, 28-35.https://linkinghub.elsevier.com/retrieve/pii/S0926669015300509 [Internet].
[http://dx.doi.org/10.1016/j.indcrop.2015.04.042]

[96]   Devappa, R.K.; Makkar, H.P.S.; Becker, K. Jatropha Diterpenes: a Review. *J. Am. Oil Chem. Soc.*, **2011**, *88*(3), 301-322.http://doi.wiley.com/10.1007/s11746-010-1720-9 [Internet].
[http://dx.doi.org/10.1007/s11746-010-1720-9]

[97]   Berrue, F.; Kerr, R.G. Diterpenes from gorgonian corals. *Nat. Prod. Rep.*, **2009**, *26*(5), 681-710.http://xlink.rsc.org/?DOI=b821918b
[http://dx.doi.org/10.1039/b821918b] [PMID: 19387501]

[98]   Vasas, A.; Hohmann, J. Euphorbia diterpenes: isolation, structure, biological activity, and synthesis (2008-2012). *Chem. Rev.*, **2014**, *114*(17), 8579-8612.https://pubs.acs.org/doi/10.1021/cr400541j
[http://dx.doi.org/10.1021/cr400541j] [PMID: 25036812]

[99]   Li, R.; Morris-Natschke, S.L.; Lee, K.H. Clerodane diterpenes: sources, structures, and biological activities. *Nat. Prod. Rep.*, **2016**, *33*(10), 1166-1226.http://xlink.rsc.org/?DOI=C5NP00137D
[http://dx.doi.org/10.1039/C5NP00137D] [PMID: 27433555]

[100]  Zhao, DL; Yang, LJ; Shi, T; Wang, CY; Shao, CL; Wang, CY Potent Phytotoxic Harziane Diterpenes from a Soft Coral-Derived Strain of the Fungus Trichoderma harzianum XS-20090075. Sci Rep [Internet]. Springer US; **2019**, *9*(1), 1-9.
[http://dx.doi.org/10.1038/s41598-019-49778-7]

[101]  Cimmino, A.; Andolfi, A.; Zonno, M.C.; Avolio, F.; Santini, A.; Tuzi, A.; Berestetskyi, A.; Vurro, M.; Evidente, A. Chenopodolin: a phytotoxic unrearranged ent-pimaradiene diterpene produced by Phoma chenopodicola, a fungal pathogen for Chenopodium album biocontrol. *J. Nat. Prod.*, **2013**, *76*(7), 1291-1297.https://pubs.acs.org/doi/10.1021/np400218z
[http://dx.doi.org/10.1021/np400218z] [PMID: 23786488]

[102] Macías, F.A.; Lacret, R.; Varela, R.M.; Nogueiras, C.; Molinillo, J.M.G. Isolation and phytotoxicity of terpenes from Tectona grandis. *J. Chem. Ecol.,* **2010**, *36*(4), 396-404. http://link.springer.com/10.1007/s10886-010-9769-3
[http://dx.doi.org/10.1007/s10886-010-9769-3] [PMID: 20237951]

[103] Evidente, A.; Sparapano, L.; Fierro, O.; Bruno, G.; Giordano, F.; Motta, A. Sphaeropsidins B and C, phytotoxic pimarane diterpenes from Sphaeropsis sapinea f. sp. Cupressi and Diplodia mutila. *Phytochemistry,* **1997**, *45*(4), 705-713. https://linkinghub.elsevier.com/retrieve/pii/S0031942-29700006X [Internet].
[http://dx.doi.org/10.1016/S0031-9422(97)00006-X]

[104] Bisio, A; Damonte, G; Fraternale, D; Giacomelli, E; Salis, A; Romussi, G Phytotoxic clerodane diterpenes from Salvia miniata Fernald (Lamiaceae). Phytochemistry [Internet]. Elsevier Ltd; **2011**, *72*(2-3), 265-75.
[http://dx.doi.org/10.1016/j.phytochem.2010.11.011]

[105] Klein Hendges, A.P.P.; dos Santos, E.F.; Teixeira, S.D.; Santana, F.S.; Trezzi, M.M.; Batista, A.N.L.; Batista, J.M., Jr; de Lima, V.A.; de Assis Marques, F.; Maia, B.H.L.N.S. Phytotoxic Neocassane Diterpenes from *Eragrostis plana. J. Nat. Prod.,* **2020**, *83*(12), 3511-3518. https://pubs.acs.org/doi/10.1021/acs.jnatprod.0c00324
[http://dx.doi.org/10.1021/acs.jnatprod.0c00324] [PMID: 33201703]

[106] Favaretto, A.; Cantrell, C.L.; Fronczek, F.R.; Duke, S.O.; Wedge, D.E.; Ali, A.; Scheffer-Basso, S.M. New Phytotoxic Cassane-like Diterpenoids from *Eragrostis plana. J. Agric. Food Chem.,* **2019**, *67*(7), 1973-1981.https://pubs.acs.org/doi/10.1021/acs.jafc.8b06832
[http://dx.doi.org/10.1021/acs.jafc.8b06832] [PMID: 30685966]

[107] Islam, A.K.M.M.; Ohno, O.; Suenaga, K.; Kato-Noguchi, H. Two novel phytotoxic substances from Leucas aspera. *J. Plant Physiol.,* **2014**, *171*(11), 877-883. https://linkinghub.elsevier.com/retrieve/pii/S0176161714000583
[http://dx.doi.org/10.1016/j.jplph.2014.03.003] [PMID: 24913044]

[108] Qin, B.; Perry, L.G.; Broeckling, C.D.; Du, J.; Stermitz, F.R.; Paschke, M.W.; Vivanco, J.M. Phytotoxic Allelochemicals From Roots and Root Exudates of Leafy Spurge ( *Euphorbia esula* L.). *Plant Signal. Behav.,* **2006**, *1*(6), 323-327.http://www.tandfonline.com/doi/abs/10.4161/psb.1.6.3563
[http://dx.doi.org/10.4161/psb.1.6.3563] [PMID: 19517003]

[109] Evidente, A.; Sparapano, L.; Bruno, G.; Motta, A. Sphaeropsidins D and E, two other pimarane diterpenes, produced *in vitro* by the plant pathogenic fungus Sphaeropsis sapinea f. sp. cupressi. *Phytochemistry,* **2002**, *59*(8), 817-823. https://linkinghub.elsevier.com/retrieve/pii/S0031942-202000158
[http://dx.doi.org/10.1016/S0031-9422(02)00015-8] [PMID: 11937160]

**CHAPTER 10**

# Applications of Natural Terpenoids as Food Additives

**Fernanda Wariss Figueiredo Bezerra**[1,*], **Giselle Cristine Melo Aires**[1], **Lucas Cantão Freitas**[1], **Marielba de Los Angeles Rodriguez Salazar**[1], **Rafael Henrique Holanda Pinto**[1], **Jorddy Neves da Cruz**[2] and **Raul Nunes de Carvalho Junior**[1]

[1] *LABEX/PPGCTA (Extraction Laboratory / Graduate Program in Food Science and Technology), Federal University of Pará, Rua Augusto Corrêa S/N, 66075-900, Belém, Pará, Brazil*

[2] *Adolpho Ducke Laboratory, Botany Coordination, Museu Paraense Emílio Goeldi, Av. Perimetral, 1900, 66077-530, Belém, Pará, Brazil*

**Abstract:** Food additives are widely used in the food industry in order to ensure the quality of products during processing, storage, packaging and subsequent reaching the consumer's table. The growing concern and doubt of the consumer market regarding artificial additives and their possible harmful effects on public health and safety have caused the demand for the use of natural additives to increase. Consequently, these natural additives have been increasingly sought by the food industry and consumers due to health, safety and sustainability issues. In this framework, terpenoids have great potential to be used with this function because they are a very extensive class of compounds, with wide chemical diversity and several proven applications in foods, mainly as anti-oxidants, anti-microbials, dyes, flavors, sweeteners and nutraceuticals. Therefore, this paper aims to make a literature search on the use of terpenoids as food additives, highlighting the main compounds used and the benefits associated with their use, ranging from the raw material to its extraction and subsequent application in food products.

**Keywords:** Secondary Metabolites, Additives, Anti-Microbials, Anti-Oxidants, Dyes, Food Industry, Food Chemistry, Flavorings, Food Preservatives, Healthy Life, Natural Additives, Natural Products, Nutraceuticals, Nutritional Fortification, Shelf Life, Sweeteners, Terpenoids, Terpenes.

---
* **Corresponding author Fernanda Wariss Figueiredo Bezerra:** LABEX/PPGCTA (Extraction Laboratory / Graduate Program in Food Science and Technology), Federal University of Pará, Rua Augusto Corrêa S/N, 66075-900, Belém, Pará, Brazil; Email: fernandawarissf@gmail.com

**Mozaniel Santana de Oliveira & Antônio Pedro da Silva Souza Filho (Eds.)**
**All rights reserved-© 2022 Bentham Science Publishers**

# INTRODUCTION

The food industry has been using additives for decades in order to give positive attributes to its products, such as longer shelf life and better sensory characteristics. The European Union database represents about 330 authorized compounds, and in the list of Substances Added to Food of FDA (Food and Drug Administration), there are more than 3000. The main additives reported in the composition of ultra-processed foods are nitrates and nitrites, (di/tri/poly) phosphates, sweeteners, monosodium glutamate, sorbate, bixin, caramel, titanium dioxide, tartrazine, butylated hydroxyanisole (BHA) and butylated hydroxytoluene (BHT), which can be applied as stabilizers, emulsifiers, dyes, flavorings, preservatives, sweeteners, gelling agents, anti-oxidants, nutrients, among others. Despite the regulations and the benefits generated to the products, researches *in vivo, in vitro* and *in silico* have been showing the harmful effect to the health (allergic, carcinogenic and mutagenic) that the additives, especially the synthetic ones, can bring [1–7].

The interest in products that are sources of bioactive substances of natural origin has increased due to the growing awareness of the consumer market about food safety, healthy eating and the possible damage to health associated with the use of synthetic additives. Thus, natural additives from animals, microorganisms and vegetables have been shown to be an alternative to synthetics and can be used to maintain and prolong food safety. Additives derived from vegetables, such as herbs, spices and their extracts or isolated compounds, contain components that can act in foods as anti-microbial agents, anti-oxidants, flavorings, dyes, and nutraceuticals, among others [8–11].

Terpenoids are the most numerous secondary metabolite group (around 80,000 compounds) and are structurally diverse, being classified according to the number and structural organization of carbons in the linear arrangement. The compounds can be present in natural sources such as plants, animals, microorganisms, insects, plant pathogens and endophytes. They have several functions that can be added to food, cosmetic and pharmaceutical products in the form of food additives, flavorings, fragrances, drug excipients and others. In the food industry, they can be used with different functions, as shown in Table **1**, which presents some terpenes approved by the FDA and the European Union [12–16].

Table 1. Some terpenes approved as food additives on FDA and European Union lists [17, 18].

| Function | Compound |
|---|---|
| Color | Carotenes, bixin, norbixin, capsanthin, capsorubin, lycopene, lutein, canthaxanthin, α-terpineol, caryophyllene. |

*(Table 1) cont.....*

| Function | Compound |
|---|---|
| Antioxidant | α-, β- and δ-tocopherol, β-carotene. |
| Antimicrobial | Bisabolene. |
| Masticatory substance | Terpene resin. |
| Flavor or adjuvant | Lemon terpenes, cedarwood oil terpenes, menthol, α-terpineol, α-terpinene, γ-terpinene, β-terpineol, terpinolene, terpinyl acetate, α-terpinyl anthranilate, terpinyl butyrate, terpinyl cinnamate, terpinyl formate, terpinyl isobutyrate, terpinyl isovalerate, terpinyl propionate, caryophyllene, thymol, carvacrol, eugenol, iso eugenol, phytol, pinene, limonene, tomato lycopene, tocopherols, bisabolene. |
| Humectant | Natural and synthetic terpene resin. |
| Solvent or vehicle | Terpene resin. |
| Nutrient supplement | Carotene, tocopherols. |

# DIVERSITY AND CHARACTERISTICS OF TERPENOIDS IN FOOD SYSTEMS

Terpenes have great chemical diversity and various applications in the pharmaceutical, food and fine chemical industries [19]. These compounds are produced in nature by some animals, microorganisms and mainly by plants, responsible for aromas and flavors characteristic of fruits and spices [20]. One of the great challenges of the food industry is to mask the unpleasant taste and odor of anti-oxidants, vitamins, minerals and other substances present in nutraceuticals and fortified food [21]. For this, several strategies are used, among them, the use of compounds of natural origin that give the food a more pleasant aroma and flavor [22].

Essential oils obtained from leaves are mainly found in monoterpenes α- and β-pinene, limonene, 3-carene, α-phellandrene and myrcene. Monoterpenes with floral and fruity aromas are most commonly found in seeds and flowers [23, 24]. Woody and balsamic aromas are characteristic of sesquiterpenes and sesquiterpenols found in woody oils [25]. Carotenoids have great anti-oxidant activity because they are able to suppress free radicals produced by chemical reactions in the human body. Other terpenes, such as lutein, γ-carotene, lycopene and β-carotene, have been linked to the fight against breast, colorectal, prostate, lung and uterine cancers [26, 27]. In these studies, carotenoids were investigated for these properties due to the protective power of human tissue when ingested in food or drinks. In addition to these functions, they help to protect the skin against UV rays and improve the immune response. Carotenoids can be found in fruits and vegetables such as sweet potatoes, squash, beets, papaya, mango, broccoli and spinach [28, 29].

Phytosterols have a vegetable origin and have a chemical structure similar to cholesterol. They compete for the same absorption site and reduce the absorbed cholesterol, preventing it from reaching the bloodstream. These terpenoids can be found in nuts, peanuts, chestnuts, sesame seeds, vegetables, fruits and grains [30–32]. Other examples and applications of terpenoids can be found in Table 2.

Table 2. Terpenoids and their applications in food chemistry.

| Terpenoids | Where to Find | Application | Refs. |
|---|---|---|---|
| Menthol | *Mentha arvensis* | Menthol aroma | [33] |
| *D*-carvone | *Carum carvi* | Spicy and bread-like aroma | [34] |
| *D*-limonene | *Cymbopogon citratus* | Fresh orange peel aroma | [35] |
| 1,8-cineole | *Eucalyptus globulus* | Camphoraceous cool aroma | [36] |
| α-Ionone | Present in many fruits and plants | Violet aroma | [37] |
| Safranal | *Crocus sativus, C. longa, Osmanthus fragrans, Ilex paraguariensis* | Woody, spicy, phenolic, camphoreous, medicinal aroma | [38] |

## POSSIBLE APPLICATIONS OF THE TERPENOIDS AS FOOD ADDITIVES

### Colorants

Dyes intensify and standardize the appearance of the product, which is one of the most relevant attributes that the consumer observes when purchasing or ingesting a product. Natural dyes are commonly applied in the food industry both in dry form and in the form of concentrated extracts obtained with specific solvents [39–41]. In this sense, contemporary consumers are increasingly focused on improving their health and consequently the quality of food consumed. Thus, the demand for good quality food produced from ingredients of natural origin has been gaining attention worldwide [42, 43].

Tetraterpenes are natural pigments present in most plants, fruits, vegetables, algae, photosynthetic bacteria, fungi and insects. The color spectrum of these compounds ranges from yellow to red, and can be classified as carotenoids, which are subdivided into carotenes (when formed only by hydrocarbons), and xanthophylls (which are oxygenated derivatives). They also may present in its structure carbonyls, carboxylic acids, hydroxyls and epoxides (Fig. 1) [44, 45]. Among the functions of wide importance that the class of terpenes and their terpenoid derivatives play in nature, we can list the following ones: *i.* they are considered accessory pigments of photosynthesis in plants, presenting maximum absorption in the range of ultraviolet and blue; ***ii.*** some of them are precursors of vitamin A in animal tissues; and ***iii.*** certain carotenoids have significant anti-oxidant activity.

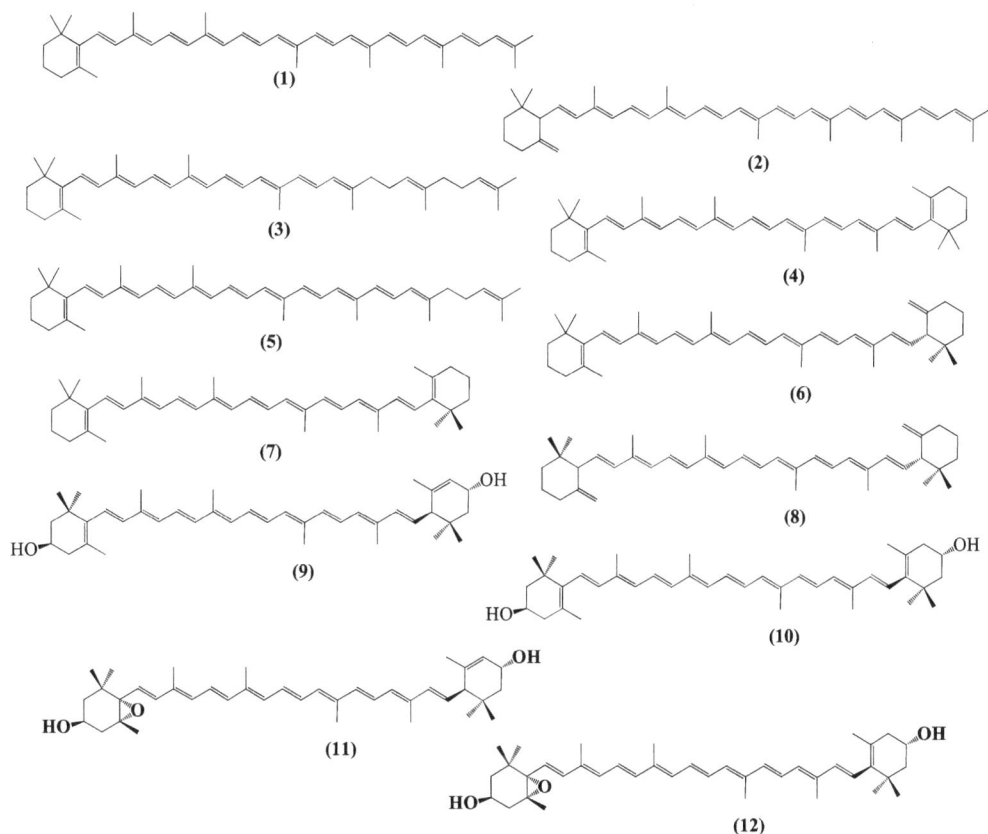

**Fig. (1).** Structure of carotenoids and xanthophylls. (1) Torulene, (2) γ-ψ-carotene, (3) β-zeacarotene (4) β-carotene, (5) β- ψ-carotene, (6) β- γ -carotene, (7) β-cryptoxanthin, (8) γ- γ-carotene, (9) Lutein,(10) Zeaxanthin, (11) Lutein epoxid and (12) Anteraxanthin.

The trend of applying natural dyes in the most diversified products has become a successful strategy that large multinationals have used as a differential, attributing greater success to their brands [46, 47]. Natural dyes can be applied in the food industry not only to add color, but can also be linked to other functions, such as packaging, flavor enhancers, nutritional fortification, beverages, the development of new food products and even pet foods [48–51].

The natural extracts most used industrially come from annatto (*Bixa orellana –* bixin), where its powder has been tested on the quality and internal and external stability of eggs in order to obtain a color acceptable to the market [52]; from carrots (*Daucus carota*), where its β-carotene can be used for the fortification and development of functional foods, pharmaceuticals and medicines [53]. The β-carotene present in palm oil (*Elaeis guineensis*) has functional and technological properties, justifying its use in the food industry [54, 55]. Products such as turmeric (*Crocus sativus* – crocetina), tomatoes (*Lycopersicum esculentum* – lycopene) and paprika (*Capsicum annuum* – β-cryptoxanthin), among others, can also produce important natural additives for the food industry, moving the global market for pigments and natural dyes [56].

**Flavoring Agent**

Different cultures have historically used vegetable raw materials to obtain products with a pleasant taste, longer shelf life and therapeutic potential. Cloves, cinnamon barks and leaves and West Indian bay leaves are world-renowned botanical raw materials for their flavoring properties, considered culinary spices that add flavor to food. The phytochemical screening of these vegetables shows the predominance of compounds from the hydroxyalkylbenzenes group and derivatives of hydroxypropenylbenzene (with emphasis on eugenol), followed by the group of aliphatic and aromatic hydrocarbons and other terpenoid components such as β-karyophylene and β-myrcene [57].

Other species rich in terpenoids, such as rosemary, ginger, lime, orange and pepper, have long been cultivated for food purposes. According to the literature, rosemary essential oil (EO) is composed of α-pinene, myrcene, 1,8-cineole, borneol and camphor [58]. In ginger, EO is found sesquiterpenes (zingiberene, AR-curcumene, β-sesquifelandreno and bisabolene), monoterpenes (camphene, β-felandrene) and monoterpenoid (1,8-cineol) [59]. Regarding lime OE, the major compounds are *d*-limonene, β-pinene and α-pinene [60], while in orange OE, the presence of l-limonene is highlighted [61]. The sabinene monoterpene is predominant in the pepper essential oil [59].

Monoterpenes and sesquiterpenes, with lower molecular weight terpenic structures, have high volatility. This characteristic is of great importance for the aroma of natural products, particularly aromatic herbs, spices, condiments and citrus fruits [59]. The sensory characteristics of terpenoids enhance applications in the food industry, especially in the production of non-alcoholic beverages (soft drinks, juices and syrups) and bakery products. Separation and purification methods of essential oils enable the production of terpenes. Conventionally, the botanical raw materials are dehydrated, ground and processed by steam distillation or recirculation of organic solvents with subsequent evaporation [57]. The use of supercritical carbon dioxide in the extraction of essential oils has been investigated in order to replace the conventional methods [62]. This technique enhances new perspectives in the production of essential oils for applications in the food industry.

## Anti-Oxidants

Oxidation reactions are common in biological systems and food. Some of these reactions can cause negative effects such as the degradation of lipids, vitamins, pigments, loss of nutritional value and undesirable flavors. Therefore, it is common to add anti-oxidant compounds to foods [63]. The most commonly used anti-oxidants are phenolic substances, due to their ability to prevent or delay oxidative stress, acting as free radical scavengers (FRSs) [64, 65]. Anti-oxidants act by donating protons to free radicals, reacting with them, stabilizing them and producing less reactive anti-oxidant radicals. In this category, we highlight the butylated hydroxyanisol (BHA), butylated hydroxytoluene (BHT), terc butyl hydroquinone (TBHQ) and propyl gallate, which are synthetic phenolic anti-oxidants approved for use in food [66, 67].

Synthetic additives have been investigated due to their possible harmful effects on health. Thus, natural anti-oxidant additives are presented as an alternative [68, 69]. Many products of natural origin contain anti-oxidant substances that prevent oxidation. Among these substances, we can highlight the terpenoids such as hemiterpenes, monoterpenes, sesquiterpenes, diterpenes and triterpenes, and especially the tetraterpenoids or carotenoids that are frequently reported as anti-oxidant agents. Tetraterpenoids or carotenoids are compounds found in plants, animals and microorganisms, and can be classified into two types: carotenes and xanthophylls [70, 71]. These compounds play a provitamin (β-carotene) activity, have an action against chronic diseases attributed to their anti-oxidant activity and are frequently used as dyes in foods in the natural or synthetic form [72].

Table **3** shows the main tetraterpenes or carotenoids reported in foods, making it possible to have a view of some plant matrices that can be used to obtain them, the possible applications as natural additives and the main *in vitro* methods for determining the anti-oxidant capacity of these compounds such as: ferric reducing / anti-oxidant power (FRAP), 2,2-diphenyl-1-picrilhhydrazyl (DPPH), oxygen radical absorption capacity (ORAC), cupric reducing anti-oxidant capacity (CUPRAC), Trolox equivalent anti-oxidant capacity (TEAC) and 2,2-azinobis-(3-ethylbenzothiazoline-6-sulfonate) (ABTS$^{\cdot+}$).

The monoterpenes (*e.g.*, α-pinene, limonena, myrcene, geraniol, linalool, nerol e terpineol) also have the anti-oxidant capacity, being predominant in wines, which can be potential sources of anti-oxidant agents [82]. The anti-oxidant activity of monoterpenes, common in essential oils, is attributed to α-pinene [83]. It was also identified triterpenes with anti-oxidant property in chamomile extracts (*Commelina benghalensis*) [84]. Table **4** shows some anti-oxidants obtained from fruits, vegetables and leaves that have been added to food products. From these studies, it is possible to suggest that tetraterpenoids such as lycopene, lutein and β-carotene could be used as anti-oxidant compounds for pasta preparation [85], in bakery products [86, 87], in meat products [68, 88] and in the protection against the oxidation of vegetable oils and fatty foods [69, 89].

Table 3. Main tetraterpenes or carotenoids in foods and their anti-oxidant capacity.

| Sample | Carotenoid Content | Determination Method/ Antioxidant Capacity | Refs. |
|---|---|---|---|
| *Citrullus lanatus in natura* | Lycopene 6.80 mg/100g<br>Total carotenoids 8.06 mg/100g | CUPRAC: 84.05 μmol TE/100g | [73] |
| Dried *Prunus persica* peach peel | Lutein 12.02 μg/100 g<br>Lycopene 36.02 μg/100 g<br>β-carotene 2730.95 μg/100 g<br>β-cryptoxanthin 814.45 μg/100 g<br>Total carotenoid 3594.35 μg/100 g | DPPH: 41.32 $EC_{50}$ (mg/ kg)<br>ABTS 43.46 $EC_{50}$ (mg/ kg) | [74] |
| *Citrus sinensis in natura* | Phytoene 13.31 mg/100 g<br>Total carotenoid 16.51 mg/100 g | DPPH: 50-75 mgTE/100g<br>ABTS 125-150 mgTE /100g | [75] |
| Dried *Solanum lycopersicum cerasiforme* pulp | Neurosporeno 7.30 mg/100 g<br>Lutein 2.03 mg/100 g<br>13-15-cis-β-carotene 2.86 mg/100 g<br>*Trans*-β-carotene 4.27 mg/100 g<br>Lycopene 11.17 mg/100 g<br>Total carotenoid 27.63 mg/100 g | FRAP: 0.28 μM $Fe^{2+}$/mg | [76] |
| *Ipomoea batatas in natura* | all-*trans*-β-carotene 79.1 mg/100g<br>13-cis-β-carotene 9.3 mg/100g<br>9-cis-β-carotene 4.9 mg/100g<br>5,6 epoxy-β-carotene 7.0 mg/100g | DPPH: 16-20 Mm Trolox/100g<br>ABTS 36-40 M Trolox/100g | [77] |
| Dried *Daucus carota* L. peel | β-carotene 1.14 mg/g<br>Total carotenoid 2.02 mg/g | ABTS: 4.71 mM TE/g CUPRAC: 19 mM TE/g | [78] |
| *Cucurbita moschata Duch in natura* | all-E-β-carotene 244.22 μg/g;<br>9-Z-β-carotene 2.34 μg/g;<br>13-Z-β-carotene 3.67 μg/g<br>α-carotene 67.06 μg/g<br>Total carotenoid 404.98 μg/g | | [79] |
| | Total carotenoid 24.2 mg/100 g | DPPH: 357.8 μg Trolox/100g | [80] |
| *Zea Mays* L. grains dried | Lutein 49.6 μg/100 g<br>Zeaxantin 218.2 μg/100 g<br>α-Criptoxanthin 68.3 μg/100 g<br>β-Criptoxanthin 5.6 μg/100 g<br>α-carotene 4.9 μg/100 g<br>β-carotene 35.7 μg/100 g | ORAC: 1827 μmol TE/ 100g | [81] |

**Table 4. Applications of tetraterpenes or carotenoids as anti-oxidant additives in food products.**

| Food additive | Type and amount of additive | Application | Improved features | Ref. |
|---|---|---|---|---|
| *Solanum lycopersicum* waste extract | Lycopene (300.85 mg/100g) Total carotenoid (654.76 mg/100g) | Cake/Biscuit | Improves anti-oxidant activity, color and sensory properties | [86] |
| Dried *Urtica dioica* L. leaves | Lutein (51.03 µg/g) β-carotene (3.51 µg/g) | Egg pasta | Enrichment and bioaccessibility of carotenoids | [85] |
| *Zea mays* | - | Pan bread | Improves the bioavailability of dietary fiber, calcium content and anti-oxidant capacity | [87] |
| *Prunus salicina* peel fiber microparticles | β-carotene (4.9 mg/kg) Lutein (5.1 mg/kg) | Breast chicken patties | Efficiency in reducing lipid oxidation | [68] |
| *Lycopersicon esculentum* peel extract | - | Steamed and Grilled Turkey Breast | Improves anti-oxidant capacity and color parameters | [90] |
| *Solanum lycopersicum* residue extract | Lycopene (218.74 ppm) | Jordanian traditional sheep ghee (*Samen Baladi*) | Improves anti-oxidant activity, oxidative and hydrolytic rancidity | [88] |
| *Solanum tuberosum* and *Lycopersicon esculentum* residue extract | Lycopene (0.65 – 12.0 mg/kg) Total carotenoid (3.60 – 64.65 mg/kg) | Canola oil | Oxidation prevention | [89] |
| *Solanum lycopersicum* peel powder extract | Lycopene (615 mg/100g) Total carotenoid (720 mg/100g) | Sunflower oil | Oxidation prevention | [69] |

## Anti-Microbial

In recent years, there has been a trend towards the use of natural anti-microbial additives due to the increasingly negative perception that the consumer market has shown in relation to synthetics, increasing the need to investigate alternative natural compounds that can perform this same function [91, 92]. Essential oils are recognized as substitutes for synthetic preservatives and can be an alternative to control or inhibit bacteria and fungi, acting on the preservation of food products due to their low toxic nature and serving as green preservatives in the food industry [93, 94]. Its anti-microbial action is associated with the ability to penetrate through the membrane into the cell, exhibiting inhibitory effects due to its lipophilic characteristics [95, 96].

Essential oils are natural volatile fractions obtained from various aromatic plant matrices such as flowers, leaves, seeds, stems, fruits, roots and barks [97, 98], consisting of fatty acids esters, monoterpenes, sesquiterpenes, phenylpropanoids, aldehyde alcohols and, in some cases, aliphatic hydrocarbons [96, 99]. The terpenoids are the class of compounds most abundant in natural products, being the main constituents of most essential oils [100]. The application of essential oils as anti-microbial food additives has some disadvantages due to their strong odor and taste, which can affect the sensory properties of the product as they are easily degraded by oxidation, volatility, heating and exposure to light, in addition to being difficult to disperse in hydrophilic media. However, these limitations can be overcome through techniques such as encapsulation in particles, emulsions and films [101–104]. Many essential oils are generally recognized as safe (GRAS) and are approved for use in food by the Food and Drug Administration [18]. Table **5** shows applications of essential oils as anti-microbial additives.

**Table 5. Essential oils as anti-microbial additives in food products.**

| Type of Food Additive | Main Terpenoids | Applied Product | Purpose | Refs. |
|---|---|---|---|---|
| Ethanol and hexane extracts of *Ocimum gratissimum* leaves | γ-terpinene, caryophyllene, humulene, squalene, thymol and *p*-cymene | Cucumber fresh cut (*Cucumis sativus*) | Reduce microbial load and improve shelf life | [105] |
| Ethanol and hexane extracts of *Ocimum gratissimum* leaves | γ-terpinene | Watermelon fresh cut (*Citrullus lanatus*) | Reduce microbial load and improve shelf life | [106] |
| *Cymbopogon citratus* essential oil | Linalol, caryophyllene, neral, geraniol and geranial | Stored corn kernels | Reduces microbial load and improves quality during storage | [107] |
| *Cymbopogon citratus* hydroalcoholic extract | Citral and myrcene | Chicken sausages | Reduces lipid oxidation, microbial load and quality maintenance during storage | [108] |
| Mixture of *Syzygium aromaticum*, *Ocimum sanctum* and *Thymus zygis* oils | α-linalool, eugenol e β-cariofilileno | Chicken sausages | Improve oxidative stability and shelf life | [97] |
| *Thymus zygis* essential oil | Thymol, *p*-cymene, α-pinene and linalool | Fish and seaweed burgers | Improve shelf life and reduce oxidation | [109] |
| *Citrus x limon* and *Citrus x sinensis* essential oils | Limonene and dihydrocarvone | Cake | Improves texture, sensory characteristics and extends shelf life | [95] |

*(Table 5) cont.....*

| Type of Food Additive | Main Terpenoids | Applied Product | Purpose | Refs. |
|---|---|---|---|---|
| *Eucalipto globulus* essential oil | 1,8-cineole, α-pinene, β-pinene, myrcene and γ-terpinene | Orangina fruit juices | Improves anti-oxidant capacity and shelf life | [93] |
| *Piper guineense, Xylopica aethiopica* and *Tetrapleura tetraptera* essential oils | Linalool | Mixed fruit juice | Improves anti-oxidant capacity and shelf life | [92] |
| *Thymus algeriensis* essential oil | Carvacrol, linalool, β-caryophyllene and *p*-cymene | Soft cheese | Improvement of sensory characteristics and shelf life | [110] |
| *Origanum vulgare* essential oil | carvacrol, thymol, cymene and terpinene | Minas Padrão cheese | Inhibit fungal growth | [104] |
| *Copaifera officinalis* oil | β-caryophyllene | Mangarito starch-based films | Inhibits microbial growth | [111] |

## Nutraceutical

Nutraceuticals are foods or parts of foods that offer proven health benefits, including disease prevention and treatment [112]. From a chemical point of view, nutraceuticals are compounds belonging to different classes, such as conjugated linoleic acids (CLAs), polyunsaturated omega-3 (PUFA omega 3), polyphenols, terpenoids, alkaloids, carotenoids and others [113, 114]. Currently, the main focus of research is to clarify the effectiveness of the biological action mechanisms of these compounds in preventing diseases (*i.e.*, heart diseases, stroke and type 2 diabetes) and that they should not be seen as drugs, but as therapeutic supplements in order to help the conventional treatments [115–117].

Human metabolism is not able to synthesize several compounds essential to health, such as carotenoids and xanthophylls. Therefore, they must be ingested through the consumption of food or through supplementation [118, 119]. Thus, some higher molecular weight terpenoids have a precursor action to certain vitamins, such as β-carotene, which has provitamin A action, in addition to vitamins E and K that have terpenoid units in their composition. Chlorophyll, present in plants, is formed by phytol diterpene. Thus, these phytochemicals must be ingested through the diet and, for this reason, they are called essential [120, 121].

The tetraterpenes from various foods are inserted in several uses considered nutraceuticals. Thus, the food supplement industries make use of several tetraterpenic carotenoids such as β-carotene, lutein, lycopene, α-carotene and astaxanthin, which are the most known and used [122, 123]. Lycopene is a potent target nutraceutical due to numerous reports of its benefits in important diseases. Its potential is mainly linked to its action in neuroprotection, preventing neurodegenerative diseases and toxin-induced neurotoxicity, and providing protection to the brain against toxicity induced by a diet rich in lipids. In addition, it improves cognitive and psychomotor deficiencies, neuronal oxidative and inflammatory stress, mitochondrial dysfunction, apoptosis and oral submucosal fibrosis [124–126].

However, when processed for application, these compounds may suffer consequences in their functional properties due to their low thermal stability or high sensitivity to light and oxygen. In this sense, techniques for protecting these compounds against degradation factors have been widely studied [127].

Astanxanthin is an example of a compound with low water solubility and high chemical instability in processing. However, when microencapsulated in an oil/water emulsion using ionic membranes, they can improve its stability and enhance its application as a functional ingredient in food [128, 129]. Thus, like astaxanthin and lutein, the encapsulation and protection of β-carotene are widely studied due to its high use as an anti-oxidant, dye and nutraceutical compound, mainly in the pharmaceutical and food industries [130–132].

There is also a multitude of studies on the effects of lutein stability, which is considered an important nutraceutical that helps to prevent macular degeneration in fetuses [133]. Studies demonstrate that the encapsulation of lutein using spraying of microencapsulated emulsions, zein/polysaccharide systems or combinations of polysaccharides as wall material or encapsulated in liposomes using supercritical carbon dioxide, beneficially affect the stability of these nutraceutical compounds [101, 134–136].

Regarding the effects of nutraceutical tetraterpenes, numerous reports in the literature support their benefits. For instance, lutein-loaded nanoparticles have been shown to confer a neuroprotective effect on locomotor activity by pointing to nanoencapsulated lutein as a therapeutic target for the development of drugs capable of protecting Parkinson's disease, allowing the identification of effective pharmacological strategies for treatments. It has also been shown that the use of lutein with complementary unsaturated fatty acids influences the functional improvement of chaperone α-critalin, which is the protein involved in maintaining eye clarity in mammals that can prevent and reduce risk factors for cataracts [102, 137].

# CONCLUDING REMARKS

The consumer market demand for natural additives has led the food industry to search for ingredients that would meet this need in a safe and efficient manner, leading to in-depth studies and the insertion of these compounds in respected databases of food products in the European Union and the United States of America (FDA). Terpenoids stand out in this context as compounds with wide industrial applicability, since they have great potential as food additives with several proven benefits for human health. Therefore, the importance of terpenoids knowledge in chemical and processing terms is emphasized. Thus, their different sources must be widely studied in order to insert them into different products, bringing more benefits to consumers and overcoming their limitations of use.

# CONSENT FOR PUBLICATION

Not applicable.

# CONFLICT OF INTEREST

The author declares no conflict of interest, financial or otherwise.

# ACKNOWLEDGEMENTS

This study was financed in part by the Coordenação de Aperfeiçoamento de Pessoal de Nível Superior – Brasil (CAPES) – Finance Code 001.

# REFERENCES

[1]     Rukosueva, E.A.; Aliyarova, G.R.; Tikhomirova, T.I.; Apyari, V.V.; Nesterenko, P.N. Simultaneous determination of synthetic food dyes using a single cartridge for preconcentration and separation followed by photometric detection. *Int. J. Anal. Chem.,* **2020**, *2020*, 1-6.
        [http://dx.doi.org/10.1155/2020/2409075]

[2]     Gatidou, G.; Vazaiou, N.; Thomaidis, N.S.; Stasinakis, A.S. Biodegradability assessment of food additives using OECD 301F respirometric test. *Chemosphere,* **2020**, *241*, 125071.
        [http://dx.doi.org/10.1016/j.chemosphere.2019.125071] [PMID: 31683420]

[3]     Tortosa, V.; Pietropaolo, V.; Brandi, V.; Macari, G.; Pasquadibisceglie, A.; Polticelli, F. Computational methods for the identification of molecular targets of toxic food additives. Butylated hydroxytoluene as a case study. *Molecules,* **2020**, *25*(9), 2229.
        [http://dx.doi.org/10.3390/molecules25092229] [PMID: 32397407]

[4]     Boutillier, S.; Fourmentin, S.; Laperche, B. Food additives and the future of health: An analysis of the ongoing controversy on titanium dioxide. *Futures,* **2020**, *122*, 102598.
        [http://dx.doi.org/10.1016/j.futures.2020.102598]

[5]     Chazelas, E.; Deschasaux, M.; Srour, B.; Kesse-Guyot, E.; Julia, C.; Alles, B.; Druesne-Pecollo, N.; Galan, P.; Hercberg, S.; Latino-Martel, P.; Esseddik, Y.; Szabo, F.; Slamich, P.; Gigandet, S.; Touvier, M. Food additives: distribution and co-occurrence in 126,000 food products of the French market. *Sci. Rep.,* **2020**, *10*(1), 3980.
        [http://dx.doi.org/10.1038/s41598-020-60948-w] [PMID: 32132606]

[6]     Miao, P.; Chen, S.; Li, J.; Xie, X. Decreasing consumers' risk perception of food additives by knowledge enhancement in China. *Food Qual. Prefer.,* **2020**, *79*, 103781.
        [http://dx.doi.org/10.1016/j.foodqual.2019.103781]

[7]     Sun, L.; Xin, F.; Alper, H.S. Bio-synthesis of food additives and colorants-a growing trend in future food. *Biotechnol. Adv.,* **2021**, *47*, 107694.
        [http://dx.doi.org/10.1016/j.biotechadv.2020.107694] [PMID: 33388370]

[8]     Mira-Sánchez, M.D.; Castillo-Sánchez, J.; Morillas-Ruiz, J.M. *Comparative study of rosemary extracts and several synthetic and natural food antioxidants. Relevance of carnosic acid/carnosol ratio,* **2020**.
        [http://dx.doi.org/10.1016/j.foodchem.2019.125688]

[9]     Gokoglu, N. Novel natural food preservatives and applications in seafood preservation: a review. *J. Sci. Food Agric.,* **2019**, *99*(5), 2068-2077.
        [http://dx.doi.org/10.1002/jsfa.9416] [PMID: 30318589]

[10]    Konuk, H.B.; Ergüden, B. Phenolic –OH group is crucial for the antifungal activity of terpenoids *via* disruption of cell membrane integrity. *Folia Microbiol. (Praha),* **2020**, *65*(4), 775-783.
        [http://dx.doi.org/10.1007/s12223-020-00787-4] [PMID: 32193708]

[11]    Mitterer-Daltoé, M.; Bordim, J.; Lise, C.; Breda, L.; Casagrande, M.; Lima, V. Consumer awareness of food anti-oxidants. Synthetic *vs.* Natural. *Food Sci. Technol.,* **2020**, *2061*, 1-5.
        [http://dx.doi.org/10.1590/fst.15120]

[12]    Carsanba, E.; Pintado, M.; Oliveira, C. Fermentation strategies for production of pharmaceutical terpenoids in engineered yeast. *Pharmaceuticals (Basel),* **2021**, *14*(4), 295.
        [http://dx.doi.org/10.3390/ph14040295] [PMID: 33810302]

[13]    Kallscheuer, N. Engineered microorganisms for the production of food additives approved by the European Union-A systematic analysis. *Front. Microbiol.,* **2018**, *9*, 1746.
        [http://dx.doi.org/10.3389/fmicb.2018.01746] [PMID: 30123195]

[14]    Lima, P.S.S.; Lucchese, A.M.; Araújo-Filho, H.G.; Menezes, P.P.; Araújo, A.A.S.; Quintans-Júnior, L.J.; Quintans, J.S.S. Inclusion of terpenes in cyclodextrins: Preparation, characterization and pharmacological approaches. *Carbohydr. Polym.,* **2016**, *151*, 965-987.
        [http://dx.doi.org/10.1016/j.carbpol.2016.06.040] [PMID: 27474645]

[15]    Perveen, S. Introductory Chapter: Terpenes and Terpenoids *erpenes and Terpenoids, IntechOpen,* **2018**, 1-12.
        [http://dx.doi.org/10.5772/intechopen.79683]

[16]    Ludwiczuk, A.; Skalicka-Woźniak, K.; Georgiev, M.I. Terpenoids. *Pharmacognosy: Fundamentals, Applications and Strategy.,* **2017**, , 233-266.
        [http://dx.doi.org/10.1016/B978-0-12-802104-0.00011-1]

[17]    European Parliament and the Concil of the European Union. **2008**.

[18]    Substances added to food (formerly EAFUS). **2021**. https://www.cfsanappsexternal.fda.gov/ scripts/fdcc/index.cfm?set=FoodSubstances&sort=Sortterm_ID&order=ASC&showAll=true&type=ba sic&search=

[19]    Diniz do Nascimento, L.; Moraes, A.A.B.; Costa, K.S.; Pereira Galúcio, J.M.; Taube, P.S.; Costa, C.M.L.; Neves Cruz, J.; de Aguiar Andrade, E.H.; Faria, L.J.G. Bioactive natural compounds and anti-oxidant activity of essential oils from spice plants: New findings and potential applications. *Biomolecules,* **2020**, *10*(7), 988.
        [http://dx.doi.org/10.3390/biom10070988]

[20]    Ben Salha, G.; Abderrabba, M.; Labidi, J. A status review of terpenes and their separation methods. *Rev. Chem. Eng.,* **2021**, *37*(3), 433-447.
        [http://dx.doi.org/10.1515/revce-2018-0066]

[21]  Hofmann, T.; Krautwurst, D.; Schieberle, P. Current Status and Future Perspectives in Flavor Research: Highlights of the 11th Wartburg Symposium on Flavor Chemistry & Biology. *J. Agric. Food Chem.,* **2018**, *66*(10), 2197-2203.
      [http://dx.doi.org/10.1021/acs.jafc.7b06144] [PMID: 29298062]

[22]  Bel-Rhlid, R.; Berger, R. G.; Blank, I. *Bio-mediated generation of food flavors – Towards sustainable flavor production inspired by nature,* **2018**.
      [http://dx.doi.org/10.1016/j.tifs.2018.06.004]

[23]  Bezerra, F.W.F. Extraction of bioactive compounds. *Green Sustainable Process for Chemical and Environmental Engineering and Science.,* **2020**, , 149-167.
      [http://dx.doi.org/10.1016/B978-0-12-817388-6.00008-8]

[24]  Noriega, P. Terpenes in Essential Oils: Bioactivity and Applications **2020**.

[25]  Lo, C.M.; Han, J.; Wong, E.S.W. Chemistry in Aromatherapy – Extraction and Analysis of Essential Oils from Plants of *Chamomilla recutita, Cymbopogon nardus, Jasminum officinale* and *Pelargonium graveolens. Biomed. Pharmacol. J.,* **2020**, *13*(3), 1339-1350.
      [http://dx.doi.org/10.13005/bpj/2003]

[26]  Mezzomo, N.; Ferreira, S.R.S. Carotenoids Functionality, Sources, and Processing by Supercritical Technology: A Review. *J. Chem.,* **2016**, *2016*, 1-16.
      [http://dx.doi.org/10.1155/2016/3164312]

[27]  Bhatt, T.; Patel, K. Natural Products and Bioprospecting. Springer, **2020**; 10, pp. (3)109-117.
      [http://dx.doi.org/10.1007/s13659-020-00244-2]

[28]  Meléndez-Martínez, A.J. A comprehensive review on carotenoids in foods and feeds: status quo, applications, patents, and research needs. *Critical Reviews in Food Science and Nutrition,* **2020**.
      [http://dx.doi.org/10.1080/10408398.2020.1867959]

[29]  Kiokias, S.; Proestos, C.; Varzakas, T. *A review of the structure, biosynthesis, absorption of carotenoids-analysis and properties of their common natural extracts,* **2016**.
      [http://dx.doi.org/10.12944/CRNFSJ.4.Special-Issue1.03]

[30]  Ling, W.H.; Jones, P.J.H. Dietary phytosterols: A review of metabolism, benefits and side effects. *Life Sci.,* **1995**, *57*(3), 195-206.
      [http://dx.doi.org/10.1016/0024-3205(95)00263-6] [PMID: 7596226]

[31]  Cabral, C.E.; Klein, M.R.S.T. "Phytosterols in the treatment of hypercholesterolemia and prevention of cardiovascular diseases," *Arquivos Brasileiros de Cardiologia,* vol. 109, no. 5. *Arq. Bras. Cardiol.,* **2017**, *109*(5), 475-482.
      [http://dx.doi.org/10.5935/abc.20170158] [PMID: 29267628]

[32]  Kritchevsky, D.; Chen, S.C. Phytosterols—health benefits and potential concerns: a review. *Nutr. Res.,* **2005**, *25*(5), 413-428.
      [http://dx.doi.org/10.1016/j.nutres.2005.02.003]

[33]  Harris, B. Menthol: A review of its thermoreceptor interactions and their therapeutic applications. *International Journal of Aromatherapy,* **2006**, *16*(3-4), 117-131.
      [http://dx.doi.org/10.1016/j.ijat.2006.09.010]

[34]  Alexander, J. Scientific Opinion on the safety assessment of carvone, considering all sources of exposure. *EFSA J.,* **2014**, *12*(7), 3806.
      [http://dx.doi.org/10.2903/j.efsa.2014.3806]

[35]  Ravichandran, C.; Badgujar, P. C.; Gundev, P.; Upadhyay, A. *Review of toxicological assessment of d-limonene, a food and cosmetics additive,* **2018**.
      [http://dx.doi.org/10.1016/j.fct.2018.07.052]

[36]  Cai, Z. M. *1,8-Cineole: a review of source, biological activities, and application,* **2020**.
      [http://dx.doi.org/10.1080/10286020.2020.1839432]

[37] Lalko, J.; Lapczynski, A.; Politano, V.T.; McGinty, D.; Bhatia, S.; Letizia, C.S.; Api, A.M. Fragrance material review on α-ionone. *Food Chem. Toxicol.*, **2007**, *45*(1), S235-S240.
[http://dx.doi.org/10.1016/j.fct.2007.09.046] [PMID: 18037212]

[38] Rezaee, R.; Hosseinzadeh, H. *Safranal: From an aromatic natural product to a rewarding pharmacological agent*, **2013**.
[http://dx.doi.org/10.22038/ijbms.2013.244]

[39] Ahmed, A.B.A. Saffron as a natural food colorant and its applications. *Saffron.*, **2021**, , 221-239.
[http://dx.doi.org/10.1016/B978-0-12-821219-6.00006-3]

[40] Latos-Brozio, M.; Masek, A. The application of natural food colorants as indicator substances in intelligent biodegradable packaging materials. *Food Chem. Toxicol.*, **2020**, *135*, 110975.
[http://dx.doi.org/10.1016/j.fct.2019.110975] [PMID: 31747619]

[41] Andretta, R.; Luchese, C.L.; Tessaro, I.C.; Spada, J.C. Development and characterization of pH-indicator films based on cassava starch and blueberry residue by thermocompression. *Food Hydrocoll.*, **2019**, *93*, 317-324.
[http://dx.doi.org/10.1016/j.foodhyd.2019.02.019]

[42] Luzardo-Ocampo, I.; Ramírez-Jiménez, A.K.; Yañez, J.; Mojica, L.; Luna-Vital, D.A. Technological applications of natural colorants in food systems: A review. *Foods*, **2021**, *10*(3), 634.
[http://dx.doi.org/10.3390/foods10030634] [PMID: 33802794]

[43] Gebhardt, B.; Sperl, R.; Carle, R.; Müller-Maatsch, J. Assessing the sustainability of natural and artificial food colorants. *J. Clean. Prod.*, **2020**, *260*, 120884.
[http://dx.doi.org/10.1016/j.jclepro.2020.120884]

[44] Britton, G. Carotenoid research: History and new perspectives for chemistry in biological systems. *Biochim. Biophys. Acta Mol. Cell Biol. Lipids*, **2020**, *1865*(11), 158699.
[http://dx.doi.org/10.1016/j.bbalip.2020.158699] [PMID: 32205211]

[45] Maoka, T.; Kawase, N.; Hironaka, M.; Nishida, R. Carotenoids of hemipteran insects, from the perspective of chemo-systematic and chemical ecological studies. *Biochem. Syst. Ecol.*, **2021**, *95*, 104241.
[http://dx.doi.org/10.1016/j.bse.2021.104241]

[46] Aggarwal, S. Indian dye yielding plants: Efforts and opportunities. *Nat. Resour. Forum*, **2021**, *45*(1), 63-86.
[http://dx.doi.org/10.1111/1477-8947.12214]

[47] Mansour, R. Natural dyes and pigments: Extraction and applications. In: *Handbook of Renewable Materials for Coloration and Finishing*; Scrivener Publishing LLC, **2018**; pp. 75-102.
[http://dx.doi.org/10.1002/9781119407850.ch5]

[48] Thompson, A. Ingredients: where pet food starts. *Top. Companion Anim. Med.*, **2008**, *23*(3), 127-132.
[http://dx.doi.org/10.1053/j.tcam.2008.04.004] [PMID: 18656839]

[49] Jang, G.W.; Choi, S.I.; Choi, S.H.; Han, X.; Men, X.; Kwon, H.Y.; Choi, Y.E.; Lee, O.H. Method validation of 12 kinds of food dye in chewing gums and soft drinks, and evaluation of measurement uncertainty for soft drinks. *Food Chem.*, **2021**, *356*, 129705.
[http://dx.doi.org/10.1016/j.foodchem.2021.129705] [PMID: 33836361]

[50] Etxabide, A.; Maté, J.I.; Kilmartin, P.A. Effect of curcumin, betanin and anthocyanin containing colourants addition on gelatin films properties for intelligent films development. *Food Hydrocoll.*, **2021**, *115*, 106593.
[http://dx.doi.org/10.1016/j.foodhyd.2021.106593]

[51] Gupta, S.; Khan, S.; Muzafar, M.; Kushwaha, M.; Yadav, A.K.; Gupta, A.P. Encapsulation: entrapping essential oil/flavors/aromas in food. In: *Encapsulations*; Elsevier Inc., **2016**; pp. 229-268.
[http://dx.doi.org/10.1016/B978-0-12-804307-3.00006-5]

[52] Martínez, Y.; Orozco, C.E.; Montellano, R.M.; Valdivié, M.; Parrado, C.A. Use of achiote (*Bixa orellana* L.) seed powder as pigment of the egg yolk of laying hens. *J. Appl. Poult. Res.,* **2021**, *30*(2), 100154.
[http://dx.doi.org/10.1016/j.japr.2021.100154]

[53] Šeregelj, V.; Vulić, J.; Ćetković, G.; Čanadanović-Brunet, J.; Tumbas Šaponjac, V.; Stajčić, S. Natural bioactive compounds in carrot waste for food applications and health benefits. *Studies in Natural Products Chemistry.,* **2021**, *Vol. 67*, 307-344.

[54] Bezerra, F.W.F.; Costa, W.A.; Oliveira, M.S.; Aguiar Andrade, E.H.; Carvalho, R.N. Transesterification of palm pressed-fibers (*Elaeis guineensis* Jacq.) oil by supercritical fluid carbon dioxide with entrainer ethanol. *J. Supercrit. Fluids,* **2018**, *136*, 136-143.
[http://dx.doi.org/10.1016/j.supflu.2018.02.020]

[55] Mozzon, M.; Foligni, R.; Mannozzi, C. Current knowledge on interspecific hybrid palm oils as food and food ingredient. *Foods,* **2020**, *9*(5), 631.
[http://dx.doi.org/10.3390/foods9050631] [PMID: 32422962]

[56] Farkas, Á.; Bencsik, T.; Deli, J. Carotenoids as food additives. *Pigments from Microalgae Handbook.,* **2020**, , 421-447.
[http://dx.doi.org/10.1007/978-3-030-50971-2_17]

[57] Gooderham, N.J.; Cohen, S.M.; Eisenbrand, G.; Fukushima, S.; Guengerich, F.P.; Hecht, S.S.; Rietjens, I.M.C.M.; Rosol, T.J.; Davidsen, J.M.; Harman, C.L.; Murray, I.J.; Taylor, S.V. FEMA GRAS assessment of natural flavor complexes: Clove, cinnamon leaf and West Indian bay leaf-derived flavoring ingredients. *Food Chem. Toxicol.,* **2020**, *145*, 111585.
[http://dx.doi.org/10.1016/j.fct.2020.111585] [PMID: 32702506]

[58] Alizadeh, L.; Abdolmaleki, K.; Nayebzadeh, K.; Shahin, R. Effects of tocopherol, rosemary essential oil and *Ferulago angulata* extract on oxidative stability of mayonnaise during its shelf life: A comparative study. *Food Chem.,* **2019**, *285*, 46-52.
[http://dx.doi.org/10.1016/j.foodchem.2019.01.028] [PMID: 30797371]

[59] Felipe, L.O.; Bicas, J.L. Terpenos, aromas e a química dos compostos naturais. *Química Nov. na Esc.,* **2017**, *39*(2), 120-130.
[http://dx.doi.org/10.21577/0104-8899.20160068]

[60] Atti-Santos, A.C.; Rossato, M.; Serafini, L.A.; Cassel, E.; Moyna, P. Extraction of essential oils from lime (*Citrus latifolia* Tanaka) by hydrodistillation and supercritical carbon dioxide. *Braz. Arch. Biol. Technol.,* **2005**, *48*(1), 155-160.
[http://dx.doi.org/10.1590/S1516-89132005000100020]

[61] Benelli, P.; Riehl, C.A.S.; Smânia, A., Jr; Smânia, E.F.A.; Ferreira, S.R.S. Bioactive extracts of orange (*Citrus sinensis* L. Osbeck) pomace obtained by SFE and low pressure techniques: Mathematical modeling and extract composition. *J. Supercrit. Fluids,* **2010**, *55*(1), 132-141.
[http://dx.doi.org/10.1016/j.supflu.2010.08.015]

[62] de Oliveira, M.S.; da Costa, W.A.; Pereira, D.S.; Botelho, J.R.S.; de Alencar Menezes, T.O.; de Aguiar Andrade, E.H.; da Silva, S.H.M.; da Silva Sousa Filho, A.P.; de Carvalho, R.N. Chemical composition and phytotoxic activity of clove ( *Syzygium aromaticum* ) essential oil obtained with supercritical $CO_2$. *J. Supercrit. Fluids,* **2016**, *118*, 185-193.
[http://dx.doi.org/10.1016/j.supflu.2016.08.010]

[63] Damodaran, S.; Parkin, K.L. *Química de alimentos de Fennema,* 5th ed; Artmed Editora, **2019**.

[64] Tauchen, J.; Bortl, L.; Huml, L.; Miksatkova, P.; Doskocil, I.; Marsik, P.; Villegas, P.P.P.; Flores, Y.B.; Damme, P.V.; Lojka, B.; Havlik, J.; Lapcik, O.; Kokoska, L. Phenolic composition, antioxidant and anti-proliferative activities of edible and medicinal plants from the Peruvian Amazon. *Rev. Bras. Farmacogn.,* **2016**, *26*(6), 728-737.
[http://dx.doi.org/10.1016/j.bjp.2016.03.016]

[65] Khoddami, A.; Wilkes, M.; Roberts, T. Techniques for analysis of plant phenolic compounds. *Molecules,* **2013**, *18*(2), 2328-2375.
[http://dx.doi.org/10.3390/molecules18022328] [PMID: 23429347]

[66] Dergal, S.B. *Química de los alimentos,* 5th ed; Pearson, **2013**.

[67] Shahidi, F. *Handbook of anti-oxidants for food preservation*; Woodhead Publishing, **2015**.

[68] Basanta, M.F.; Rizzo, S.A.; Szerman, N.; Vaudagna, S.R.; Descalzo, A.M.; Gerschenson, L.N.; Pérez, C.D.; Rojas, A.M. Plum (*Prunus salicina*) peel and pulp microparticles as natural antioxidant additives in breast chicken patties. *Food Res. Int.,* **2018**, *106*, 1086-1094.
[http://dx.doi.org/10.1016/j.foodres.2017.12.011] [PMID: 29579902]

[69] Risk, E. M.; Bedier, S. H. *Egypt. J. Agric. Res.,* **2014**, *92*(1), 309-321.
[http://dx.doi.org/10.21608/ejar.2014.154836]

[70] Kehili, M.; Kammlott, M.; Choura, S.; Zammel, A.; Zetzl, C.; Smirnova, I.; Allouche, N.; Sayadi, S. Supercritical $CO_2$ extraction and antioxidant activity of lycopene and β-carotene-enriched oleoresin from tomato ( *Lycopersicum esculentum* L.) peels by-product of a Tunisian industry. *Food Bioprod. Process.,* **2017**, *102*, 340-349.
[http://dx.doi.org/10.1016/j.fbp.2017.02.002]

[71] Rodrigues, E.; Mariutti, L.R.B.; Chisté, R.C.; Mercadante, A.Z. Development of a novel micro-assay for evaluation of peroxyl radical scavenger capacity: Application to carotenoids and structure–activity relationship. *Food Chem.,* **2012**, *135*(3), 2103-2111.
[http://dx.doi.org/10.1016/j.foodchem.2012.06.074] [PMID: 22953962]

[72] Mérillon, J-M.; Ramawat, K.G. Ramawat, Bioactive molecules in food - Reference series in phytochemistry, no. September 2018. Springer **2019**.

[73] Choudhary, B.R.; Haldhar, S.M.; Maheshwari, S.K.; Bhargava, R.; Sharma, S.K. Phytochemicals and anti-oxidants in watermelon (*Citrullus lanatus*) genotypes under hot arid region. *Indian J. Agric. Sci.,* **2015**, *85*(3), 414-421.

[74] Dabbou, S.; Maatallah, S.; Castagna, A.; Guizani, M.; Sghaeir, W.; Hajlaoui, H.; Ranieri, A. Carotenoids, phenolic profile, mineral content and anti-oxidant properties in flesh and peel of *Prunus persica* fruits during two maturation stages. *Plant Foods Hum. Nutr.,* **2017**, *72*(1), 103-110.
[http://dx.doi.org/10.1007/s11130-016-0585-y] [PMID: 27812831]

[75] Zacarías-García, J.; Rey, F.; Gil, J.V.; Rodrigo, M.J.; Zacarías, L. Antioxidant capacity in fruit of Citrus cultivars with marked differences in pulp coloration: Contribution of carotenoids and vitamin C. *Food Sci. Technol. Int.,* **2021**, *27*(3), 210-222.
[http://dx.doi.org/10.1177/1082013220944018] [PMID: 32727209]

[76] Campestrini, L.H.; Melo, P.S.; Peres, L.E.P.; Calhelha, R.C.; Ferreira, I.C.F.R.; Alencar, S.M. A new variety of purple tomato as a rich source of bioactive carotenoids and its potential health benefits. *Heliyon,* **2019**, *5*(11), e02831.
[http://dx.doi.org/10.1016/j.heliyon.2019.e02831] [PMID: 31763483]

[77] Donado-Pestana, C.M.; Salgado, J.M.; de Oliveira Rios, A.; dos Santos, P.R.; Jablonski, A. Stability of carotenoids, total phenolics and *in vitro* antioxidant capacity in the thermal processing of orange-fleshed sweet potato (Ipomoea batatas Lam.) cultivars grown in Brazil. *Plant Foods Hum. Nutr.,* **2012**, *67*(3), 262-270.
[http://dx.doi.org/10.1007/s11130-012-0298-9] [PMID: 22802046]

[78] Lau, W.K.; Van Chuyen, H.; Vuong, Q.V. Physical properties, carotenoids and anti-oxidant capacity of carrot (*Daucus carota* L.) peel as influenced by different drying treatments. *Int. J. Food Eng.,* **2018**, *14*(3), 1-13.
[http://dx.doi.org/10.1515/ijfe-2017-0042]

[79] de Carvalho, L.M.J.; Gomes, P.B.; Godoy, R.L.O.; Pacheco, S.; do Monte, P.H.F.; de Carvalho, J.L.V.; Nutti, M.R.; Neves, A.C.L.; Vieira, A.C.R.A.; Ramos, S.R.R. Total carotenoid content, α-

carotene and β-carotene, of landrace pumpkins (*Cucurbita moschata* Duch): A preliminary study. *Food Res. Int.,* **2012**, *47*(2), 337-340.
[http://dx.doi.org/10.1016/j.foodres.2011.07.040]

[80]     Priori, D.; Valduga, E.; Villela, J.C.B.; Mistura, C.C.; Vizzotto, M.; Valgas, R.A.; Barbieri, R.L. Characterization of bioactive compounds, antioxidant activity and minerals in landraces of pumpkin (*Cucurbita moschata*) cultivated in Southern Brazil. *Food Sci. Technol. (Campinas),* **2016**, *37*(1), 33-40.
[http://dx.doi.org/10.1590/1678-457x.05016]

[81]     Micheletti A, M.S.; Masciangelo, S.; Micheletti, A.; Ferretti, G. Carotenoids, phenolic compounds and anti-oxidant capacity of five local italian corn (*Zea Mays* L.) kernels. *J. Nutr. Food Sci.,* **2013**, *3*(6), 1000237.
[http://dx.doi.org/10.4172/2155-9600.1000237]

[82]     Wang, C.Y.; Chen, Y.W.; Hou, C.Y. Antioxidant and antibacterial activity of seven predominant terpenoids. *Int. J. Food Prop.,* **2019**, *22*(1), 230-238.
[http://dx.doi.org/10.1080/10942912.2019.1582541]

[83]     Wojtunik, K.A.; Ciesla, L.M.; Waksmundzka-Hajnos, M. Model studies on the antioxidant activity of common terpenoid constituents of essential oils by means of the 2,2-diphenyl-1-picrylhydrazyl method. *J. Agric. Food Chem.,* **2014**, *62*(37), 9088-9094.
[http://dx.doi.org/10.1021/jf502857s] [PMID: 25152006]

[84]     Khatun, A.; Rahman, M.; Rahman, M.S.; Hossain, M.K.; Rashid, M.A. Terpenoids and phytosteroids isolated from *Commelina benghalensis* Linn. with antioxidant activity. *J. Basic Clin. Physiol. Pharmacol.,* **2020**, *31*(1), 20180218.
[http://dx.doi.org/10.1515/jbcpp-2018-0218] [PMID: 31770097]

[85]     Marchetti, N.; Bonetti, G.; Brandolini, V.; Cavazzini, A.; Maietti, A.; Meca, G.; Mañes, J. Stinging nettle (*Urtica dioica* L.) as a functional food additive in egg pasta: Enrichment and bioaccessibility of Lutein and β-carotene. *J. Funct. Foods,* **2018**, *47*, 547-553.
[http://dx.doi.org/10.1016/j.jff.2018.05.062]

[86]     Eletr, A.A.; Siliha, H.A.E.; Elshorbagy, G.A.; Galal, G.A.; Agric, F. Evaluation of lycopene extracted from tomato processing waste as a natural anti-oxidant in some bakery products. *Zagazig Journal of Agricultural Research,* **2017**, *44*(4), 1389-1401.
[http://dx.doi.org/10.21608/zjar.2017.52942]

[87]     Acosta-Estrada, B.A.; Lazo-Vélez, M.A.; Nava-Valdez, Y.; Gutiérrez-Uribe, J.A.; Serna-Saldívar, S.O. Improvement of dietary fiber, ferulic acid and calcium contents in pan bread enriched with nejayote food additive from white maize (Zea mays). *J. Cereal Sci.,* **2014**, *60*(1), 264-269.
[http://dx.doi.org/10.1016/j.jcs.2014.04.006]

[88]     Abdullah, M.A.; Al Dajah, S.; Abu Murad, A.; El-Salem, A.M.; Khafajah, A.M. Extraction, purification, and characterization of lycopene from jordanian vine tomato cultivar, and study of its potential natural anti-oxidant effect on samen baladi. *Curr. Res. Nutr. Food Sci.,* **2019**, *7*(2), 532-546.
[http://dx.doi.org/10.12944/CRNFSJ.7.2.22]

[89]     Robles-Ramírez, M. D. C.; Monterrubio-López, R.; Mora-Escobedo, R. *Arch. Latinoam. Nutr.,* **2016**, *66*(1), 66-73.

[90]     Skiepko, N.; Chwastowska-Siwiecka, I.; Kondratowicz, J.; Mikulski, D. The effect of lycopene addition on the chemical composition, sensory attributes and physicochemical properties of steamed and grilled turkey breast. *Brazilian J. Poult. Sci,* **2016**, *18*(2), 319-330.

[91]     Valdivieso-Ugarte, M.; Gomez-Llorente, C.; Plaza-Díaz, J.; Gil, Á. Antimicrobial, anti-oxidant, and immunomodulatory properties of essential oils: A systematic review. *Nutrients,* **2019**, *11*(11), 2786.
[http://dx.doi.org/10.3390/nu11112786] [PMID: 31731683]

[92]     Ogueke, C.C.; Nnadi, N.B.; Owuamanam, C.I.; Ojukwu, M.; Nwachukwu, I.N.; Ibeabuchi, C.J.; Bede, E.N. Preservative potentials of essential oils of three Nigerian spices in mixed fruit juice and their

antioxidant capacity. *Afr. J. Biotechnol.,* **2018**, *17*(35), 1099-1110.
[http://dx.doi.org/10.5897/AJB2018.16548]

[93]   Boukhatem, M.N.; Boumaiza, A.; Nada, H.G.; Rajabi, M.; Mousa, S.A. *Eucalyptus globulus* essential oil as a natural food preservative: Anti-oxidant, antibacterial and antifungal properties *in vitro* and in a real food matrix (orangina fruit juice). *Appl. Sci. (Basel),* **2020**, *10*(16), 5581.
[http://dx.doi.org/10.3390/app10165581]

[94]   Ramawat, K.; Merillon, J. *Phytochemistry, botany and metabolism of alkaloids, phenolics and terpenes*; Springer, **2013**.

[95]   Amer, T.A.M. Effect of lemon and orange oils on shelf life of cake. *Middle East J. Appl.,* **2018**, *8*(4), 1364-1374.

[96]   de Oliveira, M.S. Potential of medicinal use of essential oils from aromatic plants. In: *Potential of Essential Oils*; IntechOpen, **2018**; pp. 1-20.
[http://dx.doi.org/10.5772/intechopen.78002]

[97]   Sharma, H.; Mendiratta, S.K.; Agarwal, R.K.; Gurunathan, K. Bio-preservative effect of blends of essential oils: natural anti-oxidant and anti-microbial agents for the shelf life enhancement of emulsion based chicken sausages. *J. Food Sci. Technol.,* **2020**, *57*(8), 3040-3050.
[http://dx.doi.org/10.1007/s13197-020-04337-1] [PMID: 32624606]

[98]   Mahizan, N.A.; Yang, S.K.; Moo, C.L.; Song, A.A.L.; Chong, C.M.; Chong, C.W.; Abushelaibi, A.; Lim, S.H.E.; Lai, K.S. Terpene derivatives as a potential agent against anti-microbial resistance (AMR) pathogens. *Molecules,* **2019**, *24*(14), 2631.
[http://dx.doi.org/10.3390/molecules24142631] [PMID: 31330955]

[99]   EL-Gioushy, S.F.; Baiea, M.H.M. Impact of gelatin, lemongrass oil and peppermint oil on storability and fruit quality of Samany date palm under cold storage. *Bull. Natl. Res. Cent.,* **2020**, *44*(1), 14.
[http://dx.doi.org/10.1186/s42269-019-0255-y]

[100]  Yang, W.; Chen, X.; Li, Y.; Guo, S.; Wang, Z.; Yu, X. Advances in pharmacological activities of terpenoids. *Nat. Prod. Commun.,* **2020**, *15*(3), 1934578X2090355.
[http://dx.doi.org/10.1177/1934578X20903555]

[101]  Wang, X.; Ding, Z.; Zhao, Y.; Prakash, S.; Liu, W.; Han, J.; Wang, Z. Effects of lutein particle size in embedding emulsions on encapsulation efficiency, storage stability, and dissolution rate of microencapsules through spray drying. *Lebensm. Wiss. Technol.,* **2021**, *146*, 111430.
[http://dx.doi.org/10.1016/j.lwt.2021.111430]

[102]  Fernandes, E.J.; Poetini, M.R.; Barrientos, M.S.; Bortolotto, V.C.; Araujo, S.M.; Santos Musachio, E.A.; De Carvalho, A.S.; Leimann, F.V.; Gonçalves, O.H.; Ramborger, B.P.; Roehrs, R.; Prigol, M.; Guerra, G.P. Exposure to lutein-loaded nanoparticles attenuates Parkinson's model-induced damage in *Drosophila melanogaster*: Restoration of dopaminergic and cholinergic system and oxidative stress indicators. *Chem. Biol. Interact.,* **2021**, *340*(January), 109431.
[http://dx.doi.org/10.1016/j.cbi.2021.109431] [PMID: 33716020]

[103]  Albuquerque, G.A.; Bezerra, F.W.F.; de Oliveira, M.S.; da Costa, W.A.; de Carvalho Junior, R.N.; Joele, M.R.S.P. Supercritical $CO_2$ impregnation of *Piper divaricatum* essential oil in fish (*Cynoscion acoupa*) skin gelatin films. *Food Bioprocess Technol.,* **2020**, *13*(10), 1765-1777.
[http://dx.doi.org/10.1007/s11947-020-02514-w]

[104]  Bedoya-Serna, C.M.; Dacanal, G.C.; Fernandes, A.M.; Pinho, S.C. Antifungal activity of nanoemulsions encapsulating oregano (*Origanum vulgare*) essential oil: *in vitro* study and application in Minas Padrão cheese. *Braz. J. Microbiol.,* **2018**, *49*(4), 929-935.
[http://dx.doi.org/10.1016/j.bjm.2018.05.004] [PMID: 30145265]

[105]  Oluwasola, O.; Maroyi, A.; Jide Afola, A. Effects of leaf extracts of *Ocimum gratissimum* L. on quality of fresh cut *Cucumis sativus* L. *Asian J. Plant Pathol.,* **2017**, *11*(4), 174-184.
[http://dx.doi.org/10.3923/ajppaj.2017.174.184]

[106]  Ebabhi, A.M.; Adeogun, O.O.; Adekunle, A.A.; Onafeko, A.O. Bio-preservation potential of leaf extracts of *Ocimum gratissimum* L. on fresh-cut fruits of *Citrullus lanatus* (Thunb). *J. Appl. Sci. Environ. Manag.,* **2019**, *23*(7), 1383-1389.
        [http://dx.doi.org/10.4314/jasem.v23i7.30]

[107]  Oliveira, F.S.; Teodoro, C.E.S.; Berbert, P.A.; Martinazzo, A.P. Evaluation of the antifungal potential of Cymbopogon citratus essential oil in the control of the fungus Aspergillus brasiliensis. *Research, Society and Development,* **2020**, *9*(7), e691974697.
        [http://dx.doi.org/10.33448/rsd-v9i7.4697]

[108]  Boeira, C.P.; Piovesan, N.; Flores, D.C.B.; Soquetta, M.B.; Lucas, B.N.; Heck, R.T.; Alves, J.S.; Campagnol, P.C.B.; dos Santos, D.; Flores, E.M.M.; da Rosa, C.S.; Terra, N.N. Phytochemical characterization and antimicrobial activity of *Cymbopogon citratus* extract for application as natural antioxidant in fresh sausage. *Food Chem.,* **2020**, *319*, 126553.
        [http://dx.doi.org/10.1016/j.foodchem.2020.126553] [PMID: 32197214]

[109]  Dolea, D.; Rizo, A.; Fuentes, A.; Barat, J.M.; Fernández-Segovia, I. Effect of thyme and oregano essential oils on the shelf life of salmon and seaweed burgers. *Food Sci. Technol. Int.,* **2018**, *24*(5), 394-403.
        [http://dx.doi.org/10.1177/1082013218759364] [PMID: 29436857]

[110]  Bukvicki, D.; Giweli, A.; Stojkovic, D.; Vujisic, L.; Tesevic, V.; Nikolic, M.; Sokovic, M.; Marin, P.D. Short communication: Cheese supplemented with *Thymus algeriensis* oil, a potential natural food preservative. *J. Dairy Sci.,* **2018**, *101*(5), 3859-3865.
        [http://dx.doi.org/10.3168/jds.2017-13714] [PMID: 29477526]

[111]  Rodrigues, G.M.; Filgueiras, C.T.; Garcia, V.A.S.; Carvalho, R.A.; Velasco, J.I.; Fakhouri, F.M. Anti-microbial activity and GC-MS profile of copaiba oil for incorporation into *Xanthosoma mafaffa* schott starch-based films. *Polymers (Basel),* **2020**, *12*(12), 2883.
        [http://dx.doi.org/10.3390/polym12122883] [PMID: 33271855]

[112]  Scicchitano, P.; Cameli, M.; Maiello, M.; Modesti, P.A.; Muiesan, M.L.; Novo, S.; Palmiero, P.; Saba, P.S.; Pedrinelli, R.; Ciccone, M.M. Nutraceuticals and dyslipidaemia: Beyond the common therapeutics. *J. Funct. Foods,* **2014**, *6*(1), 11-32.
        [http://dx.doi.org/10.1016/j.jff.2013.12.006]

[113]  Dima, C.; Assadpour, E.; Dima, S.; Jafari, S.M. Bioavailability of nutraceuticals: Role of the food matrix, processing conditions, the gastrointestinal tract, and nanodelivery systems. *Compr. Rev. Food Sci. Food Saf.,* **2020**, *19*(3), 954-994.
        [http://dx.doi.org/10.1111/1541-4337.12547] [PMID: 33331687]

[114]  Mudila, H.; Prasher, P.; Khati, B.; Kumar, S.; Punetha, H. Nutraceuticals for healthy sporting. *Nutraceuticals and innovative food products for healthy living and preventive care.,* **2018**, , 79-107.
        [http://dx.doi.org/10.4018/978-1-5225-2970-5.ch004]

[115]  Santini, A.; Novellino, E. Nutraceuticals - shedding light on the grey area between pharmaceuticals and food. *Expert Rev. Clin. Pharmacol.,* **2018**, *11*(6), 545-547.
        [http://dx.doi.org/10.1080/17512433.2018.1464911] [PMID: 29667442]

[116]  Durazzo, A.; Lucarini, M.; Santini, A. Nutraceuticals in human health. *Foods,* **2020**, *9*(3), 370.
        [http://dx.doi.org/10.3390/foods9030370] [PMID: 32209968]

[117]  Dubey, N.K.; Singh, A.K.; Dubey, R.; Deng, W.P. Nutraceutical encapsulation and delivery system for type 2 diabetes mellitus. In: *Biopolymer-Based Formulations: Biomedical and Food Applications*; Elsevier Inc., **2020**; pp. 353-363.
        [http://dx.doi.org/10.1016/B978-0-12-816897-4.00015-1]

[118]  Asefy, Z.; Tanomand, A.; Hoseinnejhad, S.; Ceferov, Z.; Oshaghi, E.A.; Rashidi, M. Unsaturated fatty acids as a co-therapeutic agents in cancer treatment. *Mol. Biol. Rep.,* **2021**, *48*(3), 2909-2916.
        [http://dx.doi.org/10.1007/s11033-021-06319-8] [PMID: 33821440]

[119] Gupta, R.; Gaur, S. Production of polyunsaturated fatty acids by fungal biofactories and their application in food industries. *Fungi in Sustainable Food Production.,* **2021**, , 117-128.
[http://dx.doi.org/10.1007/978-3-030-64406-2_7]

[120] Raal, A.; Orav, A.; Püssa, T.; Valner, C.; Malmiste, B.; Arak, E. Content of essential oil, terpenoids and polyphenols in commercial chamomile (*Chamomilla recutita* L. Rauschert) teas from different countries. *Food Chem.,* **2012**, *131*(2), 632-638.
[http://dx.doi.org/10.1016/j.foodchem.2011.09.042]

[121] Yener, I. Determination of antioxidant, cytotoxic, anticholinesterase, antiurease, antityrosinase, and antielastase activities and aroma, essential oil, fatty acid, phenolic, and terpenoid-phytosterol contents of *Salvia poculata. Ind. Crops Prod.,* **2020**, *155*, 112712.
[http://dx.doi.org/10.1016/j.indcrop.2020.112712]

[122] Mokgehle, T.M.; Madala, N.; Gitari, W.M.; Tavengwa, N.T. *Advances in the development of biopolymeric adsorbents for the extraction of metabolites from nutraceuticals with emphasis on Solanaceae and subsequent pharmacological applications,* **2021**.
[http://dx.doi.org/10.1016/j.carbpol.2021.118049]

[123] D'Onofrio, F.; Raimo, S.; Spitaleri, D.; Casucci, G.; Bussone, G. Usefulness of nutraceuticals in migraine prophylaxis. *Neurol. Sci.,* **2017**, *38*(S1) Suppl. 1, 117-120.
[http://dx.doi.org/10.1007/s10072-017-2901-1] [PMID: 28527067]

[124] Song, X.; Luo, Y.; Ma, L.; Hu, X.; Simal-Gandara, J.; Wang, L.S.; Bajpai, V.K.; Xiao, J.; Chen, F. Recent trends and advances in the epidemiology, synergism, and delivery system of lycopene as an anti-cancer agent. *Semin. Cancer Biol.,* **2021**, *73*(March), 331-346.
[http://dx.doi.org/10.1016/j.semcancer.2021.03.028] [PMID: 33794344]

[125] Gupta, N.; Kalaskar, A.; Kalaskar, R. Efficacy of lycopene in management of Oral Submucous Fibrosis– A systematic review and meta-analysis. *J. Oral Biol. Craniofac. Res.,* **2020**, *10*(4), 690-697.
[http://dx.doi.org/10.1016/j.jobcr.2020.09.004] [PMID: 33072506]

[126] Paul, R.; Mazumder, M.K.; Nath, J.; Deb, S.; Paul, S.; Bhattacharya, P.; Borah, A. Lycopene - A pleiotropic neuroprotective nutraceutical: Deciphering its therapeutic potentials in broad spectrum neurological disorders. *Neurochem. Int.,* **2020**, *140*, 104823.
[http://dx.doi.org/10.1016/j.neuint.2020.104823] [PMID: 32827559]

[127] Kehili, M.; Sayadi, S.; Frikha, F.; Zammel, A.; Allouche, N. Optimization of lycopene extraction from tomato peels industrial by-product using maceration in refined olive oil. *Food Bioprod. Process.,* **2019**, *117*, 321-328.
[http://dx.doi.org/10.1016/j.fbp.2019.08.004]

[128] Sánchez, C.A.O.; Zavaleta, E.B.; García, G.R.U.; Solano, G.L.; Díaz, M.P.R. Krill oil microencapsulation: Anti-oxidant activity, astaxanthin retention, encapsulation efficiency, fatty acids profile, *in vitro* bioaccessibility and storage stability. *Sci. Total Environ.,* **2019**, *13*, 135907.
[http://dx.doi.org/10.1016/j.lwt.2021.111476]

[129] Morales, E.; Burgos-Díaz, C.; Zúñiga, R.N.; Jorkowski, J.; Quilaqueo, M.; Rubilar, M. Influence of O/W emulsion interfacial ionic membranes on the encapsulation efficiency and storage stability of powder microencapsulated astaxanthin. *Food Bioprod. Process.,* **2021**, *126*, 143-154.
[http://dx.doi.org/10.1016/j.fbp.2020.12.014]

[130] Jiménez-Escobar, M.P.; Pascual-Pineda, L.A.; Vernon-Carter, E.J.; Beristain, C.I. *Enhanced β-carotene encapsulation and protection in self-assembled lyotropic liquid crystal structures,* **2021**.
[http://dx.doi.org/10.1016/j.lwt.2020.110056]

[131] Guedes Silva, K.C.; Feltre, G.; Dupas Hubinger, M.; Kawazoe Sato, A.C. Protection and targeted delivery of β-carotene by starch-alginate-gelatin emulsion-filled hydrogels. *J. Food Eng.,* **2021**, *290*, 110205.
[http://dx.doi.org/10.1016/j.jfoodeng.2020.110205]

[132] López-Monterrubio, D.I.; Lobato-Calleros, C.; Vernon-Carter, E.J.; Alvarez-Ramirez, J. Influence of β-carotene concentration on the physicochemical properties, degradation and antioxidant activity of nanoemulsions stabilized by whey protein hydrolyzate-pectin soluble complexes. *Lebensm. Wiss. Technol.,* **2021**, *143*(January), 111148.
[http://dx.doi.org/10.1016/j.lwt.2021.111148]

[133] Steiner, B.M.; McClements, D.J.; Davidov-Pardo, G. Encapsulation systems for lutein: A review. *Trends Food Sci. Technol.,* **2018**, *82*(May), 71-81.
[http://dx.doi.org/10.1016/j.tifs.2018.10.003]

[134] Yuan, B.; Yang, S.; Wang, M.; Jiang, X.; Bai, S. Preparation of Ag foam catalyst based on *in-situ* thermally induced redox reaction between polyvinyl alcohol and silver nitrate with supercritical $CO_2$ foaming technology. *Polymer (Guildf.),* **2020**, *206*(May), 122858.
[http://dx.doi.org/10.1016/j.polymer.2020.122858]

[135] Li, S.; Wang, C.; Fu, X.; Li, C.; He, X.; Zhang, B.; Huang, Q. Encapsulation of lutein into swelled cornstarch granules: Structure, stability and *in vitro* digestion. *Food Chem.,* **2018**, *268*, 362-368.
[http://dx.doi.org/10.1016/j.foodchem.2018.06.078] [PMID: 30064770]

[136] Zhao, L.; Temelli, F.; Curtis, J.M.; Chen, L. Encapsulation of lutein in liposomes using supercritical carbon dioxide. *Food Res. Int.,* **2017**, *100*(Pt 1), 168-179.
[http://dx.doi.org/10.1016/j.foodres.2017.06.055] [PMID: 28873676]

[137] Padmanabha, S.; Vallikannan, B. Fatty acids influence the efficacy of lutein in the modulation of α-crystallin chaperone function: Evidence from selenite induced cataract rat model. *Biochem. Biophys. Res. Commun.,* **2020**, *529*(2), 425-431.
[http://dx.doi.org/10.1016/j.bbrc.2020.06.021] [PMID: 32703446]

# CHAPTER 11

# Potential Use of Terpenoids for Control of Insect Pests

**Murilo Fazolin**[1,*], **Humberto Ribeiro Bizzo**[2] and **André Fábio Medeiros Monteiro**[1]

[1] *Agroforestry Research Center of Acre, Brazilian Agricultural Research Corporation (Embrapa/CPAFAC), Rodovia BR 364 Km 14 Rio Branco-Porto Velho- Zona Rural, CEP 69900-970, Rio Branco, AC, Brazil*

[2] *National Center for Research on Agroindustrial Food Technology (CTAA), Av. das Américas, nº 29.501, Guaratiba, RJ, CEP 23020-470, Brazil*

**Abstract:** Essential oils (EOs) have diverse chemical compositions depending on the plant species used, but the most common constituents present in EOs are mono- and sesquiterpenoids. Such volatile terpenoids have different functions in plant ecology, acting, for example, as chemical defenses against fungi, bacteria, and insects, attracting pollinators, inhibiting germination, and mediating intra- and interspecific plant communication. Mainly terpenoids present the ability to inhibit the main families of detoxifying enzymes of insects, allowing the formulation of botanical insecticides, and using blends of EO compounds considered synergists among themselves. In this case, both combinations of essential oils from different plants and the enrichment of essential oils and/or their fractions with compounds with proven synergistic effects can be considered. This chapter presents research results that indicate synergistic, additive, and antagonistic interactions between terpenoids, indicating that this is one of the main properties considered when formulating insecticides based on commercially available EOs. Considerable advances are still necessary for large-scale production, and limitations related to raw material supply, registration, and, mainly, adequacy of formulations for the control of different targets without phytotoxic effects, are the main challenges to be overcome in the short-term.

**Keywords:** Additivism, Antagonism, Agrochemical Industry, Aromatic Plants, Bioinsecticides, Biological Interference, Botanical Insecticides, Cytochrome P450, Enzyme Inhibition, Esterases, Essential Oils, Glutathione S-Transferase, Insecticide Formulations, Integrated Pest Management, Insect Toxicology, Insecticidal Plants, Microsomal Monooxygenases, Pest Control, Synergism, Terpenoid Blends.

* **Corresponding author Murilo Fazolin:** Agroforestry Research Center of Acre, Brazilian Agricultural Research Corporation, Rio Branco, AC, Brazil; Email: murilo.fazolin@embrapa.br

Mozaniel Santana de Oliveira & Antônio Pedro da Silva Souza Filho (Eds.)

# INTRODUCTION

Essential oils (EOs) are products obtained from plants by dry distillation, steam distillation, or, in the specific case of citrus, fruit pressing [1]. Their chemical composition varies greatly depending on the plant species used, but the most common constituents present in EOs are mono- and sesquiterpenoids. These volatile terpenoids have different functions in plant ecologies, such as chemical defenders against fungi, bacteria, and insects, pollinator attractors, germination inhibitors, and mediators of intra- and interspecific plant-plant communication [2].

Isolation, identification, and synthesis techniques lead to the obtaining of several volatile terpenoids in their pure form, allowing the investigation and use of specific metabolites originally present in EOs. These substances exhibit several applications and have drawn attention due to their potential use as alternative pesticides [3]. The toxicity of terpenoids and essential oils is reported against many pest insects of agricultural importance, such as *Diabrotica undecimpunctata* howardi Barber (Coleoptera: Chrysomelidae) [4]; *Spodoptera litura* (Fabricius) (Lepidoptera: Noctuidae) [5]; *Rhyzopertha dominica* (Fabricius) (Coleoptera, Bostrichidae); *Tribolium castaneum* (Herbst) (Coleoptera: Tenebrionidae); *Sitophilus oryzae* (Linnaeus) (Coleoptera: Curculionidae) [6]; and *Ceratitis capitata* (Wiedemann) (Diptera: Tephritidae) [7].

Volatile terpenoids interfere in several physiological and behavioral processes of insects, and their insecticidal action is widely reported in the literature [8 - 10]. Insectistastical properties, such as repellency, feeding inhibition, and growth reduction, are more frequent than insecticidal effects in the more than 2,000 bioactive plant species used to control pest arthropods [11].

EOs can also cause toxicity by contact or fumigation, but generally, they do not have a specific mode of action [2]. These compounds can cause cytotoxic effects due to their ability to damage cell membranes [8]. Against insects, commonly, the toxicity of essential oils and terpenoids may be related to neurotoxic effects and growth-regulating action [2, 12].

Many compounds present in essential oils, including terpenoids, are able to inhibit the main families of detoxifying enzymes in insects, and can be used in synergistic formulations with chemical insecticides in order to increase their ability to control pests, using lower doses of active ingredients with proven resistance evolution [13].

Due to these properties, so-called botanical insecticides can also be formulated by blending EO compounds considered synergistic with each other. In this case, both combinations of essential oils from distinct plants and enrichment of EOs and/or their fractions with compounds known to have proven synergistic effects can be considered.

Particularly in the USA, the development and marketing of insecticides with active ingredients obtained from EOs are facilitated and encouraged due to the relative rapidity of registration compared to conventional synthetic insecticides [14]. In fact, many formulations have been developed by American and European companies for the control of household and garden pests, and ectoparasitic mites in bees [15].

Considerable advances are still required for the large-scale production of EO-based insecticides. Limitations related to raw material production, registration, and especially adequacy of formulations to control different targets without negatively affecting host plants, are the main challenges to be overcome in the short term.

Some essential oils, although containing lower proportions of terpenoids, still cause phytotoxicity to plants [16].

## MECHANISMS OF INSECTICIDAL ACTION OF TERPENOIDS

In recent years, the use of essential oils obtained from aromatic plants as low-risk insecticides, has increased considerably due to consumer demands and market restrictions. The main plant families from which EOs can be extracted are Apiaceae, Asteraceae, Cupressaceae, Hypericaceae, Lamiaceae, Lauraceae, Myrtaceae, Pinaceae, Piperaceae, Rutaceae, Santalaceae, and Zingiberaceae [17].

Essential oils containing terpenoids as major compounds can exhibit insecticidal, repellent, and growth-regulating effects on various pest insects, effectively controlling pre- and post-harvest phytophagous species. They can also present a repellent effect on disease-causing pathogen vectors such as mosquitoes, household insects, and pests of ornamental plants. With few exceptions, their toxicity in mammals is low, with short persistence in the environment [18].

Few studies have evaluated in detail the toxicology of the major compounds of EOs. However, there is evidence of the negative effects of terpenoids on neurological processes in insects, making their use promising for insect pest control.

# Binding of GABA (Gamma-Aminobutyric Acid) Neurotransmitter to Receptors

One of the sites of action of the insecticidal activity of oxygenated monoterpenes such as carvacrol, pulegone, and thymol, is related to interaction with GABA (γ-aminobutyric acid), the main inhibitory neurotransmitter in the central and peripheral nervous systems of insects. Such compounds act as modulators of these receptors and increase the influx of chloride into neurons, leading the insect to death by inhibition of the nervous system [12, 19]. Phenolic hydroxyl, present in carvacrol and thymol, is considered one of the main structural components involved in the positive modulation of GABA receptors in insects [12]. Synthetic insecticides of cyclodienes and phenylpyrazoles are examples of commercial products that are used for pest control and have interference with GABA.

As a consequence of such terpenoid interactions, it has been suggested that thymol may interfere with the functioning of muscles responsible for the flight of blowflies *via* neuromuscular junctions, as well as in the central nervous system, by imitating the action of GABA, causing insect death [20].

## Binding to the Nicotinic Acetylcholine Receptor

Regarding interferences on the nicotinic acetylcholine receptor (a cholinergic agent that forms ion channels in the plasma membranes of certain neurons in the central nervous system of insects), it has been noted that in houseflies, carvacrol can bind to acetylcholine, acting as a modulator for this receptor [21]. However, its inhibition may not be a primary action of carvacrol because of the high doses required for such an effect [22]. Examples of chemical groups of commercial insecticides that interfere with nicotinic receptors include neonicotinoids, nicotine, sulfoxaflor, and nereistoxin analogues [23].

Although having no proven activity against arthropods, some monoterpenoids, such as borneol and camphor, have been shown to be non-competitive inhibitors of nicotinic acetylcholine receptors in mammals, indicating a potential effect on insects [24, 25].

## Inhibition of Transient Receptor Potential (TRP) Channels

Transient receptor potential (TRP) channels are defined as essential components of biological sensors. In insects, they are useful in sensing environmental changes in response to numerous stimuli (temperature variations, presence of chemical compounds, and mechanical stimuli). Terpenoids carvacrol, thymol, and menthol are all inhibitors of the transient receptor potential-like (TRPL) channel, a receptor responsible for the physiological perception of light in arthropods [26].

They also present the potential to act indirectly as insecticides by interfering with insect behavior.

The insecticidal activity related to the negative interference of terpenoids in the insect nervous system was best described for octopamine, an excitatory neurotransmitter present in the central and peripheral nervous systems of insects, and also for its immediate precursor, tyramine.

Both octopamine and tyramine occur in significant quantities in the nervous system of insects and crustaceans [27]. They present a wide spectrum of biological functions, acting as neurotransmitters, neurohormones, and neuromodulators [28, 29]. In fact, the involvement of octopamine in the regulation of heartbeat in insects [30], expression of luminescence [31], functioning of flight muscles [32] and respiratory system [33], cuticle formation, and functioning of pheromone glands [34] can be highlighted. On the other hand, tyramine is associated with interference in olfaction functions [35], as well as in the ovipositor system of insects [36].

## Activity on Octopamine and Tyramine Receptors

The octopaminergic system is considered a target for some isomeric phenolic terpenoids, such as thymol and carvacrol [37]. The partial insecticidal activity of α-terpineol is still considered, when compared to phenolic terpenoids, highlighting the importance of an electronegative group such as hydroxyls, in position 2 or 3 of the benzene ring, since they are necessary for expression of the complete toxic activity *via* interference on tyramide [37].

The toxic mode of action of these compounds is comparable to synthetic formamidine-based insecticides, which act as octopamine agonists by binding to their receptors and increasing the excited state of the target organism. Additionally, they act as behavioral disruptors, an important characteristic for the control of caterpillars, mites, and ticks [38].

## Inhibition of Detoxification Enzymes

The increased metabolic detoxification of insects is mainly due to the action of detoxification enzymes involved in phase I reactions (corresponding to processes with no synthesis), in which a reactive polar group (OH, SH, $NH_2$, COOH) is injected into a toxic molecule to make it more soluble in water. The most important effect is transforming the molecule into a substrate more suitable for phase II reactions (corresponding to biosyntheses or conjugations) [39, 40].

The main enzymatic groups involved in xenobiotic detoxification in insect are cytochrome P450-dependent monooxygenases (CYPs) and esterases (ESTs), which predominate in phase I reactions, and glutathione S-transferases (GSTs), which prevail in phase II reactions [41, 42].

The capacity of an organism to tolerate doses of xenobiotics that would be lethal to the majority of the population is called metabolic resistance [43].

Some essential oils present compounds that can inhibit enzyme groups responsible for insect detoxification. Therefore, theoretically, these EOs could be employed as insecticides in their integral form (the whole oil), rectified, or in combinations with other terpenoids.

It was possible to verify, so far, the occurrence of 37 species belonging to 13 of the main families of volatile terpenoid-producing plants, which show evidence of insecticidal effect by inhibition of detoxification enzymes (Table **1**).

**Table 1. Major compounds of essential oils from terpenoid-producing plants show evidence of insecticidal effect by inhibiting detoxification enzymes.**

| Species | Family | Major Compounds | Inhibited Enzymes | References |
|---|---|---|---|---|
| **(Popular names)** | | | | |
| *Agastache foeniculum* (Pursh) Kuntze (anise hyssop) | Lamiaceae | Estragol | GSTs and ESTs | [47, 88, 117]. |
| *Anethum graveolens* L. (dill) | Apiaceae | Carvone | ESTs | [21, 47, 84, 93, 97]. |
| *Aristolochia trilobata* L. (milhomem) | Aristolochiaceae | Sulcatyl acetate, limonene, *p*-cymene, and linalool | * | [105] |
| *Artemisia princeps* Pamp (Korean wormwood) | Aristolochiaceae | 1,8-Cineole | ESTs | [53, 64, 84, 109]. |
| *Calea serrata* Less. (snakeherb) | Asteraceae | Precocene II, germacrene D, and β-selinene | ESTs and GSTs | [108] |
| *Chenopodium ambrosioides* L. (Brazilian epazote) | Amaranthaceae | α-Terpinene, *p*-cymene, and ascaridole | ESTs and GSTs | [19, 72, 83, 87, 107, 116]. |
| Cinnamomum *camphora* (camphor) | Lauraceae | Linalool | P450, ESTs, and GSTs | [47, 53, 95, 109]. |
| *C. zeylanicum* (cinnamon) | Lauraceae | Cinnamaldehyde (FP), beta-caryophyllene, and linalool | ESTs | [13, 47, 95, 102, 103, 109, 111] |
| *Citrus aurantiifolia* (Christm.) Swingle (lime) | Rutaceae | Limonene and (Z)-β- ocimene | * | [71, 72, 84, 86, 87, 100, 110] |

*(Table 1) cont.....*

| Species | Family | Major Compounds | Inhibited Enzymes | References |
|---|---|---|---|---|
| *C. aurantium subsp. bergamia* (bergamot) | Rutaceae | Limonene | P450, ESTs, and GSTs | [71, 72, 84, 86, 87, 9, 3, 110] |
| *C. paradisi* (grapefruit) | Rutaceae | Limonene | P450, ESTs, and GSTs | [71, 72, 84, 86, 87, 93, 110] |
| *C. sinensis* (L.) (sweet orange) | Rutaceae | Limonene | * | [71, 72, 84, 86, 87, 99, 110, 112]. |
| *Cymbopogon citratus* (DC.) Stapf (lemongrass) | Poaceae | Citral, limonene, and geranyl acetate | * | [71, 72, 84, 86, 87, 99, 112, 114]. |
| *Corymbia citriodora* (Hook.) K.D. Hill & L.A.S. Johnson (lemon-scented gum) | Myrtaceae | Limonene and citronellal | * | [72, 71, 84, 86, 87, 96, 100, 110]. |
| *Eucalyptus globulus* Labill. (eucalyptus) | Myrtaceae | 1,8-Cineole | * | [64, 84, 100, 101, 109]. |
| *Geranium maculatum* L. (wild geranium) | Geraniaceae | Citronellol, geraniol, and citronellyl formate | * | [92] |
| *Helichrysum italicum* G. Donf. (immortelle) | Asteraceae | Neryl acetate | P450 | [21] |
| *Juniperus virginiana* L. (red cedar) | Cupressaceae | Thujopsene, α-cedrene, and cedrol | P450, GSTs, and ESTs | [21] |
| *J. ashei* (ashe juniper) | Cupressaceae | Thujopsene, α-cedrene, and cedrol | P450, GSTs, and ESTs | [13, 103, 104]. |
| *Lavandula angustifolia* Mill. (lavander) | Lamiaceae | Linalool | P450, ESTs, and GSTs | [47, 90, 94, 95, 109]. |
| *Lippia sidoides* Cham. (pepper-rosmarin) | Verbenaceae | Thymol and *p*-cymene | * | [113, 115]. |
| *Litsea cubeba* (Lour.) Pers. | Lauraceae | Geranial and neral | P450, ESTs, and GSTs | [13, 83, 87, 102, 110] |
| *Majorana hortensis* Moench (marjoram) | Lauraceae | Terpinen-4-ol and γ-terpinene | * | [103] |
| *Mentha arvensis* L. (wild mint) | Lamiaceae | Menthol and citronelal | * | [71], 72, 85, 87, 115]. |
| *M. piperita* (peppermint) | Lamiaceae | Menthol and menthyl acetate | * | [71, 72, 85]. |
| *M. pulegium* (squaw mint) | Lamiaceae | Pulegone, menthol, menthone, and isomethone | * | [71, 72, 91]. |
| *M. spicata* (garden mint) | Lamiaceae | Carvone, limonene, and (Z)-dihydrocarvone | * | [71, 72, 84, 86, 87, 91, 110]. |

*(Table 1) cont.....*

| Species | Family | Major Compounds | Inhibited Enzymes | References |
|---------|--------|-----------------|-------------------|------------|
| *Melaleuca alternifolia* Cheel (tea tree) | Myrtaceae | Terpin-4-ol and α-terpinene | * | [83, 87, 100, 115]. |
| *Ocimum basilicum* L. (basil) | Lamiaceae | Linalool | P450, ESTs, and GSTs | [13, 47, 95, 102, 103, 105]. |
| *Origanun* spp. (oregano) | Lamiaceae | Carvacrol | ESTs | [13, 102 - 104]. |
| *Pelargonium graveolens* L'Hér. ex Aiton. (geranium) | Geraniaceae | Citronellol, citronellyl formate, and 10-epi-γ-eudesmol | GSTs, P450, and ESTs | [13, 102, 103]. |
| *P. nigrum* (black pepper) | Piperaceae | β-caryophyllene, limonene, and β-pinene | P450, ESTs, and GSTs | [21, 69, 71, 72, 84, 86, 87, 98, 99, 110, 112]. |
| *Pogostemon* spp. Desf. (patchouli) | Lamiaceae | Patchoulol, δ-guayene, and α-guayene | P450 | [102 - 104]. |
| *Rosmarinus officinalis* L. (rosemary) | Lamiaceae | 1,8-Cineole and camphor | * | [64, 68, 84, 87, 109]. |
| *Salvia officinalis* L. (common sage) | Lamiaceae | α-Thujone, camphor, and 1,8-cineole | * | [64, 68, 84, 87, 106, 109]. |
| *Thymus vulgaris* L. (thyme) | Lamiaceae | Thymol | P450, ESTs, and GSTs | [89, 90, 115, 117]. |

Inhibited enzymes: ESTs - esterases; P450s - cytochrome P450-dependent monooxygenases; GSTs - glutathione S-transferases. *Synergistic effect observed without specification of the enzymes involved.

The family Lamiaceae represents 43% of the total species evaluated (16), followed by the families Rutaceae with 4 species; Myrtaceae with 3; Aristolochiaceae, Asteraceae, Cupressaceae, and Geraniaceae with 2; and Amaranthaceae, Apiaceae, Piperaceae, Poaceae, Verbenaceae, and Zingiberaceae, each one with 1 species.

There are controversies in defining a percentage value of a compound for it to be considered as the majority in the composition of essential oil. A more rigid definition considers as a majority of those compounds are present in reasonably high concentrations (20 to 70%). Thereafter, the essential oil is characterized by two or three major components [44]. However, in practice, contents greater than 10% already qualify a substance as the majority, increasing the possibility that up to five of them can be considered as such [45].

Based on evaluations (Table **1**) involving essential oils and isolated terpenoids, it can be noted that they may occur in several species of plants of the same family and, therefore, are considered typical of a certain botanical family (chemiosystematic markers) as for example, thymol for Lamiaceae of the genus

Thymus. However, the same compounds can be found in different families, such as linalool, an important constituent in both Lauraceae and Lamiaceae (Table **1**).

From Table **1**, it can also be inferred that monoterpenoids β-pinene, limonene, thujopsene, linalool, thymol, geranial, neral, citronellol, and citronellyl formate, as well as sesquiterpenoids β-caryophyllene, α-cedrene, cedrol, and 10-epi-γ-eudesmol are considered to have the greatest possibility of being used as an insecticide because they inhibit, simultaneously, the three main families of detoxification enzymes. However, due to their occurrence in small quantities, together with the high cost of their obtaining in pure form, their use as insecticides is unfeasible. Thus, only a few terpenoids are used in commercial insecticide formulations.

## INSECTICIDE FORMULATIONS USING TERPENOIDS

### Synergistic Interactions Between Terpenoids for Insecticide Formulation

In natural environments, polyphagous herbivores face complex mixtures of allelochemicals produced by host plants. Also, the toxicity of a single compound can be influenced by the chemical diversity in which the mixture is found [46].

Synergistic effects of certain terpenoids have been observed in previously published studies. For example, an increase in the inhibitory activity of acetylcholinesterase enzyme was noted, *in vitro*, when linalool and estragole were added, whereas a reversible inhibition was observed after the addition of other terpenoids, such as camphor, geraniol, carvone, fenchone, and terpinene [47]. The concentration of these compounds also influences the inhibitory activity, since fenchone, carvone, and linalool require higher concentrations than 1,8-cineole to inhibit the same enzyme group.

The use of a single active terpenoid isolated from essential oil, with only one mechanism of action, may result in the rapid selection of insect populations with metabolic resistance, similar to what occurs for other synthetic insecticides [48, 49]. However, the fact that EOs are complex mixtures of active compounds, mainly terpenoids, with different mechanisms of action, prevents or delays the process of resistance evolution, which can be pointed out as a great advantage [18, 50, 51].

Although the bioactivity of essential oils is generally attributed to a few major compounds, a synergistic interaction between their components can result in a higher level of activity compared to the components used alone [5, 52]. Studies have shown that binary mixtures of terpenoids have greater insecticidal potential when compared to the same purified compounds [53, 54].

Even before the extraction of terpenoids, the synergistic effects of complex mixtures are believed to be important in plant defenses against herbivory. Plants usually present defenses resulting from a set of compounds and, therefore, it is believed that the minor constituents can act as synergists, increasing the effect of the major compounds through a variety of mechanisms. It is reported that essential oils made up of terpenoid mixtures are considerably more effective than single compounds. The identification of such synergistic terpenoids may allow the development of more effective products. It should be noted, however, that terpenoids isolated from essential oils can exhibit both synergistic and antagonistic effects [5].

Even before the extraction of terpenoids, synergistic effects of complex mixtures are believed to be important in plant defenses. Interactions among the constituents, especially terpenoids, can have a significant interference in the overall biological efficacy of essential oils. Thus, new essential oil-based (bio) insecticides, whose formulations are standardized mixtures with known efficacy, may be developed [5, 55, 56].

Using binary combinations of terpenoids, interactions among 22 compounds present in essential oils, with proven acute toxicity to lepidopteran and dipteran larvae, were evaluated (Table **2**). At least ten synergistic binary interactions were found among such compounds considering both biological targets [5, 44, 56].

**Table 2a. Synergistic, antagonistic, and additive interactions between terpenoids present in essential oils. Results against *Culex quinquafasciatus* larvae and *Spodoptera littoralis* and *Spodoptera litura* caterpillars.**

| Terpenoid | 1,8-Cineole | Borneol | Canphene | Camphor | Carvone | Carvracrol | Citronellal | Estragole | Limonene | Linalool | Menthone |
|---|---|---|---|---|---|---|---|---|---|---|---|
| 1,8-Cineole | - | S | A/S | S | - | S | I/S | I/S | I/S | S | S/A |
| Borneol | S | - | S | S | S/I | S | S | S | S | S | S/I |
| Camphene | A/S | S | - | S | S/I | S | A/S | - | I/S | I/S | - |
| Camphor | S | S | S | - | S/I | S | I/S | - | S | S | - |
| Carvone | - | S/I | S/I | S/I | - | S/A | A/I | S/A | S | - | S/A |
| Carvracrol | S | S | S | S | S/A | - | I/S | S | S | S | S/A |
| Citronellal | I/S | S | A/S | I/S | A/I | I/S | - | S | - | - | S/I |
| Estragole | I/S | S | - | - | S/A | S | S | - | S | I | - |
| Limonene | I/S | S | I/S | S | S | S | - | S | - | S | S |
| Linalool | S | S | I/S | S | - | S | - | I | S | - | S/I |
| Menthone | S/A | S/I | - | - | S/A | S/A | S/I | - | S | S/I | - |
| Myrcene | S/I | S | - | - | S/A | S/A | A/S | - | S | S/I | S |
| p-Cimene | S | S | A/I | I/S | - | S | - | S | - | - | S |
| Terpinolene | A | S | - | - | S/A | S | A/S | I | S/I | I/S | S/A |
| Thymol | S | S | S | S | S/A | s | I/S/AD | S | I/S | A/S | S/A |
| (E)-anethole | I/S | S | S | - | S/A | S | I/S | S | S | I/S | S/I |
| α-Pinene | I/S | S | S | S/I | A | S/A | - | S/I | A/S | S | S/I |

*(Table 2a) cont.....*

| Terpenoid | 1,8-Cineole | Borneol | Canphene | Camphor | Carvone | Carvracrol | Citronellal | Estragole | Limonene | Linalool | Menthone |
|---|---|---|---|---|---|---|---|---|---|---|---|
| α-Terpinene | A/S | S | I/S | S | S/A | S/A | A/S | S/I | S | A/S | S/A |
| α-Terpineol | S/I | S | - | - | S/A | S | I/S | - | S | S/I | S |
| β-Citronellol | S | S | S | S | S/A | S/I | S | S/I | A/S | S | S/A |
| β-Pinene | S/I | S/A | A/I | S/I | S/A | - | A/I | S/I | S | A/I | S/I |
| γ-Terpinene | S | I/S | A/S | A/S | - | S/I | - | S | I/S | I/S | S |

**Table 2b. Synergistic, antagonistic, and additive interactions between terpenoids present in essential oils. Results against *Culex quinquafasciatus* larvae and *Spodoptera littoralis* and *Spodoptera litura* caterpillars.**

| Myrcene | p-Cimene | Terpinolene | Thymol | (E)-anethole | α-Pinene | α-Terpinene | α-Terpineol | β-Citronellol | β-Pinene | γ-Terpinene |
|---|---|---|---|---|---|---|---|---|---|---|---|
| S/I | S | A | S | I/S | I/S | A/S | S/I | S | S/I | S |
| S | S | S | S | S | S | S | S | S | S/A | I/S |
| - | A/I | - | S | S | S | I/S | - | S | A/I | A/S |
| - | I/S | - | S | - | S/I | S | - | S | S/I | A/S |
| S/A | - | S/A | S/A | S/A | A | S/A | S/A | S/A | S/A | - |
| S/A | S | S | S | S | S/A | - | S | S/I | - | S/I |
| A/S | - | A/S | I/S/AD | I/S | - | A/S | I/S | S | A/I | - |
| - | S | I | S | S | S/I | S/I | - | S/I | S/I | S |
| S | - | S/I | I/S | S | A/S | S | S | A/S | S | I/S |
| S/I | - | I/S | A/S | I/S | S | A/S | S/I | S | A/I | I/S |
| S | S | S/A | S/A | S/I | S/I | S/A | S | S/A | S/I | S |
| - | I/S | I/A | S/A | S/I | A | S/I | I/S | I | I A | I/S |
| I/S | - | S | S | S | A/I | S | I/S | I/S | I | - |
| - | S | - | I | S | A | S/A | I/A | S/I | A/I | S |
| S/A | S | I | - | S | S/I | A/I | S/I/AD | S | I | S |
| - | S | S | S | - | S | S | S | S | I | S |
| A | A/I | A | S/I | S | - | A | S/I | A/I | A | A/S |
| - | S | S/A | A/I | S | A | - | S/A | I | - | S |
| - | I/S | I/A | S/I/AD | S | S/I | S/A | - | S/I | I/S | I/S |
| I | I/S | S/I | S | S | A/I | I | S/I | - | I | A/S |
| I A | I | A/I | I | I | A | - | I/S | I | - | I |
| I/S | - | S | S | S | A/S | S | I/S | A/S | I | - |

S: synergism; A: antagonism; I: ineffective or no effect: AD: additivism. Based on the works by Hummelbrunner and Isman (2001) [5] and Pavela (2014; 2015 a) [56, 45]. The first capital letter is for compound interaction for *Culex quinquafasciatus* larvae, and the second capital letter is for compound interaction for *Spodoptera littoralis* and *S. litura*. Single letters are for classification for both species.

It should also be noted that borneol and camphor, when used individually, caused low mortality even at the highest concentrations tested. In synergistic combinations with each other, they also caused low mortality at a 1:2 ratio (borneol/camphor), while at 2:1 and 1:1 ratios, a significant increase in efficacy was observed, expressed by the values of estimated median lethal dose ($LD_{50}$) [56]. Such a result shows the synergistic interaction between these two terpenoids, observed in Table **2**.

Borneol and camphor, along with two arylpropanoids estragole and *trans-*anethole, were the ones that presented only one antagonistic interaction when combined with the other 21 compounds, demonstrating the potential of being used in insecticide combinations. This occurs mainly because borneol presented a synergistic effect on all compounds evaluated, and camphor on 14 of them (Table **2**).

Still, regarding synergistic interactions in combinations with camphor, an increase in the efficacy of this compound was observed when associated with 1,8-cineole. This compound promoted an increase in its penetration in the cuticle of *Trichoplusia ni* larvae (Hübner, 1803) (Lepidoptera: Noctuidae) [57]. Probably, 1,8-cineole not only interacted with the wax layer of the insect integument but also with camphor [57].

The increase in penetration of xenobiotics for the control of pest insects is of fundamental importance since, during environmental variations, there might be interference in the allele frequencies of genes involved in insecticide resistance. These genes are linked to other insect characteristics, such as resistance to dehydration and greater tolerance to thermal changes [58 - 60].

As insects adapt to drier environments (as a result of climate changes), variations in the composition and amount of their waxy layer should also occur, which could lead to decreased cuticular permeability, and consequent reduction or loss of insecticide effectiveness. As the increase in production of the waxy layer in insects is related to resistance to dehydration, probably selecting this characteristic would also include insects with such resistance mechanisms [60, 61].

The use of insecticides with different action mechanisms, associated with biological and cultural control can also be pointed out as an important strategy for resistance management in the face of climate change [60, 62].

Based on this assumption, terpenoids such as 1,8-cineole have the potential to be used in insecticide formulations, allowing an increase in the effectiveness of xenobiotics by the ease of penetration.

The total of 19 synergistic interactions of terpenoids carvacrol, limonene, and 1,8-cineole, along with arylpropanoid anethole, can be considered a high value. In addition, only one antagonistic interaction was observed, except for carvacrol which antagonized five different terpenoids (Table **2**).

Terpenoids were noticed to have a high number of synergistic interactions also associated with antagonistic ones, depending on the combination considered. Carvone showed 15 synergistic and 13 antagonistic interactions, similar to β-pinene, which presented 8 and 7 interactions of such types, respectively. These results reveal the need for further studies on the use of terpenoid combinations with insecticidal purposes.

A combined analysis of the results shows that 70.2% of the interactive effects observed were synergistic, 21.7% were antagonistic, whereas in 30.4% of the cases, no significant interaction occurred. It should be noted that some terpenoids, when used in different combinations, may present different interactive responses depending on the target evaluated, confirming the observations previously made [63].

Table **2** shows the effect inversion, from synergistic to antagonistic and *vice versa*, of carvone and α-terpinene on 11 and 7 compounds, respectively. It depended exclusively on whether the target was Diptera or Lepidoptera larvae. Thus, the order of the insect to be controlled, when a formulation with such compounds is made, must be considered. Effect inversions could be observed in at least one interaction, except for *p*-cymene and camphor.

When the best synergistic combinations were selected (based on χ2 values of concentration/dose-response curves), it was concluded that camphor, borncol, and limonene were the compounds that showed significant synergistic interactions when combined with the other constituents [56].

Additional evaluations involving other compounds, mainly sesquiterpenes and monoterpenes, indicated synergistic and/or additive effects on most interactions (Table **3**) [5, 63 - 66]. Therefore, it can be concluded that the ability to promote synergistic effects in combinations of terpenoids present in essential oils depends on: i) the respective chemical structures, ii) the type of combination, and iii) the appropriate proportion of substances in the mixture. Thus, apparently inactive compounds may play an important role in complex mixtures, similar to compounds biosynthesized in plant metabolism in order to provide plant defense mechanisms against herbivorous insects. By accepting this hypothesis, the same principle could be used in the development of new insecticide formulations.

**Table 3. Other terpenoids present in essential oils present interactions with each other. Results compiled from the works by Hummelbrunner e Isman (2001) [5], Savelev *et al.* (2003) [64], Attia *et al.* (2011) [65], Chaubey (2012) [66], and Chang *et al.* (2014).**

| Compound | 1,8-Cineole | Citronellal | Terpinolene | Thymol | (*E*)-Anethole | α-Humulene | α-Pinene | α-Terpinene | β-Caryophyllene |
|---|---|---|---|---|---|---|---|---|---|
| **Monoterpene** | | | | | | | | | |
| *Trans*-anethole | - | S | S | S | - | - | - | S | - |

*(Table 3) cont.....*

| Compound | 1,8-Cineole | Citronellal | Terpinolene | Thymol | (E)-Anethole | α-Humulene | α-Pinene | α-Terpinene | β-Caryophyllene |
|---|---|---|---|---|---|---|---|---|---|
| **Sesquiterpene** | | | | | | | | | |
| (±)-*Trans*-nerolidol | - | - | - | - | | S | S | - | S |
| Caryophyllene oxide | S | - | - | - | - | - | - | - | - |
| β-Caryophyllene | - | - | - | - | - | - | S | - | S |

A: antagonism; AD: additivism; S: synergism; - no information.

## Production of Blends From Synergistic Terpenoids Present in Essential Oils and Development of Commercial Products

Effective results of essential oil blends or even artificial combinations of some of their isolated constituents have highlighted the importance of the absence and presence of certain major and/or minor components. Artificial terpene mixtures (blends) have been developed in recent years and serve as a reference for the development of new insecticide formulations, including blends of synergistic terpenoids.

The interactions among the main components of the essential oil of *R. officinalis* (rosemary) were evaluated regarding toxicity to spider mite, *Tetranychus urticae* Koch (Acari: Tetranychidae) [67]. Based on the results, the authors concluded that when 2 of the 3 major terpenoids (1,8-cineole or α-pinene) were absent, there was a significant decrease in the toxicity of the mixture against spider mite. The removal of minority terpenoids such as *p*-cymene, α-terpineol, or bornyl acetate also led to a significant reduction in toxicity, but much less than that caused by the absence of the majority compounds. Considering that *p*-cymene, when individually evaluated, presented non-significant toxicity, it can be concluded that this type of minority constituent presents a synergistic effect on the majority compounds and, although it is not active alone, its presence is necessary to promote total toxicity against *T. urticae*.

Taking as reference the interaction results presented in Table **2**, it can be noted that the effect of 1,8-cineole on all compounds (camphene, β-pinene, camphor, *p*-cymene, borneol, limonene, α-terpineol, and α-pinene) was synergistic. The same result was not observed for α-pinene, showed an antagonistic effect on β-pinene and *p*-cymene, and synergistic effect in combination with the other compounds.

When the target insect was *T. ni*, in bioassays that evaluated the removal of compounds from the original essential oil, the response indicated 1,8-cineole as the main insecticidal component of rosemary oil (*Rosmarinus officinalis*) [68]. Using its LD95 as a reference of insecticidal efficacy, there was a significant decrease in lethality when 1,8-cineole was eliminated from the synthetic blend.

Regarding the fumigation assay, the most impactful effects of removals or absences on the toxicity of the artificial blends were those of 1,8-cineole and camphor, demonstrating that these two compounds were more active than all other constituents [68]. This indicated that such components should not be removed from an insecticide formulation based on this essential oil, as they were the most important for the expression of overall toxicity. On the other hand, α - terpineol, besides having the lowest $LD_{50}$ value, presented relatively low content in the oil (1.7%), and its elimination from the mixture did not lead to decreased lethality. The lack of interaction of this compound in combinations with other terpenoids could be noted by the high number of terpene blends with no effect (11) (Table 2). This suggests that the contribution of this constituent to the total toxicity is limited, and thus, α-terpineol can be removed in the standardization of certain blends.

The authors [68] also evaluated interactions among 4 major constituents of *R. officinalis* oil, testing blends or combinations of 2, 3 or 4 compounds. The most significant synergy was in the blend of camphor + α-pinene + camphene. A mixture with the 10 most abundant compounds, corresponding to 98.5% of the total composition of the essential oil of *Deverra scoparia* Coss. & Durieu (Apiaceae), was evaluated for the control of *T. urticae*. This blend also included myrcene, which corresponded to 3.5% of the total oil. Even with the low concentration of myrcene in the mixture, its presence allowed reaching a maximum mite lethality rate (83%), suggesting that this minority constituent presented a significant contribution to the overall toxicity of the oil, possibly by synergistic effect [65].

Another interesting result of this study was related to the artificial combination of the major terpenoids (α-pinene, 3-carene, and terpinen-4-ol) of *D. scoparia* essential oil, responsible for 65% of the lethality of *T. urticae* nymphs, a value equivalent to the that caused by the individual application of each of these compounds. Thus, these active constituents would have no additional effect on each other, because their level of toxicity was similar when tested alone. Furthermore, the removal of α-pinene from the mixture caused a significant decrease in toxicity for *T. urticae* (about 80%), which led to the conclusion that this compound is the most important for the oil toxicity. However, in combination with α-pinene, 43% of the compounds evaluated in other studies (Table 2) showed antagonistic effects, depending on the target insect evaluated, which suggests expressive ambiguity or uncertainty of the action of α-pinene in combinations. Therefore, the toxicity effects are species-specific, and it is necessary to test pure oils or terpenoids on different species that are desired to control.

Yeom *et al.* (2012) [69] evaluated 11 essential oils from plants of the Apiaceae family: *Anethum graveolens* L. (dill), *Carum carvi* L. (caraway), and *Cuminum cyminum* L. (cumin), as effective fumigants for the control of *Blatella germanica* (L.) (Blattaria: Ectobiidae) [69]. Artificial blends of these oils were produced, and the absence of carvone from *A. graveolens* and *C. carvi* essential oils caused a significant reduction in their toxicity. These results indicated that carvone is the main fumigant agent of dill and caraway oils. Coincidentally, carvone is the majority compound of these 2 essential oils (with relative compositions of 41 and 56%, respectively). However, for *A. graveolens,* the combined elimination of carvone and limonene negatively influenced the toxicity for this pest species, suggesting synergy between them and corroborating the hypothesis raised and discussed previously (Table **2**).

Evidence of synergistic interaction between compounds can also be observed in the essential oil of *C. cyminum*, since the absence of *p*-cymene, γ-terpinene, and cuminaldehyde in the artificial blend caused a significant decrease in toxicity on *B. germanica* [69]. The hypothesis is that a synergistic effect occurs between *p*-cymene and γ-terpinene (Table **2**), in addition to interaction of each compound with cuminaldehyde.

The complexity of interactions between compounds increases when the chemical profile of the essential oil is diversified (regarding chemical groups). Evaluations of the toxicity of compounds present in the essential oil of 2 species of *Ocotea glomerata* (Nees) Mez (Lauraceae) against *T. Urticae* were performed [70]. The majority compounds, in this case, were sequiterpenes aromadendrene and β-caryophyllene, composing the blends with presence and absence of the terpenoids α-pinene, β-pinene, *p*-cymene, limonene, terpinolene, terpinen-4-ol, α-terpineol, and α-humulene.

The authors concluded that the removal of α-pinene did not affect the toxicity of the artificial blend, and the residual toxicity of the oil was not reduced by the removal of limonene or terpinen-4-ol. However, removing aromadendrene led to a drastic reduction in the lethality of exposed mites, indicating that this sesquiterpene was the one that most contributed to the toxicity of the artificial blend made from *O. glomerata* essential oil, followed by β-caryophyllene and *p*-cymene.

The individual relative toxicities of the constituents showed that aromadendrene (major compound) exhibited the same level of toxicity as *p*-cymene, regardless of the type of exposure (fumigation or residual contact), whereas the second most abundant constituent, β-caryophyllene, was 170- and 50-fold more toxic than aromadendrene in the residual contact and fumigation bioassays, respectively. These results suggested that the contribution of a chemical constituent in a complex blend is not predictable, based solely on its individual toxicity or the proportion in which it is found in the blend.

As essential oils obtained from plants vary chemically based on numerous biotic and abiotic factors, the chemical profile of these oils may not be uniform. To overcome such limitations, essential oils from different sources can be intentionally blended to obtain a composition with fewer variations [68]. Thus, the importance of understanding the contributions of individual constituents and their combinations for the toxicity of the resulting derivatives is emphasized, enabling the use of this information in the standardization and quality control of commercial formulations.

Some evaluations have made important contributions to the understanding of the mechanisms involved in increased cell penetration favored by synergistic compounds, demonstrated by reduced surface tension, increased solubility, and interactions between the synergistic substance and the lipid layer of the insect cuticle [57]. Knowing the role of each constituent and its interactions in the composition of essential oil is a prime factor for the evaluation of lethal and sublethal toxicity on a certain arthropod species, which allows selecting and obtaining compounds for application in agriculture, and/or in the formulation of new botanical insecticides/acaricides.

In practice, the limiting factor for the formulation of commercial products, based on this knowledge, is the cost of the purified compounds. Therefore, the use of essential oil blends containing major active compounds in expressive concentrations is the most economically feasible. If it is not possible to naturally obtain these essential oils with the desirable characteristics, purification or fractionation processes can be performed. It is important to point out that in these processes, it is possible to eliminate compounds considered antagonists, remaining only those that add synergistic characteristics to the blend or to part of it.

It is also worth mentioning that blends of compounds often act as inhibitors of detoxification enzymes, enabling their use as insecticides and synergists of commercial insecticides. In this case, as the amount of product needed to be added into a blend is smaller, there is the possibility of using less purified compounds.

## Essential Oil-Based Products with High Levels of Terpenoids

Many commercial products, for instance, Buzz Away®, contain a blend of oils from citronella (*Cymbopogon nardus* (L.)), cedarwood (*Juniperus cedrus* Webb & Berthel), eucalyptus (*Eucalyptus globulus* Labill.), and lemon balm (*Melissa officinalis* L.), whereas Green Ban® presents oil blends from citronella (*C.*

*nardus*), Melaleuca spp, lavender (*Lavandula angustifolia* Mill.), sassafras (*Ocotea odorifera* (Vellozo) Rohwer) without safrole, peppermint (*Mentha piperita*), and bergamot oil (*Citrus aurantium* subsp. bergamia (Risso & Poit.) Wight & Arn. ex Engl.) without bergapten. Similarly, American and European companies have developed many other formulations, which contain mint oil for home and garden pest control, or menthol for control of ectoparasitic bee mites [15].

In Italy, formulations such as ApilifeVar®, containing menthol and smaller amounts of cineole and camphor, have been developed to control ectoparasitic bee mites. Studies have been conducted to develop new formulations and new technologies for the storage and application of these products, especially in laboratories in Canada, France, and Israel. However, factors such as scarcity of raw materials, the need for chemical standardization and quality control, and the difficulties of registration in many countries have hindered the commercialization of these pesticides [15].

In the 2000s, essential oil-based pesticides, rich in terpenoids, were introduced into the North American market by a company named *EcoSmart*. Several botanical insecticides were available for many purposes, using mainly essential oils of rosemary (*R. officinalis*), peppermint (*Mentha piperita*), citronella (*C. nardus*), and thyme (*T. vulgaris*) [68].

Table **4** compiles information on commercially available products intended for pest arthropod control. In addition to the products marketed by *EcoSmart*, other companies can be pointed out in this field, such as *Arysta LifeScience*, *Brandt*, *Key Plex*, and *Zoecon*. Of the 22 products listed, 68% have in their formulation R. *officinalis* essential oil (rosemary), 50% *Mentha piperita* (mint), and 27% of them have *T. vulgaris* (thyme).

**Table 4. Main commercial products are formulated from terpenoids and essential oils, and their applications.**

| Commercial Products (Manufacturers) | Composition of Formulations Containing Essential Oils | Applications |
|---|---|---|
| Akabrown (Green Corp Biorganiks) | Cinnamon (1.5%), clove (1%), mint (1%), and oregano (0.5%) | Acaricide for agricultural pests |
| Ant & Roach Killer (EcoSmart) | Rosemary (1.5%) and mint (1.5%) | Insecticide for urban pests (cockroaches and ants) |
| ApiLife Var (Chemicals LAIF) | Thymol (74.1%), eucalyptus (cineole type) (16%), and L-menthol (3.73%) | Acaricide for pests associated with bees |

*(Table 4) cont.....*

| Commercial Products (Manufacturers) | Composition of Formulations Containing Essential Oils | Applications |
|---|---|---|
| Natural Insect Repellent (Babyganics) | Rosemary (1.5%), citronella (0.95%), geranium (0.75%), juniper (0.7%), mint (0.6%), and lemongrass (0.25%) | Repellent for human use (mosquitoes) |
| Bed Bug Killer (EcoSmart) | Rosemary (1.5%) and mint (1.5%) | Insecticide for urban pests (bed bugs) |
| Bed Bug Repellent (EcoSmart) | Rosemary (1.5%) and mint (1.5%) | Repellent for human use (bed bugs) |
| BioBlock (Homs) | Juniper (1.5%) and lemongrass (0.5%) | Repellent and broad-spectrum insecticide for urban pests |
| Biomite (Arysta LifeScience) | Citronellol (0.42%), geraniol (0.42%), nerolidol (0.42%), and farnesol (0.17%) | Acaricide for agricultural and garden pests |
| Bite Blocker Xtreme (Homs) | Geranium (6%) | Repellent for human use (mosquitoes) |
| Botanical Insect Control (EcoShield) | Juniper (2%) | Insecticide for agricultural, veterinary, and garden pests |
| Ebioluzion Plus vO (Green Corp Biorganiks) | Clove (2%), cinnamon (1.24%), and black pepper (1.24%) | Broad-spectrum insecticide for agricultural pests |
| Eco PCO WP- X (Zoecon) | Mint (5%) and permethrin (0.5%) | Broad-spectrum insecticide for urban and garden pests |
| Eco-oil (Organic Crops Protection) | Fractions of *Melaleuca* sp. and *Eucalyptus* sp. (2%) | Acaricide and insecticide for agricultural pests |
| Ecotec Plus (Brandt) | Rosemary (10%), geraniol (5%), and mint (2%) | Acaricide and insecticide for agricultural and garden pests |
| Ecotrol G (Key Plex) | Thyme (2.7%), clove (2%), and cinnamon (1%) | Insecticide for agricultural pests |
| Ecotrol Plus (Key Plex) | Rosemary (10%), geraniol (5%), and mint (2%) | Broad-spectrum acaricide and insecticide for agricultural and garden pests |
| Equine & Livestock Fly Spray (Homs) | Geranium (8%), juniper (1%), and lemongrass (0.5%) | Insecticide for veterinary pests (horses and cattle) |
| Essentria All Purpose (Zoecon) | Rosemary (10%) and mint (2%) | Broad-spectrum insecticide for agricultural and veterinary pests |
| Essentria G (Zoecon) | Eugenol (2.9%) and thyme (0.6%) | Broad-spectrum insecticide for urban pests |
| Essentria IC 3 (Zoecon) | Rosemary (10%) and geraniol (5%) | Broad-spectrum insecticide for urban pests |
| Essentria Wasp & Hornet (Zoecon) | Rosemary (1.5%) | Broad-spectrum insecticide for urban pests |

*(Table 4) cont.....*

| Commercial Products (Manufacturers) | Composition of Formulations Containing Essential Oils | Applications |
|---|---|---|
| Flying InsectKiller (EcoSmart) | Rosemary (1%) and geraniol (0.5%) | Insecticide for urban pests (mosquitoes and moths) |
| Garden Insect (EcoSmart) | Rosemary (0.5%) and mint (0.5%) | Insecticide and acaricide for garden pests |
| Herbal Insect Repellent (Burst's Bees) | Rosemary (3.77%), lemongrass (2.83%), juniper (0.94%), mint (0.76%), citronella (0.57%), cinnamon (0.38%), and geranium (0.19%) | Repellent for human use (mosquitoes) |
| Home Pest Control (EcoSmart) | Rosemary (0.5%), clove (0.5%), and thyme (0.25%) | Insecticide and arachnicide for urban pests |
| Home Pest Control (Orange Guard) | *D*-Limonene (5.8%) | Repellent and broad-spectrum insecticide for urban and garden pests |
| Insect Killer (EcoSmart) | Clove (0.5%) and thyme (0.25%) | Insecticide for urban pests |
| Insect Repellent (EcoSmart) | Geraniol (1%), rosemary (0.5%), cinnamon (0.5%), and citronella (0.5%) | Repellent for human use (mosquitoes and ticks) |
| Lawn Insect Killer (EcoSmart) | Clove (1%) and thyme (1%) | Insecticide for urban and garden pests (ants and cockroaches) |
| Mosquito & Tick Control (EcoSmart) | Mint (1.5%), rosemary (1%), clove (0.7%), and thyme (0.5%) | Insecticide, acaricide, and arachnicide for garden pests |
| Mosquito Repellent Spray (Kinven) | Mint (1.6%), citronella (0.6%), geranium (1.1%), rosemary (0.6%), and cinnamon (1.1%) | Repellent for human use (mosquitoes) |
| Mosquito Repellent Incense (Murphy's) | Rosemary (4.5%), mint (2%), citronella (1.5%), lemongrass (1%), and juniper (1%) | Repellent for human use (mosquitoes) |
| Mosquito Fogger (EcoSmart) | Geraniol (3%), rosemary (2%), and mint (0.4%) | Insecticide and fumigant repellent for urban pests (mosquitoes and flies) |
| PetFresh Insect Repellent (Homs) | Geranium (6%), juniper (1%), and lemongrass (0.5%) | Insect repellent for use on dogs and horses |
| Plant Based Lemon Eucalyptus Insect Repellent (Repel) | Citriodiol (19.5%) | Repellent for human use (mosquitoes) |
| Prev-Am (Oro Agri) | Orange (5-6%) | Acaricide and broad-spectrum insecticide for agricultural pests |
| Requiem (Bayer) | α-Terpinene (59.7%), *p*-cimene (22.4%), and *D*-limonene (17.9%) | Acaricide and insecticide for agricultural pests |
| Spider Blaster (EcoSmart) | Rosemary (5%) | Arachnicide for urban pests (spiders) |

(Table 4) cont.....

| Commercial Products (Manufacturers) | Composition of Formulations Containing Essential Oils | Applications |
|---|---|---|
| Sporan EC2 (Keyplex) | Rosemary (16%), cinnamon (10%), thyme (10%), and mint (2%) | Fungicide |
| TetraCURB (Kemin) | Rosemary (50%), clove (3%), and mint (1.95%) | Acaricide for agricultural pests |
| Wasp & Hornet (EcoSmart) | Mint (0.4%) | Insecticides and fumigant repellent for urban pests (wasps) |

Rosemary: EO from *Rosmarinus officinalis* L.; *Cinnamon* : EO from *Cinnamomum zeylanicum* Blume; *Lemongrass* : EO from Cymbopogon citratus (DC.) Stapf.; Clove: EO from *Syzygium aromaticum* (L.) Merr. & L.M. Perry; *Citronella* : EO from *Cymbopogon nardus* (L.) Rendle; Juniper: EO from *Juniperus virginiana* L.; Orange: EO from *Citrus sinensis* (L.) Osbeck; Mint: EO from *Mentha piperita* ; *Geranium* : EO from *Pelargonium graveolens* L'Hér. ex Aiton; Oregano: EO from *Origanum* spp.; Black pepper: EO from *Piper nigrum* L.; Thyme: EO from *Thymus vulgaris* L.

Except for Biomate® (*Arysta Lifescience*), the products are combinations of 2 or more essential oils that have as major compounds terpenoids that also present synergistic properties, as exemplified in Table **2**. When the pesticide is formulated with only 1 essential oil, it is usually *R. officinalis*, which has as major components 1,8-cineole (32%), α-pinene (18%), and camphor (20%), proven to be toxic to arthropods and synergistic among each other.

Among the products listed in Table **4**, 8 of them presented at least 1 essential oil and 1 isolated compound, usually geraniol. It is also noteworthy that the percentage of essential oils was low, not exceeding 10%, while the isolated compounds did not exceed 5% of the total composition. However, the cost of production should influence the mixture components, since isolated compounds are more expensive than crude essential oils.

In the formulation of pyrethrin-based insecticide Eco PCO WP-X®, menthol and menthone, the main terpenoids of mint essential oil, are considered activators of cytochrome P450-dependent monooxygenase enzymes and, at the same time, inhibitors of acetylcholinesterase [71, 72]. In this case, it was observed that the synergistic property of the essential oil was crucial for the toxic effect of the insecticide. In addition, the ratio of synergist (mint oil) to insecticidal active ingredient pyrethrin was 10:1 (v/v), similar to that used in chemical insecticides with synthetic piperonyl butoxide (PBO). With the prohibition of PBO for organic crops in Europe [73 - 75], the use of natural synergists in the form of isolated compounds or synergistic blends could be an option in the formulation of pyrethrum-based products.

The most positive aspect of using essential oils and/or their terpenoids for pest control is their low toxicity to mammals, since many of them are commonly obtained from culinary herbs and spices. These new pesticides are usually a combination of small amounts of essential oils from 2 to 4 different plant species.

Some pesticides promote the death of insects and arachnids while others repel them. The marketing of these products is diversified, not limited to domestic use for garden pest control, as seen in Table **4**. In addition, many commercial products use essential oils that are in the FDA (Food and Drug Administration) list of GRAS (Generally Recognized as Safe) regulated by the Environmental Protection Agency (EPA), including their use as additives in food and beverages [76]. Thus, they do not require extensive regulatory approval and can be available more quickly on the market than a conventional pesticide. This puts US companies at an advantage in terms of innovation in this market segment compared to those in other countries.

Essential oils and active substances, usually terpenoids, with this special "status" in the USA can be employed without costly and time-consuming requirements, which include necessary toxicological and ecotoxicological tests for commercial product registration [77].

In general, the European market is more regulated and less developed. However, the company Pranarom (France) sells blends of mosquito repellent for children and pregnant women (Pranarom Anti-Mosquito Spray). This type of product is also marketed for the control of house dust mites, lice, and skin irritations due to mosquito bites. For such blends, the proportion of essential oils (4 ~ 9 oils) in the formulations is not disclosed.

Although essential oils have a promising future in the biopesticide market, progress in facilitating the registration process must be achieved before they occupy a significant share of the global pesticide market [18]. Therefore, the barrier to regulatory approval is a constraint to marketing essential oil-based products and will continue to be until a generalized policy of exemption makes their use viable. Few countries are exempt from registering products derived from essential oils such as pesticides or repellents.

In the US, some products are defined as minimum-risk pesticides by the Federal Insecticide, Fungicide, and Rodenticide Act (FIFRA), section 25(b), because their active and inert ingredients are low-risk and generally recognized as safe (GRAS list). Consequently, they are exempt from FIFRA regulatory requirements and can be used on any crop described on the product label, as well as in a non-agricultural environment, as they do not need to be registered as pesticides. For use in organic farming, these substances need to be listed by OMRI (Organic Materials Review Institute) [78].

## Challenges to the Production of Commercial Insecticides Based on Essential Oils and Terpenoids

The development of commercial pesticides containing terpenoid-rich essential oils can have four industrialization paths depending on the origin of the active ingredients: a blend of essential oils; a single essential oil or constituent derived from it; a blend of synthetically produced compounds that imitate its original composition; or an innovative blend obtained from essential oil constituents from distinct botanical species. This must be done by overcoming the main factors that limit the use of essential oils for biopesticide production, such as availability, price, and the chance of regulatory approval [79].

Some of the terpenoids purified from essential oils are moderately toxic to mammals, however, their high volatility promotes reduced persistence under field conditions. There is a small probability that after treatment, predators and parasitoids might be intoxicated by residues, as is often the case with conventional insecticides. Furthermore, there are some specificities regarding their neurotoxic action, which result in different levels of toxicity for mammals and invertebrates exposed to the same xenobiotic compound [80].

Thus, in the formulation of products based on essential oils, the synergy between their components and the addition of others should be considered in order to increase the control efficacy through synergism. Other aspects that need to be evaluated are the method of application, the solvent used in the formulation, the composition of the target organism membrane, the product ability to penetrate the cuticle, the development stage of the target arthropod, and the mechanism of action of the insecticide. All of these factors can affect its bioactivity and, more specifically, its physicochemical properties that may be partially responsible for the interspecific differences observed in toxicity [55, 50, 68].

Recently, nanoemulsions have been developed, such as a nanopesticide option to replace traditional emulsifiable concentrates (oils), by reducing the use of organic solvents and increasing the dispersion and penetration properties of the droplets. The advantages of using oil-in-water microemulsions in formulations containing essential oils are improvement of biological efficacy and reduction of the dose used, a suitable strategy for commercial development of green pesticides on a commercial scale [81].

The development of suitable formulations that use essential oils, and particularly terpenoids, are of great importance to mitigate the negative effects of phytotoxicity that most of these compounds cause when applied directly to plants.

Considerable advances in the understanding of target sites and mode of action of terpenoids have been achieved in recent years. Thus, the cytotoxic action of monoterpenes on animal and plant tissues has been well explained, demonstrating that they cause a drastic reduction in the number of mitochondria and Golgi complexes, thus impairing respiration and photosynthesis, and decreasing cell membrane permeability [15]. In this context, some essential oils, although containing lower proportions of terpenoids, cause phytotoxicity to plants due to the high concentration they require to promote pest control [16].

The effects of EO phytotoxicity on plant leaves can be acute or chronic, which might include scorching and darkening of leaves, reduced flowering, and stunting of growth. Phytotoxicity may also be associated with plant stress, environmental temperature and humidity, oil concentration, and type of formulation. It can vary according to the plant species and cultivars applied. In addition, the phytotoxicity caused by essential oils on plants is lower than that caused by mineral oils [15].

## Insect Resistance to Commercial Terpenoid-Based Insecticides

In evaluating the insecticidal effect of the major compounds of 17 different essential oils on Choristoneura rosaceana (Harris) (Lepidoptera: Tortricidae), Dysaphis plantaginea (Passerini) (Hemiptera: Aphididae), Myzus persicae (Sulzer) (Hemiptera: Aphididae), and *T. ni*, it was observed that the young species recovered, even though they initially appeared to show biological disturbances. This occurred especially when they were subjected to treatments with low concentrations of the compounds, suggesting the role of detoxification enzymes [82]. However, due to the limited information on this topic, the authors stated that under field conditions, resistance was unlikely to be a concern because an EO-based insecticide or synergist would probably consist of more than one compound, mostly volatile terpenoids, with unrelated modes of action. It would require insects to develop cross-resistance or even multiple resistance to the active ingredients, which would also demand multiple genetic mutations, with negative impacts on the growth and development of the species involved. This concept reinforces the idea of developing synergistic commercial products, by combining compounds that present this action individually in order to ensure a final blend with chemical standardization and biological effects on the target pests to be controlled.

## CONCLUDING REMARKS

Meeting the global trend toward the consumption of healthy, residue-free, and ecologically sustainable food, less toxic products with low environmental impact are required to control insect pests. Insecticides based on essential oils rich in terpenoids may be able to supply this demand in the short term. Therefore, knowledge advances are necessary to develop adequate insecticide formulations taking as reference the existing information and thus, encouraging discoveries from aromatic plant species, combined with the study of properties and modes of action of terpenoids present in EOs.

## CONSENT FOR PUBLICATION

Not applicable.

## CONFLICT OF INTEREST

The author declares no conflict of interest, financial or otherwise.

## ACKNOWLEDGEMENT

Declared none.

## REFERENCES

[1]     ISO (2013)- International Organization for Standardization, ISO 9235:2013: aromatic natural raw materials—vocabulary. **2013**. https://www.iso.org/standard/51017.html

[2]     Isman, M.B. Plant essential oils for pest and disease management. *Crop Prot.,* **2000**, *19*(8-10), 603-608.
[http://dx.doi.org/10.1016/S0261-2194(00)00079-X]

[3]     Zwenger, S.; Basu, C. Plant terpenoids: applications and future potentials. *Biotechn. Mol. Biol. Ver.,* **2008**, *3*, 1-7.

[4]     Rice, P.J.; Coats, J.R. Insecticidal properties of several monoterpenoids to the house fly (Diptera: Muscidae), red flour beetle (Coleoptera: Tenebrionidae), and southern corn rootworm (Coleoptera: Chrysomelidae). *J. Econ. Entomol.,* **1994**, *87*(5), 1172-1179.
[http://dx.doi.org/10.1093/jee/87.5.1172] [PMID: 7962947]

[5]     Hummelbrunner, L.A.; Isman, M.B. Acute, sublethal, antifeedant, and synergistic effects of monoterpenoid essential oil compounds on the tobacco cutworm, Spodoptera litura (Lep., Noctuidae). *J. Agric. Food Chem.,* **2001**, *49*(2), 715-720.
[http://dx.doi.org/10.1021/jf000749t] [PMID: 11262018]

[6]     Rozman, V.; Kalinovic, I.; Liska, A. Insecticidal activity of some aromatic plants from Croatia against granary weevil ( *Sitophilus granarius* L.) on stored wheat. *Cereal Res. Commun.,* **2006**, *34*(1), 705-708.
[http://dx.doi.org/10.1556/CRC.34.2006.1.176]

[7]     Papachristos, D.P.; Kimbaris, A.C.; Papadopoulos, N.T.; Polissiou, M.G. Toxicity of citrus essential oils against *Ceratitis capitata* (Diptera: Tephritidae) larvae. *Ann. Appl. Biol.,* **2009**, *155*(3), 381-389.
[http://dx.doi.org/10.1111/j.1744-7348.2009.00350.x]

[8]     Bakkali, F.; Averbeck, S.; Averbeck, D.; Idaomar, M. Biological effects of essential oils – A review. *Food Chem. Toxicol.,* **2008**, *46*(2), 446-475.
[http://dx.doi.org/10.1016/j.fct.2007.09.106] [PMID: 17996351]

[9]     Zoubiri, S.; Baaliouamer, A. Potentiality of plants as source of insecticide principles. *J. Saudi Chem. Soc.,* **2014**, *18*(6), 925-938.
[http://dx.doi.org/10.1016/j.jscs.2011.11.015]

[10]    Ebadollahi, A.; Jalali Sendi, J. A review on recent research results on bio-effects of plant essential oils against major Coleopteran insect pests. *Toxin Rev.,* **2015**, *34*(2), 76-91.

[http://dx.doi.org/10.3109/15569543.2015.1023956]

[11]   Dimetry, N.Z. Different plant families as bioresource for pesticides.*Advances in plant biopesticides*; Singh, D., Ed.; Springer: New Delhi, India, **2014**, pp. 1-20.
[http://dx.doi.org/10.1007/978-81-322-2006-0_1]

[12]   Tong, F.; Coats, J.R. Effects of monoterpenoid insecticides on [3H]-TBOB binding in house fly GABA receptor and 36Cl– uptake in American cockroach ventral nerve cord. *Pestic. Biochem. Physiol.,* **2010**, *98*(3), 317-324.
[http://dx.doi.org/10.1016/j.pestbp.2010.07.003]

[13]   Gross, A.D.; Norris, E.J.; Kimber, M.J.; Bartholomay, L.C.; Coats, J.R. Essential oils enhance the toxicity of permethrin against *Aedes aegypti* and *Anopheles gambiae. Med. Vet. Entomol.,* **2017**, *31*(1), 55-62.
[http://dx.doi.org/10.1111/mve.12197] [PMID: 27800630]

[14]   Isman, M.B. Botanical insecticides, deterrents, and repellents in modern agriculture and an increasingly regulated world. *Annu. Rev. Entomol.,* **2006**, *51*(1), 45-66.
[http://dx.doi.org/10.1146/annurev.ento.51.110104.151146] [PMID: 16332203]

[15]   Rathore, H.S. Green pesticides for organic farming: Occurrence and properties of essential oils for use in pest control.*Green pesticides handbook: Essential oils for pest control*; Nollet, L.M.; Rathore, H.S., Eds.; CRC Press: Washington, DC, USA, **2017**, pp. 3-26.
[http://dx.doi.org/10.1201/9781315153131-1]

[16]   Volpe, H.X.L.; Fazolin, M.; Magnani, R.F.; Garcia, R.B.; Barbosa, J.C.; Miranda, M.P. *Eficácia do óleo essencial de Piper aduncum L. (Piperaceae) para o controle de Diaphorina citri Kuwayama*; Embrapa Acre: Rio Branco, Acre, **2018**.

[17]   Hunter, M. *Essential oils: Art, agriculture, science, industry and entrepreneurship*; Nova Science Publishers Inc.: New York, USA, **2009**.

[18]   Regnault-Roger, C.; Vincent, C.; Arnason, J.T. Essential oils in insect control: low-risk products in a high-stakes world. *Annu. Rev. Entomol.,* **2012**, *57*(1), 405-424.
[http://dx.doi.org/10.1146/annurev-ento-120710-100554] [PMID: 21942843]

[19]   Waliwitiya, R.; Belton, P.; Nicholson, R.A.; Lowenberger, C.A. Effects of the essential oil constituent thymol and other neuroactive chemicals on flight motor activity and wing beat frequency in the blowfly *Phaenicia sericata. Pest Manag. Sci.,* **2010**, *66*(3), 277-289.
[http://dx.doi.org/10.1002/ps.1871] [PMID: 19890946]

[20]   Waliwitiya, R.; Kennedy, C.J.; Lowenberger, C.A. Larvicidal and oviposition-altering activity of monoterpenoids, *trans* -anithole and rosemary oil to the yellow fever mosquito *Aedes aegypti* (Diptera: Culicidae). *Pest Manag. Sci.,* **2009**, *65*(3), 241-248.
[http://dx.doi.org/10.1002/ps.1675] [PMID: 19086001]

[21]   Tong, F.; Bloomquist, J.R. Plant essential oils affect the toxicities of carbaryl and permethrin against Aedes aegypti (Diptera: Culicidae). *J. Med. Entomol.,* **2013**, *50*(4), 826-832.
[http://dx.doi.org/10.1603/ME13002] [PMID: 23926781]

[22]   Anderson, J.A.; Coats, J.R. Acetylcholinesterase inhibition by nootkatone and carvacrol in arthropods. *Pestic. Biochem. Physiol.,* **2012**, *102*(2), 124-128.
[http://dx.doi.org/10.1016/j.pestbp.2011.12.002]

[23]   http://www.irac-br.org/

[24]   Park, T.J.; Seo, H.K.; Kang, B.J.; Kim, K.T. Noncompetitive inhibition by camphor of nicotinic acetylcholine receptors. *Biochem. Pharmacol.,* **2001**, *61*(7), 787-793.
[http://dx.doi.org/10.1016/S0006-2952(01)00547-0] [PMID: 11274963]

[25]   Park, T.J.; Park, Y.S.; Lee, T.G.; Ha, H.; Kim, K.T. Inhibition of acetylcholine-mediated effects by borneol. *Biochem. Pharmacol.,* **2003**, *65*(1), 83-90.
[http://dx.doi.org/10.1016/S0006-2952(02)01444-2] [PMID: 12473382]

[26]    Parnas, M.; Peters, M.; Dadon, D.; Lev, S.; Vertkin, I.; Slutsky, I.; Minke, B. Carvacrol is a novel inhibitor of Drosophila TRPL and mammalian TRPM7 channels. *Cell Calcium,* **2009**, *45*(3), 300-309. [http://dx.doi.org/10.1016/j.ceca.2008.11.009] [PMID: 19135721]

[27]    Roeder, T. Octopamine in invertebrates. *Prog. Neurobiol.,* **1999**, *59*(5), 533-561. [http://dx.doi.org/10.1016/S0301-0082(99)00016-7] [PMID: 10515667]

[28]    Orchard, I. Octopamine in insects: neurotransmitter, neurohormone, and neuromodulator. *Can. J. Zool.,* **1982**, *60*(4), 659-669. [http://dx.doi.org/10.1139/z82-095]

[29]    Hollingworth, R.M.; Johnstone, E.M.; Wright, N. *ACS Symposium Series No, 255*DC**1984**, , pp. 103-125.

[30]    Prier, K.R.; Beckman, O.H.; Tublitz, N.J. Modulating a modulator: biogenic amines at subthreshold levels potentiate peptide-mediated cardioexcitation of the heart of the tobacco hawkmoth Manduca sexta. *J. Exp. Biol.,* **1994**, *197*(1), 377-391. [http://dx.doi.org/10.1242/jeb.197.1.377] [PMID: 7852910]

[31]    Christensen, T.A.; Sherman, T.G.; McCaman, R.E.; Carlson, A.D. Presence of octopamine in firefly photomotor neurons. *Neuroscience,* **1983**, *9*(1), 183-189. [http://dx.doi.org/10.1016/0306-4522(83)90055-6] [PMID: 6410304]

[32]    Orchard, I.; Ramirez, J.M.; Lange, A.B. multifunctional role for octopamine in locust flight. *Annu. Rev. Entomol.,* **1993**, *38*(1), 227-249. [http://dx.doi.org/10.1146/annurev.en.38.010193.001303]

[33]    Zeng, H.; Loughton, B.G.; Jennings, K.R. Tissue specific transduction systems for octopamine in the locust (Locusta migratoria). *J. Insect Physiol.,* **1996**, *42*(8), 765-769. [http://dx.doi.org/10.1016/0022-1910(96)00013-3]

[34]    Rafaeli, A.; Gileadi, C. Modulation of the PBAN-stimulated of pheromonotropic activity in Helicoverpa armigera. *Insect Biochem. Mol. Biol.,* **1995**, *25*(7), 827-834. [http://dx.doi.org/10.1016/0965-1748(95)00019-R]

[35]    Kutsukake, M.; Komatsu, A.; Yamamoto, D.; Ishiwa-Chigusa, S. A tyramine receptor gene mutation causes a defective olfactory behavior in Drosophila melanogaster. *Gene,* **2000**, *245*(1), 31-42. [http://dx.doi.org/10.1016/S0378-1119(99)00569-7] [PMID: 10713442]

[36]    Donini, A.; Lange, A.B. Evidence for a possible neurotransmitter/neuromodulator role of tyramine on the locust oviducts. *J. Insect Physiol.,* **2004**, *50*(4), 351-361. [http://dx.doi.org/10.1016/j.jinsphys.2004.02.005] [PMID: 15081828]

[37]    Enan, E.E. Molecular response of Drosophila melanogaster tyramine receptor cascade to plant essential oils. *Insect Biochem. Mol. Biol.,* **2005**, *35*(4), 309-321. [http://dx.doi.org/10.1016/j.ibmb.2004.12.007] [PMID: 15763467]

[38]    Guedes, R.; Vilela, E. *Produtos que agem na fisiologia dos insetos. Novos produtos para o manejo integrado de pragas*; ABEAS: Brasília, Brasil, **1991**.

[39]    Dauterman, W.C. Metabolism of toxicants: Phase II reactions. In: *Introduction to biochemical toxicology*; Hodgson, E; Levi, PE, Eds.; Appleton & Lange, Norwalk: CT, USA, **1994**; pp. 126-132.

[40]    Hodgson, E.; Levi, P.E. Metabolism of toxicants: Phase I reactions. In: *Introduction to biochemical toxicology*; Appleton & Lange, Norwalk: CT, USA, **1994**; pp. 75-93.

[41]    Hemingway, J. The molecular basis of two contrasting metabolic mechanisms of insecticide resistance. *Insect Biochem. Mol. Biol.,* **2000**, *30*(11), 1009-1015. [http://dx.doi.org/10.1016/S0965-1748(00)00079-5] [PMID: 10989287]

[42]    Perry, T.; Batterham, P.; Daborn, P.J. The biology of insecticidal activity and resistance. *Insect Biochem. Mol. Biol.,* **2011**, *41*(7), 411-422. [http://dx.doi.org/10.1016/j.ibmb.2011.03.003] [PMID: 21426939]

[43]   Li, X.; Schuler, M.A.; Berenbaum, M.R. Molecular mechanisms of metabolic resistance to synthetic and natural xenobiotics. *Annu. Rev. Entomol.,* **2007**, *52*(1), 231-253.
[http://dx.doi.org/10.1146/annurev.ento.51.110104.151104] [PMID: 16925478]

[44]   Pavela, R. Acute toxicity and synergistic and antagonistic effects of the aromatic compounds of some essential oils against Culex quinquefasciatus Say larvae. *Parasitol. Res.,* **2015**, *114*(10), 3835-3853.
[http://dx.doi.org/10.1007/s00436-015-4614-9] [PMID: 26149532]

[45]   Pavela, R. Essential oils for the development of eco-friendly mosquito larvicides: A review. *Ind. Crops Prod.,* **2015**, *76*, 174-187.
[http://dx.doi.org/10.1016/j.indcrop.2015.06.050]

[46]   Berenbaum, M.C. The expected effect of a combination of agents: the general solution. *J. Theor. Biol.,* **1985**, *114*(3), 413-431.
[http://dx.doi.org/10.1016/S0022-5193(85)80176-4] [PMID: 4021503]

[47]   López, M.D.; Pascual-Villalobos, M.J. Mode of inhibition of acetylcholinesterase by monoterpenoids and implications for pest control. *Ind. Crops Prod.,* **2010**, *31*(2), 284-288.
[http://dx.doi.org/10.1016/j.indcrop.2009.11.005]

[48]   Ranson, H.; Abdallah, H.; Badolo, A.; Guelbeogo, W.M.; Kerah-Hinzoumbé, C.; Yangalbé-Kalnoné, E.; Sagnon, N.F.; Simard, F.; Coetzee, M. Insecticide resistance in Anopheles gambiae: data from the first year of a multi-country study highlight the extent of the problem. *Malar. J.,* **2009**, *8*(1), 299.
[http://dx.doi.org/10.1186/1475-2875-8-299] [PMID: 20015411]

[49]   http://whqlibdoc.who.int/Publications/2012/9789241564472_eng.pdf**2019**.

[50]   Rattan, R.S. Mechanism of action of insecticidal secondary metabolites of plant origin. *Crop Prot.,* **2010**, *29*(9), 913-920.
[http://dx.doi.org/10.1016/j.cropro.2010.05.008]

[51]   Sutthanont, N.; Choochote, W.; Tuetun, B.; Junkum, A.; Jitpakdi, A.; Chaithong, U.; Riyong, D.; Pitasawat, B. Chemical composition and larvicidal activity of edible plant-derived essential oils against the pyrethroid-susceptible and -resistant strains of Aedes aegypti (Diptera: Culicidae). *J. Vector Ecol.,* **2010**, *35*(1), 106-115.
[http://dx.doi.org/10.1111/j.1948-7134.2010.00066.x] [PMID: 20618656]

[52]   Gillij, Y.G.; Gleiser, R.M.; Zygadlo, J.A. Mosquito repellent activity of essential oils of aromatic plants growing in Argentina. *Bioresour. Technol.,* **2008**, *99*(7), 2507-2515.
[http://dx.doi.org/10.1016/j.biortech.2007.04.066] [PMID: 17583499]

[53]   Liu, P.T.; Stenger, S.; Li, H.; Wenzel, L.; Tan, B.H.; Krutzik, S.R.; Ochoa, M.T.; Schauber, J.; Wu, K.; Meinken, C.; Kamen, D.L.; Wagner, M.; Bals, R.; Steinmeyer, A.; Zügel, U.; Gallo, R.L.; Eisenberg, D.; Hewison, M.; Hollis, B.W.; Adams, J.S.; Bloom, B.R.; Modlin, R.L. Toll-like receptor triggering of a vitamin D-mediated human antimicrobial response. *Science,* **2006**, *311*(5768), 1770-1773.
[http://dx.doi.org/10.1126/science.1123933] [PMID: 16497887]

[54]   Singh, B.; Sharma, D.K.; Kumar, R.; Gupta, A. Controlled release of the fungicide thiram from starch–alginate–clay based formulation. *Appl. Clay Sci.,* **2009**, *45*(1-2), 76-82.
[http://dx.doi.org/10.1016/j.clay.2009.03.001]

[55]   Pavela, R. Acute and synergistic effects of some monoterpenoid essential oil compounds on the house fly (Musca domestica L.). *J. Essent. Oil-Bear. Plants,* **2008**, *11*(5), 451-459.
[http://dx.doi.org/10.1080/0972060X.2008.10643653]

[56]   Pavela, R. Acute, synergistic and antagonistic effects of some aromatic compounds on the Spodoptera littoralis Boisd. (Lep., Noctuidae) larvae. *Ind. Crops Prod.,* **2014**, *60*, 247-258.
[http://dx.doi.org/10.1016/j.indcrop.2014.06.030]

[57]   Tak, J.H.; Isman, M.B. Enhanced cuticular penetration as the mechanism for synergy of insecticidal constituents of rosemary essential oil in Trichoplusia ni. *Sci. Rep.,* **2015**, *5*(1), 12690.

[http://dx.doi.org/10.1038/srep12690] [PMID: 26223769]

[58] Turner, T.L.; Levine, M.T.; Eckert, M.L.; Begun, D.J. Genomic analysis of adaptive differentiation in Drosophila melanogaster. *Genetics,* **2008**, *179*(1), 455-473.
[http://dx.doi.org/10.1534/genetics.107.083659] [PMID: 18493064]

[59] Schmidt, J.M.; Good, R.T.; Appleton, B.; Sherrard, J.; Raymant, G.C.; Bogwitz, M.R.; Martin, J.; Daborn, P.J.; Goddard, M.E.; Batterham, P.; Robin, C. Copy number variation and transposable elements feature in recent, ongoing adaptation at the Cyp6g1 locus. *PLoS Genet.,* **2010**, *6*(6), e1000998.
[http://dx.doi.org/10.1371/journal.pgen.1000998] [PMID: 20585622]

[60] Pu, J.; Wang, Z.; Chung, H. Climate change and the genetics of insecticide resistance. *Pest Manag. Sci.,* **2020**, *76*(3), 846-852.
[http://dx.doi.org/10.1002/ps.5700] [PMID: 31793168]

[61] Balabanidou, V.; Kampouraki, A.; MacLean, M.; Blomquist, G.J.; Tittiger, C.; Juárez, M.P.; Mijailovsky, S.J.; Chalepakis, G.; Anthousi, A.; Lynd, A.; Antoine, S.; Hemingway, J.; Ranson, H.; Lycett, G.J.; Vontas, J. Cytochrome P450 associated with insecticide resistance catalyzes cuticular hydrocarbon production in *Anopheles gambiae. Proc. Natl. Acad. Sci. USA,* **2016**, *113*(33), 9268-9273.
[http://dx.doi.org/10.1073/pnas.1608295113] [PMID: 27439866]

[62] Sparks, T.C.; Nauen, R. IRAC: Mode of action classification and insecticide resistance management. *Pestic. Biochem. Physiol.,* **2015**, *121*, 122-128.
[http://dx.doi.org/10.1016/j.pestbp.2014.11.014] [PMID: 26047120]

[63] Chang, K.S.; Shin, E.H.; Yoo, D.H.; Ahn, Y.J. Enhanced toxicity of binary mixtures of Bacillus thuringiensis subsp. israelensis and three essential oil major constituents to wild Anopheles sinensis (Diptera: Culicidae) and Aedes albopictus (Diptera: Culicidae). *J. Med. Entomol.,* **2014**, *51*(4), 804-810.
[http://dx.doi.org/10.1603/ME13128] [PMID: 25118412]

[64] Savelev, S.; Okello, E.; Perry, N.S.L.; Wilkins, R.M.; Perry, E.K. Synergistic and antagonistic interactions of anticholinesterase terpenoids in Salvia lavandulaefolia essential oil. *Pharmacol. Biochem. Behav.,* **2003**, *75*(3), 661-668.
[http://dx.doi.org/10.1016/S0091-3057(03)00125-4] [PMID: 12895684]

[65] Attia, S.; Grissa, K.L.; Lognay, G.; Heuskin, S.; Mailleux, A.C.; Hance, T. Chemical composition and acaricidal properties of Deverra scoparia essential oil (Araliales: Apiaceae) and blends of its major constituents against Tetranychus urticae (Acari: Tetranychidae). *J. Econ. Entomol.,* **2011**, *104*(4), 1220-1228.
[http://dx.doi.org/10.1603/EC10318] [PMID: 21882686]

[66] Chaubey, M.K. Acute, lethal and synergistic effects of some terpenes against Tribolium castaneum Herbst (Coleoptera: Tenebrionidae). *Ecol. Balk.,* **2012**, *4*, 53-62.

[67] Miresmailli, S.; Bradbury, R.; Isman, M.B. Comparative toxicity ofRosmarinus officinalis L. essential oil and blends of its major constituents againstTetranychus urticae Koch (Acari: Tetranychidae) on two different host plants. *Pest Manag. Sci.,* **2006**, *62*(4), 366-371.
[http://dx.doi.org/10.1002/ps.1157] [PMID: 16470541]

[68] Tak, J.H.; Jovel, E.; Isman, M.B. Comparative and synergistic activity of *Rosmarinus officinalis* L. essential oil constituents against the larvae and an ovarian cell line of the cabbage looper, *Trichoplusia ni* (Lepidoptera: Noctuidae). *Pest Manag. Sci.,* **2016**, *72*(3), 474-480.
[http://dx.doi.org/10.1002/ps.4010] [PMID: 25809531]

[69] Yeom, H.J.; Kang, J.S.; Kim, G.H.; Park, I.K. Insecticidal and acetylcholine esterase inhibition activity of Apiaceae plant essential oils and their constituents against adults of German cockroach (Blattella germanica). *J. Agric. Food Chem.,* **2012**, *60*(29), 7194-7203.
[http://dx.doi.org/10.1021/jf302009w] [PMID: 22746406]

[70]   Moraes, M.M.D.; Camara, C.A.G.D.; Silva, M.M.C.D. Comparative toxicity of essential oil and blends of selected terpenes of Ocotea species from Pernambuco, Brazil, against Tetranychus urticae Koch. *An. Acad. Bras. Cienc.,* **2017**, *89*(3), 1417-1429.
       [http://dx.doi.org/10.1590/0001-3765201720170139] [PMID: 28767894]

[71]   Yu, S.J.; Berry, R.E.; Terriere, L.C. Host plant stimulation of detoxifying enzymes in a phytophagous insect. *Pestic. Biochem. Physiol.,* **1979**, *12*(3), 280-284.
       [http://dx.doi.org/10.1016/0048-3575(79)90113-5]

[72]   Lee, S.E.; Lee, B.H.; Choi, W.S.; Park, B.S.; Kim, J.G.; Campbell, B.C. Fumigant toxicity of volatile natural products from Korean spices and medicinal plants towards the rice weevil,Sitophilus oryzae (L). *Pest Manag. Sci.,* **2001**, *57*(6), 548-553.
       [http://dx.doi.org/10.1002/ps.322] [PMID: 11407032]

[73]   DiMatteo, K. Pesticides and organic agriculture. *Environ. Health Perspect.,* **2004**, *112*(15), A865-A865.
       [http://dx.doi.org/10.1289/ehp.112-a865b] [PMID: 15531416]

[74]   Canyon, D.V.; Heukelbach, J.; Shaalan, E.S.; Speare, R. Piperonyl butoxide. In: *The use of antibiotics*; Grayson, ML, Ed.; Hodder Arnold: London, England, **2010**; pp. 2306-2309.
       [http://dx.doi.org/10.1201/b13787-235]

[75]   Bioland,        R.F.        **2019**.*Verband        für        organisch-biologischen        Landbau.,*
       https://www.bioland.de/suche/search

[76]   http://www.epa.gov/oppsrrd1/ REDs/factsheets/ 4097fact.pdf

[77]   Isman, M.B. Pesticides based on plant essential oils: Phytochemical and practical considerations.*Medicinal and aromatic crops: Production, phytochemistry, and utilization*; Zheljazkov, V.D.; Cantrell, C.L., Eds.; American Chemical Society: Washington, DC, USA, **2016**, pp. 13-26.
       [http://dx.doi.org/10.1021/bk-2016-1218.ch002]

[78]   Marrone, P.G. Pesticidal natural products – status and future potential. *Pest Manag. Sci.,* **2019**, *75*(9), 5433.
       [http://dx.doi.org/10.1002/ps.5433] [PMID: 30941861]

[79]   Isman, M.B. Commercial development of plant essential oils and their constituents as active ingredients in bioinsecticides. *Phytochem. Rev.,* **2020**, *19*(2), 235-241.
       [http://dx.doi.org/10.1007/s11101-019-09653-9]

[80]   Koul, O.; Walia, S.; Dhaliwal, G.S. Essential oils as green pesticides: Potential and constraints. *Biopestic. Int. (Jalandhar),* **2008**, *4*, 63-84.

[81]   Bullangpoti, V. Essential oils and synthetic pesticides.*Green pesticides handbook: Essential oils for pest control*; Nollet, L.M.; Rathore, H.S., Eds.; CRC Press: New York, USA, **2017**, pp. 449-478.
       [http://dx.doi.org/10.1201/9781315153131-24]

[82]   Machial, C.M. **2010**.

[83]   Abbassy, M.A.; Abdelgaleil, S.A.M.; Rabie, R.Y.A. Insecticidal and synergistic effects of *Majorana hortensis* essential oil and some of its major constituents. *Entomol. Exp. Appl.,* **2009**, *131*(3), 225-232.
       [http://dx.doi.org/10.1111/j.1570-7458.2009.00854.x]

[84]   Abdelgaleil, S.A.M.; Mohamed, M.I.E.; Badawy, M.E.I.; El-arami, S.A.A. Fumigant and contact toxicities of monoterpenes to Sitophilus oryzae (L.) and Tribolium castaneum (Herbst) and their inhibitory effects on acetylcholinesterase activity. *J. Chem. Ecol.,* **2009**, *35*(5), 518-525.
       [http://dx.doi.org/10.1007/s10886-009-9635-3] [PMID: 19412756]

[85]   Akhtar, Y.; Pages, E.; Stevens, A.; Bradbury, R.; da CAMARA, C.A.G.; Isman, M.B. Effect of chemical complexity of essential oils on feeding deterrence in larvae of the cabbage looper. *Physiol. Entomol.,* **2012**, *37*(1), 81-91.

[http://dx.doi.org/10.1111/j.1365-3032.2011.00824.x]

[86] Brattsten, L.B.; Wilkinson, C.F.; Eisner, T. Herbivore-plant interactions: mixed-function oxidases and secondary plant substances. *Science,* **1977**, *196*(4296), 1349-1352.
[http://dx.doi.org/10.1126/science.196.4296.1349] [PMID: 17831753]

[87] De-Oliveira, A.C.; Ribeiro-Pinto, L.F.; Paumgartten, J.R. *In vitro* inhibition of CYP2B1 monooxygenase by β-myrcene and other monoterpenoid compounds. *Toxicol. Lett.,* **1997**, *92*(1), 39-46.
[http://dx.doi.org/10.1016/S0378-4274(97)00034-9] [PMID: 9242356]

[88] Ebadollahi, A.; Davari, M.; Razmjou, J.; Naseri, B. Separate and combined effects of Mentha piperata and Mentha pulegium essential oils and a pathogenic fungus Lecanicillium muscarium against Aphis gossypii (Hemiptera: Aphididae). *J. Econ. Entomol.,* **2017**, *110*(3), 1025-1030.
[http://dx.doi.org/10.1093/jee/tox065] [PMID: 28334238]

[89] Erler, F.; Erdemir, T.; Ceylan, F.O.; Toker, C. Fumigant toxicity of three essential oils and their binary and tertiary mixtures against the pulse beetle, Callosobruchus maculatus F. (Coleoptera: Bruchidae). *Fresenius Environ. Bull.,* **2009**, *18*, 975-981.

[90] Faraone, N.; Hillier, N.K.; Cutler, G.C. Plant essential oils synergize and antagonize toxicity of different conventional insecticides against Myzus persicae (Hemiptera: Aphididae). *PLoS One,* **2015**, *10*(5), e0127774.
[http://dx.doi.org/10.1371/journal.pone.0127774] [PMID: 26010088]

[91] Franzios, G.; Mirotsou, M.; Hatziapostolou, E.; Kral, J.; Scouras, Z.G.; Mavragani-Tsipidou, P. Insecticidal and genotoxic activities of mint essential oils. *J. Agric. Food Chem.,* **1997**, *45*(7), 2690-2694.
[http://dx.doi.org/10.1021/jf960685f]

[92] Gallardo, A.; Picollo, M.I.; González-Audino, P.; Mougabure-Cueto, G. Insecticidal activity of individual and mixed monoterpenoids of geranium essential oil against Pediculus humanus capitis (Phthiraptera: Pediculidae). *J. Med. Entomol.,* **2012**, *49*(2), 332-335.
[http://dx.doi.org/10.1603/ME11142] [PMID: 22493851]

[93] Joffe, T. Evaluation of potential pyrethrum sinergists on agriculturally significant inset species, School of Agricultural Science/TIAR, University of Tasmania. **2011**.

[94] Jyoti, N.; Singh, N.K.; Singh, H.; Mehta, N.; Rath, S.S. *In vitro* assessment of synergistic combinations of essential oils against Rhipicephalus (Boophilus) microplus (Acari: Ixodidae). *Exp. Parasitol.,* **2019**, *201*, 42-48.
[http://dx.doi.org/10.1016/j.exppara.2019.04.007] [PMID: 31034814]

[95] Keane, S.; Ryan, M.F. Purification, characterisation, and inhibition by monoterpenes of acetylcholinesterase from the waxmoth, Galleria mellonella (L.). *Insect Biochem. Mol. Biol.,* **1999**, *29*(12), 1097-1104.
[http://dx.doi.org/10.1016/S0965-1748(99)00088-0]

[96] Khairul, F.K.; Harbant, S.; Hunter, M.; Ahmad, M.N. A novel mosquitoes repellent soap based on Azadirachta indica and Eucalyptus citriodora oil. *J. Pen. Ut. Mal.,* **2005**, *2*, 77-81.

[97] Lichtenstein, E.P.; Liang, T.T.; Schulz, K.R.; Schnoes, H.K.; Carter, G.T. Insecticidal and synergistic components isolated from dill plants. *J. Agric. Food Chem.,* **1974**, *22*(4), 658-664.
[http://dx.doi.org/10.1021/jf60194a037] [PMID: 4842328]

[98] Liu, Y.; Xue, M.; Zhang, Q.; Zhou, F.; Wei, J. Toxicity of β-caryophyllene from Vitex negundo (Lamiales: Verbenaceae) to Aphis gossypii Glover (Homoptera: Aphidiae) and its action mechanism. *Acta Entomol. Sin.,* **2010**, *53*, 396-404.

[99] Martins, G.D.S.O.; Zago, H.B.; Costa, A.V.; Araujo Junior, L.M.D.; Carvalho, J.R.D. CHEMICAL COMPOSITION AND TOXICITY OF CITRUS ESSENTIAL OILS ON Dysmicoccus brevipes (HEMIPTERA: PSEUDOCOCCIDAE). *Rev. Caatinga,* **2017**, *30*(3), 811-817.

[http://dx.doi.org/10.1590/1983-21252017v30n330rc]

[100] Melo, J.P.R. *Tese Doutorado. Produtos formulados à base de óleos essenciais para o manejo de populações de traça-das-crucíferas Plutella xylostella (L.) (Lepidoptera: Plutellidae) resistentes ao ingrediente ativo deltametrina*; Universidade Federal de Pernambuco, **2017**.

[101] Meshram, M.D.; Kulkarni, M.D.; Pawalkar, D.A.; Ghumare, B.C.; Muluk, P.K.; Pawar, A.J. Acaricidal effect of a herbal spray formulation of tobacco extract (nicotine) and eucalyptus oil combination in Holstein Friesian crossbred cattle. *An. Science Rep.,* **2012**, *61*, 31-35.

[102] Norris, E.J.; Gross, A.D.; Dunphy, B.M.; Bessette, S.; Bartholomay, L.; Coats, J.R. Comparison of the insecticidal characteristics of commercially available plant essential oils against Aedes aegypti and Anopheles gambiae (Diptera: Culicidae). *J. Med. Entomol.,* **2015**, *52*(5), 993-1002.
[http://dx.doi.org/10.1093/jme/tjv090] [PMID: 26336230]

[103] Norris, E.; Johnson, J.; Gross, A.; Bartholomay, L.; Coats, J. Plant essential oils enhance diverse pyrethroids against multiple strains of mosquitoes and inhibit detoxification enzyme processes. *Insects,* **2018**, *9*(4), 132.
[http://dx.doi.org/10.3390/insects9040132] [PMID: 30287743]

[104] Norris, E.J.; Gross, A.D.; Bartholomay, L.C.; Coats, J.R. Plant essential oils synergize various pyrethroid insecticides and antagonize malathion in *AEDES AEGYPTI. Med. Vet. Entomol.,* **2019**, *33*(4), 453-466.
[http://dx.doi.org/10.1111/mve.12380] [PMID: 31102301]

[105] Oliveira, I.M. Dissertação (Mestrado Profissional em Defesa Sanitária Vegetal), Resistência de artrópodos de importância agrícola ao controle químico, Universidade Federal de Viçosa. **2017**.

[106] O'Neal, S.T.; Johnson, E.J.; Rault, L.C.; Anderson, T.D. Vapor delivery of plant essential oils alters pyrethroid efficacy and detoxification enzyme activity in mosquitoes. *Pestic. Biochem. Physiol.,* **2019**, *157*, 88-98.
[http://dx.doi.org/10.1016/j.pestbp.2019.03.007] [PMID: 31153481]

[107] Pavela, R.; Maggi, F.; Lupidi, G.; Mbuntcha, H.; Woguem, V.; Womeni, H.M.; Barboni, L.; Tapondjou, L.A.; Benelli, G. Clausena anisata and Dysphania ambrosioides essential oils: from ethnomedicine to modern uses as effective insecticides. *Environ. Sci. Pollut. Res. Int.,* **2018**, *25*(11), 10493-10503.
[http://dx.doi.org/10.1007/s11356-017-0267-9] [PMID: 28965298]

[108] Ribeiro, V.L.S. *Tese Doutorado. Estudo do mecanismo de ação da atividade acaricida de Calea serrata (Asteraceae) em Rhipicephalus (Boophilus) microplus e da sua toxicidade em roedores*; Universidade Federal do Rio Grande do Sul, **2012**.

[109] Ruttanaphan, T.; Pluempanupat, W.; Aungsirisawat, C.; Boonyarit, P.; Goff, G.L.; Bullangpoti, V. Effect of plant essential oils and their major constituents on cypermethrin tolerance associated detoxification enzyme activities in Spodoptera litura (Lepidoptera: Noctuidae). *J. Econ. Entomol.,* **2019**, *112*(5), 2167-2176.
[http://dx.doi.org/10.1093/jee/toz126] [PMID: 31139824]

[110] Ryan, M.F.; Byrne, O. Plant-insect coevolution and inhibition of acetylcholinesterase. *J. Chem. Ecol.,* **1988**, *14*(10), 1965-1975.
[http://dx.doi.org/10.1007/BF01013489] [PMID: 24277106]

[111] Saad, M.M.G.; Abou-Taleb, H.K.; Abdelgaleil, S.A.M. Insecticidal activities of monoterpenes and phenylpropenes against Sitophilus oryzae and their inhibitory effects on acetylcholinesterase and adenosine triphosphatases. *Appl. Entomol. Zool.,* **2018**, *53*(2), 173-181.
[http://dx.doi.org/10.1007/s13355-017-0532-x]

[112] Shafiei, S.; Talebi, J.K.; Sabahi, Q. Effects of mixture of orange peel essential oil with two chemical acaricides, spirodiclofen and propargite against adult females of Tetranychus urticae (Acari: Tetranychidae). *Plant Pest Res.,* **2016**, *5*, 39-50.

[113] Soares, A.M.S.; Penha, T.A.; Araújo, S.A.; Cruz, E.M.O.; Blank, A.F.; Costa-Junior, L.M. Assessment of different Lippia sidoides genotypes regarding their acaricidal activity against Rhipicephalus (Boophilus) microplus. *Rev. Bras. Parasitol. Vet.,* **2016**, *25*(4), 401-406.
[http://dx.doi.org/10.1590/s1984-29612016087] [PMID: 27982301]

[114] Tak, J.H.; Jovel, E.; Isman, M.B. Synergistic interactions among the major constituents of lemongrass essential oil against larvae and an ovarian cell line of the cabbage looper, Trichoplusia ni. *J. Pest Sci.,* **2017**, *90*(2), 735-744.
[http://dx.doi.org/10.1007/s10340-016-0827-7]

[115] Waliwitiya, R.; Nicholson, R.A.; Kennedy, C.J.; Lowenberger, C.A. The synergistic effects of insecticidal essential oils and piperonyl butoxide on biotransformational enzyme activities in Aedes aegypti (Diptera: Culicidae). *J. Med. Entomol.,* **2012**, *49*(3), 614-623.
[http://dx.doi.org/10.1603/ME10272] [PMID: 22679869]

[116] Wei, H.; Liu, J.; Li, B.; Zhan, Z.; Chen, Y.; Tian, H.; Lin, S.; Gu, X. The toxicity and physiological effect of essential oil from Chenopodium ambrosioides against the diamondback moth, Plutella xylostella (Lepidoptera: Plutellidae). *Crop Prot.,* **2015**, *76*, 68-74.
[http://dx.doi.org/10.1016/j.cropro.2015.06.013]

[117] Youssef, N.S. Toxic and synergistic properties of several volatile oils against larvae of the house fly, Musca domestica vicina Maquart (Diptera: Muscidae). *J. Egy. Ger. Soc. Zool.,* **1997**, *22*, 131-150.

# CHAPTER 12

# Potential Antimicrobial Activities of Terpenoids

**Hamdy A. Shaaban**[1,*] and **Amr Farouk**[1]

[1] *Chemistry of Flavor and Aroma Department, National Research Center, Dokki, Cairo, Egypt*

**Abstract:** The antimicrobial effect of essential oils and their main constituents, the terpenoids, has been generally reviewed in this article, with a comparative investigation of the structure-activity relationship. Terpenoids are widespread metabolites in plants belonging to different chemical classes, whereas oxygenated derivatives constitute the predominates. They could be classified as diterpenes, triterpenes, tetraterpenes, or hemiterpenes and sesquiterpenes. As crude materials, terpenoids are also broadly utilized in drug, food, and beauty care product ventures. Terpenoids have antitumor, anti-inflammatory, antibacterial, antiviral, antimalarial effects, promote transdermal absorption, prevent and treat cardiovascular diseases, and hypoglycemic activities. Moreover, terpenoids have many critical uses as insecticides, immunoregulators, antioxidants, antiaging, and neuroprotection agents. Terpenoids have a complicated construction with assorted impacts and various components of activity. Using plants – containing – terpenoids as neutraceuticals in the nutrition of humans and animals also constitutes a potential issue as natural inhibitors for microbes. These phytochemicals are generally conveyed in soil products and are particularly helpful in food protection as microbial development inhibitors.

**Keywords:** Terpenoids, Essential oils, Antimicrobial activities, Mode of action, Progress of research.

## INTRODUCTION

Terpenoids are one of the natural bioactive classes classified according to the isoprene units. According to the number of isoprenes, terpenoids could be categorized into monoterpene ($C_{10}$), sesquiterpene ($C_{15}$), diterpene ($C_{20}$), triterpene ($C_{30}$), tetraterpene ($C_{40}$), and polyterpenoid (C > 40). Another well-known classification in the literature is based on oxygenated derivatives like carboxylic acids, esters, aldehydes, alcohols, and glycosides. The interconvertible of $C_5$ precursors isopentenyl diphosphateproduced *via* mevalonate (MVA) and themethylerythritol phosphate (MEP) pathways is responsible for the natural

---

* **Corresponding author Hamdy A. Shaaban:** Chemistry of Flavor and Aroma Department, National Research Center, Dokki, Cairo, Egypt; E-mail: hamdy.shaaban64@gmail.com

**Mozaniel Santana de Oliveira & Antônio Pedro da Silva Souza Filho (Eds.)**

synthesis of terpenoids. The MVA pathway exists in the cytosol with the formation of metabolites such as sesquiterpenes, sterols, and triterpenes, while the MEP pathway is primarily present in plastids through many enzymes that lead to the generation of monoterpenes, diterpenes, and tetraterpenes [1].

Terpenoids represent the major bioactive constituents of the oils found in higher medicinal plants belonging to families like composite, Ranunculaceae, Araliaceae, Oleaceae, Magnoliaceae, Lauraceae, Aristolochiaceae, Rutaceae, Labiatae, Pinaceae, Apiaceae, Celastraceae, Acanthaceae, Taxaceae, and so on. Monoterpenes and sesquiterpenes are predominantly found in essential oils of the medicinal plant, while higher terpenes as triterpene, are primarily found in amber and gum.

The terpenoids are bioactive classes with antimicrobial properties against many microorganisms (Fig. **1**) [2]. Generally, terpenoids showed higher antimicrobial activity than terpenes. For example, the terpenoid fraction of *Helichrysum italicum* essential oil was higher than its terpene fraction against *S. aureus* and *Candida albicans* development [3]. Functional groups of the terpenoids structure play a key role in their antimicrobial activity [2, 4], whereas alcoholic and aldehydic terpenoids, *e.g.*, such as terpinene-4-ol and cinnamaldehyde, have a crucial role in their antimicrobial activity, higher antimicrobial efficiency than others containing carbonyl group only. Moreover, geranyl acetic acid has been shown to have a higher antimicrobial activity than geraniol cause of carbonyl and hydroxyl moieties in its structure [2]. Eugenol and cinnamaldehyde are popular terpenoids widespread in the essential oils of many plants with remarkable bioactivity against a broad spectrum of microbes. After a survey of 30 strains of *H. pylori,* a significant human microbe associated with gastric and duodenal ulcers, Ali *et al.* [5] revealed that eugenol and cinnamaldehyde could inhibit H. pylori strains development without any reinforcement. Eugenol additionally has shown striking bioactivity against enterotoxins and biofilms of methicillin-resistant *Staphylococcus aureus* (MRSA) and methicillin-susceptible *S. Aureus* (MSSA) clinical strains [6]. According to Yadav *et al.* [6], eugenol depresses biofilm development, interferes with cell correspondence, destroys the pre-setup biofilms, and kills the microorganisms in biofilms, similarly to MRSA and MSSA mechanisms. These eugenol effects were due to the hindrance of the bacterial cell film and the spillage of the cell substance. In the study of Rathinam *et al.* [7], eugenol displayed practically identical impacts on biofilm development and the harmfulness factor combination of *P. aeruginosa.* A review on the cinnamaldehyde activity against *E. coli* and *S. aureus* using electron microscopy showed damage to the integrity of the bacterial membrane decreased the membrane potential and affected the metabolic activity, thus inhibiting bacterial growth [8]. The hydrogen bonding parameters and the solubility of terpenoids proved to affect their antimicrobial activity by Griffin and colleagues [9] during their study against *P. aeruginosa*, *E. coli*, *S. aureus*, and *C. albicans*.

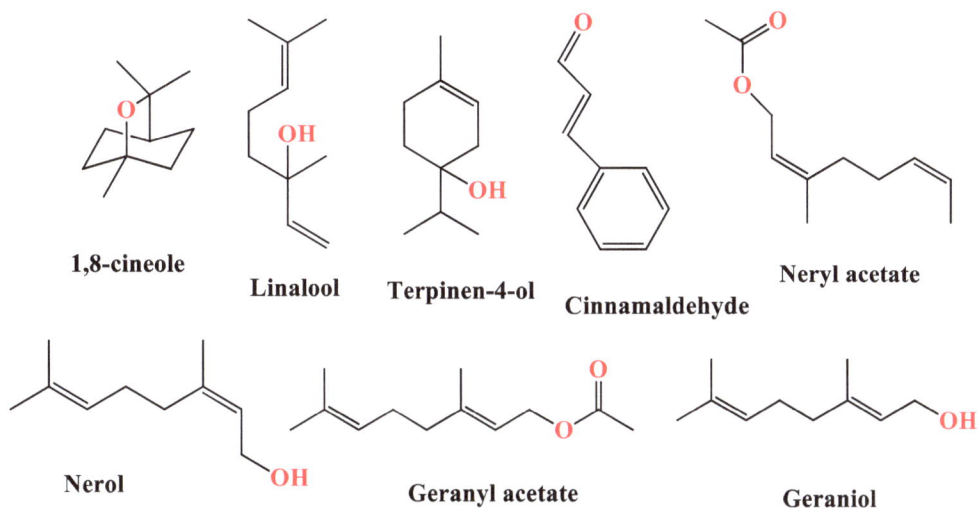

**Fig. (1).** Chemical structure ofmain antimicrobial constituents (terpenoids) in essential oils.

## Antibacterial Activity

Other higher terpenoids like labdane diterpenoid (andrographolide) and pentacyclic triterpenoid (oleanolic acid) were showed a strong antibacterial effect and used as a therapeutic agent against diseases like tuberculosis [10 - 14] (Table **1**, Fig. **2**). Mentha family members are rich in monoterpenoids, which have a strong antimicrobial effect [15]. For example, menthol showed critical inhibitory action of biofilm on *Candida albicans* [16 - 20]. Patchouli liquor (PA) is a tricyclic sesquiterpenoid compound found in *Pogostemon cablin* (Blanco) Benth revealed an antioxidant efficiency against *Helicobacter pylori* activity *in vitro* and *in vivo* [21]. The exploratory information shows that the bactericidal impact of PA is time, pH, and concentration-dependent, whereas the minimal bactericidal concentrations were 25-75 µg/mL [21]. Many researchers discovered that *Artemisia annua* L. oil and extracts have diverse antibacterial activities against anaerobic microscopic organisms, facultative anaerobic microorganisms, microaerophilic microbes, and high-impact microorganisms [22 - 24].

The common use of antibiotics may lead to a lower efficiency toward clinically deadly pathogens like *Pseudomonas aeruginosa*. Cheng *et al*. [25] revealed the impressive inhibitory impact of andrographolide on the biofilm of P.aeruginosa and its synergistic antibacterial effect with azithromycin. Again, Banerjee *et al*. [26] showed an antibacterial activity for the labdane diterpenoid against the significant gram-positive microorganisms, among which *S. aureus* with a negligible MIC of 100 µg/mL. Meanwhile, oleanolic acid showed an inhibitory impact on *S. aureus*, methicillin-safe *S. aureus*, and *Streptococcus mutans* [27] and the ability to kill microorganisms like listeria monocytogenes, *Enterococcus faecium*, and *E. faecalis* at MICs of 16-32 µg/mL for L. monocytogenes and 32-64 µg/mL for *E. faecium* and *E. faecalis*, and bacterial cell openness at 2×MIC of oleanolic acid [28, 29]. Several terpenoid subsidiaries likewise displayed antimycobacterial movement as perhaps the primary pathogen. A rundown of such regular items with the accentuate on M. tuberculosis was accounted for by Copp, including sandaracopimaric corrosive, (+)-Totarol, Agelasine F, elisapterosin B, costunolide, parthenolide, 1,10-poxycostunolide, santamarine, reynosin, alantolactone, puupehenone, elatol, deschloroelatol, debromolaurinterol, allolaurinterol, and aureol. The higher lipophilic structure of terpenoid subsidiaries is responsible for their antimycobacterial activity, which depends on the invasion of the mycobacterial cell divider [30].

**Table 1. The Antibacterial Activity of Terpenoids.**

| Classification | Compound | Antibacterial Range | References |
|---|---|---|---|
| Monoterpene | 1,8- eucalyptus | *Staphylococcus aureus, Bacillus subtilis, Escherichia coli*, and *Streptococcus, Listeria, Bacillus cereus* | [10, 11] |
| | Limonene, Geranialdehyde | *B. subtilis, S. aureus, Streptococcus mutans, E. coli, Candida albicans*, methicillin-resistant *S. aureus* and *Saccharomyces cerevisiae* | [20] |
| | Sabinene | *S. aureus* (Gram-positive) and *E. coli* (Gram-negative) | [21] |
| | Menthol | *S. aureus, Streptococcus pneumonia, Streptococcus pyogenes, Haemophilus influenzae* | [15 - 19] |
| | Sabinol | Oral bacteria | [12] |
| | Carvone | *S. aureus, E. coli, B. subtilis, Salmonella typhimurium, Listeria* | [13, 14] |
| Sesquiterpene | Patchouli Alcohol | *Helicobacter pylori* | [22] |
| Diterpene | Artemisinin | Variouspathogenssuch as *B. subtilis, E. coli, Pseudomonas aeruginosa, S. cerevisiae, S. aureus,Mycobacteriumtuberculosis* | [23 - 25] |
| | Andrographolide | *P. aeruginosa* | [26, 27] |
| Triterpene | Oleanolic acid | *S. aureus*, methicillin-resistant *S. aureus, S. mutans*and*Listeria monocytogenes, Enterococcusfaecium, Enterococcus faecalis* | [28, 29] |

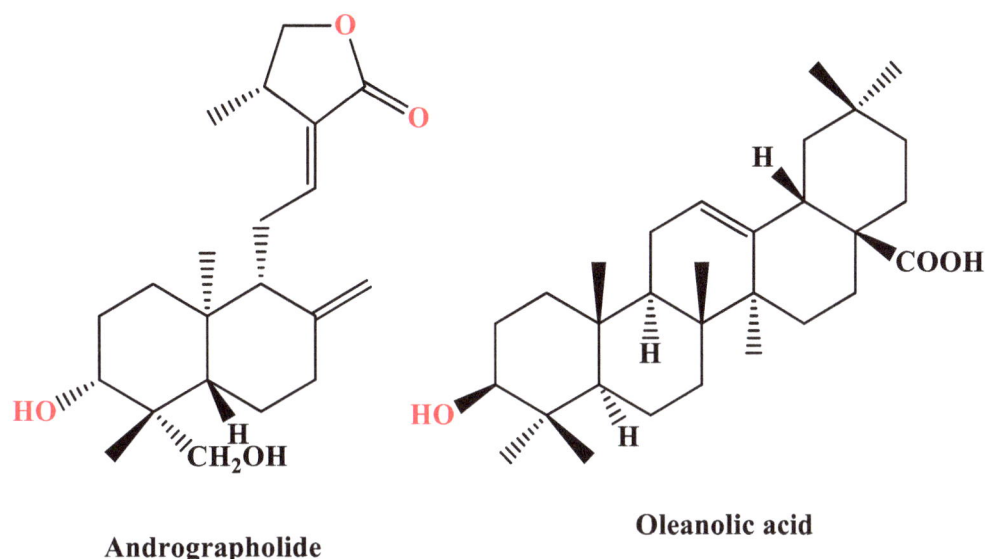

**Androgapholide**

**Oleanolic acid**

**Fig. (2).** The structures of common terpenoids with antibacterial activities.

## Antiviral Effect

Many studies revealed terpenoids are a promising antimicrobial agent against infections, microscopic organisms, parasites, and protozoa [31]. The structures of terpenoids with antiviral effects are shown in Fig. (**3**). Monoterpenoids like isoborneol and borneol proved a great activity against herpes simplex viruses (HSV-1). Isoborneol is a monoterpenoid found widely as a principal constituent in many plants' essential oils. It showed a complete control on HSV-1 replication at a concentration of 0.06% [32]. Other monoterpenoids, such as α-terpinene, γ-terpinene, thymol, 1,8-cineole, α-terpineol, and citral isolated from tea tree, eucalyptus, and thyme showed antiviral activity against HSV-1 by about>80% in comparison to the oils which inhibited the virus by >96% [33]. Brezani *et al.* [34] extracted 12 pure compounds and a mixture of 2 isomers from the leaves and twigs of *Eucalyptus globulus*, and studied their antiviral activities against HSV-1 and HSV-2. The results showed that tereticornate A ($IC_{50}$: 0.96 μg/mL; selectivity list $CC_{50}/IC_{50}$: 218.8) showed the highest activity against HSV-1 compared with acyclovir ($IC_{50}$: 1.92 μg/mL; selectivity file $CC_{50}/IC_{50}$: 109.4), the standard antiviral medication. Moreover, the compound showed moderate antimicrobial and anti-inflammatory impacts during the same study. Putranjivain A, a diterpenoid extracted from *Euphorbia joking*, exhibited an antiviral efficiency against HSV-2 *in vitro* with an $IC_{50}$ 6.3 μM [35]. The two triterpenoid components, betulonic acid and moronic acid, extracted from the plant *Rhus javanica*, showed *in vitro* a robust inhibitory effect on HSV-1 with $EC_{50}$ values of 4.1 and 2.9μg/mL, respectively [36]. Saponins like Notoginsenoside ST-4, extracted from steam-treated not sensing, the famous traditional Chinese medicine, showed *in vitro* surprising inhibitory effects against HSV-1 and HSV-2 with EC$_{50}$ 16.4 and 19.44 µM, respectively [37]. The after-effects of the inhibitory movement against HSV-1 of glycyrrhizin (GR) demonstrated that GR wiped out the expanded weakness of thermally harmed mice to HSV infectivity during the acceptance of CD4+contra suppressor T cells [38].

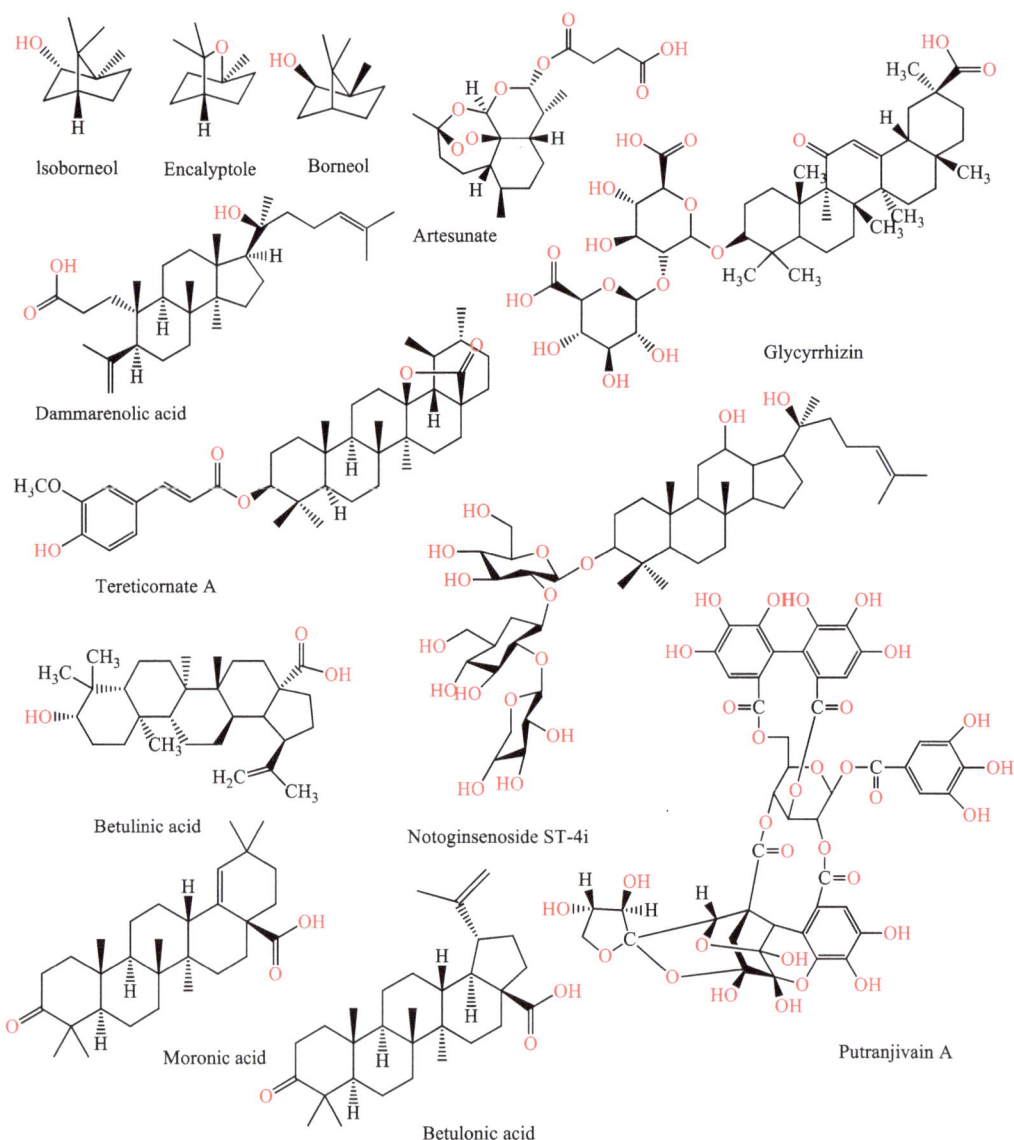

**Fig. (3).** Structures of significant terpenoids with antiviral effects.

Artemisinin derived from *Artemisia annual* showed prominent inhibitory activity on the hepatitis B virus (HBV) and its semisynthetic derivative; artesunate even has a better inhibitory effect on hepatitis surface antigen (HBsAg). A synergistic effect was observed by combining artesunate with lamivudine (a drug for treating HBV) [39].

Recent studies have found that andrographolide has a critical inhibitory impact on the replication cycle of the chikungunya infection (CHIKV) (a mosquito-borne alphavirus) [40]. Andrographolide derived from *A. paniculata* showed a potential inhibition of CHIKV with a reduction in virus production by roughly $3_{log10}$ with a50% $EC_{50}$ of 77 μM without cytotoxicity. Time of addition and RNA transfection studies showed that andrographolide influenced CHIKV replication with cell-type independent activity. Efforts have been made to overcome dengue fever, the most widely recognized viral illness sent by arthropods in people, without effective medicine against it. Panraksa showed that andrographolide displayed dramatic cytotoxicity against DENV in HepG2 and HeLa cell lines with half $EC_{50}$ 21.304 μM and 22.739 μM, respectively [41].

Harada *et al.* [42] discovered that glycyrrhizin (a triterpenoid saponin) could suppress HIV, influenza A, and vesicular stomatitis viruses, but not by poliovirus through lower plasma membrane fluidity. Betulinic acid is a pentacyclic triterpenoid compound extracted from *Syzygium claviflorum* and known as an inhibitor for HIV replication in lymphocyte cells [43, 44]. Oleanolic acid, dammarenolic acid, and ursolic acid also have anti-HIV activity *in vitro* and in lymphocyte cells [45 - 47].

## Terpenoids and Essential Oils as Antimicrobial Agents in Food Preservation

Essential oils are mixtures of many components with different polarities and may have been found at higher percentages (20-70%) of the total constituents [48]. Generally, essential oils components can be classified into two main groups with different biosynthesis pathways; terpenes (mono,- and sesquiterpenes) and terpenoids (monoterpenoids) [49]. Aliphatic mixtures of aldehydes, phenols, and methoxy derivatives may be found in essential oils but as minors [50]. Therapeutic plant parts like roots, leaves, branches, stems, bark, blossoms, and organic products are usually wealthy in terpenes, such as carvacrol and citral (a characteristic combination of geranial and neral), linalool, geraniol, and numerous others [51]. Among the broad spectrum of components that constitute oils and extracts of plants, a few proved successful biological activity against microscopic organisms [52], viruses [53], and fungi [54] (Table **1**, Fig. **1**). For example, carvacrol (a monoterpenoid phenol) (Fig. **2**) is one of the most active components [50]; however, other common terpenes, such as *p*-cymene (Fig. **2**), lack intense antimicrobial activity, and many *in vitro* tests have shown that some terpenes are inactive as antimicrobials when used alone [55]. The antimicrobial activity of terpenoids is based on their structure. In particular, electrons delocalization in phenolic terpenoids due to their hydroxyl groups is responsible for the structure-activity relation [50]. For example, methyl ether branches cannot show the same antimicrobial action as hydroxyl groups due to the difference in hydrophobicity and the mechanism of their interaction with the microbial cell wall. Another unique function for the hydroxyl group is their ability as monovalent cation transporter across layers, conveying $H^+$ into the cell cytoplasm and shipping $K^+$ back out [56]. The synergistic effect among the different constituents of the essential oils is a potential factor in the antimicrobial activity of oils or extracts. A great model is a synergistic cooperation between carvacrol and *p*-cymene [57]. Carvacrol (a monoterpenoid) and *p*-cymene (monoterpene) are found in oregano and thyme separately and can be utilized as common additives when both are being used together [58]. Carvacrol has antimicrobial action against a broad scope of bacteria, $\rho$-cymene barely inhibits microbial growth but can improve the antimicrobial properties of carvacrol by a synergistic process (Table **1**). Carvacrol /*p*-cymene synergistic system has been examined in carrot juice, where *Vibrio cholerae* growth was inhibited when both terpenes were spiked together into the carrot juice. However, this bacterial strain was not affected when these active compounds were applied separately [59]. According to Ultee *et al.* [56], *p*-cymene acts synergistically with carvacrol by expanding the membrane, which destabilizes the membrane. The previous mechanism supports the possible role of carvacrol as an exchanger of cations [56].

*Rosmarinus officinalis* essential oil contains mainly monoterpenes such as myrcene, α-pinene, borneol, β-pinene, verbenone, and camphor (Fig. **2**), phenolic diterpenes like carnosol and carnosic acid, and rosmarinic acid as a polyphenol. Most of these compounds were reported as antimicrobial agents disrupting bacterial membrane integrity [60]. Different studies have demonstrated their potential activity against Gram-negative (*E. coli* and *Klebsiella pneumonia*) and Gram-positive (*S. aureus* and *Bacillus subtilis*) bacteria [61]. Feeding animals with adequate amounts of rosemary or phenolic terpenoids utilized antimicrobial impacts in their meat products [62]. For example, a decrease in bacterial waste was observed in sheep after applying 200 mg/kg dry of *rosemary* in their nutrition to improve meat preservation, whereas the polyphenols content of *rosemary* influenced bacterial layers, hence hindering bacterial multiplication rates [63]. [63].

Oregano essential oil (rich in thymol and carvacrol) showed more significant antimicrobial activity against Gram-positive bacteria *S. aureus* than Gram-negative bacteria *E. coli* and *Pseudomonas aeruginosa* [64, 65]. The addition of

*Satureja horvatii* essential oil, rich in *p*-cymene (33.14%) and thymol (26.11%), to pork meat, inhibited *L. monocytogenes* growth. In this case, an improvement in treated food's flavor and odor was noticed after four days of storage compared to controls [66].

With the increasing demand for organic food, the necessity for resistant crops without synthetic chemicals has become an important issue. In the USA, organic food production is regulated by the United States Department of Agriculture (USDA) National Organic Program, whereas chemical antimicrobials approved for post-harvest treatment of organic products are very limited. Therefore, developing new GRAS additives such as essential oil is necessary, which can be used interestingly in ready-to-eat vegetables such as packaged salads [67]. In the same context, some polymeric films containing oregano essential oil, mainly carvacrol and citral, showed a drastic reduction in spoilage microbiota and inhibited the growth of some pathogens in spiked salads like *E. coli*, *Salmonella enterica*, and *Listeria spp.*, with a more significant effect on Gram-negative bacteria. Additionally, such essential oils enhanced the sensory qualities of the food products [68]. The mixture of carvacrol and nisin effectively eliminated *L. monocytogenes* in ready-to-eat carrots [69].In other studies, citron oil was more effective against *E. coli, Salmonella* spp., and *L. monocytogenes* in fruit-based salads [70]; oregano essential oil (carvacrol) was effective against *E. coli* O157:H7, *Campylobacter jejuni, S. enterica* and *L. monocytogenes* in apple juice [71]; and packaging films that incorporated carvacrol were able to reduce *Salmonella* populations in spiked bagged leaves [67].

Formulatingnanoemulsion could be an efficient and convincing strategy to build food-grade materials with the following goals; preservation of both the food product and the oil itself, emulsification of the oil to be applied in all media, avoid of using synthetic chemicals [72].

## The Site of Influence of Terpenoids and Essential Oils

Terpenoids have shown an expansive range of inhibitory potential against different Gram-positive and Gram-negative pathogenic microbes. Being lipophilic, they effectively saturate through the cell divider and cell layer. Interruption of layer trustworthiness and potential, spillage of cell substance, denaturation of cytoplasmic proteins, and inactivation of cell catalysts lead to bacterial cell passing [73]. The movement of terpenoids equals their synthetic designs; thus, the most bioactive ones will generally have a phenolic structure where the length of the aliphatic chains of terpene alcohols and the presence of twofold bonds are urgent. Bioactivity is expanded by the company of an aldehyde or ketone bunch; acetic acid derivation moiety (*e.g.*, geraniol and geranyl acetic acid derivation), or the situation of the hydroxyl bunch/s. The action of terpenoids with a nonphenolic structure relies upon the sort of alkyl substituent.

Additionally, α-isomers are less dynamic contrasted with β-isomers (pinene) and cis-isomers contrasted with trans isomers (*e.g.*, geraniol and nerol) [74]. Recently, investigations have shown that the site of action of terpenoids [75, 76] and tea tree oil [77 - 79] is at the cell membrane. In particular, tea tree oil was found to cause $K^+$ leakage in *E. coli* [77], at levels known to inhibit reducing the growth of this organism and stimulate autolysis [78]. Electron micrographs indicated the autolysis of stationary and exponential growth phase cells, which showed loss of electron-dense material, coagulation of cell cytoplasm, and formation of extracellular blebs. It has been proposed that the ability of tea tree oil to disrupt the permeability barrier of cell membrane structures and the accompanying loss of chemiosmotic control is the most likely source of its destructive action at minimum inhibitory levels [79].

Carvacrol and thymol have likewise been displayed to act straightforwardly against bacterial layers. The two mixtures have compelling external film deteriorating properties, as demonstrated by their improving impact on 1-N-phenylnapththlamine take-up and LPS discharge just as sharpening to cleansers. These compounds also inhibited bacterial growth at concentrations similar to those required for outer membrane disintegration and increased the permeability of the cytoplasmic membrane to ATP. In contrast, the same study showed that cinnamaldehyde, found to have an equal activity to carvacrol and thymol, and (+)-carvone, which was less active, exhibited none of these outer membrane disintegrating characteristics, nor did they affect levels of intracellular ATP. The mechanism by which thymol and carvacrol attack the outer membrane has yet to be clarified. However, observations that $MgCl_2$ does not affect the disintegrating ability of the two compounds suggest a mechanism other than chelation of divalent metal cations from the outer membrane exists [76].

In addition to their membrane disintegrating capability, terpenes such as 1,8-cineole, limonene, and nerolidol are known to increase the permeation of drugs such as 5-fluorouracil and estradiol through the human epidermis [80, 81], and model systems developed to mimic the stratum corneum [82 - 84]. These terpenes also impacted the structure of both the model matrix and actual stratum corneum by affecting lipid bilayers' inter and intra arrangements [83, 85].

β-pinene has also been shown to act at the membrane causing $K^+$, and $H^+$ leakage in yeast [86]. In the same study,β-pinene was also found to impact isolated yeast mitochondria by de-energizing the organelle followed by inhibition of respiration. This influence on respiration was attributed to effects on the cytochrome b region of the electron transport chain. Likewise, limonene, cyclohexane, and $\beta$–pinene were found to inhibit respiration and other energy-dependent processes associated with the membrane of the yeast *S. Cerevisiae* [87, 88].

The cyclic hydrocarbon tetralin [89], and the hydrocarbon terpenes α-pinene, limonene, and ɤ-terpinene [90], affected the structural and functional properties of artificial membranes. These compounds were shown to permeabilize the membranes making them swell and increase membrane fluidity. Additionally, inhibited respiratory enzymes lead to a partial dissipation of the pH gradient and electrical potential due to the more permeability to $H^+$ ions [89, 90]. Each of these Gradients is crucial to the energy system in a cell. The effect of many terpenoids on microbial oxygen uptake and oxidative phosphorylation has also been studied [91 - 93]. A large portion of the terpenoids tried were found to repress the two cycles. Specifically, the phenolic and non-phenolic alcohols showed the most grounded inhibitory impacts, trailed by aldehydes and ketones. The monoterpene hydrocarbons showed almost no movement by any means. It was proposed that the free ‾OH bunch moved by the alcohols might be a key to their action [91].

Terpenoid compounds also inhibit electron transport, proton translocation, phosphorylation steps, and other enzyme-dependent reactions in membranes prepared from *Pseudomonas aeruginosa*. Even in lower concentrations, specific terpenoids with functional groups such as alcohols, aldehydes, or phenolics interfere with membrane-integrated or associated enzyme proteins [94].

## Factors Affecting Antimicrobial Activity

Despite the mode group of an antimicrobial compound, two significant boundaries should be survived. These are its accessibility to the cell (for example, will adequately measures of the compound or the mixture be carried into contact with the cell?), and its ability to reach its site of action (*i.e.* can the compound pass through the physical barriers presented by the cell?). Since the cell envelope is known to be the site of action of terpenoids and regulates the flow of molecules in and out of the cell, it is necessary to include it in determining the factors affecting antimicrobial activity.

## CONCLUDING REMARKS

The constructions of the most encouraging terpenoids that could be utilized as layouts for additional examinations were recorded in this section. The energy research results show that distinctive terpenoids have been displayed to have critical sickness anticipation and therapy impacts, antitumor, calming, antibacterial, antiviral, and antimalarial effects, and likely capacities in a safe guideline, neuroprotection, and antiallergy. The movement displayed by terpenoids plays a significant part in advancing new medications and enhancements in existing treatment choices. Simultaneously, the natural exercises of numerous terpenoids still need top to bottom and careful exploration.

The terpenoids are plentiful in helpful plants, and terpenoids are used for the most part and have tremendous application potential and broad progression prospects. Terpenoids cannot exclusively be re-extracted and segregated from plants, yet, in addition, they can be acquired dependent on metabolic designing, engineered science, and biotransformation, which are new wellsprings of the amalgamation of terpenoids and address the deficiency of terpenoids. Fundamentally improved terpenoids can be ready in vast amounts by engineered science to address the issues of innovative medication work.

As of now, the system of activity of numerous terpenoids has not yet been clarified. The blend of "omics" innovation and sub-atomic organization pharmacology can be utilized to further review terpenoids' component and construction movement relationship. Screening the movement of terpenoids is a vital stage in the advancement of new medications. The mixtures with higher action can be straightforwardly formed into new drugs or altered as lead mixtures, and afterward screening, new combinations with critical activity. It is an effective method for the innovative work of the medication, and it is likewise a problem area in the field of typical item research. The new dosage form of terpenoids can be developed in combination with the recent progress of pharmaceutics to maximize its pharmacological activity. Terpenoids can be added as additives to health care products and cosmetics, with broad market prospects and economic benefits. Also, these types of scientific data, together with increasing consumer perception of the need for safer and more natural food-processing techniques and additives, have resulted in recent years in growing efforts at scientific and industrial levels devoted to the use of plant metabolites such as terpenoids among others, as food additives. The antibacterial exercises for a portion of these plant nutraceuticals have been tried *in vitro*, in the put-away food lattice, and sometimes in creature disease models, revealing insight into the mind-boggling field of food added substances and additives.

## CONSENT FOR PUBLICATION

Not applicable.

## CONFLICT OF INTEREST

The author declares no conflict of interest, financial or otherwise.

# ACKNOWLEDGEMENT

Declared none.

# REFERENCES

[1]    Liao, P.; Hemmerlin, A.; Bach, T.J.; Chye, M.L. The potential of the mevalonate pathway for enhanced isoprenoid production. *Biotechnol. Adv.,* **2016**, *34*(5), 697-713.
[http://dx.doi.org/10.1016/j.biotechadv.2016.03.005] [PMID: 26995109]

[2]    Dorman, H.J.D.; Deans, S.G. Antimicrobial agents from plants: antibacterial activity of plant volatile oils. *J. Appl. Microbiol.,* **2000**, *88*(2), 308-316.
[http://dx.doi.org/10.1046/j.1365-2672.2000.00969.x] [PMID: 10736000]

[3]    Mastelic, J.; Politeo, O.; Jerkovic, I.; Radosevic, N. Composition and antimicrobial activity of *Helichrysumitalicum*essential oils and its terpene and the terpenoid fractions. *Chem. Nat. Compd.,* **2005**, *41*(1), 35-40.
[http://dx.doi.org/10.1007/s10600-005-0069-z]

[4]    Vuuren, S.F.; Viljoen, A.M. Antimicrobial activity of limonene enantiomers and 1,8-cineole alone and in combination. *Flavour Fragrance J.,* **2007**, *22*(6), 540-544.
[http://dx.doi.org/10.1002/ffj.1843]

[5]    Ali, S.M.; Khan, A.A.; Ahmed, I.; Musaddiq, M.; Ahmed, K.S.; Polasa, H.; Rao, L.V.; Habibullah, C.M.; Sechi, L.A.; Ahmed, N. Antimicrobial activities of Eugenol and Cinnamaldehyde against the human gastric pathogen Helicobacter pylori. *Ann. Clin. Microbiol. Antimicrob.,* **2005**, *4*(1), 20.
[http://dx.doi.org/10.1186/1476-0711-4-20] [PMID: 16371157]

[6]    Yadav, M.K.; Chae, S.W.; Im, G.J.; Chung, J.W.; Song, J.J. Eugenol: a phyto-compound effective against methicillin-resistant and methicillin-sensitive Staphylococcus aureus clinical strain biofilms. *PLoS One,* **2015**, *10*(3), e0119564.
[http://dx.doi.org/10.1371/journal.pone.0119564] [PMID: 25781975]

[7]    Rathinam, P.; Vijay Kumar, H.S.; Viswanathan, P. Eugenol exhibits anti-virulence properties by competitively binding to quorum sensing receptors. *Biofouling,* **2017**, *33*(8), 624-639.
[http://dx.doi.org/10.1080/08927014.2017.1350655] [PMID: 28792229]

[8]    Zhang, Y.B.; Liu, X.Y.; Jiang, P.P.; Li, W.D.; Wang, Y.F. Mechanism and antibacterial activity of cinnamaldehyde against Escherichia coli and *Staphylococcus aureus. Xiandai Shipin Keji,* **2015**, *31*(5), 31-35.

[9]    Griffin, S.G.; Wyllie, S.G.; Markham, J.L.; Leach, D.N. The role of structure and molecular properties of terpenoids in determining their antimicrobial activity. *Flavour Fragrance J.,* **1999**, *14*(5), 322-332.
[http://dx.doi.org/10.1002/(SICI)1099-1026(199909/10)14:5<322::AID-FFJ837>3.0.CO;2-4]

[10]   Sebei, K.; Sakouhi, F.; Herchi, W.; Khouja, M.; Boukhchina, S. Chemical composition and antibacterial activities of seven Eucalyptus species essential oils leaves. *Biol. Res.,* **2015**, *48*(1), 7-5.
[http://dx.doi.org/10.1186/0717-6287-48-7] [PMID: 25654423]

[11]   Dogan, G.; Kara, N.; Bagci, E.; Gur, S. Chemical composition and biological activities of leaf and fruit essential oils from *Eucalyptus camaldulensis. Z. Naturforsch. C J. Biosci.,* **2017**, *72*(11-12), 483-489.
[http://dx.doi.org/10.1515/znc-2016-0033] [PMID: 28640755]

[12]   Wang, T.H.; Hsia, S.M.; Wu, C.H.; Ko, S.Y.; Chen, M.Y.; Shih, Y.H.; Shieh, T.M.; Chuang, L.C.; Wu, C.Y. Evaluation of the antibacterial potential of liquid and vapor phase phenolic essential oil compounds against oral microorganisms. *PLoS One,* **2016**, *11*(9), e0163147.
[http://dx.doi.org/10.1371/journal.pone.0163147] [PMID: 27681039]

[13]   Chan, Y.W.; Siow, K.S.; Ng, P.Y.; Gires, U.; Yeop Majlis, B. Plasma polymerized carvone as an antibacterial and biocompatible coating. *Mater. Sci. Eng. C,* **2016**, *68*, 861-871.
[http://dx.doi.org/10.1016/j.msec.2016.07.040] [PMID: 27524089]

[14]   Shahbazi, Y. Chemical Composition and *in vitro* Antibacterial Activity of *Mentha spicata* Essential Oil against Common Food-Borne Pathogenic Bacteria. *J. Pathogens,* **2015**, *2015*(2), 916305.
       [PMID: 26351584]

[15]   Park, Y.J.; Baskar, T.B.; Yeo, S.K.; Arasu, M.V.; Al-Dhabi, N.A.; Lim, S.S.; Park, S.U. Composition of volatile compounds and *in vitro* antimicrobial activity of nine *Mentha* spp. *Springerplus,* **2016**, *5*(1), 1628.
       [http://dx.doi.org/10.1186/s40064-016-3283-1] [PMID: 27722047]

[16]   Pattnaik, S.; Subramanyam, V.R.; Bapaji, M.; Kole, C.R. Antibacterial and antifungal activity of aromatic constituents of essential oils. *Microbios,* **1997**, *89*(358), 39-46.
       [PMID: 9218354]

[17]   Osawa, K.; Saeki, T.; Yasuda, H.; Hamashima, H.; Sasatsu, M.; Arai, T. The antibacterial activities of peppermint oil and green tea polyphenols, alone and in combination, against enterohemorrhagic *Escherichia coli. Biocontrol Sci.,* **1999**, *4*(1), 1-7.
       [http://dx.doi.org/10.4265/bio.4.1]

[18]   Inouye, S.; Takizawa, T.; Yamaguchi, H. Antibacterial activity of essential oils and their major constituents against respiratory tract pathogens by gaseous contact. *J. Antimicrob. Chemother.,* **2001**, *47*(5), 565-573.
       [http://dx.doi.org/10.1093/jac/47.5.565] [PMID: 11328766]

[19]   Trombetta, D.; Castelli, F.; Sarpietro, M.G.; Venuti, V.; Cristani, M.; Daniele, C.; Saija, A.; Mazzanti, G.; Bisignano, G. Mechanisms of antibacterial action of three monoterpenes. *Antimicrob. Agents Chemother.,* **2005**, *49*(6), 2474-2478.
       [http://dx.doi.org/10.1128/AAC.49.6.2474-2478.2005] [PMID: 15917549]

[20]   Raut, J.S.; Shinde, R.B.; Chauhan, N.M.; Mohan Karuppayil, S. Terpenoids of plant origin inhibit morphogenesis, adhesion, and biofilm formation by *Candida albicans. Biofouling,* **2013**, *29*(1), 87-96.
       [http://dx.doi.org/10.1080/08927014.2012.749398] [PMID: 23216018]

[21]   Xu, Y.F.; Lian, D.W.; Chen, Y.Q.; Cai, Y.F.; Zheng, Y.F.; Fan, P.L.; Ren, W.K.; Fu, L.J.; Li, Y.C.; Xie, J.H.; Cao, H.Y.; Tan, B.; Su, Z.R.; Huang, P. *in vitro* and in *vivo* antibacterial activities of patchouli alcohol, a naturally occurring tricyclicsesquiterpene, against Helicobacter pylori infection. *Antimicrob. Agents Chemother.,* **2017**, *61*(6), e00122-17.
       [http://dx.doi.org/10.1128/AAC.00122-17] [PMID: 28320722]

[22]   Appalasamy, S.; Lo, K.Y.; Ch'ng, S.J.; Nornadia, K.; Othman, A.S.; Chan, L.K. Antimicrobial activity of artemisinin and precursor derived from *in vitro* plantlets of Artemisia annua L. *BioMed Res. Int.,* **2014**, *2014*(31), 215872.
       [PMID: 24575401]

[23]   Juteau, F.; Masotti, V.; Bessière, J.M.; Dherbomez, M.; Viano, J. Antibacterial and antioxidant activities of Artemisia annua essential oil. *Fitoterapia,* **2002**, *73*(6), 532-535.
       [http://dx.doi.org/10.1016/S0367-326X(02)00175-2] [PMID: 12385883]

[24]   Kim, W.S.; Choi, W.J.; Lee, S.; Kim, W.J.; Lee, D.C.; Sohn, U.D.; Shin, H.S.; Kim, W. Anti-inflammatory, Antioxidant and Antimicrobial Effects of Artemisinin Extracts from *Artemisia annua* L. *Korean J. Physiol. Pharmacol.,* **2014**, *19*(1), 21-27.
       [http://dx.doi.org/10.4196/kjpp.2015.19.1.21] [PMID: 25605993]

[25]   Cheng, H.J.; Liu, J.; Zhang, G. The anti–bacterial effect of andrographolide against Pseudomonas aeruginosa biofilm and azithromycin. *Chinese J Microecology,* **2012**, *24*, 120-123.

[26]   Banerjee, M.; Parai, D.; Chattopadhyay, S.; Mukherjee, S.K. Andrographolide: antibacterial activity against common bacteria of human health concern and possible mechanism of action. *Folia Microbiol. (Praha),* **2017**, *62*(3), 237-244.
       [http://dx.doi.org/10.1007/s12223-017-0496-9] [PMID: 28097636]

[27]   Zhao, L.X.; Wang, Z.W.; Zheng, C.J. synthesis, characterization and antibacterial activity of oleanolic

acid C–28 modification Liaoning Normal University, Natural Science Edition. **2013**, *36*, 529-533.

[28]    Kim, S.; Lee, H.; Lee, S.; Yoon, Y.; Choi, K.H. Antimicrobial action of oleanolic acid on Listeria monocytogenes, Enterococcus faecium, and Enterococcus faecalis. *PLoS One,* **2015**, *10*(3), e0118800.
[http://dx.doi.org/10.1371/journal.pone.0118800] [PMID: 25756202]

[29]    Kurek, A.; Nadkowska, P.; Pliszka, S.; Wolska, K.I. Modulation of antibiotic resistance in bacterial pathogens by oleanolic acid and ursolic acid. *Phytomedicine,* **2012**, *19*(6), 515-519.
[http://dx.doi.org/10.1016/j.phymed.2011.12.009] [PMID: 22341643]

[30]    Copp, B.R. Antimycobacterial natural products. *Nat. Prod. Rep.,* **2003**, *20*(6), 535-557.
[http://dx.doi.org/10.1039/b212154a] [PMID: 14700198]

[31]    Hassan, S.T.S.; Masarčíková, R.; Berchová, K. Bioactive natural products with anti-herpes simplex virus properties. *J. Pharm. Pharmacol.,* **2015**, *67*(10), 1325-1336.
[http://dx.doi.org/10.1111/jphp.12436] [PMID: 26060043]

[32]    Armaka, M.; Papanikolaou, E.; Sivropoulou, A.; Arsenakis, M. Antiviral properties of isoborneol, a potent inhibitor of herpes simplex virus type 1. *Antiviral Res.,* **1999**, *43*(2), 79-92.
[http://dx.doi.org/10.1016/S0166-3542(99)00036-4] [PMID: 10517310]

[33]    Astani, A.; Reichling, J.; Schnitzler, P. Comparative study on the antiviral activity of selected monoterpenes derived from essential oils. *Phytother. Res.,* **2010**, *24*(5), 673-679.
[http://dx.doi.org/10.1002/ptr.2955] [PMID: 19653195]

[34]    Brezáni, V.; Leláková, V.; Hassan, S.; Berchová-Bímová, K.; Nový, P.; Klouček, P.; Maršík, P.; Dall'Acqua, S.; Hošek, J.; Šmejkal, K. Anti-infectivity against herpes simplex virus and selected microbes and antiInflammatory activities of compounds isolated from *Eucalyptus globulus* Labill. *Viruses,* **2018**, *10*(7), 360-18.
[http://dx.doi.org/10.3390/v10070360] [PMID: 29986399]

[35]    Cheng, H.Y.; Lin, T-C.; Yang, C-M.; Wang, K-C.; Lin, L-T.; Lin, C-C. Putranjivain A from Euphorbia jolkini inhibits both virus entry and late stage replication of herpes simplex virus type 2 *in vitro. J. Antimicrob. Chemother.,* **2004**, *53*(4), 577-583.
[http://dx.doi.org/10.1093/jac/dkh136] [PMID: 14998984]

[36]    Kurokawa, M.; Basnet, P.; Ohsugi, M.; Hozumi, T.; Kadota, S.; Namba, T.; Kawana, T.; Shiraki, K. Anti-herpes simplex virus activity of moronic acid purified from Rhus javanica *in vitro* and *in vivo. J. Pharmacol. Exp. Ther.,* **1999**, *289*(1), 72-78.
[PMID: 10086989]

[37]    Pei, Y.; Du, Q.; Liao, P.Y.; Chen, Z.P.; Wang, D.; Yang, C.R.; Kitazato, K.; Wang, Y.F.; Zhang, Y.J. Notoginsenoside ST-4 inhibits virus penetration of herpes simplex virus *in vitro. J. Asian Nat. Prod. Res.,* **2011**, *13*(6), 498-504.
[http://dx.doi.org/10.1080/10286020.2011.571645] [PMID: 21623512]

[38]    Utsunomiya, T.; Kobayashi, M.; Herndon, D.N.; Pollard, R.B.; Suzuki, F.; Glycyrrhizin, S.F. Glycyrrhizin (20β-carboxy-11-oxo-30-norolean-12-en-3β-yl-2-O-β-d-glucopyranuronosyl-α-d-glu-copyranosiduronic acid) improves the resistance of thermally injured mice to opportunistic infection of herpes simplex virus type 1. *Immunol. Lett.,* **1995**, *44*(1), 59-66.
[http://dx.doi.org/10.1016/0165-2478(94)00183-R] [PMID: 7721345]

[39]    Wohlfarth, C.; Efferth, T. Natural products as promising drug candidates for the treatment of hepatitis B and C. *Acta Pharmacol. Sin.,* **2009**, *30*(1), 25-30.
[http://dx.doi.org/10.1038/aps.2008.5] [PMID: 19060918]

[40]    Wintachai, P.; Kaur, P.; Lee, R.C.H.; Ramphan, S.; Kuadkitkan, A.; Wikan, N.; Ubol, S.; Roytrakul, S.; Chu, J.J.H.; Smith, D.R. Activity of andrographolide against chikungunya virus infection. *Sci. Rep.,* **2015**, *5*(1), 14179.
[http://dx.doi.org/10.1038/srep14179] [PMID: 26384169]

[41]    Panraksa, P.; Ramphan, S.; Khongwichit, S.; Smith, D.R. Activity of andrographolide against dengue

virus. *Antiviral Res.,* **2017**, *139*, 69-78.
[http://dx.doi.org/10.1016/j.antiviral.2016.12.014] [PMID: 28034742]

[42]     Harada, S. The broad anti-viral agent glycyrrhizin directly modulates the fluidity of plasma membrane and HIV-1 envelope. *Biochem. J.,* **2005**, *392*(1), 191-199.
[http://dx.doi.org/10.1042/BJ20051069] [PMID: 16053446]

[43]     Fujioka, T.; Kashiwada, Y.; Kilkuskie, R.E.; Cosentino, L.M.; Ballas, L.M.; Jiang, J.B.; Janzen, W.P.; Chen, I.S.; Lee, K.H. Anti-AIDS agents, 11. Betulinic acid and platanic acid as anti-HIV principles from Syzigium claviflorum, and the anti-HIV activity of structurally related triterpenoids. *J. Nat. Prod.,* **1994**, *57*(2), 243-247.
[http://dx.doi.org/10.1021/np50104a008] [PMID: 8176401]

[44]     Cichewicz, R.H.; Kouzi, S.A. Chemistry, biological activity, and chemotherapeutic potential of betulinic acid for the prevention and treatment of cancer and HIV infection. *Med. Res. Rev.,* **2004**, *24*(1), 90-114.
[http://dx.doi.org/10.1002/med.10053] [PMID: 14595673]

[45]     Zhu, Y.M.; Shen, J.K.; Wang, H.K.; Cosentino, L.M.; Lee, K.H. Synthesis and anti-HIV activity of oleanolic acid derivatives. *Bioorg. Med. Chem. Lett.,* **2001**, *11*(24), 3115-3118.
[http://dx.doi.org/10.1016/S0960-894X(01)00647-3] [PMID: 11720855]

[46]     Esimone, C.O.; Eck, G.; Nworu, C.S.; Hoffmann, D.; Überla, K.; Proksch, P.; Charles, O.E.; Gero, E.; Chukwuemeka, S.N.; Dennis, H.; Klaus, U.; Peter, P. Dammarenolic acid, a secodammarane triterpenoid from Aglaia sp. shows potent anti-retroviral activity *in vitro. Phytomedicine,* **2010**, *17*(7), 540-547.
[http://dx.doi.org/10.1016/j.phymed.2009.10.015] [PMID: 19962871]

[47]     Ma, C.; Nakamura, N.; Miyashiro, H.; Hattori, M.; Shimotohno, K. Inhibitory effects of constituents from *Cynomorium songaricum* and related triterpene derivatives on HIV-1 protease. *Chem. Pharm. Bull. (Tokyo),* **1999**, *47*(2), 141-145.
[http://dx.doi.org/10.1248/cpb.47.141] [PMID: 10071849]

[48]     Bakkali, F.; Averbeck, S.; Averbeck, D.; Idaomar, M. Biological effects of essential oils – A review. *Food Chem. Toxicol.,* **2008**, *46*(2), 446-475.
[http://dx.doi.org/10.1016/j.fct.2007.09.106] [PMID: 17996351]

[49]     Breitmaier, E. *Terpenes: importance, general structure, and biosynthesis. Wein- heim*; Wiley-VCH Verlag: Germany, **2006**, pp. 1-9.
[http://dx.doi.org/10.1002/9783527609949]

[50]     Hyldgaard, M.; Mygind, T.; Meyer, R.L. Essential oils in food preservation: mode of action, synergies, and interactions with food matrix components. *Front. Microbiol.,* **2012**, *3*, 12.
[http://dx.doi.org/10.3389/fmicb.2012.00012] [PMID: 22291693]

[51]     Ortega-Ramirez, L.A.; Rodriguez-Garcia, I.; Leyva, J.M.; Cruz-Valenzuela, M.R.; Silva-Espinoza, B.A.; Gonzalez-Aguilar, G.A.; Siddiqui, M.W.; Ayala-Zavala, J.F. Potential of medicinal plants as antimicrobial and antioxidant agents in food industry: a hypothesis. *J. Food Sci.,* **2014**, *79*(2), R129-R137.
[http://dx.doi.org/10.1111/1750-3841.12341] [PMID: 24446991]

[52]     Schelz, Z.; Molnar, J.; Hohmann, J. Antimicrobial and antiplasmid activities of essential oils. *Fitoterapia,* **2006**, *77*(4), 279-285.
[http://dx.doi.org/10.1016/j.fitote.2006.03.013] [PMID: 16690225]

[53]     Duschatzky, C.B.; Possetto, M.L.; Talarico, L.B.; García, C.C.; Michis, F.; Almeida, N.V.; de Lampasona, M.P.; Schuff, C.; Damonte, E.B. Evaluation of chemical and antiviral properties of essential oils from South American plants. *Antivir. Chem. Chemother.,* **2005**, *16*(4), 247-251.
[http://dx.doi.org/10.1177/095632020501600404] [PMID: 16130522]

[54]     Pawar, V.C.; Thaker, V.S. *in vitro* efficacy of 75 essential oils against Aspergillus niger. *Mycoses,* **2006**, *49*(4), 316-323.

[http://dx.doi.org/10.1111/j.1439-0507.2006.01241.x] [PMID: 16784447]

[55]   Koutsoudaki, C.; Krsek, M.; Rodger, A. Chemical composition and antibacterial activity of the essential oil and the gum of Pistacia lentiscus Var. chia. *J. Agric. Food Chem.*, **2005**, *53*(20), 7681-7685.
[http://dx.doi.org/10.1021/jf050639s] [PMID: 16190616]

[56]   Ultee, A.; Bennik, M.H.J.; Moezelaar, R. The phenolic hydroxyl group of carvacrol is essential for action against the food-borne pathogen Bacillus cereus. *Appl. Environ. Microbiol.*, **2002**, *68*(4), 1561-1568.
[http://dx.doi.org/10.1128/AEM.68.4.1561-1568.2002] [PMID: 11916669]

[57]   Šarac, Z.; Matejić, J.S.; Stojanović-Radić, Z.Z.; Veselinović, J.B.; Džamić, A.M.; Bojović, S.; Marin, P.D. Biological activity of Pinus nigra terpenes—Evaluation of FtsZ inhibition by selected compounds as contribution to their antimicrobial activity. *Comput. Biol. Med.*, **2014**, *54*, 72-78.
[http://dx.doi.org/10.1016/j.compbiomed.2014.08.022] [PMID: 25217763]

[58]   Burt, S. Essential oils: their antibacterial properties and potential applications in foods—a review. *Int. J. Food Microbiol.*, **2004**, *94*(3), 223-253.
[http://dx.doi.org/10.1016/j.ijfoodmicro.2004.03.022] [PMID: 15246235]

[59]   Rattanachaikunsopon, P.; Phumkhachorn, P. Assessment of factors influencing antimicrobial activity of carvacrol and cymene against Vibrio cholerae in food. *J. Biosci. Bioeng.*, **2010**, *110*(5), 614-619.
[http://dx.doi.org/10.1016/j.jbiosc.2010.06.010] [PMID: 20638331]

[60]   Santoyo, S.; Cavero, S.; Jaime, L.; Ibañez, E.; Señoráns, F.J.; Reglero, G. Chemical composition and antimicrobial activity of Rosmarinus officinalis L. essential oil obtained *via* supercritical fluid extraction. *J. Food Prot.*, **2005**, *68*(4), 790-795.
[http://dx.doi.org/10.4315/0362-028X-68.4.790] [PMID: 15830672]

[61]   Okoh, O.O.; Sadimenko, A.P.; Afolayan, A.J. Comparative evaluation of the antibacterial activities of the essential oils of Rosmarinus officinalis L. obtained by hydrodistillation and solvent free microwave extraction methods. *Food Chem.*, **2010**, *120*(1), 308-312.
[http://dx.doi.org/10.1016/j.foodchem.2009.09.084]

[62]   Moñino, I.; Martínez, C.; Sotomayor, J.A.; Lafuente, A.; Jordán, M.J. Polyphenolic transmission to Segureno lamb meat from ewes' diet supplemented with the distillate from rosemary (Rosmarinus officinalis) leaves. *J. Agric. Food Chem.*, **2008**, *56*(9), 3363-3367.
[http://dx.doi.org/10.1021/jf7036856] [PMID: 18422334]

[63]   Ortuño, J.; Serrano, R.; Jordán, M.J.; Bañón, S. Shelf life of meat from lambs given essential oil-free rosemary extract containing carnosic acid plus carnosol at 200 or 400mgkg−1. *Meat Sci.*, **2014**, *96*(4), 1452-1459.
[http://dx.doi.org/10.1016/j.meatsci.2013.11.021] [PMID: 24412737]

[64]   Aguirre, A.; Borneo, R.; León, A.E. Antimicrobial, mechanical and barrier properties of triticale protein films incorporated with oregano essential oil. *Food Biosci.*, **2013**, *1*, 2-9.
[http://dx.doi.org/10.1016/j.fbio.2012.12.001]

[65]   Pelissari, F.M.; Grossmann, M.V.E.; Yamashita, F.; Pineda, E.A.G. Antimicrobial, mechanical, and barrier properties of cassava starch-chitosan films incorporated with oregano essential oil. *J. Agric. Food Chem.*, **2009**, *57*(16), 7499-7504.
[http://dx.doi.org/10.1021/jf9002363] [PMID: 19627142]

[66]   Bukvički, D.; Stojković, D.; Soković, M.; Vannini, L.; Montanari, C.; Pejin, B.; Savić, A.; Veljić, M.; Grujić, S.; Marin, P.D. Satureja horvatii essential oil: *in vitro* antimicrobial and antiradical properties and *in situ* control of Listeria monocytogenes in pork meat. *Meat Sci.*, **2014**, *96*(3), 1355-1360.
[http://dx.doi.org/10.1016/j.meatsci.2013.11.024] [PMID: 24342186]

[67]   Zhu, L.; Olsen, C.; McHugh, T.; Friedman, M.; Jaroni, D.; Ravishankar, S. Apple, carrot, and hibiscus edible films containing the plant antimicrobials carvacrol and cinnamaldehyde inactivate Salmonella Newport on organic leafy greens in sealed plastic bags. *J. Food Sci.*, **2014**, *79*(1), M61-M66.

[http://dx.doi.org/10.1111/1750-3841.12318] [PMID: 24460771]

[68]   Muriel-Galet, V.; Cerisuelo, J.P.; López-Carballo, G.; Lara, M.; Gavara, R.; Hernández-Muñoz, P. Development of antimicrobial films for microbiological control of packaged salad. *Int. J. Food Microbiol.,* **2012**, *157*(2), 195-201.
[http://dx.doi.org/10.1016/j.ijfoodmicro.2012.05.002] [PMID: 22633535]

[69]   Ndoti-Nembe, A.; Vu, K.D.; Doucet, N.; Lacroix, M. Antimicrobial effects of essential oils, nisin, and irradiation treatments against Listeria monocytogenes on ready-to-eat carrots. *J. Food Sci.,* **2015**, *80*(4), M795-M799.
[http://dx.doi.org/10.1111/1750-3841.12832] [PMID: 25807882]

[70]   Belletti, N.; Lanciotti, R.; Patrignani, F.; Gardini, F. Antimicrobial efficacy of citron essential oil on spoilage and pathogenic microorganisms in fruit-based salads. *J. Food Sci.,* **2008**, *73*(7), M331-M338.
[http://dx.doi.org/10.1111/j.1750-3841.2008.00866.x] [PMID: 18803716]

[71]   Friedman, M.; Henika, P.R.; Levin, C.E.; Mandrell, R.E. Antibacterial activities of plant essential oils and their components against Escherichia coli O157:H7 and Salmonella enterica in apple juice. *J. Agric. Food Chem.,* **2004**, *52*(19), 6042-6048.
[http://dx.doi.org/10.1021/jf0495340] [PMID: 15366861]

[72]   Donsì, F.; Annunziata, M.; Sessa, M.; Ferrari, G. Nanoencapsulation of essential oils to enhance their antimicrobial activity in foods. *Lebensm. Wiss. Technol.,* **2011**, *44*(9), 1908-1914.
[http://dx.doi.org/10.1016/j.lwt.2011.03.003]

[73]   Raut, JS; Karuppayil, SM *A status review on the medicinal properties ofessential oils.,* **2014**.

[74]   Patra, A.K. An Overview of Antimicrobial Properties of Different Classes of Phytochemicals. In: *Dietary Phytochemicals and Microbes*; Patra, A.K., Ed.; Springer Netherlands: Dordrecht, **2012**; pp. 1-32.
[http://dx.doi.org/10.1007/978-94-007-3926-0_1]

[75]   Sikkema, J.; de Bont, J.A.; Poolman, B. Mechanisms of membrane toxicity of hydrocarbons. *Microbiol. Rev.,* **1995**, *59*(2), 201-222.
[http://dx.doi.org/10.1128/mr.59.2.201-222.1995] [PMID: 7603409]

[76]   Helander, I.M.; Alakomi, H.L.; Latva-Kala, K.; Mattila-Sandholm, T.; Pol, I.; Smid, E.J.; Gorris, L.G.M.; von Wright, A. Characterization of the action of selected essential oil components on Gram-negative bacteria. *J. Agric. Food Chem.,* **1998**, *46*(9), 3590-3595.
[http://dx.doi.org/10.1021/jf980154m]

[77]   Cox, S.D.; Gustafson, J.E.; Mann, C.M.; Markham, J.L.; Liew, Y.C.; Hartland, R.P.; Bell, H.C.; Warmington, J.R.; Wyllie, S.G. Tea tree oil causes K $^+$ leakage and inhibits respiration in *Escherichia coli*. *Lett. Appl. Microbiol.,* **1998**, *26*(5), 355-358.
[http://dx.doi.org/10.1046/j.1472-765X.1998.00348.x] [PMID: 9674165]

[78]   Gustafson, J.E.; Liew, Y.C.; Chew, S.; Markham, J.; Bell, H.C.; Wyllie, S.G.; Warmington, J.R. Effects of tea tree oil on Escherichia coli. *Lett. Appl. Microbiol.,* **1998**, *26*(3), 194-198.
[http://dx.doi.org/10.1046/j.1472-765X.1998.00317.x] [PMID: 9569708]

[79]   Cox, S.D.; Mann, C.M.; Markham, J.L.; Bell, H.C.; Gustafson, J.E.; Warmington, J.R.; Wyllie, S.G. The mode of antimicrobial action of the essential oil of Melaleuca alternifolia (tea tree oil). *J. Appl. Microbiol.,* **2000**, *88*(1), 170-175.
[http://dx.doi.org/10.1046/j.1365-2672.2000.00943.x] [PMID: 10735256]

[80]   Williams, A.C.; Barry, B.W. Terpenes and the lipid-protein-partitioning theory of skin penetration enhancement. *Pharm. Res.,* **1991**, *8*(1), 17-24.
[http://dx.doi.org/10.1023/A:1015813803205] [PMID: 2014203]

[81]   Williams, A.C.; Barry, B.W. The enhancement index concept applied to terpene penetration enhancers for human skin and model lipophilic (oestradiol) and hydrophilic (5-fluorouracil) drugs. *Int. J. Pharm.,* **1991**, *74*(2-3), 157-168.

[http://dx.doi.org/10.1016/0378-5173(91)90232-D]

[82]   Moghimi, H.R.; Williams, A.C.; Barry, B.W. A lamellar matrix model for stratum corneum intercellular lipids. V. Effects of terpene penetration enhancers on the structure and thermal behaviour of the matrix. *Int. J. Pharm.,* **1997**, *146*(1), 41-54.
[http://dx.doi.org/10.1016/S0378-5173(96)04766-7]

[83]   Moghimi, H.; Williams, A.C.; Barry, B.W. A lamellar matrix model for stratum corneum intercellular lipids IV. Effects of terpene penetration enhancers on the permeation of 5-fluorouracil and oestradiol through the matrix. *Int. J. Pharm.,* **1996**, *145*(1-2), 49-59.
[http://dx.doi.org/10.1016/S0378-5173(96)04716-3]

[84]   Cornwell, P.A.; Barry, B.W.; Bouwstra, J.A.; Gooris, G.S. Modes of action of terpene penetration enhancers in human skin; Differential scanning calorimetry, small-angle X-ray diffraction and enhancer uptake studies. *Int. J. Pharm.,* **1996**, *127*(1), 9-26.
[http://dx.doi.org/10.1016/0378-5173(95)04108-7]

[85]   Yamane, M.A.; Williams, A.C.; Barry, B.W. Terpene penetration enhancers in propylene glycol/water co-solvent systems: effectiveness and mechanism of action. *J. Pharm. Pharmacol.,* **2011**, *47*(12A), 978-989.
[http://dx.doi.org/10.1111/j.2042-7158.1995.tb03282.x] [PMID: 8932680]

[86]   Uribe, S.; Ramirez, J.; Peña, A. Effects of beta-pinene on yeast membrane functions. *J. Bacteriol.,* **1985**, *161*(3), 1195-1200.
[http://dx.doi.org/10.1128/jb.161.3.1195-1200.1985] [PMID: 3156123]

[87]   Uribe, S.; Rangel, P.; Espínola, G.; Aguirre, G. Effects of cyclohexane, an industrial solvent, on the yeast Saccharomyces cerevisiae and on isolated yeast mitochondria. *Appl. Environ. Microbiol.,* **1990**, *56*(7), 2114-2119.
[http://dx.doi.org/10.1128/aem.56.7.2114-2119.1990] [PMID: 2202257]

[88]   Uribe, S.; Pena, A. Toxicity of allelopathic monoterpene suspensions on yeast dependence on droplet size. *J. Chem. Ecol.,* **1990**, *16*(4), 1399-1408.
[http://dx.doi.org/10.1007/BF01021035] [PMID: 24263736]

[89]   Sikkema, J; Poolman, B; Konings, WN; de Bont, JA Effects of the membrane action of tetralin on the functional and structural properties of artificial and bacterial membranes. J Bacteriol. May **1992**, *174*(9), 2986-2992.

[90]   Sikkema, J; de Bont, JA; Poolman, B Interactions of cyclic hydrocarbons with biological membranes. J Biol Chem. Mar 18 **1994**, *269*(11), 8022-8028.

[91]   Knobloch, K.; Weigand, H.; Weis, N.; Schwärm, H.; Vigenschow, H. Action of terpenoids on energy metabolism. *Progress in Essential Oil Research.,* **1986**, , 429-445.

[92]   Knobloch, K.; Weis, N.; Weigand, H. Mechanism of antimicrobial activity of essential oils. *Planta Med.,* **1986**, *52*(6), 556.
[http://dx.doi.org/10.1055/s-2007-969370] [PMID: 17345526]

[93]   Weis, N.; Weigand, H.; Knobloch, K. On the enfluenceofterpene and phenylpropane derivatives on bacterial respiration and oxidative phosphorylation. *Biol. Chem.,* **1985**, *366*, 866.

[94]   Knobloch, E.; Pauli, A.; Iberl, B.; Wies, N.; Weigand, H. Mode of action of essential oil components on whole cells of bacteria and fungi in plate tests. *Bioflavour'87.,* **1998**, , 287-299.

# CHAPTER 13

# Terpenoids in Propolis and Geopropolis and Applications

**Jorddy Neves Cruz[1,2,*], Mozaniel Santana de Oliveira[2], Lindalva Maria de Meneses Costa Ferreira[3], Daniel Santiago Pereira[1], João Paulo de Holanda Neto[4], Aline Carla de Medeiros[5], Patrício Borges Maracajá[6] and Antônio Pedro da Silva Souza Filho[1]**

[1] *Laboratory of Agro-Industry, Embrapa Eastern Amazon, Belem, Pará, Brazil*

[2] *Adolpho Ducke Laboratory, Paraense Emílio Goeldi Museum, Belém, Pará, Brazil*

[3] *Laboratório de Nanotecnologia Farmacêutica, Faculdade de Farmácia, Universidade Federal do Pará, Brazil*

[4] *Belém, Pará, Brazil*

[5] *Federal Institute of education, Science and Technology of Sertão Pernambucano, Oricuri, Pernambuco, Brasil*

[6] *Federal University of Campina Grande, Paraiba, Brasil*

**Abstract:** Propolis is a resin, which comes from from bee colonies and is considered a natural antibiotic, without serious side effects, compared to synthetic treatments, and has several pharmacological properties. Geopropolis is a mixture of clay and propolis produced by species of stingless bees of the genus Melipona, hence the name geopropolis. It is formed in the same way as propolis produced by other bee species. In this review, we aim to address general aspects related to terpenoids present in propolis and geopropolis. Here, we report the main terpenoids, their chemical structure, and pharmacological and food industry applications.

**Keywords:** Bess, Food Industry, Pharmaceutical Properties, Stingless Bees, Terpenoids.

## INTRODUCTION

Propolis is composed of approximately 50-60% of resins and aromatic balsams, 30-40% of waxes, 5-10% of essential oils, and up to 5% of other substances. Microelements such as aluminum, calcium, strontium, iron, copper, manganese,

---

*Corresponding author Jorddy Neves Cruz: Laboratory of Agro-Industry, Embrapa Eastern Amazon, Belem, Pará, Brazil. E-mail: jorddynevescruz@gmail.com

**Mozaniel Santana de Oliveira & Antônio Pedro da Silva Souza Filho (Eds.)**

magnesium, silicon, titanium, bromine, zinc, and vitamins B1, B2, B6, C, and E are also present [1].

Its insoluble portion is composed of organic matter, plant tissues, pollen grains, and other substances. The soluble constituents of propolis, obtained using organic solvents, are divided into waxy materials (~30%), balsams, essential oils, and phenolic derivatives (~60%) [2].

According to Sforcin (2007) [3], bees use propolis to seal holes in their hives, smooth out the inner walls, and cover the remains of intruders, who have died inside the hive in order to prevent their decomposition. Propolis protects the colony from diseases due to its antiseptic efficacy and antimicrobial properties. Studies report that propolis's chemical composition can vary according to regional seasonality, which can influence its potential action [4].

The first records of the use of propolis by humankind date back to ancient Egypt (1700 B.C.), and it used to be employed as one of the materials to embalm the dead [1]. The Greeks, including Hippocrates, used it for internal and external cicatrization. The Roman historian Pliny refers to propolis as a medicine capable of reducing swelling and relieving pain. The term propolis was already described in the 16th century in France; and in 1908, the first scientific paper on its properties and chemical composition was published. Later in 1968, the first patent using Romanian propolis for the production of bath lotions was presented. Both works were indexed in *Chemical Abstracts* [2].

In South Africa, the war that occurred at the end of the 19th century was widely used due to its healing properties, and in World War II, it was used in several Soviet clinics. In the former USSR (Union of Soviet Socialist Republics), propolis received special attention in human and veterinary medicine, with applications in the treatment of tuberculosis, observing the regression of lung problems and recovery of appetite [5].

Propolis composition is mainly determined by the phytogeographical characteristics around the hive. However, it also varies seasonally in the same locality. The probable plant source, compared to its chemical composition, is the best indicator of the botanical origin of propolis [6].

Not only the chemical composition of propolis is determined by the vegetation characteristics but also by the pollen and honey deposits. As a consequence of this different chemical composition, there is also a variation in its pharmacological activities [7].

Propolis extract is a mixture of different components in different proportions, and it is not clear how these constituents interact and promote their effects on other organisms. Additionally, there is considerable variation in the composition of propolis extracts according to certain plant species and seasonality [8].

Hernandez *et al.,* (2010) [9] infer that at least one plant species contributes to the production of Cuban propolis. Therefore, although it is a product of animal origin, some chemical compounds of propolis are derived from the botanical source used by bees, especially those with biological action.

In Brazil, some types of propolis have already been characterized and classified by their coloration. According to Daugsch *et al.,* (2007) [10], a new type of red-colored propolis was verified in beehives found along the coast and rivers of Northeastern Brazil, showing physicochemical and biological characteristics different from the others already studied. However, this classification is still underestimated since bees can collect resin from a wide variety of plants.

Most papers in the literature refer to green propolis, and only in recent years has red propolis begun to be studied. Brazilian red propolis has new bioactive compounds never before found in the products already evaluated. It is an important source of substances with biological properties, including antioxidant activity [11].

The global interest in propolis has two justifications: the first is due to its panacea characteristics. In a certain way, these features also hinder its acceptance since doctors and other professionals tend to distrust its efficacy because dozens of biological activities are simultaneously attributed to it. The second reason is its high added-value, as a bottle of the alcoholic extract purchased in Brazil can cost up to 30 times as much in Tokyo. This high added value may justify, in part, their interest in propolis, especially the Brazilian propolis. Although Brazil produces 10 to 15% of the world's production, it supplies about 80% of the Japanese demand for propolis [12].

## TERPENOIDS PRESENT IN PROPOLIS FROM *APIS MELLIFERA* BEES

Propolis contains a variety of different constituents, which include phenolic acids, esters, flavonoids, other phenolic molecules, terpenes, ketones, aromatic aldehydes and alcohols, proteins, fatty acids, waxy acids, amino acids, steroids, sugars, vitamins, minerals, and even enzymes [13 - 15]. Propolis has been studied for several applications, such as in human medicine, quality of life, cosmetics and food industries, aquaculture, and livestock, due to its antioxidant and antimicrobial properties [12].

Terpenes are volatile constituents present in the essential oils of citrus fruits, cherries, mints, and herbs that contain only carbon, hydrogen, and oxygen atoms [16]. They can be chemically classified as alcohols, hydrocarbons, ketones, and epoxies. Physiologically, terpenes function primarily as chemoattractants or chemorepellents and are largely responsible for the characteristic fragrances of many plants [17].

The potential of terpenes as skin permeation enhancers for drug delivery systems has received considerable attention, especially because they are naturally occurring substances with low skin irritation [16]. Terpenes deserve to be highlighted as the biologically important constituents in propolis. They are volatile and mostly non-polar compounds, being found in essential oils. Terpenes have been described mainly for tropical propolis, which is rarer than poplar-type and Mediterranean propolis. Among these substances, sesquiterpenes are biologically the most important [14, 18, 19].

The lipophilic potential of terpenes is of extreme importance. Studies on their interactions with lipid membranes have become more relevant. The knowledge from these studies can be used for the development of new cancer suppressants and antimicrobial drugs, as well as for many other uses [20]. Since terpenes are not very water-soluble and rarely present in the cellular environment, their interactions with membranes still need further studies [14, 18, 21].

Limonene is one of the least polar terpenes (cyclic monoterpene), the only one that does not have at least one OH group in its constitution (Fig. **1**) [20]. Due to its hydrophobicity, limonene permeates rapidly through membranes. Studies have shown that limonene has toxic effects on fibroblast and erythrocyte membranes [16]. However, sesquiterpenes are more cytotoxic to membranes than monoterpenes, although both made membranes more fluid [22].

Terpenes lupeol, α-amyrin, β-amyrin, and geraniol present one OH group in their structure, while isocupressic acid has two (Fig. **1**). Members of the triterpene class (α-amyrin, β-amyrin, and lupeol) have at least four C6 rings. Geraniol (monoterpene) is the only one that does not present a carbon ring, while isopressic acid has two C6 rings and a long acyl chain [19, 23]. Geraniol, the only non-cyclic terpene, fluidizes saturated *Candida albicans* membranes, and decreases their surface potential. Due to its strong interactions with acyl lipid chains, it expands saturated membranes and makes them more permeable to other molecules [19, 23].

Isocupressic acid is the most polar compound among terpenes. It permeates membranes and is mainly intercalated in the inner layer of erythrocyte bilayers. Such substance does not have the strong effects on membranes as seen for other terpenes [20]. The main reason is that isocupressic acid is intercalated closer to the polar groups than the other terpenes. Its effects on membranes are also concentration-dependent, thus, higher concentrations have greater effects [20].

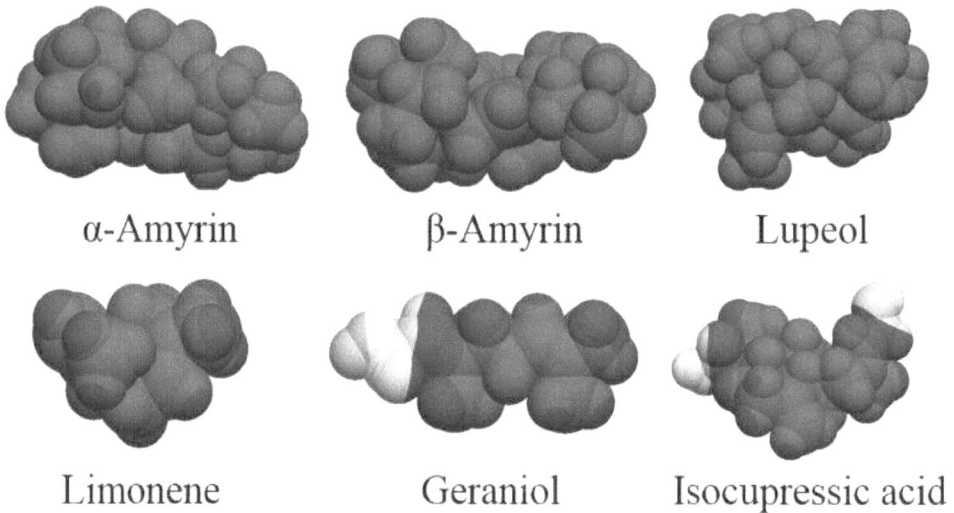

**Fig. (1).** Lipophilic potential of some terpenes from propolis. The darker the color, the more is the region of the hydrophobic molecule [20].

Terpenoids play an important role in distinguishing premium propolis from inferior or false propolis. Although volatiles represents only 10% of propolis constituents, they confer its characteristic odor and contribute to the pharmacological effects of propolis [14]. Propolis has a variable chemical composition that depends on the biodiversity of each region visited by bees [13, 19, 24, 25]. This diversity of propolis types and compositions also generates distinct biological properties [26].

Research related to the bioactivity of propolis volatiles are relatively scarce and most are about antimicrobial properties [18]. Studies have confirmed the activity of propolis volatiles against several microorganisms. Among them are gram-positive bacteria, such as *Staphylococcus aureus, Staphylococcus epidermidis, Micrococcus glutamicus, Bacillus subtilis, Bacillus cereus, Sarcina lutea, Streptococcus pyogenes, Streptococcus mutans,* and*Streptococcus faecalis* [27 - 32], as well as gram-negative bacteria, such as *Escherichia coli, Enterobacter cloacae, Klebsiella pneumonie,* and *Pseudomonas aeruginosa* [29 - 32].

Propolis composition can be influenced by temperature due to the seasonality of each region where bees collect it [18]. Regardless of its chemical composition, propolis has common biological properties. As a result, it has been studied for applications in several areas, such as food, pharmaceuticals, and livestock [12]. Among the terpenes most commonly found in propolis, monoterpenes, sesquiterpenes, diterpenes, and triterpenes stand out [33, 34].

The monoterpenes isolated from propolis include acyclic, monocyclic, and bicyclic monoterpenes and their derivatives (Table **1**). The primary acyclic and monocyclic monoterpenes are myrcenes, *p*-menthanes, and cineoles, respectively. The dicyclic monoterpenes in propolis are classified into five groups: thujanes, caranes, pinanes, fenchanes, and camphenes. Sesquiterpenes are the most abundant chemical components in propolis [14]. Simionatto *et al.* (2012) [30] identified one monoterpene (*trans*-β-terpineol) and three sesquiterpenes (γ-elemene, α-ylangene, valencene) in Brazilian propolis with valuable biological activities.

Table 1. Monoterpenes identified in propolis (*Apis mellifera*) in Greece, Brazil, and Iran since the year 2000.

| Monoterpenes | Geographic Location | References |
|---|---|---|
| Camphor | Iran | [35] |
| Linalool | Brazil | [30] |
| *Trans*-β-Terpineol | Greece | [29] |

According to the number of rings, sesquiterpenes (Table **2**) are divided into four categories: acyclic, monocyclic, bicyclic, and tricyclic. There are four types of monocyclic sesquiterpenes, five types of bicyclic sesquiterpenes, and ten types of tricyclic sesquiterpenes in propolis [14]. In Turkish propolis, Kartal *et al.* (2002) reported the presence of sesquiterpenes 8-βH-Cedran-8-ol,4-βH,5α-Ere-ophil-1(10)-ene,α-Bisabolol,α-Eudesmol, and α-Cadinol [36].

Table 2. Sesquiterpenes identified in propolis (*Apis mellifera*) in Greece, Brazil, and Turkey since the year 2000.

| Sesquiterpenes | Geographic Location | References |
|---|---|---|
| α-Bisabolol | Turkey | [29] |
| α-Cadinol | Turkey | [37] |
| α-Eudesmol | Turkey | [37] |
| α-Ylangene | Brazil | [30] |
| γ-Elemene | Brazil | [30] |
| Junipene | Greece | [29] |
| Valencene | Brazil | [30] |

(Table 2) cont.....

| Sesquiterpenes | Geographic Location | References |
|---|---|---|
| 4-βH,5α-Eremophil-1(10)-ene | Turkey | [36] |
| 8-βH-Cedran-8-ol | Turkey | [36] |

Cembrane, labdane, abietane, pimarane, and totarane are reported as the main diterpenes in propolis (Table **3**), and some of them are proven to have a broad spectrum of pharmacological properties [14]. Popova *et al.* (2009) [19] identified the "Mediterranean" diterpenes in samples from Greece, along with some diterpenes that are considered to be characteristic oleoresin components of different *Coniferae* plants (mainly *Pinaceae* and *Cupressaceae*). Their plant source was considered to be *Cupressaceae* because Greek propolis contained ferruginol, totarol, oxygenated ferruginol, totarol derivatives, and sempervirol, which are typically found in Cupressaceae plants, but not in Pinaceae.

**Table 3. Diterpenes identified in propolis (*Apis mellifera*) from Greece since the year 2000.**

| Diterpenes | References |
|---|---|
| Abietic acid Greece | [38] |
| Diterpenic acid | [38] |
| Dihydroxyabieta-8,11,13-triene | [38] |
| Dehydroabietic acid | [38] |
| Ferruginol Greece | [38] |
| Ferruginolone Greece | [38] |
| Hydroxydehydroabietic acid | [38] |
| Imbricataloic acid | [38] |
| Imbricatoloic acid | [38] |
| Junicedric acid | [19] |
| Labda-8(17),12,13-triene | [38] |
| Manoyl oxide | [38] |
| Neoabietic acid | [38] |
| Oleoyl isocupressic acid | [19] |
| Palmitoyl isocupressic acid | [19] |
| Pimaric acid | [19] |
| Sempervirol | [38] |
| Totarolone | [19] |
| *Tran*-Communal | [19] |
| 2-Hydroxyferruginol | [38] |

*(Table 3) cont.....*

| Diterpenes | References |
|---|---|
| 6/7-Hydroxyferruginol | [38] |
| 13(14)-Dehydrojunicedric acid | [38] |
| 13-Hydroxy-8(17),14-labdadien-19-oic acid | [19] |
| 14,15-Dinor-13-oxo-8(17)-labden-19-oic acid | [19] |
| 15-Oxolabda-8(17),13(E)-dien-19-oic acid | [19] |
| 18-Succinyloxyabietadiene | [38] |
| 18-Succinyloxyhydroxyabietatriene | [38] |
| 18-Hydroxyabieta-8,11,13-triene | [38] |

The tetracyclic triterpenes in propolis are lanostane and cycloartane, and the pentacyclic triterpenes are oleanane, ursane, and lupane (Table **4**) [14]. Some triterpenes belonging to lupane (lupeol alkanoates, lupeol, lupeol acetate) [9, 39], lanostane (lanosterol acetate, lanosterol) [9, 40], oleanane (germanicol acetate, germanicol,β-amyrin acetate), ursane (β-amyrone, α-amyrin acetate,α-amyrone) [9], and other types (methylene-9,19-ciclolanostan-3β-ol; (22Z,24E)-3-oxocyclo-art-22,24-dien-26-oic acid; (24E)-3-oxo-27,28-dihydroxycycloart-24-en-26-oic acid; 3,4-seco-cycloart-12-hydroxy-4(28),24-dien-3-oicacid;cycloart-3,7-dihydroxy-24-en-28-oic acid; and 3-oxo-triterpenic acid methyl ester) were found for the first time in Brazilian, Cuban, Greek, Burmese, and Egyptian propolis [19, 39, 41, 42].

**Table 4. Triterpenes identified in propolis (*Apis mellifera*) from Greece, Brazil, Cuba, Egypt, and Burma since the year 2000.**

| Triterpenes | Origin | References |
|---|---|---|
| α-Amyrin acetate | Cuba | [9] |
| α-Amyrone | Cuba | [9] |
| β-Amyrin acetate | Cuba | [9] |
| β-Amyrone | Cuba | [9] |
| Cycloart-3,7-dihydroxy-24-en-28-oic acid | Greece | [19] |
| Germanicol acetate | Cuba | [9] |
| Germanicol | Cuba | [9] |
| Lanosterol | Cuba | [9] |
| Lanosterol acetate | Egypt | [40] |
| Lupeol acetate | Cuba | [9] |
| Lupeol | Brazil | [39] |
| Lupeol alkanoates | Brazil | [39] |

*(Table 4) cont.....*

| Triterpenes | Origin | References |
|---|---|---|
| 3,4-seco-Cycloart-12-hydroxy-4(28),24-dien-3-oicacid | Greece | [19] |
| 3-Oxo-triterpenic acid methyl ester | Egypt | [42] |
| (22Z,24E)-3-Oxocycloart-22,24-dien-26-oic acid | Burma | [41] |
| 24-Methylene-9,19-ciclolanostan-3β-ol | Brazil | [39] |
| (24E)-3-Oxo-27,28-dihydroxycycloart-24-en-26-oic acid | Burma | [41] |

Pereira *et al.* (2002) [39] studied the crude dichloromethane extract of propolis (*Apis mellifera*) collected from eucalyptus trees and native plants of the Brazilian cerrado, and identified four pentacyclic triterpenes:α-amyrin, β-amyrin, lupeol, lupenone, and for the first time,lupeol alkanoates. In propolis samples from the eastern Mediterranean region, such as Greece, Crete, and Turkey, diterpenes such as isocupressic acid were found. These diterpenes belong to species of the Cupressaceae family, which are widely cultivated in those regions [38]. They demonstrated activity against HT-29 human colon adenocarcinoma cells, without affecting normal cells, with diterpene manool being the most active compound [43]. In the southern region of Brazil, *Araucaria angustifolia* trees are considered to be possible sources of diterpenes [44].

In Brazilian propolis, Banskota *et al.* (2001) [45] found isocupressic and agathic acids (diterpenes), which demonstrated hepatoprotective activity. In green propolis, Matsuno *et al.* (1997) [46] isolated 13Z-symphyoreticulic acid, a clerodane-type diterpene, which showed anticarcinogenic activity. In green propolis, triterpenes that occur widely in plants, such as pentacyclic lupeol and procrims, were found [47 - 49]. In samples of Minamar propolis, thirteen cycloartane triterpenoids were found, and among these, 27-dihydroxycycloart-24E-en-26-oic acid showed the most potent cytotoxic activity against murine melanoma cells [50].

El-Guendouz *et al.* (2019) [12] updated information on the chemical composition of propolis from all Mediterranean countries, as well as their biological properties and applications. Diterpenes were present in the non-volatile fraction of propolis in some countries such as Sicily, Croatia, Malta, Crete, Turkey, Cyprus, Egypt, Libya, Algeria, and Morocco.

There is little information on the composition of volatiles of propolis from Mediterranean countries when compared to the non-volatile profile. In poplar-type propolis, sesquiterpenes, particularly β-eudesmol, are always present in relatively high amounts, while monoterpenes such as α-pinene are mainly present in Mediterranean propolis species. This is perhaps due to the plant source used by coniferous bees, such as *Cupressus sempervirens* [12].

Bankova *et al.* (2014) [18]reported the presence of isocupressic acid, imbricatoloic acid, communic acid, agathadiol, totarol, 13-epi-cupressic acid, lupeol, α-amyrin, lupeol acetate, lanosterol, and cycloartenol in Moroccan propolis (Fig. **2**).

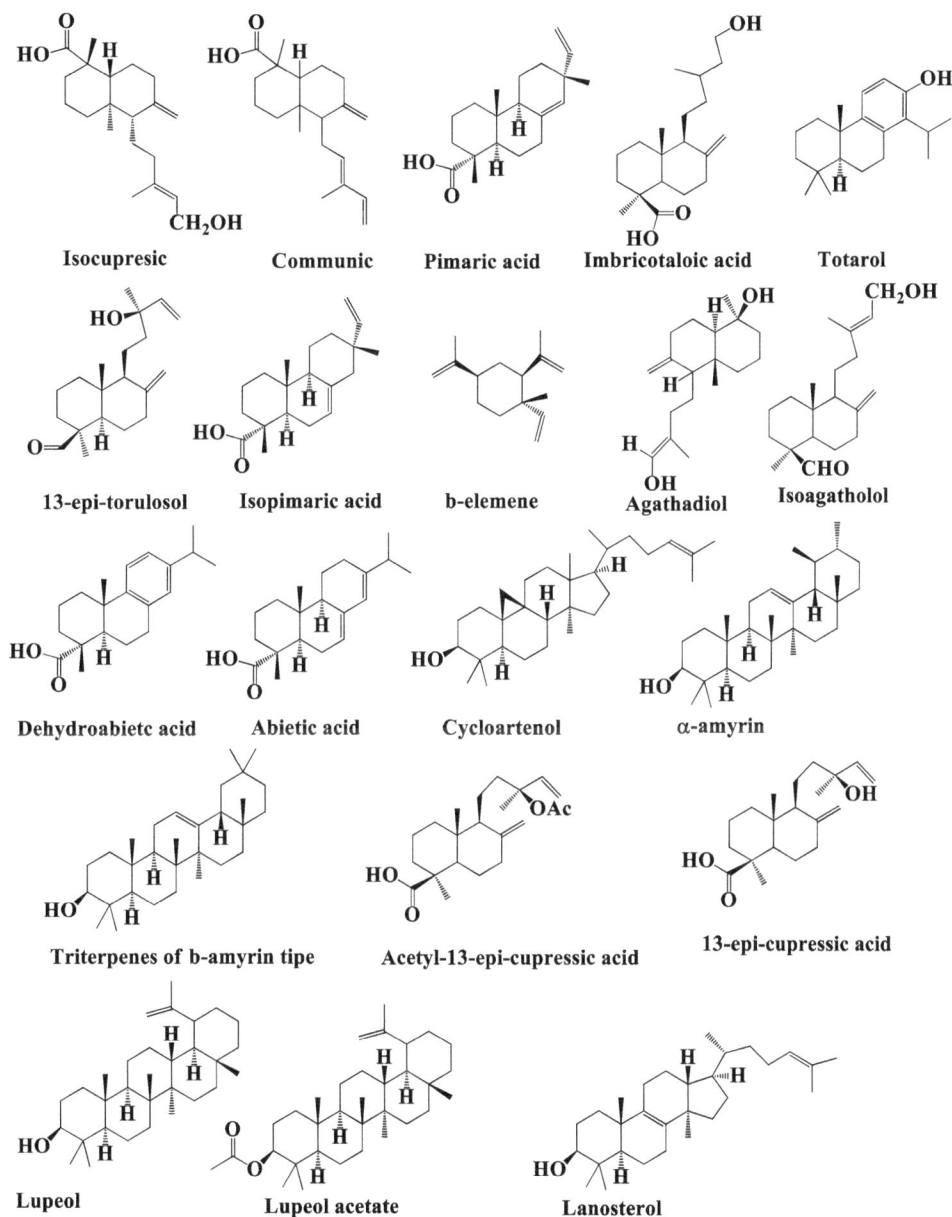

**Fig. (2).**  Terpenes and their derivatives present in Mediterranean *Apis melliferapropolis* [12].

Castro *et al.* (2007) [26] investigated ethanolic extracts of two types of *Apis mellifera* propolis obtained by maceration, in order to evaluate the influence of seasonality (northeastern and southeastern regions)on their production. Terpene β-caryophyllene was identified by gas chromatography coupled to mass spectrometry (GC-MS) in the northeastern propolis. In the propolis from the southeastern region, terpenes α-pinene, β-pinene, and farnesol were found. The authors concluded that throughout the bee harvest periods evaluated, seasonality influenced the antibacterial activity of propolis from both regions, probably due to changes in the concentration of bioactive compounds in the plant sources.

A frouzan *et al.* (2018) [51] analyzed the chemical composition and antimicrobial activity of four samples of *Apis mellifera* propolis from different areas of Iran (Chenaran, Taleghan, Morad Beyg, and Kalaleh) with diversified climates and flora, and identified several terpene derivatives. They performed the extraction with 70% ethanol and dichloromethane, and verified by GC-MS, the presence of terpene derivatives, such as: 3α,4α-4-methyl-stigmast-22-en-3-ol (1.93%) and dimethyl-1,3,5,6-tetramethyl-[1,3-($^{13}C_2$)]bicyclo[5.5.0]dodeca-1,3,5,6,8,10-hexa-ene-9,10-dicarboxylate (2.13%). In the ethanolic extract of Taleghan propolis, terpene derivatives were identified, such as germanicol (3.50%), dimethyl-1,3,5,6-tetramethyl-[1,3-($^{13}C_2$)]bicyclo[5.5.0]dodeca-1,3,5,6,8,10-hexaene-9,10-dicarbo-xylate (1.88%), and spiro[benzocyclopenta-3,4-cyclobuta-1,2-ccycloheptene]-8(5H),2'-1,3dioxane 6,7,7b,10a-tetrahydro-1.

In the ethanolic extract of Chenaran propolis, they [51] found terpene derivatives, such as (3α,4α)-4-methyl-stigmast-22-en-3-ol (1.09%) and dimethyl-1,3,5,6-tetra-methyl-[1,3-($^{13}C_2$)]bicyclo[5.5.0]dodeca-1,3,5,6,8,10-hexaene-9,10-dicarboxylate (2.51%). And in theethanolic extract of Kalaleh propolis, they identified 1,7,11-trimethyl-4-(1-methylethyl)cyclotetradecane (0.12%) and dibutyl phthalate (0.89%) [51]. All extracts ofIranian *Apis mellifera* propolis showed antimicrobial activity against *C. albicans, E. coli*, and *S. aureus,* perhaps due to the presence of such terpene derivatives [51].

Silici and Kutluca (2005) [37] investigated the chemical composition and antibacterial activity of three types of propolis collected from three different races of *Apis mellifera* bees in the same apiary (*Apis mellifera caucasica, Apis mellifera anatolica,* and *Apis mellifera carnica*). They were analyzed by GC/MS and terpenes, such as α-cadinol, chrysophanol, α-bisabolol, β-eudesmol, and α-eudesmol (sesquiterpenes) were identified.

The authors [37] also evaluated their antimicrobial activity against *Staphylococcus aureus, Escherichia coli, Pseudomonas aeruginosa,* and *Candida albicans.* The ethanolic extracts of propolis showed high antibacterial activity against Gram-positive cocci (*Staphylococcus aureus*), but had weak activity against Gram-negative bacteria (*Escherichia coli* and *Pseudomonas aeruginosa*) and the yeast *Candida albicans*. Propolis collected from *Apis melliferacaucasica* showed higher antibacterial activity than those collected from *Apis mellifera anatolica* and *Apis mellifera carnica*.

Propolis essential oil has been of interest to researchers because it contains large amounts of terpenes. Several essential oils from *Apis mellifera* propolis have been analyzed and substances such as terpenoids have been reported [52]. The chemical composition of propolis essential oils presented qualitative and quantitative variations due to the influence of local soil conditions and seasonal harvesting periods [52].

Bankova *et al.* (1998) [52] investigated the essential oils from African and European *Apis mellifera* obtained byhydrodistillation and extracted with n-pentane-diethyl ether mixture (1:1). After analysis by GC-MS, sesquiterpenes (mainly β-eudesmol) in addition to (2,6 E)-farnesol and spathulenol were identified.

Oliveira *et al.* (2010) [30]studied the chemical composition of the essential oil of *Apis mellifera* propolis from Rio de Janeiro (Brazil), as well as its ability to inhibit the growth *in vitro* of *Staphylococcus aureus, Staphylococcus epidermides, Staphylococcus pyogenes,* and *Escherichia coli* after extraction by steam distillation using a Clevenger apparatus. The compounds identified by GC-MS represented 67.5% of the oil. The most abundant were β-caryophyllene (12.7%), acetophenone (12.3%), linalool (6.47%), γ-elemene (6.25%), γ-cadinene (5.86%), and γ-muurolene (3.61%).

## TERPENOIDS PRESENT IN PROPOLIS AND GEOPROPOLIS OF STINGLESS BEES

Stingless bees also collect plant resin and store it in large deposits inside their hives. These deposits can be used in a similar way to other types of propolis. In Brazil, stingless bee propolis is a traditional medicine for wound healing, gastritis, and hemorrhoids [53]. Studies reported that in most cases, the sources of propolis used by *A. mellifera* and stingless bees are different. They also revealed various types of *Meliponinae* propolis, according to the compound predominant in the alcoholic extracts (diterpenic and triterpenic substances) [54]. They found that triterpenes comprised more than 35% of the total amount of *Apis mellifera* and *T. angustula* propolis samples from the same location, and that their chemical profiles were almost identical [55].

Velikova *et al.* (2000) [54] analyzed by GC-MS the propolis ethanolic extract produced by 12 different *Meliponinae* species, and identified some diterpenes and triterpenes (Table **5**). In the study, most samples showed weak or no activity against *E. coli* and weak action against *Candida albicans*. However, some of them showed significant activity against *S. Aureus,* possibly due to the high concentration of diterpene acids and high cytotoxic activity.

There are few studies on the volatile oils from stingless bees. However, these works demonstrated that their chemical composition is different from those of *Apis Mellifera* bees from the same region because they use different plant sources. Patricio *et al.* (2002) [56]noted that some of the samples studied were rich in monoterpenes; others, in sesquiterpenes, and the bee species could also be important for the botanical source selection.

Table 5. Chemical composition of *Meliponinae* propolis in Brazil, Mexico, and Venezuela.

| Bee Species | Origin | Compound Type | References |
|---|---|---|---|
| *Friesomelitta silvestrii* | Goias, Brazil | Monoterpenes, Sesquiterpenes | [56] |
| *Friesomelitta silvestrii languida* | Minas Gerais, Brazil | Monoterpenes, Sesquiterpenes, Diterpenes, Triterpenes | [56] |
| *Friesomelitta varia* | São Paulo. Brazil | Monoterpenes, Sesquiterpenes, Triterpenes | [56] |
| *Lestrimellata spp.* | Paraná, Brazil | Diterpenes, Triterpenes | [54] |
| *Melipona beechei* | Yucatan, Mexico | Monoterpenes, Sesquiterpenes | [57] |
| *Melipona compressipes* | Piauí, Brazil | Sesquiterpenes, Diterpenes, Triterpenes | [27] |
| *Melipona favora orlinge* | Mato Grosso do Sul, Brazil | Diterpenes, Triterpenes | [54] |
| *Melipona favosa* | Venezuela | Triterpenes | [58] |
| *Melipona quadrifasciata* | Paraná, Brazil | Monoterpenes, Sesquiterpenes, Diterpenes, Diterpenic acids | [59] |
| *Nanotrigona testacularis* | Minas Gerais, Brazil | Diterpenes, Triterpenes | [54] |
| *Plebeia remota* | Paraná, Brazil | Diterpenes, Triterpenes | [54] |

| Bee Species | Origin | Compound Type | References |
|---|---|---|---|
| *Plebeia spp.* | Paraná, Brazil | Diterpenes, Triterpenes | [54] |

Although terpenoids are the main components of stingless bee propolis, their biological activities have not been fully quantified and investigated [60]. *Heterotrigona itama* propolis has abundant terpenoids, however, little information is provided regarding its extraction optimization. Zhang *et al.* (2020) [60] optimized the extraction process of terpenoids from *Heterotrigona itama* propolis, using response surface methodology, and obtained satisfactory parameters. The authors demonstrated its strong anti-inflammatory activity by decreasing the inflammatory mediators iNOS, IL - 1β,and IL – 10, however, its antioxidant activity was low.

In the *Melipona quadrifasciata anthidioides* propolis, the only individual compounds isolated were the kaurene-type diterpenic acids:ent-15β-hydroxy-16-kauren-19-oic acid and ent-15β-3-methylbutanoyloxy-16-kauren-19-oic acid [59]. Pino *et al.* (2006) [57] evaluated *Apis mellifera* and *Melipona beecheii* samples, and identified 94 volatile compounds. In particular, the common constituents, such as α-pinene, β-pinene, *trans*-verbenol, α-copaene, β-bourbonene, β-caryophyllene, spatulenol, and caryophyllene oxide, were found in greater amounts.

Dutra *et al.* (2019) [61] investigated the antileishmanial activity, cytotoxicity, and chemical composition of geopropolis produced by *M. fasciculata* in order to find new products with activity against Leishmania infection. They identified triterpenes β-amyrin, lupenone, and oleanolic acid by GC-MS.

Smith (2020) [62] evaluated the antioxidant, anti-inflammatory, and cytotoxic activities of the hydroethanolic extract of geopropolis produced by *M. fasciculata*, and identified β-amyrin that belongs to the triterpene class. He correlated the compounds with the detected biological activities through *in silico* assays, and concluded that the study could contribute to the development of new products because *M. fasciculata* geopropolis presents antitumor activity.

Triterpenes are among the main constituents found in *Melipona fasciculata* geopropolis [63, 64]. Araújo *et al.* [63] studied the pharmacological activity of *M.fasciculata* geopropolis and observed that the product alone showed no antibacterial activity against *Staphylococcus aureus* and *Escherichia coli*, but its combination with chloramphenicol presented a stronger action against *S. aureus,* actually greater than chloramphenicol alone. It also exhibited cytostatic action towards human laryngeal epidermoid carcinoma cells and stimulated tumor necrosis factor-alpha and interleukin-10 production by human monocytes, showing an activating profile for human monocytes.

The main activities studied in *Meliponinae* propolis are antibacterial and antimycotic [53]. In*Melipona quadrifasciata* propolis, diterpenic acids have been identified as antibacterial components. However, data on other types of activities are unfortunately quite scarce [53]. Velikova (2000) [54] analyzed the potential cytotoxicity of stingless bee propolis from *Mellipona quadrifasciata* and *Nanotrigona testacularis*, which showed very promising toxicity. It is worth noting that all samples have diterpene-type compounds.

## RECENT APPLICATIONS OF PROPOLIS AND GEOPROPOLIS IN THE FOOD AND PHARMACEUTICAL INDUSTRIES

The scientific literature has reported the biological properties of propolis, such as bacteriostatic and bactericidal, fungistatic and fungicidal, virustatic and virucidal, antioxidant, antitumor, cicatrizing, tissue repair, anesthetic, against intestinal and blood parasites, antimutagenic, and against cardiovascular and respiratory diseases [65, 66].

Since the 1980s, this product has been widely used in food supplements and drinks, as a preventive therapy for diseases, and in topical applications. In the mid-1980s, propolis became an important product in complementary medicine. Currently, in many parts of the world, propolis is commercialized by the pharmaceutical industry as alternative medicine. The biological properties of propolis are obviously directly related to its chemical composition, and this is possibly the biggest problem for the use of propolis in "phytotherapy", since it varies according to the flora and harvesting season, the technique used, as well as the bee species [67].

In Brazil, the first paper published on propolis, in 1984 presented a comparative study of the effect of propolis and antibiotics on the inhibition of *Staphylococcus aureus*. The Brazilian propolis studied showed more activity than several of the antibiotics tested [68].

The occurrence of several diseases, such as cardiovascular, rheumatic, neurological, and psychiatric diseases, premature aging, neoplasms, osteoporosis, diabetes, and inflammation, is related to increasing levels of free radicals in the human organism. The possibility of using plants containing known polyphenols with antioxidant properties, for the control and prevention of the diseases mentioned above has been growing in recent years [69].

Besides polyphenols, propolis contains a wide range of other compounds that are capable of removing the excess free radicals from the human organism. Several research groups have reported this property, and many of them have isolated several compounds that would be responsible for such antioxidant activity [70].

Free radical scavenging could be an antioxidant mechanism of propolis, which would result in an ultimate anti-inflammatory activity. Propolis possesses anti-inflammatory properties that have been described mainly against diseases of the muscular system, infections, rheumatisms, sprains, and other types of inflammation [71].

Some researchers report the viability of propolis extract with bioactive substances responsible for antibiotic and cytotoxic activities that preserve its organoleptic characteristics. According to them [72], propolis has a very relevant synergistic action and may constitute a therapeutic alternative for microbial resistance.

Besides its antibacterial activity, propolis stands out for its antifungal action. Fungal infections have become an important health problem in recent decades due to the significant increase in the number of immune-compromised patients. These infections have a wide variety and diffusion, presenting diverse clinical manifestations [73].

Several other properties of propolis are described in several works, such as hepatoprotective, analgesic, estrogenic, antiangiogenic, and cartilage and bone regenerative activities, through the stimulation of chondrocyte proliferation. However, such regeneration properties, such as ulcer and wound healing and hepatoprotection, are possibly related to the antioxidant activity of propolis. When free radicals are produced, they hinder and/or prevent the regeneration of the local cells. Their removal by propolis flavonoids would allow the diseased organ or tissue to regenerate normally [74].

Prevention of lipid oxidation is one of the goals of the meat industry since its products are undesirable not only because of the production of offensive odors and flavors as a result of lipid breakdown, but also because of the destruction of essential constituents, causing decreased nutritional value and formation of toxic compounds during its processing [75].

## CONCLUDING REMARKS

The propolis can be an important source of volatile bioactive compounds, especially terpenoids and terpenoids; with this, it can be targeted by various research to analyze chemical variability. In addition, propolis has demonstrated major pharmacological applications that may be related to its chemical composition and geographical origin over the years.

## CONSENT FOR PUBLICATION

Not applicable.

## CONFLICT OF INTEREST

The author declares no conflict of interest, financial or otherwise.

## ACKNOWLEDGEMENTS

We would like to thank the authors for their contributions and commitment to writing this scientific work and the editors of this book for the invitation and vote of confidence. The author Dr Mozaniel Santana de Oliveira, thank the *Programa de Capacitação Institucional* (PCI), linked and supervised by the *Ministério da Ciência, Tecnologia e Inovações* – MCTI, for the scholarship (process number 302050/2021-3).

## REFERENCES

[1]     de Freitas, M.C.D.; de Miranda, M.B.; de Oliveira, D.T.; Vieira-Filho, S.A.; Caligiorne, R.B.; de Figucircdo, S.M. Biological Activities of Red Propolis: A Review. *Recent Pat. Endocr. Metab. Immune Drug Discov.,* **2018**, *11*(1), 3-12.
[http://dx.doi.org/10.2174/1872214812666180223120316] [PMID: 29473533]

[2]     Santos, L. M.; Fonseca, M. S.; Sokolonski, A. R. "Propolis: types, composition, biological activities, and veterinary product patent prospecting," Journal of the Science of Food and Agriculture, *100*(4), 1369-1382.

[3]     Sforcin, J. M. "Propolis and the immune system: a review," Journal of Ethnopharmacology, vol. 113, no. 1. J Ethnopharmacol, , 1-14.

[4]     Machado, G.M.C.; Leon, L.L.; Castro, S.L. Activity of Brazilian and Bulgarian propolis against different species of Leishmania. *Mem. Inst. Oswaldo Cruz,* **2007**, *102*(1), 73-77.
[http://dx.doi.org/10.1590/S0074-02762007000100012] [PMID: 17294003]

[5]     Cauich-Kumul, R.; Segura Campos, M.R. Bee Propolis: Properties, Chemical Composition, Applications, and Potential Health Effects. *Bioactive Compounds: Health Benefits and Potential Applications.,* **2018**, , 227-243.

[6]     Andrade, J.K.S.; Denadai, M.; de Oliveira, C.S.; Nunes, M.L.; Narain, N. Evaluation of bioactive compounds potential and antioxidant activity of brown, green and red propolis from Brazilian northeast region. *Food Res. Int.,* **2017**, *101*, 129-138.
[http://dx.doi.org/10.1016/j.foodres.2017.08.066] [PMID: 28941675]

[7]     Toreti, V.C.; Sato, H.H.; Pastore, G.M.; Park, Y.K. "Recent progress of propolis for its biological and chemical compositions and its botanical origin," *Evidence-based Complement. Altern. Med.,* **2013**, *2013*, 13.

[8]     Lavinas, F. C.; Macedo, E. H. B. C.; Sá, G. B. L. "Brazilian stingless bee propolis and geopropolis: promising sources of biologically active compounds," Brazilian Journal of Pharmacognosy, **2019**, *29*(3), 389-399.

[9]    Márquez Hernández, I.; Cuesta-Rubio, O.; Campo Fernández, M.; Rosado Pérez, A.; Montes de Oca Porto, R.; Piccinelli, A.L.; Rastrelli, L. Studies on the constituents of yellow Cuban propolis: GC-MS determination of triterpenoids and flavonoids. *J. Agric. Food Chem.,* **2010**, *58*(8), 4725-4730.
[http://dx.doi.org/10.1021/jf904527n] [PMID: 20230059]

[10]   Daugsch, A.; Moraes, C.S.; Fort, P.; Park, Y.K. Brazilian red propolis--chemical composition and botanical origin. *Evid. Based Complement. Alternat. Med.,* **2008**, *5*(4), 435-441.
[http://dx.doi.org/10.1093/ecam/nem057] [PMID: 18955226]

[11]   Silva, B.B.; Rosalen, P.L.; Cury, J.A.; Ikegaki, M.; Souza, V.C.; Esteves, A.; Alencar, S.M. Chemical composition and botanical origin of red propolis, a new type of brazilian propolis. *Evid. Based Complement. Alternat. Med.,* **2008**, *5*(3), 313-316.
[http://dx.doi.org/10.1093/ecam/nem059] [PMID: 18830449]

[12]   El-Guendouz, S.; Lyoussi, B.; Miguel, M.G. Insight on Propolis from Mediterranean Countries: Chemical Composition, Biological Activities and Application Fields. *Chem. Biodivers.,* **2019**, *16*(7), e1900094.
[http://dx.doi.org/10.1002/cbdv.201900094] [PMID: 31099458]

[13]   Marcucci, M.C. Propolis: chemical composition, biological properties and therapeutic activity. *Apidologie (Celle),* **1995**, *26*(2), 83-99.
[http://dx.doi.org/10.1051/apido:19950202]

[14]   Huang, S.; Zhang, C.P.; Wang, K.; Li, G.; Hu, F.L. Recent advances in the chemical composition of propolis. *Molecules,* **2014**, *19*(12), 19610-19632.
[http://dx.doi.org/10.3390/molecules191219610] [PMID: 25432012]

[15]   Zhang, C.P.; Liu, G.; Hu, F.L. Hydrolysis of flavonoid glycosides by propolis β-glycosidase. *Nat. Prod. Res.,* **2012**, *26*(3), 270-273.
[http://dx.doi.org/10.1080/14786419.2010.541877] [PMID: 21851328]

[16]   Mendanha, S.A.; Moura, S.S.; Anjos, J.L.V.; Valadares, M.C.; Alonso, A. Toxicity of terpenes on fibroblast cells compared to their hemolytic potential and increase in erythrocyte membrane fluidity. *Toxicol. In Vitro,* **2013**, *27*(1), 323-329.
[http://dx.doi.org/10.1016/j.tiv.2012.08.022] [PMID: 22944593]

[17]   Hasler, C.M.; Blumberg, J.B. Introduction. *J. Nutr.,* **1999**, *129*(3), 756S-757S.
[http://dx.doi.org/10.1093/jn/129.3.756S] [PMID: 10082785]

[18]   Bankova, V.; Popova, M.; Trusheva, B. Propolis volatile compounds: chemical diversity and biological activity: a review. *Chem. Cent. J.,* **2014**, *8*(1), 28.
[http://dx.doi.org/10.1186/1752-153X-8-28] [PMID: 24812573]

[19]   Popova, M.P.; Chinou, I.B.; Marekov, I.N.; Bankova, V.S. Terpenes with antimicrobial activity from Cretan propolis. *Phytochemistry,* **2009**, *70*(10), 1262-1271.
[http://dx.doi.org/10.1016/j.phytochem.2009.07.025] [PMID: 19698962]

[20]   Šturm, L.; Poklar Ulrih, N. Propolis flavonoids and terpenes, and their interactions with model lipid membranes: a review. *Advances in Biomembranes and Lipid Self-Assembly,* **2020**, *32*, 25-52.
[http://dx.doi.org/10.1016/bs.abl.2020.04.003]

[21]   Martins, M.A.R.; Silva, L.P.; Ferreira, O.; Schröder, B.; Coutinho, J.A.P.; Pinho, S.P. Terpenes solubility in water and their environmental distribution. *J. Mol. Liq.,* **2017**, *241*, 996-1002.
[http://dx.doi.org/10.1016/j.molliq.2017.06.099]

[22]   Witzke, S.; Duelund, L.; Kongsted, J.; Petersen, M.; Mouritsen, O.G.; Khandelia, H. Inclusion of terpenoid plant extracts in lipid bilayers investigated by molecular dynamics simulations. *J. Phys. Chem. B,* **2010**, *114*(48), 15825-15831.
[http://dx.doi.org/10.1021/jp108675b] [PMID: 21070035]

[23]   S. CLARKE, "Families of compounds that occur in essential oils," Essent. Chem. Aromather., pp. 41–77, 2008. **2008**, 41-77.

[24]  Velikova, M.; Bankova, V.; Sorkun, K.; Houcine, S.; Tsvetkova, I.; Kujumgiev, A. Propolis from the Mediterranean region: chemical composition and antimicrobial activity. *Z. Naturforsch. C J. Biosci.,* **2000**, *55*(9-10), 790-793.
[http://dx.doi.org/10.1515/znc-2000-9-1019] [PMID: 11098832]

[25]  Banskota, A.H.; Tezuka, Y.; Prasain, J.K.; Matsushige, K.; Saiki, I.; Kadota, S. Chemical constituents of Brazilian propolis and their cytotoxic activities. *J. Nat. Prod.,* **1998**, *61*(7), 896-900.
[http://dx.doi.org/10.1021/np980028c] [PMID: 9677271]

[26]  Castro, M.L.; Cury, J.A.; Rosalen, P.L.; Alencar, S.M.; Ikegaki, M.; Duarte, S.; Koo, H. Própolis do sudeste e nordeste do Brasil: influência da sazonalidade na atividade antibacteriana e composição fenólica. *Quim. Nova,* **2007**, *30*(7), 1512-1516.
[http://dx.doi.org/10.1590/S0100-40422007000700003]

[27]  Kujumgiev, A.; Tsvetkova, I.; Serkedjieva, Y.; Bankova, V.; Christov, R.; Popov, S. Antibacterial, antifungal and antiviral activity of propolis of different geographic origin. *J. Ethnopharmacol.,* **1999**, *64*(3), 235-240.
[http://dx.doi.org/10.1016/S0378-8741(98)00131-7] [PMID: 10363838]

[28]  Bankova, V.S.; Christov, R.; Tejera, A.D. Lignans and other constituents of propolis from the canary islands. *Phytochemistry,* **1998**, *49*(5), 1411-1415.
[http://dx.doi.org/10.1016/S0031-9422(98)00108-3]

[29]  Melliou, E.; Stratis, E.; Chinou, I. Volatile constituents of propolis from various regions of Greece – Antimicrobial activity. *Food Chem.,* **2007**, *103*(2), 375-380.
[http://dx.doi.org/10.1016/j.foodchem.2006.07.033]

[30]  Oliveira, A.; França, H.; Kuster, R.; Teixeira, L.; Rocha, L. Received: June 5, 2009 Accepted: July 30, 2009. *J. Venom. Anim. Toxins,* **2010**, *16*(1), 121-130.
[http://dx.doi.org/10.1590/S1678-91992010005000007]

[31]  Simionatto, E.; Facco, J.T.; Morel, A.F.; Giacomelli, S.R.; Linares, C.B. Chiral analysis of monoterpenes in volatile oils from propolis. *J. Chil. Chem. Soc.,* **2012**, *57*(3), 1240-1243.
[http://dx.doi.org/10.4067/S0717-97072012000300009]

[32]  e, E.H.K.; Betul, D.; Atac, U.; Fatih, D. Volatile composition of Anatolian propolis by headspace-solid-phase microextraction (HS-SPME), antimicrobial activity against food contaminants and antioxidant activity. *J. Med. Plants Res.,* **2013**, *7*(28), 2140-2149.
[http://dx.doi.org/10.5897/JMPR2013.4470]

[33]  Salatino, A.; Teixeira, É.W.; Negri, G.; Message, D. "Origin and chemical variation of Brazilian propolis," *Evidence-based Complement. Evid. Based Complement. Alternat. Med.,* **2005**, *2*(1), 33-38.
[http://dx.doi.org/10.1093/ecam/neh060] [PMID: 15841276]

[34]  Taniguchi, M.; Kataoka, T.; Suzuki, H.; Uramoto, M.; Ando, M.; Arao, K.; Magae, J.; Nishimura, T.; Õtake, N.; Nagai, K. Costunolide and dehydrocostus lactone as inhibitors of killing function of cytotoxic T lymphocytes. *Biosci. Biotechnol. Biochem.,* **1995**, *59*(11), 2064-2067.
[http://dx.doi.org/10.1271/bbb.59.2064] [PMID: 8541643]

[35]  Trusheva, B.; Todorov, I.; Ninova, M.; Najdenski, H.; Daneshmand, A.; Bankova, V. Antibacterial mono- and sesquiterpene esters of benzoic acids from Iranian propolis. *Chem. Cent. J.,* **2010**, *4*(1), 8.
[http://dx.doi.org/10.1186/1752-153X-4-8] [PMID: 20350297]

[36]  Kartal, M.; Kaya, S.; Kurucu, S. GC-MS analysis of propolis samples from two different regions of Turkey. *Z. Naturforsch. C J. Biosci.,* **2002**, *57*(9-10), 905-909.
[http://dx.doi.org/10.1515/znc-2002-9-1025] [PMID: 12440732]

[37]  Silici, S.; Kutluca, S. Chemical composition and antibacterial activity of propolis collected by three different races of honeybees in the same region. *J. Ethnopharmacol.,* **2005**, *99*(1), 69-73.
[http://dx.doi.org/10.1016/j.jep.2005.01.046] [PMID: 15848022]

[38]  Popova, M.P.; Graikou, K.; Chinou, I.; Bankova, V.S. GC-MS profiling of diterpene compounds in

Mediterranean propolis from Greece. *J. Agric. Food Chem.,* **2010**, *58*(5), 3167-3176.
[http://dx.doi.org/10.1021/jf903841k] [PMID: 20112913]

[39]   Pereira, A.S.; Nascimento, E.A.; De Aquino Neto, F.R. "Lupeol alkanoates in Brazilian propolis,"
       *Zeitschrift fur Naturforsch. - Sect. C J. Biosci.,* **2002**, *57*(7–8), 721-726.

[40]   Hegazi, A.G.; El Hady, F.K.A. Egyptian propolis: 3. Antioxidant, antimicrobial activities and
       chemical composition of propolis from reclaimed lands. *Z. Naturforsch. C J. Biosci.,* **2002**, *57*(3-4),
       395-402.
       [http://dx.doi.org/10.1515/znc-2002-3-432] [PMID: 12064746]

[41]   Li, F.; Awale, S.; Zhang, H.; Tezuka, Y.; Esumi, H.; Kadota, S. Chemical constituents of propolis
       from Myanmar and their preferential cytotoxicity against a human pancreatic cancer cell line. *J. Nat.
       Prod.,* **2009**, *72*(7), 1283-1287.
       [http://dx.doi.org/10.1021/np9002433] [PMID: 19572611]

[42]   El Hady, F.K.A.; Hegazi, A.G. Egyptian propolis: 2. Chemical composition, antiviral and
       antimicrobial activities of East Nile Delta propolis. *Z. Naturforsch. C J. Biosci.,* **2002**, *57*(3-4), 386-
       394.
       [http://dx.doi.org/10.1515/znc-2002-3-431] [PMID: 12064745]

[43]   Pratsinis, H.; Kletsas, D.; Melliou, E.; Chinou, I. Antiproliferative activity of Greek propolis. *J. Med.
       Food,* **2010**, *13*(2), 286-290.
       [http://dx.doi.org/10.1089/jmf.2009.0071] [PMID: 20132046]

[44]   Missima, F.; Filho, A.A.S.; Nunes, G.A.; Bueno, P.C.P.; De Sousa, J.P.B.; Bastos, J.K.; Sforcin, J.M.
       Effect of *Baccharis dracunculifolia* D.C (Asteraceae) extracts and its isolated compounds on
       macrophage activation. *J. Pharm. Pharmacol.,* **2010**, *59*(3), 463-468.
       [http://dx.doi.org/10.1211/jpp.59.3.0017] [PMID: 17331351]

[45]   Banskota, A.H.; Tezuka, Y.; Adnyana, I.K.; Ishii, E.; Midorikawa, K.; Matsushige, K.; Kadota, S.
       Hepatoprotective and anti-Helicobacter pylori activities of constituents from Brazilian propolis.
       *Phytomedicine,* **2001**, *8*(1), 16-23.
       [http://dx.doi.org/10.1078/0944-7113-00004] [PMID: 11292234]

[46]   Matsuno, T.; Matsumoto, Y.; Saito, M.; Morikawa, J. Isolation and characterization of cytotoxic
       diterpenoid isomers from propolis. *Z. Naturforsch. C J. Biosci.,* **1997**, *52*(9-10), 702-704.
       [http://dx.doi.org/10.1515/znc-1997-9-1020] [PMID: 9374000]

[47]   Negri, G.; Marcucci, M.C.; Salatino, A.; Salatino, M.L.F. Comb and Propolis Waxes from Brazil:
       Triterpenoids in Propolis Waxes. *J. Apic. Res.,* **2000**, *39*(1-2), 86-88.
       [http://dx.doi.org/10.1080/00218839.2000.11101026]

[48]   Furukawa, S.; Takagi, N.; Ikeda, T.; Ono, M.; Nafady, A.M.; Nohara, T.; Sugimoto, H.; Doi, S.;
       Yamada, H. Two novel long-chain alkanoic acid esters of lupeol from alecrim-propolis. *Chem. Pharm.
       Bull. (Tokyo),* **2002**, *50*(3), 439-440.
       [http://dx.doi.org/10.1248/cpb.50.439] [PMID: 11911218]

[49]   Teixeira, É.W.; Message, D.; Negri, G.; Salatino, A. Bauer-7-en-3β-yl acetate: a major constituent of
       unusual samples of Brazilian propolis. *Quim. Nova,* **2006**, *29*(2), 245-246.
       [http://dx.doi.org/10.1590/S0100-40422006000200013]

[50]   Li, F.; Awale, S.; Tezuka, Y.; Kadota, S. Cytotoxic constituents of propolis from Myanmar and their
       structure-activity relationship. *Biol. Pharm. Bull.,* **2009**, *32*(12), 2075-2078.
       [http://dx.doi.org/10.1248/bpb.32.2075] [PMID: 19952433]

[51]   Afrouzan, H.; Tahghighi, A.; Zakeri, S.; Es-haghi, A. Chemical composition and antimicrobial
       activities of Iranian Propolis. *Iran. Biomed. J.,* **2018**, *22*(1), 50-65.
       [PMID: 28558440]

[52]   Bankova, V.; Boudourova-Krasteva, G.; Popov, S.; Sforcin, J.M.; Funari, S.R.C. Seasonal variations
       in essential oil from brazilian propolis. *J. Essent. Oil Res.,* **1998**, *10*(6), 693-696.

[http://dx.doi.org/10.1080/10412905.1998.9701012]

[53] Bankova, V.; Popova, M. Propolis of stingless bees: a promising source of biologically active compounds. *Pharmacogn. Rev.,* **2007**, *1*(1), 88-92.

[54] Velikova, M.; Bankova, V.; Marcucci, M.C.; Tsvetkova, I.; Kujumgiev, A. Chemical composition and biological activity of propolis from Brazilian meliponinae. *Z. Naturforsch. C J. Biosci.,* **2000**, *55*(9-10), 785-789.
[http://dx.doi.org/10.1515/znc-2000-9-1018] [PMID: 11098831]

[55] Ibrahim, A.; Reuter, G.S.; Spivak, M. Field trial of honey bee colonies bred for mechanisms of resistance against *Varroa destructor. Apidologie (Celle),* **2007**, *38*(1), 67-76.
[http://dx.doi.org/10.1051/apido:2006065]

[56] Patricio, E.F.L.R.A.; Cruz-López, L.; Maile, R.; Tentschert, J.; Jones, G.R.; Morgan, E.D. The propolis of stingless bees: terpenes from the tibia of three Frieseomelitta species. *J. Insect Physiol.,* **2002**, *48*(2), 249-254.
[http://dx.doi.org/10.1016/S0022-1910(01)00170-6] [PMID: 12770125]

[57] Pino, J.A.; Marbot, R.; Delgado, A.; Zumárraga, C.; Sauri, E. Volatile constituents of propolis from honey bees and stingless bees from yucatán. *J. Essent. Oil Res.,* **2006**, *18*(1), 53-56.
[http://dx.doi.org/10.1080/10412905.2006.9699384]

[58] Tomás-Barberán, F.A.; García-Viguera, C.; Vit-Olivier, P.; Ferreres, F.; Tomás-Lorente, F. Phytochemical evidence for the botanical origin of tropical propolis from Venezuela. *Phytochemistry,* **1993**, *34*(1), 191-196.
[http://dx.doi.org/10.1016/S0031-9422(00)90804-5]

[59] Velikova, M.; Bankova, V.; Tsvetkova, I.; Kujumgiev, A.; Marcucci, M.C. Antibacterial ent-kaurene from Brazilian propolis of native stingless bees. *Fitoterapia,* **2000**, *71*(6), 693-696.
[http://dx.doi.org/10.1016/S0367-326X(00)00213-6] [PMID: 11077178]

[60] Zhang, W.; Cai, Y.; Chen, X.; Ji, T.; Sun, L. Optimized extraction based on the terpenoids of *Heterotrigona itama* propolis and their antioxidative and anti□inflammatory activities. *J. Food Biochem.,* **2020**, *44*(8), e13296.
[http://dx.doi.org/10.1111/jfbc.13296] [PMID: 32529646]

[61] Dutra, R.P.; Bezerra, J.L.; Silva, M.C.P.; Batista, M.C.A.; Patrício, F.J.B.; Nascimento, F.R.F.; Ribeiro, M.N.S.; Guerra, R.N.M. Antileishmanial activity and chemical composition from Brazilian geopropolis produced by stingless bee Melipona fasciculata. *Rev. Bras. Farmacogn.,* **2019**, *29*(3), 287-293.
[http://dx.doi.org/10.1016/j.bjp.2019.02.009]

[62] Smith, M. *Anti-Inflammatory Activities of Geopropolis Produced by the Stingless Bee,* **2020**.

[63] Araujo, M.; Bufalo, M.; Conti, B.; Jr, A.; Trusheva, B.; Bankova, V.; Sforcin, J. The chemical composition and pharmacological activities of geopropolis produced by Melipona fasciculata Smith in northeast Brazil. *Journal of Molecular Pathophysiology,* **2015**, *4*(1), 12.
[http://dx.doi.org/10.5455/jmp.20150204115607]

[64] Araújo, M.J.A.M.; Bosco, S.M.G.; Sforcin, J.M. Pythium insidiosum: inhibitory effects of propolis and geopropolis on hyphal growth. *Braz. J. Microbiol.,* **2016**, *47*(4), 863-869.
[http://dx.doi.org/10.1016/j.bjm.2016.06.008] [PMID: 27522931]

[65] Parreira, N.A.; Magalhães, L.G.; Morais, D.R.; Caixeta, S.C.; de Sousa, J.P.B.; Bastos, J.K.; Cunha, W.R.; Silva, M.L.A.; Nanayakkara, N.P.D.; Rodrigues, V.; da Silva Filho, A.A. Antiprotozoal, schistosomicidal, and antimicrobial activities of the essential oil from the leaves of Baccharis dracunculifolia. *Chem. Biodivers.,* **2010**, *7*(4), 993-1001.
[http://dx.doi.org/10.1002/cbdv.200900292] [PMID: 20397234]

[66] Elnakady, Y.A.; Rushdi, A.I.; Franke, R.; Abutaha, N.; Ebaid, H.; Baabbad, M.; Omar, M.O.M.; Al Ghamdi, A.A. Characteristics, chemical compositions and biological activities of propolis from Al-

Bahah, Saudi Arabia. *Sci. Rep.,* **2017**, *7*(1), 41453.
[http://dx.doi.org/10.1038/srep41453] [PMID: 28165013]

[67]    Więckiewicz, W.; Miernik, M.; Więckiewicz, M.; Morawiec, T. Does Propolis Help to Maintain Oral Health. **2013**, *Vol. 2013*

[68]    Lu, L.C.; Chen, Y.W.; Chou, C.C. Antibacterial activity of propolis against Staphylococcus aureus. *Int. J. Food Microbiol.,* **2005**, *102*(2), 213-220.
[http://dx.doi.org/10.1016/j.ijfoodmicro.2004.12.017] [PMID: 15992620]

[69]    Silva-Carvalho, R.; Baltazar, F.; Almeida-Aguiar, C. Propolis: A Complex Natural Product with a Plethora of Biological Activities That Can Be Explored for Drug Development. *Evid. Based Complement. Alternat. Med.,* **2015**, *2015*, 1-29.
[http://dx.doi.org/10.1155/2015/206439] [PMID: 26106433]

[70]    Chaillou, L.L.; Nazareno, M.A. Bioactivity of propolis from Santiago del Estero, Argentina, related to their chemical composition. *Lebensm. Wiss. Technol.,* **2009**, *42*(8), 1422-1427.
[http://dx.doi.org/10.1016/j.lwt.2009.03.002]

[71]    Silva, H.; Francisco, R.; Saraiva, A.; Francisco, S.; Carrascosa, C.; Raposo, A. The cardiovascular therapeutic potential of propolis—a comprehensive review. *Biology (Basel),* **2021**, *10*(1), 27.
[http://dx.doi.org/10.3390/biology10010027] [PMID: 33406745]

[72]    Anjum, S.I.; Ullah, A.; Khan, K.A. Composition and functional properties of propolis (bee glue): A review. *Saudi J. Biol. Sci.,* **2018**.
[PMID: 31762646]

[73]    Mohdaly, A.A.A.; Mahmoud, A.A.; Roby, M.H.H.; Smetanska, I.; Ramadan, M.F. Phenolic Extract from Propolis and Bee Pollen: Composition, Antioxidant and Antibacterial Activities. *J. Food Biochem.,* **2015**, *39*(5), 538-547.
[http://dx.doi.org/10.1111/jfbc.12160]

[74]    da Silva Frozza, C.O.; Garcia, C.S.C.; Gambato, G.; de Souza, M.D.O.; Salvador, M.; Moura, S.; Padilha, F.F.; Seixas, F.K.; Collares, T.; Borsuk, S.; Dellagostin, O.A.; Henriques, J.A.P.; Roesch-Ely, M. Chemical characterization, antioxidant and cytotoxic activities of Brazilian red propolis. *Food Chem. Toxicol.,* **2013**, *52*, 137-142.
[http://dx.doi.org/10.1016/j.fct.2012.11.013] [PMID: 23174518]

[75]    Berretta, A.A.; Arruda, C.; Miguel, F.G. Functional Properties of Brazilian Propolis: From Chemical Composition Until the Market **2017**.
[http://dx.doi.org/10.5772/65932]

# CHAPTER 14

# Terpenoids and Biotechnology

**Jorddy Neves Cruz**[1,2,*], **Fernanda Wariss Figueiredo Bezerra**[3], **Renan Campos e Silva**[4], **Mozaniel Santana de Oliveira**[1], **Márcia Moraes Cascaes**[1], **Jose de Arimateia Rodrigues do Rego**[5], **Antônio Pedro da Silva Souza Filho**[2], **Daniel Santiago Pereira**[2] and **Eloisa Helena de Aguiar Andrade**[1]

[1] *Adolpho Ducke Laboratory, Paraense Emílio Goeldi Museum, Belém, Brazil*

[2] *Laboratory of Agro-Industry, Embrapa Eastern Amazon, Belem, Pará, Brazil*

[3] *Program of Post-Graduation in Food Science and Technology, Federal University of Para, Belém, Pará, Brazil*

[4] *Program of Post-Graduation in Chemistry, Federal University of Pará, Belém, Brazil*

[5] *Institute of Technology, Federal University of Pará, Belém, Brazil*

**Abstract:** Terpenoids, or isoprenoids, represent a large and structurally diverse class of isoprene-based secondary metabolites that play a fundamental role in the organism of all living beings. In nature, terpenes are essential for the interaction of organisms with their environment, mediating antagonistic and beneficial interactions between organisms. In this chapter, we will cover the biotechnology production of terpenes, as well as their biosynthesis by micro-organisms. We will also investigate the various pharmaceutical applications of these compounds.

**Keywords:** Applications, Biosynthesis, Metabolites, Micro-organisms.

## INTRODUCTION

Green plants, particularly angiosperms, exhibit a high number of terpenoids compared to other living organisms [1. It is estimated that more than 80,000 compounds belonging to this class are known, and many more are still unknown in all existing life forms [2].

The structural diversity of terpenoids results from a natural background marked by herbivore stress and other selectivity imposed by animals, resulting in a wide range of functionalized terpenoids preselected for their potent biological activities. It is also driven by stereospecific carbocation cyclizationrearrangement,

---

[*] **Corresponding author Jorddy Neves Cruz**: Laboratory of Agro-Industry, Embrapa Eastern Amazon, Belem, Pará, Brazil. E-mail: jorddynevescruz@gmail.com

and elimination reactions that transform some universal isopentenyl diphosphate precursors into core layers of numerous structurally distinct terpenoids [3 - 5]. Furthermore, reactions catalyzed by terpene cyclases from cryptic pathways are believed to be largely responsible for the expansive chemodiversity of terpenoid natural products [6].

Biosynthesis of terpenoids occurs *via* mevalonate (MVA) or methylerythritol 4-phosphate (MEP) pathways to generate five-carbon isoprene units, dimethylallyl diphosphate (DMAPP) and isopentenyl diphosphate (IPP), which are coupled to isoprenyl diphosphates (Fig. **1**) and undergo cyclization reactions to produce a myriad of terpenoids [7 - 9].

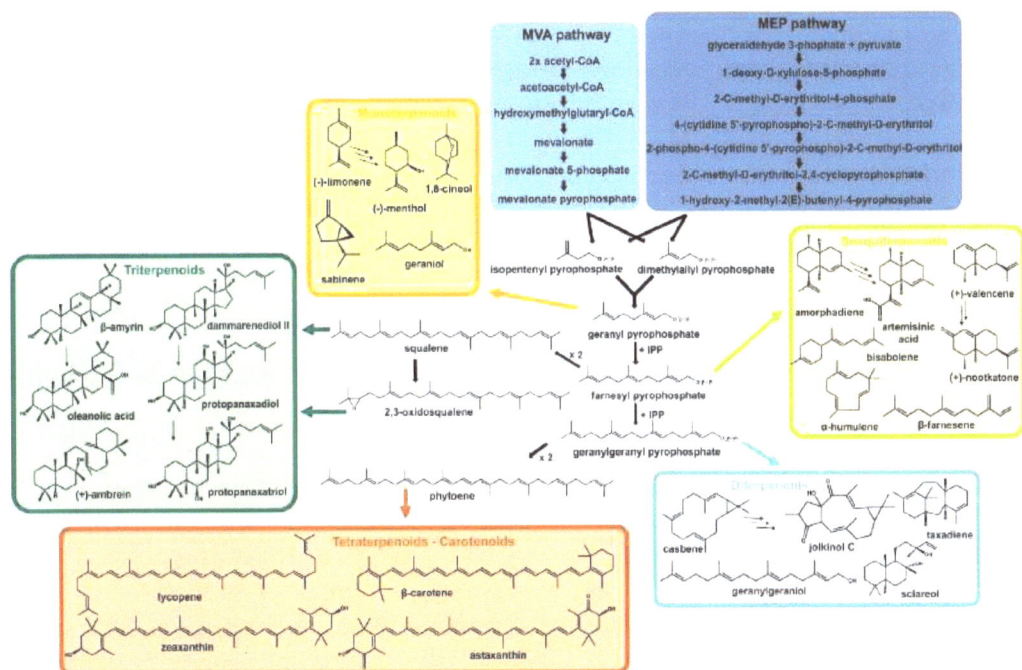

**Fig. (1).** Biosynthesis of terpenoids. Adapted from Moser; Pichler (2019).

Terpenes are responsible for defending many species of plants, animals, and micro-organisms against predators, pathogens, and competitors, and are involved in transmitting messages to co-species and mutualists about the presence of food, companions, and enemies [10]. For instance, there is much evidence that isoprenoids can act as chemical messengers that influence the expression of genes involved in mechanisms of plant defense or even influence the gene expression of neighboring plants [11].

Terpenes are of great industrial interest because they have promising pharmacological properties, which can lead to the identification of new pharmaceuticals, and can also be used in perfumery and food preservation [12]. For example, the taxol-based compound, Paclitaxel, is one of the most widely used drugs in the treatment of breast cancer [13], in addition to menthol, linalool, camphor, and limonene, which are used in the manufacture of essences, and also the natural rubber used mainly in the automotive industry [14]. Also, recent research has identified terpenes as potential materials for the production of specialty biofuels, since some compounds in this class meet current industrial and chemical requirements, including viscosity, flash and freezing points, high energy densities, and high net heat of combustion [15, 16].

Various methods of obtaining terpenes have been used, such as distillation or solvent extraction techniques, which are typically time-consuming and labor-intensive. In recent years, microextraction techniques (solid-phase microextraction - SPME and stir bar sorptive extraction - SBSE) have been developed, which aid in sample preparation and are environmentally friendly [17]. However, nowadays, supercritical $CO_2$ extraction, microwave-assisted extraction, and other solid-liquid extraction methods are the most common techniques to isolate and purify hydrophobic terpenes and other natural products from plant-derived raw materials [18].

Despite being considered a renewable source, plants normally produce low concentrations of terpenoids in their tissues. Furthermore, due to the complexity of these molecules, the chemical synthesis of terpenoids is inherently difficult, expensive, and produces relatively low yields. Thus, engineering metabolic pathways to produce large amounts of complex terpenoids in a treatable biological host represents an attractive alternative to extraction processes from environmental sources [19]. In this scenario, micro-organisms, such as *Escherichia coli* and *Saccharomyces cerevisiae* have emerged as a sustainable alternative for the production of industrially valued terpenes, by applying synthetic biology techniques. In addition, they provide a promising alternative to producing non-native terpenes because of the genetic tools available in metabolic engineering and genome editing [4].

Recently, the use of modern biotechnological techniques has increased considerably in order to achieve large-scale production of terpenes with vast structural diversity for applications in the pharmaceutical industry, using the heterologous expression method aided by metabolic engineering techniques [20]. They have been applied for agronomic purposes, producing more resistant plants and obtaining a higher yield of aromatic compounds through the manipulation of transcription factors [21]; or for biotransformation of terpenes into more powerful derivatives [22]. The use of metabolic modifications of *E. coli* has already made it possible to obtain zizaene, a khusinol-precursor sesquiterpene, which is the main constituent of *Chrysopogonzizanioides* essential oil, also applied in fine perfumery [23].

Another example is the production of sesquiterpenes from patchouli essential oil by metabolically modified *E. coli* strains. Such processes show that the use of biotechnological procedures can provide a sustainable route to produce aromatic terpenes [24]. The key to some recent advances in terpenoid metabolic engineering is insights into the structures and functions of a diverse family of enzymes known as terpene synthases (TPSs), that have emerged in nature in a chemical ecology context, and produce the multiple core structures of terpenoids found in nature [25]. Considering the importance of terpenes for all industrial sectors and their potential to offer beneficial products to society, this study aims to conduct a literature review in order to address the main advances in biochemistry, biotechnology, and methods of extraction of such metabolites.

## BIOTECHNOLOGY OF TERPENE PRODUCTION IN MICROORGANISMS

Since most terpenoids were discovered in plants, vegetables have become the main source of commercially available terpenoids. However, plant extraction is not feasible due to low yields, slow plant growth, low content, and difficult growing conditions for some species. Although biotechnological progress exists in transgenic plants, designing plants for improved isoprenoid contents remains difficult.

Microbes are fast-growing and not much dependent on soil/water resources. Thus, they have great potential for sustainable mass production of terpenoids on plants. Micro-organisms need to meet the criteria to be used in industry: (i) high metabolic potential that supports the efficient synthesis of products of interest, with robust and rapid cell growth; (ii) well-understood metabolism and well-developed genetic tools; and (iii) high ability to grow on cheap carbon sources.

The synthesis of terpenoids depends on MEP or MVA pathways for the generation of isoprene units IPP and DMAPP. Therefore, two representative microbial hosts, *E. coli* and *S. cerevisiae*, based on MEP and MVA pathways, respectively, have been used to produce various terpenoids [26]. Successful overproduction of terpenes in microbes requires several optimization steps. The main difficulties are increasing flux through the major precursor pathway, increasing the expression of enzymes that convert prenyl diphosphates into the desired compound and introducing long and complex metabolic pathways (when necessary for complex end products). In addition, it is necessary to remove or decrease the flux of competing for metabolic pathways. Also, the concentrations of the precursors must be balanced, and any toxicity of intermediates or end products must be mitigated [27].

Synthetic biology and metabolic engineering offer new approaches to induce increased production of desired molecules in the native organism and, more importantly, relocate the biosynthetic pathways to other hosts. Microbial systems are well studied, and genetic manipulations allow microbial metabolism optimization to produce common terpenoid precursors. Unprecedented advances in the large-scale production of terpenoids have been achieved in recent years through biotechnological tools. Terpenoid production can be improved by identifying limiting steps and pathway regulation, along with the design of strategies to minimize terpenoid by-products [28]. In addition, fermentation strategies are also used for the production of terpenoids of industrial importance [29]. However, terpene production using microbial platforms still relies on external sugar sources and elaborated production facilities, which significantly increases its production costs [30].

Microorganisms secrete several enzymes that act in the biotransformation of terpene substrates by catalyzing various reactions, generating new structurally diverse products [31]. A recent study showed that *Aspergillus niger* is able to produce (S)-(+)-carvone hydrate from (-)-α-pinene; also 4-hydroxy-beta-ionone, 2-hydroxy-beta-ionone,α-ionone, and dihydro-3-oxo-beta-ionol from β-ionone; and 4-hydroxyisoisophorone,3,3-dimethyl-5-oxocyclohexane-1-carboxaldehyde, 4-ketoisoisophorone, and dihydrooxoisophorone from isophorone [32]. Similarly, fungi *Cunninghamellaechinulata*and*Rhizopus oryzae*promoted reactions of hydrogenation and epoxidation on the structures of sesquiterpene lactones (+)-costunolide, (+)-cnicine, (+)-salitenolide, (-)-dehydrocostuslactone, (--lycnofolide, and (-)- eremantholide C [33]. The success of fungal biotransformation of terpenes can also be seen in the formation of (R)-(+)-α-terpineol from (R)-(+)-limonene using *Penicillium digitatum* [34].

The efficiency of this biotransformation was evident in the work of Sales *et al.* [35], in which the highest ever recorded concentration of limonene-1,2-diol (3.34-4.01 g) was obtained using R-(+)-limonene or S-(-)-limonene as substrate. Biotransformation has several advantages when compared to the corresponding chemical methods. Besides being regio- and stereospecific, many biotransformations are also enantiospecific, generating chiral products from racemic mixtures. The conditions for biotransformations are mild and, in most cases, do not require the protection of pre-existing functional groups [31].

Another method for the optimized production of terpenes *via* microorganisms is the heterologous expression by treatable hosts, such as *Escherichia coli* and *Saccharomyces cerevisiae*, which present very satisfactory results [36]. The expression of sesquiterpene hydroxylases CYP260A1 and CYP264B1 from *Sorangiumcelulosum* So ce56, using *E. coli* as the host, was performed to compare their substrate specificities and regio-selectivities. The enzyme CYP260A1 was identified as responsible for sesquiterpene conversion to eremophilane, humulane, and cedran-type structures and CYP264B1 was characterized as a highly selective sesquiterpene hydroxylase that catalyzed the conversion of the substrates zerumbone, α-humulane, β-caryophyllene, and eremophilane to produce allylic alcohols [37].

In another work, 11 terpenoids with 11-carbon chains, as well as 24 new terpenes that had not yet been described as terpene synthase products, were produced by heterologous expression of mevalonate pathway genes, an IPP isomerase gene, a GPP synthase gene, a GPPmethyltransferase gene, and a four-gene C11-TS in *E. coli*. Among them, four compounds were identified as 3,4-dimethylcoumarin, 2-methylborneol, and two diastereomers of 2-methylcitronellol [38]. Heterologous expression experiments were also conducted using *S. cerevisiae* as a host for *Botrytis cinerea* native genes bcaba1, bcaba2, bcaba3, and bcaba4, involved in the biosynthesis of abscisic acid (ABA).

The results showed that the expression of these genes was sufficient to produce ABA, an important compound for the agriculture and pharmaceutical industry [39]. *S. cerevisiae* was also used for the expression of sesquiterpene synthase genes GME3634, GME3638, and GME9210 from *Lignosusrhinocerotis,* and produced a wide variety of sesquiterpenes, highlighting (+)-torreyol and α-cadinol synthesized by the expression of GME3638 and GME3634, respectively. Both compounds showed significant cytotoxicity, showing that this method can be of great use in the search for new metabolites with medicinal properties [39].

Recently, the discovery of cytochrome P450 has also contributed to obtaining terpenes in large quantities from microorganisms, as it allows post-modifications to occur in the structure of an existing terpene. An example is a platform for the production of steviol glycosides (SGs) using designed *S. cerevisiae*. Two P450s of kaurene oxidase (KO) and kaurenoic acid hydroxylase (KAH) are required in the conversion of kaurene into steviol. This conversion was maximized by optimizing the modulation of KO-KAHCPR combinations [40].

Fukushima *et al.,* [41] used combinatorial biosynthesis of terpenes from the expression of P450s, producing bAS, CPR, CYP72A63, and CYP93E2 or CYP716A12 from *Medicago truncatula* in *S. cerevisiae*. They biosynthesized rare tryptans such as soyasapogenol B and gypsogenic acid from β-amyrin and oleanolic acid, respectively. Such metabolites were not detected in the original organism. It is usually considered that the ability to express soluble P450s in *E. coli* is limited. This is justified by the lack of an endomembrane system for binding eukaryotic P450s, since they have a hydrophobic transmembrane helices domain containing 20-30 amino acid residues at their *N*-terminal ends [4].

Thus, some researchers have succeeded in producing terpenes through the expression of P450s, such as Biggs *et al.*, [42], who optimized the production of oxygenated taxol derivatives (~570 ± 45 mg/L) mediated by the expression of P450s, reductase interactions, and *N*-terminal modifications. An unexpected docking of P450 expression and expression of upstream pathway enzymes were discovered and identified as a key obstacle for functional oxidative chemistry. In another work, a combination of improvements of a heterologous mevalonate pathway with a superior fermentation process provided a significant increase in the production of amorpha-4,11-diene, a precursor to artemisinin, an important drug in the treatment of malaria.

In this process, yeast genes for HMG-CoA synthase and HMG-CoA reductase (the second and third enzymes in the pathway) were replaced by equivalent genes from *Staphylococcus aureus*, nearly doubling the production [43]. Understanding post-modifications in terpenoid biosynthesis is of particular importance for their efficient production, because such selective reactions performed by CYPs are extremely difficult to achieve by conventional chemical routes [44].

Besides the mentioned techniques, several others have contributed to the improvement in the production and discovery of new commercially viable terpenes. Among them, the design of biosynthetic enzymes allowed obtaining 653 mg/L of 1,8-cineole and 505 mg/L of linalool, which were improved by introducing a mutation (Ser81 -Phe) in the native FPP synthase of *E. coli,* resulting in an enzyme that preferentially synthesized GPP instead of FPP [45].

The co-culture engineering has been gaining more space in recent years, since under certain conditions, a single host cell cannot provide an optimal environment for the functioning of all enzymes involved in biosynthesis. Also, the metabolic charges of overexpression of complex pathways can reduce biosynthetic efficiency. Zhou *et al.,* [46] obtained 33 mg/L of oxygenated taxanes using a co-culture composed of *E. coli* designed to perform the biosynthesis of the taxa diene intermediate of *S. cerevisiae*, the preferred host for cytochrome P450 expression.

Another method for obtaining terpenes is subcellular engineering, which allows the optimization of geraniol production, compared to strains producing this compound in the cytosol. This was achieved by directing the geraniol biosynthetic pathway into the mitochondria of *S. cerevisiae* to protect the GPP pool from consumption by the cytosolic ergosterol pathway [42]. In addition, there are also systems of cell-free biosynthesis (CFB), which are easy to use for multiple enzyme pathways from various organisms, without the need for precursor supplies, and exclude the risk of generating toxic products.

This type of system has been used by Korman *et al.*, [47] for obtaining monoterpenes on a platform consisting of 27 enzymes, using reconstituted glycolysis to generate ATP, NADPH, and acetyl-CoA, resulting in the production of 12.5 g/L of limonene and 14.9 g/L of pinene. The system converted glucose into monoterpenes at high rates and could sustain operation for five days with only one addition of glucose.

Programming the biosynthetic pathway of the precursor (as an alternative route) can also be used when aiming to obtain terpenes or other natural products. In this sense, Kang *et al.*, [48] reported an alternative IPP-bypass AMM pathway using promiscuous activities of phosphomevalonate decarboxylase and endogenous phosphatase from *E. coli*, which made isopentenol production independent on IPP generation, and increased isoprenolconcentration to 3.7 g/L. Such approaches have increased terpene production and provided resources for new industry supplies by exploiting the functionalities of these compounds.

## PHARMACEUTICAL APPLICATIONS OF TERPENOIDS PRODUCED BY BIOTECHNOLOGICAL METHODS

The ethnopharmacological use of several plant species has led to scientific studies [49 - 52], which can support this popular knowledge and obtain phytotherapeutic agents with pharmaceutical applications with efficacy and safety. Terpenoids are a very wide and diverse class, having derivatives of five-carbon units arranged in various ways, and can be classified as hemiterpenoids, monoterpenoids, iridoids, sesquiterpenoids, diterpenoids, sesterterpenoids, triterpenoids, tetraterpenoids, polyterpenoids, and irregular terpenoids. [53, 54]. The use of these compounds for pharmacological purposes is reported in several papers for the treatment of some types of cancer, viruses, fungi, tumors, cardiovascular diseases, diabetes, malaria, inflammations, *etc.* [12, 55 - 57]. Table **1** presents some examples of terpenoids and their pharmacological indications.

## BIOTECHNOLOGICAL APPLICATIONS OF TERPENES

Terpenes are a class of compounds synthesized and stored mainly by plants, and can act as substitutes for artificial compounds, generating great interest in the market due to greater sustainability and lower risk to health and the environment. The technological applications of extracts, essential oils and isolated terpenic compounds are quite varied, and can be used in food, cosmetics, biotechnological, and pharmacological areas, as previously reported.

**Table 1. Examples of pharmaceutical applications of terpenoids.**

| Terpenoid | Raw material | Effect/Use | References |
|---|---|---|---|
| β-Elemene, δ-elemene, furanodiene, furanodienone, curcumol, and germacrone | *Rhizomacurcumae* | Anticancer | [58] |
| Thymol and carvacrol | - | Antimicrobial, Antibiotic | [59] |
| *trans*-farnesol, nerolidoland α-bisabolol | - | Antibiotic | [60] |
| larixol | *Crotonmatourensis* | Neuroprotective, anti-inflammatory | [61] |
| clerodane diterpene 16α-hydroxycleroda-3, 13(14)-Z-dien-15,16-olide | Leaves of *Polyalthia longifolia* | Antibiotic | [62] |
| Parthenolide | - | Breast cancer | [63] |
| Tripchlorolide | - | Lung cancer | [64] |
| Betulinicacid | - | Human cervical cancer | [65] |
| Isoprenoids, monoterpenoids, sesquiterpenoids,germacranolide, andditerpenoids | - | Antileishmanial | [66] |
| Lactonescumanin, psilostachyin,and cordilin,and the sterol glycoside daucosterol. | *Ambrosia elatiorandA. scabra* | Antiparasitic (*Trypanosoma cruzi*and*Leishmania* sp.) | [67] |
| DiosgeninandGinsenoside Re | - | Cardioprotective | [68] |
| 3α-Hydroxy-5,6-epoxy-7-megastigmen-9-one, (+)-dehydrovomifoliol, loliolide, (6R,7E,9R)- 9-hydroxy-4,7-megastigmadien-3-one, petasol,andoplodiol | *Padinapavonia* | Anti-hyperglycemic, anti-dyslipidemic, antioxidant, and anti-inflammatory | [69] |

In the food industry, synthetic additives have become the target of an increasingly negative perception due to their potential harm to health. Thus, natural additives have been seeking to fill this gap and serve consumers with greater safety. Several terpenoids are used for this purpose, and can be applied as colorants [70, 71], flavorings [72, 73], and nutraceutical [74, 75], antioxidant [76, 77], and antimicrobial agents [78, 79].

In cosmetics, terpenes are very appreciated for their fragrance, antioxidant effect, and protection against ultraviolet radiation adding benefits to skin care products, among other health-related advantages [80, 81]. The biosynthesis of terpenes has been studied because of their potential to be used in the production of aviation fuels, missile propellants, gasoline, and diesel fuels [82 - 84].

Terpenes, such as lutein and bixin, have been investigated for use in the textile industry as replacements for synthetic dyes since they have been associated with a higher risk of allergic and toxic effects on the skin, and greater environmental impact [85, 86]. The application of terpene-based biopesticides has also been reported to control pests in agribusiness due to their low toxicity [87, 88]. Table **2** summarizes some applications of these terpene compounds and their sources.

Table 2. Terpenes and their technological applications.

| Terpenoid | Raw Material | Activity/Application | References |
|---|---|---|---|
| Bixin and norbixin | Achiote (*Bixa orellana* L.) | Pigments | [70, 71] |
| Menthone and menthol | - | Aroma compounds | [72] |
| Steviol glycosides and mogrosides | *Stevia rebaudiana* (Bertoni) and *Siraitiagrosvenorii* (Swingle) | Sweeteners | [73] |
| Lutein, lycopene and β-carotene | *Prunus persica* | Antioxidant | [77] |
| Neurosporene, lutein, 13-15-cis-β-carotene, *trans*-β-carotene, lycopene | *Solanum lycopersicum cerasiforme* | Antioxidant | [76] |
| Astaxanthin | - | Nutraceutical | [74] |
| Thymol, carvacrol, eugenol, and menthol | - | Antifungal | [78] |
| Linalol, caryophyllene, neral, geraniol and geranial | *Cymbopogon citratus* | Antifungal | [79] |
| Eugenol, *D*-limonene, and eucalyptol | - | Skin care | [80] |
| L-β-pinene, *D*-limonene, α-pinene and β-phellandrene | Essential oils of *Ligulariafischeri* and *L. fischeri*var. *Spicifoprmis* | Fragrance for perfumery products | [89]] |
| Sabinene | Microbial production | Precursor of biofuels | [83] |
| Lutein | Marigold (*Tagetes erecta*L.) flower | Textile dyeing | [86] |
| Bixin | *Bixa orellana*seeds | Textile dyeing | [85] |
| Terpene-based biopesticides | Orange oil, *Chenopodium ambrosioides* extract and neem oil | Control of aphid pests | [87] |

*(Table 2) cont.....*

| Terpenoid | Raw Material | Activity/Application | References |
|---|---|---|---|
| 4-methyl-2-pentyl acetate, α-pinene, 1.8 cineole, isopinocarveol, caryophyllene and β-humulene | *Eucalyptus globulus* essential oil | Insecticides | [88] |

## CONCLUDING REMARKS

Currently, with the search for healthier life habits, researchers have shown great interest in natural products, and biotechnological processes can be an alternative because, through genetic engineering, plant species can be improved for the production of secondary metabolites present in oils essential products of commercial interest for various applications, such as terpenoids, which over the years have demonstrated several important biological activities and can serve as a basis for the development of new drugs.

## CONSENT FOR PUBLICATION

Not applicable.

## CONFLICT OF INTEREST

The author declares no conflict of interest, financial or otherwise.

## ACKNOWLEDGEMENTS

We would like to thank the authors for their contributions and commitment to writing this scientific work, and the editors of this book for the invitation and vote of confidence. The author Dr Mozaniel Santana de Oliveira, thank the *Programa de Capacitação Institucional* (PCI), linked and supervised by the *Ministério da Ciência, Tecnologia e Inovações* – MCTI, for the scholarship (process number 302050/2021-3).

## REFERENCES

[1]    Bergman, M.E.; Davis, B.; Phillips, M.A. Medically useful plant terpenoids: Biosynthesis, occurrence, and mechanism of action. *Molecules,* **2019**, *24*(21), 3961.
[http://dx.doi.org/10.3390/molecules24213961] [PMID: 31683764]

[2]    Bergman, M.E.; Phillips, M.A. Structural diversity and biosynthesis of plant derived *p*-menthane monoterpenes. *Phytochem. Rev.,* **2021**, *20*(2), 433-459.
[http://dx.doi.org/10.1007/s11101-020-09726-0]

[3]    Pichersky, E.; Raguso, R.A. Why do plants produce so many terpenoid compounds? *New Phytol.,* **2018**, *220*(3), 692-702.
[http://dx.doi.org/10.1111/nph.14178] [PMID: 27604856]

[4]    Zhang, C.; Hong, K. Production of Terpenoids by Synthetic Biology Approaches. *Front. Bioeng. Biotechnol.,* **2020**, *8*, 347.

[http://dx.doi.org/10.3389/fbioe.2020.00347] [PMID: 32391346]

[5]     Rajčević, N.; Nikolić, B.; Marin, P. Different responses to environmental factors in terpene composition of Pinus heldreichii and P. peuce: Ecological and chemotaxonomic considerations. *Arch. Biol. Sci.,* **2019**, *71*(4), 629-637.
[http://dx.doi.org/10.2298/ABS190705045R]

[6]     He, H.; Bian, G.; Herbst-Gervasoni, C.J.; Mori, T.; Shinsky, S.A.; Hou, A.; Mu, X.; Huang, M.; Cheng, S.; Deng, Z.; Christianson, D.W.; Abe, I.; Liu, T. Discovery of the cryptic function of terpene cyclases as aromatic prenyltransferases. *Nat. Commun.,* **2020**, *11*(1), 3958.
[http://dx.doi.org/10.1038/s41467-020-17642-2] [PMID: 32769971]

[7]     Tholl, D. Biosynthesis and biological functions of terpenoids in plants. *Adv. Biochem. Eng. Biotechnol.,* **2015**, *148*, 63-106.
[http://dx.doi.org/10.1007/10_2014_295] [PMID: 25583224]

[8]     Huang, A.C.; Osbourn, A. Plant terpenes that mediate below-ground interactions: prospects for bioengineering terpenoids for plant protection. *Pest Manag. Sci.,* **2019**, *75*(9), 5410.
[http://dx.doi.org/10.1002/ps.5410] [PMID: 30884099]

[9]     Moser, S.; Pichler, H. Identifying and engineering the ideal microbial terpenoid production host. *Appl. Microbiol. Biotechnol.,* **2019**, *103*(14), 5501-5516.
[http://dx.doi.org/10.1007/s00253-019-09892-y] [PMID: 31129740]

[10]    Gershenzon, J.; Dudareva, N. The function of terpene natural products in the natural world. *Nat. Chem. Biol.,* **2007**, *3*(7), 408-414.
[http://dx.doi.org/10.1038/nchembio.2007.5] [PMID: 17576428]

[11]    S, Z.; C, B. Plant terpenoids: applications and future potentials. *Biotechnol. Mol. Biol. Rev.,* **2008**, *3*, 1-7.

[12]    Abbas, F.; Ke, Y.; Yu, R.; Yue, Y.; Amanullah, S.; Jahangir, M.M.; Fan, Y. Volatile terpenoids: multiple functions, biosynthesis, modulation and manipulation by genetic engineering. *Planta,* **2017**, *246*(5), 803-816.
[http://dx.doi.org/10.1007/s00425-017-2749-x] [PMID: 28803364]

[13]    Pattanaik, B.; Lindberg, P. Terpenoids and their biosynthesis in cyanobacteria. *Life (Basel),* **2015**, *5*(1), 269-293.
[http://dx.doi.org/10.3390/life5010269] [PMID: 25615610]

[14]    Arendt, P.; Pollier, J.; Callewaert, N.; Goossens, A. Synthetic biology for production of natural and new-to-nature terpenoids in photosynthetic organisms. *Plant J.,* **2016**, *87*(1), 16-37.
[http://dx.doi.org/10.1111/tpj.13138] [PMID: 26867713]

[15]    Mewalal, R.; Rai, D.K.; Kainer, D.; Chen, F.; Külheim, C.; Peter, G.F.; Tuskan, G.A. Plant-Derived Terpenes: A Feedstock for Specialty Biofuels. *Trends Biotechnol.,* **2017**, *35*(3), 227-240.
[http://dx.doi.org/10.1016/j.tibtech.2016.08.003] [PMID: 27622303]

[16]    Pahima, E.; Hoz, S.; Ben-Tzion, M.; Major, D.T. Computational design of biofuels from terpenes and terpenoids. *Sustain. Energy Fuels,* **2019**, *3*(2), 457-466.
[http://dx.doi.org/10.1039/C8SE00390D]

[17]    Petronilho, S.; Coimbra, M.A.; Rocha, S.M. A critical review on extraction techniques and gas chromatography based determination of grapevine derived sesquiterpenes. *Anal. Chim. Acta,* **2014**, *846*, 8-35.
[http://dx.doi.org/10.1016/j.aca.2014.05.049] [PMID: 25220138]

[18]    Janoschek, L.; Grozdev, L.; Berensmeier, S. Membrane-assisted extraction of monoterpenes: from *in silico* solvent screening towards biotechnological process application. *R. Soc. Open Sci.,* **2018**, *5*(4), 172004.
[http://dx.doi.org/10.1098/rsos.172004] [PMID: 29765654]

[19]    Martin, V.J.J.; Pitera, D.J.; Withers, S.T.; Newman, J.D.; Keasling, J.D. Engineering a mevalonate

pathway in Escherichia coli for production of terpenoids. *Nat. Biotechnol.,* **2003**, *21*(7), 796-802.
[http://dx.doi.org/10.1038/nbt833] [PMID: 12778056]

[20] Ma, C.; Zhang, K.; Zhang, X.; Liu, G.; Zhu, T.; Che, Q.; Li, D.; Zhang, G. Heterologous expression and metabolic engineering tools for improving terpenoids production. *Curr. Opin. Biotechnol.,* **2021**, *69*, 281-289.
[http://dx.doi.org/10.1016/j.copbio.2021.02.008] [PMID: 33770560]

[21] Mahmoud, S.S.; Croteau, R.B. Strategies for transgenic manipulation of monoterpene biosynthesis in plants. *Trends Plant Sci.,* **2002**, *7*(8), 366-373.
[http://dx.doi.org/10.1016/S1360-1385(02)02303-8] [PMID: 12167332]

[22] Kuriata-Adamusiak, R.; Strub, D.; Lochyński, S. Application of microorganisms towards synthesis of chiral terpenoid derivatives. *Appl. Microbiol. Biotechnol.,* **2012**, *95*(6), 1427-1436.
[http://dx.doi.org/10.1007/s00253-012-4304-9] [PMID: 22846902]

[23] Aguilar, F.; Scheper, T.; Beutel, S. Improved production and *in situ* recovery of sesquiterpene (+)-zizaene from metabolically-engineered E. Coli. *Molecules,* **2019**, *24*(18), 3356.
[http://dx.doi.org/10.3390/molecules24183356] [PMID: 31540161]

[24] Aguilar, F.; Ekramzadeh, K.; Scheper, T.; Beutel, S. Whole-cell production of patchouli oil sesquiterpenes in Escherichia coli: Metabolic engineering and fermentation optimization in solid∑liquid phase partitioning cultivation. *ACS Omega,* **2020**, *5*(50), 32436-32446.
[http://dx.doi.org/10.1021/acsomega.0c04590] [PMID: 33376881]

[25] Bohlmann, J. Terpenoid synthases--from chemical ecology and forest fires to biofuels and bioproducts. *Structure,* **2011**, *19*(12), 1730-1731.
[http://dx.doi.org/10.1016/j.str.2011.11.009] [PMID: 22153494]

[26] Wang, C.; Liwei, M.; Park, J. Bin; Jeong, S.H.; Wei, G.; Wang, Y.; Kim, S.W. Microbial platform for terpenoid production: Escherichia coli and Yeast. *Front. Microbiol.,* **2018**, *9*, 1-8.

[27] Vickers, C.; Bongers, M.; Liu, Q.; Delatte, T.; Bouwmeester, H. Metabolic engineering of volatile isoprenoids in plants and microbes. *Plant Cell Environ.,* **2014**, *37*(8), 1753-1775.
[http://dx.doi.org/10.1111/pce.12316] [PMID: 24588680]

[28] Ajikumar, P.K.; Tyo, K.; Carlsen, S.; Mucha, O.; Phon, T.H.; Stephanopoulos, G. Terpenoids: opportunities for biosynthesis of natural product drugs using engineered microorganisms. *Mol. Pharm.,* **2008**, *5*(2), 167-190.
[http://dx.doi.org/10.1021/mp700151b] [PMID: 18355030]

[29] Du, F.; Wang, Y.Z.; Xu, Y.S.; Shi, T.Q.; Liu, W.Z.; Sun, X.M.; Huang, H. Biotechnological production of lipid and terpenoid from thraustochytrids. *Biotechnol. Adv.,* **2021**, *48*, 107725.
[http://dx.doi.org/10.1016/j.biotechadv.2021.107725] [PMID: 33727145]

[30] Jiang, Z.; Kempinski, C.; Bush, C.J.; Nybo, S.E.; Chappell, J. Engineering triterpene and methylated triterpene production in plants provides biochemical and physiological insights into terpene metabolism. *Plant Physiol.,* **2016**, *170*(2), 702-716.
[http://dx.doi.org/10.1104/pp.15.01548] [PMID: 26603654]

[31] Sultana, N.; Saify, Z.S. Enzymatic biotransformation of terpenes as bioactive agents. *J. Enzyme Inhib. Med. Chem.,* **2013**, *28*(6), 1113-1128.
[http://dx.doi.org/10.3109/14756366.2012.727411] [PMID: 23046385]

[32] Çorbacı, C. Biotransformation of terpene and terpenoid derivatives by Aspergillus niger NRRL 326. *Biologia (Bratisl.),* **2020**, *75*(9), 1473-1481.
[http://dx.doi.org/10.2478/s11756-020-00459-1]

[33] Barrero, A.F.; Oltra, J.E.; Raslan, D.S.; Saúde, D.A. Microbial transformation of sesquiterpene lactones by the fungi cunninghamella echinulata and rhizopus oryzae. *J. Nat. Prod.,* **1999**, *62*(5), 726-729.
[http://dx.doi.org/10.1021/np980520w] [PMID: 10346955]

[34]   Prieto, S. G.A.; Perea V, J.A.; Ortiz L, C.C. Microbial biotransformation of (R)-(+)-limonene by penicillium digitatum dsm 62840 for producing (R)-(+)-terpineol. *Vitae, 2011, 18,* 163-172.

[35]   Sales, A.; Afonso, L.F.; Americo, J.A.; de Freitas Rebelo, M.; Pastore, G.M.; Bicas, J.L. Monoterpene biotransformation by Colletotrichum species. *Biotechnol. Lett.,* **2018**, *40*(3), 561-567.
[http://dx.doi.org/10.1007/s10529-017-2503-2] [PMID: 29288353]

[36]   Ikram, N.K.B.K.; Zhan, X.; Pan, X.W.; King, B.C.; Simonsen, H.T. Stable heterologous expression of biologically active terpenoids in green plant cells. *Front. Plant Sci.,* **2015**, *6,* 129.
[http://dx.doi.org/10.3389/fpls.2015.00129] [PMID: 25852702]

[37]   Schifrin, A.; Litzenburger, M.; Ringle, M.; Ly, T.T.B.; Bernhardt, R. New sesquiterpene oxidations with CYP260A1 and CYP264B1 from Sorangium cellulosum so ce56. *ChemBioChem,* **2015**, *16*(18), 2624-2632.
[http://dx.doi.org/10.1002/cbic.201500417] [PMID: 26449371]

[38]   Kschowak, M.J.; Wortmann, H.; Dickschat, J.S.; Schrader, J.; Buchhaupt, M. Heterologous expression of 2-methylisoborneol / 2 methylenebornane biosynthesis genes in Escherichia coli yields novel C11-terpenes. *PLoS One,* **2018**, *13*(4), e0196082.
[http://dx.doi.org/10.1371/journal.pone.0196082] [PMID: 29672609]

[39]   Otto, M.; Teixeira, P.G.; Vizcaino, M.I.; David, F.; Siewers, V. Integration of a multi-step heterologous pathway in Saccharomyces cerevisiae for the production of abscisic acid. *Microb. Cell Fact.,* **2019**, *18*(1), 205.
[http://dx.doi.org/10.1186/s12934-019-1257-z] [PMID: 31767000]

[40]   Gold, N.D.; Fossati, E.; Hansen, C.C.; DiFalco, M.; Douchin, V.; Martin, V.J.J. A Combinatorial Approach To Study Cytochrome P450 Enzymes for *De Novo* Production of Steviol Glucosides in Baker's Yeast. *ACS Synth. Biol.,* **2018**, *7*(12), 2918-2929.
[http://dx.doi.org/10.1021/acssynbio.8b00470] [PMID: 30474973]

[41]   Fukushima, E.O.; Seki, H.; Sawai, S.; Suzuki, M.; Ohyama, K.; Saito, K.; Muranaka, T. Combinatorial biosynthesis of legume natural and rare triterpenoids in engineered yeast. *Plant Cell Physiol.,* **2013**, *54*(5), 740-749.
[http://dx.doi.org/10.1093/pcp/pct015] [PMID: 23378447]

[42]   Biggs, B.W.; Lim, C.G.; Sagliani, K.; Shankar, S.; Stephanopoulos, G.; De Mey, M.; Ajikumar, P.K. Overcoming heterologous protein interdependency to optimize P450-mediated Taxol precursor synthesis in *Escherichia coli. Proc. Natl. Acad. Sci. USA,* **2016**, *113*(12), 3209-3214.
[http://dx.doi.org/10.1073/pnas.1515826113] [PMID: 26951651]

[43]   Tsuruta, H.; Paddon, C.J.; Eng, D.; Lenihan, J.R.; Horning, T.; Anthony, L.C.; Regentin, R.; Keasling, J.D.; Renninger, N.S.; Newman, J.D. High-level production of amorpha-4,11-diene, a precursor of the antimalarial agent artemisinin, in Escherichia coli. *PLoS One,* **2009**, *4*(2), e4489.
[http://dx.doi.org/10.1371/journal.pone.0004489] [PMID: 19221601]

[44]   Xiao, H.; Zhang, Y.; Wang, M. Discovery and Engineering of Cytochrome P450s for Terpenoid Biosynthesis. *Trends Biotechnol.,* **2019**, *37*(6), 618-631.
[http://dx.doi.org/10.1016/j.tibtech.2018.11.008] [PMID: 30528904]

[45]   Mendez-Perez, D.; Alonso-Gutierrez, J.; Hu, Q.; Molinas, M.; Baidoo, E.E.K.; Wang, G.; Chan, L.J.G.; Adams, P.D.; Petzold, C.J.; Keasling, J.D.; Lee, T.S. Production of jet fuel precursor monoterpenoids from engineered *Escherichia coli. Biotechnol. Bioeng.,* **2017**, *114*(8), 1703-1712.
[http://dx.doi.org/10.1002/bit.26296] [PMID: 28369701]

[46]   Zhou, K.; Qiao, K.; Edgar, S.; Stephanopoulos, G. Distributing a metabolic pathway among a microbial consortium enhances production of natural products. *Nat. Biotechnol.,* **2015**, *33*(4), 377-383.
[http://dx.doi.org/10.1038/nbt.3095] [PMID: 25558867]

[47]   Korman, T.P.; Opgenorth, P.H.; Bowie, J.U. A synthetic biochemistry platform for cell free production of monoterpenes from glucose. *Nat. Commun.,* **2017**, *8*(1), 15526.

[http://dx.doi.org/10.1038/ncomms15526] [PMID: 28537253]

[48]   Kang, A.; George, K.W.; Wang, G.; Baidoo, E.; Keasling, J.D.; Lee, T.S. Isopentenyl diphosphate (IPP)-bypass mevalonate pathways for isopentenol production. *Metab. Eng.,* **2016**, *34*, 25-35.
[http://dx.doi.org/10.1016/j.ymben.2015.12.002] [PMID: 26708516]

[49]   Leonti, M.; Casu, L. Traditional medicines and globalization: current and future perspectives in ethnopharmacology. *Front. Pharmacol.,* **2013**, *4*, 92.
[http://dx.doi.org/10.3389/fphar.2013.00092] [PMID: 23898296]

[50]   Huang, J.; Zhang, Y.; Dong, L.; Gao, Q.; Yin, L.; Quan, H.; Chen, R.; Fu, X.; Lin, D. Ethnopharmacology, phytochemistry, and pharmacology of Cornus officinalis Sieb. et Zucc. *J. Ethnopharmacol.,* **2018**, *213*, 280-301.
[http://dx.doi.org/10.1016/j.jep.2017.11.010] [PMID: 29155174]

[51]   Nabeelah Bibi, S.; Fawzi, M.M.; Gokhan, Z.; Rajesh, J.; Nadeem, N.; Pandian, ; R D D G, A.; Pandian, S.K. Ethnopharmacology, phytochemistry, and global distribution of mangroves-a comprehensive review. *Mar. Drugs,* **2019**, *17*(4), 231.
[http://dx.doi.org/10.3390/md17040231] [PMID: 31003533]

[52]   Miroddi, M.; Calapai, G.; Navarra, M.; Minciullo, P.L.; Gangemi, S. Passiflora incarnata L.: Ethnopharmacology, clinical application, safety and evaluation of clinical trials. *J. Ethnopharmacol.,* **2013**, *150*(3), 791-804.
[http://dx.doi.org/10.1016/j.jep.2013.09.047] [PMID: 24140586]

[53]   Ludwiczuk, A.; Skalicka-Woźniak, K.; Georgiev, M.I. Terpenoids. *Pharmacognosy: Fundamentals, Applications and Strategy.,* **2017**, , 233-266.
[http://dx.doi.org/10.1016/B978-0-12-802104-0.00011-1]

[54]   Zheng, X.; Li, P.; Lu, X. Research advances in cytochrome P450-catalysed pharmaceutical terpenoid biosynthesis in plants. *J. Exp. Bot.,* **2019**, *70*(18), 4619-4630.
[http://dx.doi.org/10.1093/jxb/erz203] [PMID: 31037306]

[55]   Carsanba, E.; Pintado, M.; Oliveira, C. Fermentation strategies for production of pharmaceutical terpenoids in engineered yeast. *Pharmaceuticals (Basel),* **2021**, *14*(4), 295.
[http://dx.doi.org/10.3390/ph14040295] [PMID: 33810302]

[56]   Lu, X.; Tang, K.; Li, P. Plant metabolic engineering strategies for the production of pharmaceutical terpenoids. *Front. Plant Sci.,* **2016**, *7*, 1647.
[http://dx.doi.org/10.3389/fpls.2016.01647] [PMID: 27877181]

[57]   Zhang, Y.; Nielsen, J.; Liu, Z. Engineering yeast metabolism for production of terpenoids for use as perfume ingredients, pharmaceuticals and biofuels. *FEMS Yeast Res.,* **2017**, *17*(8), 1-11.
[http://dx.doi.org/10.1093/femsyr/fox080] [PMID: 29096021]

[58]   Lu, J.J.; Dang, Y.Y.; Huang, M.; Xu, W.S.; Chen, X.P.; Wang, Y.T. Anti-cancer properties of terpenoids isolated from Rhizoma Curcumae – A review. *J. Ethnopharmacol.,* **2012**, *143*(2), 406-411.
[http://dx.doi.org/10.1016/j.jep.2012.07.009] [PMID: 22820242]

[59]   Zhang, D.; Hu, H.; Rao, Q.; Zhao, Z. Synergistic effects and physiological responses of selected bacterial isolates from animal feed to four natural antimicrobials and two antibiotics. *Foodborne Pathog. Dis.,* **2011**, *8*(10), 1055-1062.
[http://dx.doi.org/10.1089/fpd.2010.0817] [PMID: 21612425]

[60]   Gonçalves, O.; Pereira, R.; Gonçalves, F.; Mendo, S.; Coimbra, M.A.; Rocha, S.M. Evaluation of the mutagenicity of sesquiterpenic compounds and their influence on the susceptibility towards antibiotics of two clinically relevant bacterial strains. *Mutat. Res. Genet. Toxicol. Environ. Mutagen.,* **2011**, *723*(1), 18-25.
[http://dx.doi.org/10.1016/j.mrgentox.2011.03.010] [PMID: 21453784]

[61]   Bezerra, F.W.F.; Salazar, M.L.A.R.; Freitas, L.C.; de Oliveira, M.S.; dos Santos, I.R.C.; Dias, M.N.C.; Gomes-Leal, W.; Andrade, E.H.A.; Ferreira, G.C.; Carvalho, R.N., Jr Chemical composition,

antioxidant activity, anti-inflammatory and neuroprotective effect of Croton matourensis Aubl. Leaves extracts obtained by supercritical $CO_2$. *J. Supercrit. Fluids,* **2020**, *165*, 104992.
[http://dx.doi.org/10.1016/j.supflu.2020.104992]

[62] Gupta, V.K.; Tiwari, N.; Gupta, P.; Verma, S.; Pal, A.; Srivastava, S.K.; Darokar, M.P. A clerodane diterpene from Polyalthia longifolia as a modifying agent of the resistance of methicillin resistant Staphylococcus aureus. *Phytomedicine,* **2016**, *23*(6), 654-661.
[http://dx.doi.org/10.1016/j.phymed.2016.03.001] [PMID: 27161406]

[63] D'Anneo, A.; Carlisi, D.; Lauricella, M.; Puleio, R.; Martinez, R.; Di Bella, S.; Di Marco, P.; Emanuele, S.; Di Fiore, R.; Guercio, A.; Vento, R.; Tesoriere, G. Parthenolide generates reactive oxygen species and autophagy in MDA-MB231 cells. A soluble parthenolide analogue inhibits tumour growth and metastasis in a xenograft model of breast cancer. *Cell Death Dis.,* **2013**, *4*(10), e891.
[http://dx.doi.org/10.1038/cddis.2013.415] [PMID: 24176849]

[64] Chen, L.; Liu, Q.; Huang, Z.; Wu, F.; Li, Z.; Chen, X.; Lin, T. Tripchlorolide induces cell death in lung cancer cells by autophagy. *Int. J. Oncol.,* **2012**, *40*(4), 1066-1070.
[http://dx.doi.org/10.3892/ijo.2011.1278] [PMID: 22139090]

[65] Xu, T.; Pang, Q.; Zhou, D.; Zhang, A.; Luo, S.; Wang, Y.; Yan, X. Proteomic investigation into betulinic acid-induced apoptosis of human cervical cancer HeLa cells. *PLoS One,* **2014**, *9*(8), e105768.
[http://dx.doi.org/10.1371/journal.pone.0105768] [PMID: 25148076]

[66] Ogungbe, I.V.; Setzer, W.N. *In-silico* Leishmania target selectivity of antiparasitic terpenoids. **2013**, *18*

[67] Sülsen, V.P.; Cazorla, S.I.; Frank, F.M.; Laurella, L.C.; Muschietti, L.V.; Catalán, C.A.; Martino, V.S.; Malchiodi, E.L. Natural terpenoids from Ambrosia species are active *in vitro* and *in vivo* against human pathogenic trypanosomatids. *PLoS Negl. Trop. Dis.,* **2013**, *7*(10), e2494.
[http://dx.doi.org/10.1371/journal.pntd.0002494] [PMID: 24130916]

[68] Hua, F.; Shi, L.; Zhou, P. Phytochemicals as potential IKK-β inhibitor for the treatment of cardiovascular diseases in plant preservation: terpenoids, alkaloids, and quinones. *Inflammopharmacology,* **2020**, *28*(1), 83-93.
[http://dx.doi.org/10.1007/s10787-019-00640-2] [PMID: 31487001]

[69] Germoush, M.O.; Elgebaly, H.A.; Hassan, S.; Kamel, E.M.; Bin-Jumah, M.; Mahmoud, A.M. Consumption of terpenoids-rich padina pavonia extract attenuates hyperglycemia, insulin resistance and oxidative stress, and upregulates ppary in a rat model of type 2 diabetes. *Antioxidants,* **2020**, *9*, 1-24.

[70] Raddatz-Mota, D.; Pérez-Flores, L.J.; Carrari, F.; Mendoza-Espinoza, J.A.; de León-Sánchez, F.D.; Pinzón-López, L.L.; Godoy-Hernández, G.; Rivera-Cabrera, F. Achiote (*Bixa orellana* L.): a natural source of pigment and vitamin E. *J. Food Sci. Technol.,* **2017**, *54*(6), 1729-1741.
[http://dx.doi.org/10.1007/s13197-017-2579-7] [PMID: 28559632]

[71] Martínez, Y.; Orozco, C.E.; Montellano, R.M.; Valdivié, M.; Parrado, C.A. Use of achiote (Bixa orellana L.) seed powder as pigment of the egg yolk of laying hens. *J. Appl. Poult. Res.,* **2021**, *30*(2), 100154.
[http://dx.doi.org/10.1016/j.japr.2021.100154]

[72] Ades, H.; Kesselman, E.; Ungar, Y.; Shimoni, E. Complexation with starch for encapsulation and controlled release of menthone and menthol. *Lebensm. Wiss. Technol.,* **2012**, *45*(2), 277-288.
[http://dx.doi.org/10.1016/j.lwt.2011.08.008]

[73] Pawar, R.S.; Krynitsky, A.J.; Rader, J.I. Sweeteners from plants—with emphasis on Stevia rebaudiana (Bertoni) and Siraitia grosvenorii (Swingle). *Anal. Bioanal. Chem.,* **2013**, *405*(13), 4397-4407.
[http://dx.doi.org/10.1007/s00216-012-6693-0] [PMID: 23341001]

[74] Morales, E.; Burgos-Díaz, C.; Zúñiga, R.N.; Jorkowski, J.; Quilaqueo, M.; Rubilar, M. Influence of O/W emulsion interfacial ionic membranes on the encapsulation efficiency and storage stability of powder microencapsulated astaxanthin. *Food Bioprod. Process.,* **2021**, *126*, 143-154.

[http://dx.doi.org/10.1016/j.fbp.2020.12.014]

[75]    Paul, R.; Mazumder, M.K.; Nath, J.; Deb, S.; Paul, S.; Bhattacharya, P.; Borah, A. Lycopene - A pleiotropic neuroprotective nutraceutical: Deciphering its therapeutic potentials in broad spectrum neurological disorders. *Neurochem. Int.,* **2020**, *140*, 104823.
[http://dx.doi.org/10.1016/j.neuint.2020.104823] [PMID: 32827559]

[76]    Campestrini, L.H.; Melo, P.S.; Peres, L.E.P.; Calhelha, R.C.; Ferreira, I.C.F.R.; Alencar, S.M. A new variety of purple tomato as a rich source of bioactive carotenoids and its potential health benefits. *Heliyon,* **2019**, *5*(11), e02831.
[http://dx.doi.org/10.1016/j.heliyon.2019.e02831] [PMID: 31763483]

[77]    Dabbou, S.; Maatallah, S.; Castagna, A.; Guizani, M.; Sghaeir, W.; Hajlaoui, H.; Ranieri, A. Carotenoids, Phenolic Profile, Mineral Content and Antioxidant Properties in Flesh and Peel of Prunus persica Fruits during Two Maturation Stages. *Plant Foods Hum. Nutr.,* **2017**, *72*(1), 103-110.
[http://dx.doi.org/10.1007/s11130-016-0585-y]

[78]    Abbaszadeh, S.; Sharifzadeh, A.; Shokri, H.; Khosravi, A.R.; Abbaszadeh, A. Antifungal efficacy of thymol, carvacrol, eugenol and menthol as alternative agents to control the growth of food-relevant fungi. *J. Mycol. Med.,* **2014**, *24*(2), e51-e56.
[http://dx.doi.org/10.1016/j.mycmed.2014.01.063] [PMID: 24582134]

[79]    Oliveira, F.S.; Teodoro, C.E.S.; Berbert, P.A.; Martinazzo, A.P. Evaluation of the antifungal potential of Cymbopogon citratus essential oil in the control of the fungus Aspergillus brasiliensis. *Research, Society and Development,* **2020**, *9*(7), e691974697.
[http://dx.doi.org/10.33448/rsd-v9i7.4697]

[80]    Nastiti, C.M.R.R.; Ponto, T.; Mohammed, Y.; Roberts, M.S.; Benson, H.A.E. Novel nanocarriers for targeted topical skin delivery of the antioxidant resveratrol. *Pharmaceutics,* **2020**, *12*(2), 108.
[http://dx.doi.org/10.3390/pharmaceutics12020108] [PMID: 32013204]

[81]    Tetali, S.D. Terpenes and isoprenoids: a wealth of compounds for global use. *Planta,* **2019**, *249*(1), 1-8.
[http://dx.doi.org/10.1007/s00425-018-3056-x] [PMID: 30467631]

[82]    Szucs, I.; Escobar, M.; Grodzinski, B. Emerging Roles for Plant Terpenoids. Elsevier B.V, **2011**; 4, pp. 273-286.

[83]    Zhang, H.; Liu, Q.; Cao, Y.; Feng, X.; Zheng, Y.; Zou, H.; Liu, H.; Yang, J.; Xian, M. Microbial production of sabinene—a new terpene-based precursor of advanced biofuel. *Microb. Cell Fact.,* **2014**, *13*(1), 20.
[http://dx.doi.org/10.1186/1475-2859-13-20]

[84]    Mewalal, R.; Rai, D.K.; Kainer, D.; Chen, F.; Külheim, C.; Peter, G.F.; Tuskan, G.A. Plant-Derived Terpenes: A Feedstock for Specialty Biofuels. *Trends Biotechnol.,* **2017**, *35*(3), 227-240.
[http://dx.doi.org/10.1016/j.tibtech.2016.08.003] [PMID: 27622303]

[85]    Saha, P.D.; Sinha, K. Natural dye from bixa seeds as a potential alternative to synthetic dyes for use in textile industry. *Desalination Water Treat.,* **2012**, *40*(1-3), 298-301.
[http://dx.doi.org/10.1080/19443994.2012.671169]

[86]    Adeel, S.; Gulzar, T.; Azeem, M.; Fazal-ur-Rehman, ; Saeed, M.; Hanif, I.; Iqbal, N. Appraisal of marigold flower based lutein as natural colourant for textile dyeing under the influence of gamma radiations. *Radiat. Phys. Chem.,* **2017**, *130*, 35-39.
[http://dx.doi.org/10.1016/j.radphyschem.2016.07.010]

[87]    Smith, G.H.; Roberts, J.M.; Pope, T.W. Terpene based biopesticides as potential alternatives to synthetic insecticides for control of aphid pests on protected ornamentals. *Crop Prot.,* **2018**, *110*, 125-130.
[http://dx.doi.org/10.1016/j.cropro.2018.04.011]

[88]    Ghrab, S.; Balme, S.; Cretin, M.; Bouaziz, S.; Benzina, M. Adsorption of terpenes from Eucalyptus

globulus onto modified beidellite. *Appl. Clay Sci.,* **2018**, *156*, 169-177.
[http://dx.doi.org/10.1016/j.clay.2018.02.002]

[89]   Yeon, B.R.; Cho, H.M.; Yun, M.S.; Jhoo, J.W.; Jung, J.W.; Park, Y.H.; Kim, S. Comparison of fragrance and chemical composition of essential oils in Gom-chewi (Ligularia fischeri) and Handaeri Gom-chewi (Ligularia fischeri var. spicifoprmis). *Journal of the Korean Society of Food Science and Nutrition,* **2012**, *41*(12), 1758-1763.
[http://dx.doi.org/10.3746/jkfn.2012.41.12.1758]

# SUBJECT INDEX

## A

Abiotic stresses 41, 187
Abscisic acid 11
  responses 11
  signaling transduction pathway enzymes 11
Absorption 148, 153, 155, 156, 159, 161, 183,
    187, 227, 279
  oral 153
  plant mineral 183
  transdermal 279
Acetylcholine 83, 149, 249
Acetylcholinesterase 46, 52, 82, 83, 84, 112,
    113, 254, 266
  assay 83
  enzyme 254
  inhibition 112
  inhibitory activity 82
Acetyl-CoA acetyltransferase 173
*Achillea millefolium* 206
Acid(s) 18, 26, 31, 50, 54, 106, 177, 180, 186,
    203, 211, 212, 226, 233, 279, 281, 282,
    283, 285, 286, 300, 304, 305, 306, 307,
    310, 311, 312, 326
  agathic 306
  amino 54, 186, 300
  barthydrolic 211, 212
  bartsiifolic 211
  betulinic 180, 285
  betulonic 283
  blakielic 211, 212
  blakifolic 211
  butyric 186
  carboxylic 226, 279
  carnosic 286
  communic 307
  conjugated linoleic 233
  dehydroabietic 304
  dehydrojunicedric 305
  diterpenic 304, 310, 312
  fatty 18, 31, 106, 300
  ganoderenic 26

  ganoderic 26
  gypsogenic 326
  heptelidic 50
  hexadecanoic 203
  hydroxydehydroabietic 304
  imbricataloic 304
  junicedric 304
  oleanolic 26, 180, 281, 282, 285, 311, 326
  oleoyl isocupressic 304
  palmitoyl isocupressic 304
  phenolic 18, 300
  pimaric 304
  rosmarinic 286
  ursolic 180, 285
Action 80, 81, 148, 149, 150, 169, 187, 247,
    268, 311, 313
  anthelmintic 148, 149
  anthropic 169
  anti-inflammatory 80, 81
  anti-nociceptive 150
  cytostatic 311
  growth-rcgulating 247
  neurotoxic 268
  synergistic 313
  toxic 187
Active anthelmintic product 161
Activities 91, 92, 119, 120, 153, 157, 158,
    186, 190, 191, 213, 279, 280, 282, 283
  amoebicidal 92
  anthelmintic 153, 157, 158
  antiaflatoxigenic 120
  antidepressant 120
  antimalarial 119
  antimycobacterial 282
  antiplasmodial 120
  antiviral 283
  enzymatic 191
  hypoglycemic 279
  leishmanicidal 91
  metabolic 280
  microbial 186, 190
  osmotic 213

Agents 85, 170, 181, 224, 229, 327
  anti-microbial 224
  anti-oxidant 229
  effective pharmaco-therapeutic 170
  parasitic 85
  phytotherapeutic 327
Agrochemical industry 246
Agroecosystems 191
*Agrostis stolonifera* 209
Alcohols 1, 68, 176, 177, 232, 279, 289, 300,
    301
  aldehyde 232
  monoterpenic 176
Aldehydes 68, 177, 279, 285, 287, 289, 300
  aromatic 300
Algae, green 5
*Alibertia macrophylla* 46
Allelochemicals 169, 170, 181, 182, 183, 184,
    185, 186, 187, 188, 189, 190, 191, 200,
    201
  activities 190
  behavior in soil 185
  phytotoxicity 186, 191
  release 188
Allelopathic 106, 131, 185
  agent 185
  communication 106
Allelopathy 185, 189, 190, 201
*Allium cepa* 121, 210, 211
Alphonsea tonkinensis 107
Alzheimer's disease 82, 84, 94, 112
*Amblyomma* 132, 137
  *americanum* 132
  *sculptum* 132, 137
American trypanosomiasis 119
*Anaxagorea brevipes* 114, 117
Andditerpenoids 328
Andrographolide 281, 282, 285
*Anethum graveolens* 251, 261
*Annona* 106, 115, 116, 117, 120, 121
  *cherimola* 115
  *leptopetala* 117, 120
  *muricata* 106, 121
  *squamosa* 106, 115, 116
  *vepretorum* 116, 120

Anteraxanthin 227
Antibacterial 113, 114, 281, 282, 283, 308,
    311, 313
  activity 113, 114, 281, 283, 308, 311, 313
  activity of terpenoids 282
Antidiabetic activity 93, 94
Antifungal 56, 114, 313
  action 313
  activity 56
  inhibitory effects 114
Anti-inflammatory 80, 81, 82, 94, 105, 115,
    116, 313
  activity 80, 81, 82, 94, 115, 116, 313
  effects 105, 116
Antileishmanial activity 91, 92, 311
Antimicrobial 21, 39, 40, 44, 54, 56, 105, 113,
    114 115, 122, 185, 225, 279, 280, 281,
    285, 286, 289, 299, 300, 301, 302 308,
    328
  action 286
  activities 54, 113, 114, 115, 185, 279, 280,
    281, 286, 289, 308
  agents 285, 328
  control nystatin 56
  drugs 301
  effect 279
  properties 113, 280, 286, 299, 300, 302
  tests 54
Antioxidant 22, 67, 70, 75, 79, 105, 112, 121,
    122, 131, 300, 311, 312, 313, 328, 329
  activity 67, 70, 75, 79, 121, 122, 300, 311,
    313
Antiparasitic efficacy 154, 161
Antiproliferative activity 87, 88, 116, 117, 118
Anti-protozoan activity 90
Antitumor activity 311
Apis mellifera 308, 309
  anatolica 308, 309
  bees 300, 308
  carnica 308, 309
  caucasica 308
  propolis 308, 309
Apoptosis 177, 234
Arabidopsis 4, 7, 9, 150, 211
  seedlings 150

thaliana 4, 7, 9, 211
Arachnicide for urban pests 265
Aromatherapy 177
Ascomycetes 48, 49, 55
Assays 49, 52, 81, 87, 92, 93 113, 136, 137,
        138, 260
    anti-amastigote 92
    fumigation 260
    luciferase 81
    spectrophotometric 93
ATP-dependent decarboxylation 173
Autolysis 288
    stimulate 288
*Avena sativa* 206, 211
Azithromycin 49, 281

**B**

*Bacillus* 114, 282, 286, 302
    *cereus* 114, 282, 302
    *subtilis* 282, 286, 302
Bacteria 44, 54, 67, 94, 226, 231, 246, 247,
        286
    endodontic pathogenic 94
    photosynthetic 226
*Barbarea verna* 209
Basidiomycetes 55
Beverages 227, 228, 267
    non-alcoholic 228
Bicyclogermacrene 107, 108, 109, 110, 111,
        112, 113, 114, 116, 117, 118, 119, 120,
        121, 122
Bioinsecticides 246
Biomarkers, inflammatory 82
Biosynthesis 10, 12, 13, 170, 285
    brassinosteroid 10
    cytokinin 13
    glycoprotein 170
    pathways 12, 285
Biosynthetic 13, 54, 324, 327
    pathways 54, 324, 327
    terpenoid pathways enzymes 13
Biotechnology of terpene production 323
Biotransformation 153, 290, 322, 324

*Blakiella* 211, 261
    *bartsiifolia* 211
    *germanica* 261
Blood-brain barrier (BBB) 156, 157
Bocageopsis 108, 113, 115, 119
    multiflora 108, 113, 119
    pleiosperma 115
*Botryosphaeria mamane* 45
Brassinosteroid biosynthesis pathway 11
Braziliensis 49
Breast adenocarcinoma 116
Burma, acid 306
Butyrylcholinesterase 52, 83, 84, 113

**C**

*Caesalpinia* 48, 49, 50
    *echinata* 50
    *pyramidalis* 48, 49
*Callistemon citrinus* 71, 74, 77
*Calyptranthes* 90
    *grandifolia* 90
    *tricona* 90
*Campylobacter jejuni* 287
Cancer 85, 86, 88, 116, 118, 327
    cells, gastric 85, 86, 88
    ovarian 118
*Candida* 56, 114, 281, 282, 308, 310
    *albicans* 56, 114, 281, 282, 308, 310
    *parapsilosis* 114
*Cannabis sativa* 22, 25
Canola oil 231
*Capsicum annuum* 228
Carbocation 44
    transoid allylic 44
Carbotricyclic ophiobolanes 54
*Cardiopetalum calophyllum* 108, 116
*Carum carvi* 206, 207, 226, 261
Cell-free biosynthesis (CFB) 327
*Ceratitis capitata* 247
Cervical adenocarcinoma 116
Chagas disease 92, 119
*Chamomilla recutita* 206
Chemoattractants 301

Chemorepellents 301
*Chenopodium album* 206
Chikungunya 130
Cholesterol 55, 170, 179, 180, 226
   absorbed 226
   absorption 180
   biosynthesis 179
Cholinesterase 52
*Choristoneura rosaceana* 269
*Cinnamomum camphora* 251
*Cirsium arvense* 210
*Citrus limetta* 26
Commercial products, developing synergistic
   269
Compounds 80, 83, 153, 169, 180, 182, 186
  pathways release allelopathic 182
  pentacyclic triterpene 180
  phenolic 80, 169, 186
  phytochemical 153
  therapeutic 83
Conditions 132, 183, 185, 186, 200
  aerobic 186
  anaerobic 186
  environmental 183, 200
  natural 185
  nutritional 185
  toxic 132
Conjugated linoleic acids (CLAs) 233
Consumption, lower electrical 22
*Copaifera* 207, 233
   *duckei* 207
   *martii* 207
   *officinalis* oil 233
   *reticulata* 207
*Coriandrum sativum* 203
*Corymbia citriodora* 93, 252
Cosmetic industries 44
*Crassostrea rhizophorae* 83
*Crocus sativus* 226, 228
Crops, organic 266
*Cucumis sativus* 232
*Cuminum cyminum* 261
Cyanobacteria 5
Cycloartenol synthase 10
*Cymbopogon nardus* 262, 266

*Cynara cardunculus* 208
Cytokinin 2, 13
  nucleotides 13
Cytotoxicity 55, 85, 86
  activity 85, 86
  assay 55

**D**

Damage 39, 40, 139, 151, 176, 200, 247
  cell membranes 247
  economic 200
Decarboxylation 52, 171
  oxidative 52
Degradation 154, 183
  metabolic 154
  microbial 183
Dengue fever 285
*Dermacentor andersoni* 139
Dermatophytosis 152
Detoxification enzymes 250, 251, 254, 262,
  269
  inhibiting 251
Diabetes 93, 233, 312, 327
*Diaporthe* 44
   *anacardii* 44
   *foeniculaceae* 44
Diphosphate decarboxylase 5
Diseases 40, 82, 92, 93, 112, 115, 119, 130,
   132, 139, 233, 234, 279, 281, 299, 312,
   313, 327
  autoimmune 93
  cardiovascular 279, 327
  heart 233
  inflammatory 115
  parasitic 92
  protecting Parkinson's 234
  psychiatric 312
  respiratory 312
  stemming 139
DMAPP 170, 171, 173, 177
  biosynthesis 170
  condensation 177
  isomer 171

isomerases 173
Drought 5, 7, 41, 178
  stresses 5, 41
Drug(s) 41, 153, 154, 155, 156, 159, 160, 233,
    234, 279, 285, 288
  absorption 155
  anthelmintic 154
  drug interactions (DDI) 159, 160
  metabolizing enzymes 154
*Duguetia gardneriana* 118
*Dysaphis plantaginea* 269

**E**

Ecosystems 40, 42, 67, 181, 190
  terrestrial 42
Effects 120, 121, 148, 152, 154, 159, 190,
    207, 234, 247, 266, 301, 329
  anthelmintic 152, 154, 159
  antidepressant 120
  carcinogenic 121
  neuroprotective 234
  neurotoxic 247
  phytotoxicity 207
  synergic 148, 152
  synergism 152
  toxic 190, 266, 301, 329
Electron 280, 289
  microscopy 280
  transport 289
*Electrophorus electricus* 83
Emulsifiers 224
Encapsulation technique 158
Endophytic fungi 39, 40, 41, 42, 45, 51, 56, 58
Energy 17, 19, 24
  electromagnetic 19
  microwave photon 19
*Enterococcus* 282
  *faecalis* 282
  *faecium* 282
Enterotoxins 280
Environmental 40, 85, 129, 140, 169, 182,
    183, 184, 185, 267
  factors 40, 169, 182, 183, 184, 185

health problems 129
  pollution 85
  protection agency (EPA) 140, 267
Enzyme(s) 7, 9, 10, 12, 13, 43, 44, 45, 83, 84,
    154, 177, 191, 246, 247, 323, 324, 326
  acethylcholinesterase 83
  biosynthetic 326
  detoxifying 246, 247
  drug-metabolizing 154
  inhibition 246
  sesquiterpene synthases 44
Enzymatic hydrolysis 32
Epidermal growth factor (EGF) 82
*Escherichia coli* 282, 302, 308, 309, 311, 322,
    325
Essential oils (EO) 70, 75, 80, 81, 82, 83, 84,
    85, 89, 91, 92, 93, 94, 114, 115, 116,
    117, 120, 232, 247, 248, 266
  crude 266
  cytotoxicity of 85, 89
Esterases 246, 251, 253
Estragole 150, 207, 254, 257
  arylpropanoids 257
*Eucalyptus* 76, 77
  *angulosa* 76, 77
  *camaldulensis* 77
*Eugenia* 83, 87, 90
  *anomala* 90
  *arenosa* 90
  *brasiliensis* 83
  *tapacumensis* 87
Extraction processes 19, 20, 21, 27, 31, 32,
    33, 311, 322
  green sustainable 32, 33

**F**

Factors 1, 5, 40, 82, 182, 183, 184, 185, 186,
    191, 262, 263, 268
  abiotic 182, 191, 262
  biotic 184
  epidermal growth 82
  platelet-derived growth 82
  post-transcriptional 1

tumor necrosis 82
Farnesyl transferase 10
Ferruginol 304
   oxygenated 304
Fibroblast, non-human lung 87
Flavin-monooxygenase 154
FMO-dependent production 154
*Foeniculum vulgare* 206, 207
Food 140, 223, 224, 225, 226, 227, 228, 229,
      231, 232, 234, 235, 267, 298, 300
   and drug administration (FDA) 140, 224,
      232, 235, 267
      industry 223, 224, 225, 226, 227, 228, 229,
      231, 234, 235, 298, 300
Formation 50, 51, 53, 68, 75, 150, 170, 171,
      175, 280, 288, 313, 324
   allelic cation 171
   biosynthetic 51
   inhibiting microtubules 150
Free radical scavengers (FRSs) 229
Fresh orange peel aroma 226
Fumigant toxicity 87
*Fusaea longifolia* 113, 119
*Fusarium* 54, 115
   *fujikuroi* 54
   *oxysporium* 115


**G**


GABA neurotransmission 83
Ganoderma lucidum spore powder (GLSP) 25,
   26
Gas chromatography 308
Gas extraction 17, 19, 30, 32
   liquefied petroleum 17, 19
Gene(s) 1, 4, 5, 7, 8, 9, 10, 48, 257, 321, 325,
   326
   expression 1, 321
   mevalonate pathway 325
   paralogs 4
   splicing 10
   yeast 326
Genome editing 322
Genotoxic profiles 116

Geraniol 326, 327
   biosynthetic pathway 327
   production 326
Geranyl diphosphate 175, 176
Geranylgeranyl diphosphate 172
Germination inhibitors 247
GI nematodes 154
Glioblastoma 46, 116
Glutamatergic neurotransmission 150
Gonubiensis 45
GPP synthase gene 325
Green extraction method 31
Growth 40, 41, 89, 91, 121, 170, 188, 233,
      247, 269, 280, 287, 288, 309
   fungal 233
   inhibited bacterial 288
   inhibiting bacterial 280
   reduction 247
   regulation 170
Gymnosperms 9, 174


**H**


*Haemaphysalis longicornis* 140
*Haemophilus* 55, 282
   *influenzae* 282
   *impetiginosus* 55
Health, veterinary 132
HeLa, mammalian 89
*Helicobacter pylori* 282
Helminth parasites 152
Hemolysis 81, 86
   erythrocyte 81
Hepatitis B virus (HBV) 285
Herbicides 85, 201
   commercial 201
HMG-CoA 173, 177, 326
   reductase 173, 177, 326
   reduction 173
   synthase 326
Homeostasis 151
Human 81, 88, 117, 118, 328
   cervical cancer 328
   embryonic kidney 81

fibroblast 88
hepatocellular carcinoma 117, 118
Humidity deficiency 190
Hydrocarbons 43, 74, 205, 226, 301
    acyclic 43
Hydrodistillation method 27, 29
Hydrodistilled oil 22
Hydrolytic rancidity 231
Hydrophobic cavity 44
Hypothetical biosynthetic precursor 50
*Hyssopus officinalis* 207

**I**

Immobilized capillary enzyme reactor (ICER)
    52
Industrial processes 17, 18, 27
    green sustainable 17, 18
Industries 44, 70, 170, 181, 225, 322, 323, 329
    agricultural 170
    automotive 322
    cellulose 70
    chemical 44, 225
    perfume 70
    textile 329
Infections 40, 159, 283, 285, 313
    chikungunya 285
    fungal 313
Inflammation 80, 115, 116, 312, 313, 327
    carrageenan-induced acute 115
Insect detoxification 251
Insecticidal 246, 249, 250
    activity 249, 250
    plants 246
Insecticides 187, 246, 247, 248, 249, 250, 251,
    254, 255, 257, 262, 263, 264, 265, 266,
    268, 269
    botanical 246, 247, 263
    chemical 247, 266
    commercial 249, 262
    for agricultural pests 264
    for urban pests 263, 264, 265
    for veterinary pests 264
    synthetic formamidine-based 250

toxic 187
Intensity 23, 159, 188, 204
    ultrasonic 23
Interactions 19, 21, 40, 139, 140, 155, 169,
    170, 176, 183, 249, 255, 258, 259, 260,
    261, 262, 301, 320
    plant-environment 170
    plant-fungal 40
    plant-insect 176
    plant-microorganism 183
    terpenoid 249
Intestinal absorption 155
Ion exchange capacity 182, 186
IPP isomerase 8, 9, 325
    activity 8
    gene 325

**J**

Junctions, neuromuscular 249

**K**

Kaurene oxidase (KO) 325
Kinetic disposition 154, 157
*Klebsiella pneumonia* 114, 115, 286
*Kocuria rhizophila* 114

**L**

Lactones 68, 130, 155, 159, 161, 177
    macrocyclic 155, 159, 161
Larvicidal activity 89, 118, 119
*Lasiodiplodia theobromae* 45
*Lavandula angustifolia* 206, 207, 252
*Leishmania* 49, 90, 91, 120, 311
    *amazonenses* 90, 91
    *infantum* 91, 120
    infection 311
Lemna paucicostata 209, 211
*Lepidium sativum* 207, 208, 210
*Lepidoptera larvae* 258
*Leptospermum* 87

*citratum* 87
  *scoparium* 87
Leukemia 46, 117, 118
  human chronic myelocytic 117
  human promyelocytic 117, 118
Limonene synthase 177
Lipid oxidation 232, 313
Lipophilicity 18, 153
Liquefied petroleum gas 30, 32
Liquid nitrification 188
*Listeria monocytogenes* 115, 282
LPG extraction 32
Lung 87, 89, 152, 328
  cancer 87, 89, 328
  parenchyma 152
Lycopene, tomato 225
Lycopersicon esculentum 231

# M

Macrocyclic lactones (MLs) 159
*Majorana hortensis* 207
Malaria 130, 326, 327
Mass spectrometry 308
*Matricaria chamomilla* 187
Meat 286, 313
  industry 313
  preservation 286
Medicinal plants 188, 190, 280
Medicines 228, 284, 299, 312
  complementary 312
  traditional Chinese 284
  veterinary 299
Mediterranean propolis 301
*Melaleuca* 72, 79, 82, 84, 93
  *alternifolia* 79, 93
  *cajuputi* 82
  *citrina* 72, 82, 84
Melanoma 46, 85, 86, 87, 88
*Melipona beechei* 310
*Melissa officinalis* 206, 207
MEP and MVA pathways 323
MEP pathway 1, 2, 5, 6, 7, 8, 13, 171, 174, 280

genes 7
*Mercurialis annua* 210
Meroterpenoids 54, 55
Metabolic pathways 1, 48, 172, 324
  secondary 48
Metabolic processes 1, 170
  secondary 170
Metabolism 148, 153, 154, 155, 156, 157, 161, 179, 185, 323
  hepatic 161
  hepatic CYP-dependent 155
  oxidative 153
Metabolites 40, 41, 50, 132, 153, 154, 182, 185, 189, 247, 279, 280, 320, 323, 326
  of monoterpenes 153
  isoprenoid 132
  purified fungal 50
  toxic 40
Methicillin 113, 280
  resistant *Staphylococcus aureus* (MRSA) 113, 280
  susceptible *S. aureus* (MSSA) 280
Methods 21, 26
  conventional heating 21, 26
  economical 26
Mevalonate 2, 3, 4, 43, 54, 68, 172, 173
  diphosphate decarboxylase 3
  kinase (MK) 2, 4, 173
  pathway 43, 54, 68, 172
  pyrophosphate decarboxylase (MPD) 173
Microbes 40, 279, 280, 281, 323
  microaerophilic 281
Microbial-mediated allelochemical production 191
*Micrococcus glutamicus* 302
Microorganisms 39, 40, 41, 113, 152, 224, 225, 229, 280, 281, 282, 323, 324, 325
  facultative anaerobic 281
  resistant 113, 152
Microplate dilution method 87
Microscopic organisms, anaerobic 281
Microsomal monooxygenases 246
Microwave(s) 19, 21
  energy 19
  heating 19

irradiation 21
   release 21
Microwave-assisted 17, 19, 20, 21, 22, 25, 26,
      33, 322
   extraction (MAE) 17, 19, 20, 21, 25, 26,
      33, 322
   hydro-distillation (MAHD) 22, 25
Migration 80, 81
   leukocyte 80, 81
   neutrophil 80
*Mimosa pudica* 207
Mitochondrial 151, 234
   dysfunction 234
   profile 151
Molecules 9, 30, 177, 250, 300, 302
   allylic diphosphate 9
   carotenoid 30
   hydrophobic 177, 302
   phenolic 300
   toxic 250
Monocytogenes 282, 287
Monoterpene(s) 21, 25, 43, 68, 74, 106, 150,
      152, 153, 154, 155, 157, 158, 161, 175,
      176, 177, 182, 204, 205, 228, 229, 249,
      280, 289, 301, 303, 310
   dihydrojasmone 152
   hydrocarbons 21, 25, 106, 289
   monocyclic 303
   oxygenated 74, 204, 249
   production 182
Monoterpenoids 39, 43, 44, 107, 108, 109,
      110, 111, 112, 132, 175, 176, 283, 285,
      286, 327, 328
   bicyclic 132
   diterpenes 39
Moroccan propolis 307
Movement 129, 138, 183, 186, 282, 287, 289,
      290
   antimycobacterial 282
MTT 90, 91
   assay 90, 91
MTT colorimetric 85, 86, 87, 90, 91
   assay 90, 91
   method 85, 86, 87
Muscle 149, 150

contraction 149
   paralysis 150
   relaxation 149
MVA pathway 1, 2, 3, 4, 5, 7, 8, 172, 173,
      177, 179, 323
   genes expression 5
*Myrcia* 83, 84, 88
   *mollis* 83, 84
   *silvatica* 88
*Myrcia sylvatica* 85
   oil 85
*Myrrhinium atropurpureum* 90, 91

**N**

Nanotrigona testacularis 310, 312
Necrotic lesions 210
Nectria pseudotrichia 49
Nematodes 54, 148, 149, 150, 157
   intestinal 157
*Neofusicoccum* 45, 46
   *cordaticola* 45
   *parvum* 46
Nervous system 149, 249, 250
Neurodegenerative diseases 82, 83, 112
Neurohormones 250
Neuroprotection 234, 279, 289
   agents 279
Neuroprotective activity 82, 84
Neurotransmitters 83, 149, 249, 250
Nitrogen mineralization 188
NMR analyses 50
Nutrients 181, 184, 187, 188, 189, 190, 200,
      224
   mineral 187

**O**

*Ocimum* 187, 207, 232, 253
   *basilicum* 187, 207, 253
   *gratissimum* 232
*Ocotea glomerata* 261
Oils 140, 176, 179, 225, 228, 230, 269, 288,
      329

cinnamon 140
clove 140
cumin 176
mineral 269
neem 329
palm 228
tea tree 288
turpentine 179
vegetable 230
woody 225
Oral submucosal fibrosis 234
Orange 32, 329
    oil 329
    waste extracts 32
Organic 264
    crops protection 264
Organic food 287
    production 287
Organisms, taxonomical 170
*Origanum vulgare* 24, 203, 207, 233
*Oryza sativa* 206
*Osmanthus fragrans* 226
Osteoporosis 312
Oxidation 13, 121, 172, 187, 188, 229, 230,
    232
    ammonium pathway 188
    reactions 13, 229
Oxygen radical absorption capacity (ORAC)
    229, 230

**P**

*Paepalanthus chiquitensis* 54
*Papaver rhoeas* 211
Parasites 154, 157
    nematode 154
    pathogenic gastrointestinal 157
Parasitoids 268
Paromomycin 49
*Parthenium hysterophorus* 206
Pathogen vectors, disease-causing 248
Pathways 1, 2, 3, 10, 13, 68, 93, 154, 170,
    171, 173, 321, 326, 327
    cryptic 321

cytosolic ergosterol 327
    enzymatic 93
    plastid 1
Paw edema 80
    carrageenan-induced 80
Peritonitis 80
Pest control 139, 246, 249, 266, 269
Pesticides 44, 70, 200, 263, 266, 267, 268
    agricultural 200
    commercial 268
    conventional 267
    essential oil-based 263
    green 268
    natural 70
Phenolic terpenoids 250, 286
Phomopsis 44
Phosphomevalonate decarboxylase 327
Phosphorylation, oxidative 289
Photosynthesis 1, 170, 226, 269
Phylogenetic analyses 44
Phytochrome interacting factors (PIFs) 7
Phytotoxic activity 200, 201, 204, 205, 207,
    208
Phytotoxicity 186, 190, 268, 269
*Piper nigrum* 266
Plant(s) 1, 13, 40, 41, 170, 176, 186, 190, 191,
    203, 209, 232, 251, 258, 269
    communication 170
    communities 190
    defense mechanisms 40, 258
    disease 40
    essential oils 209
    growth 186, 190, 191
    hormones 1, 41
    matrices, aromatic 232
    metabolism 13, 258
    micro-ecosystems 40
    plant communication 170, 176
    terpenoid-producing 203, 251
    stress 269
Plant metabolites 1, 129, 290
    secondary 1, 129
Plasma 156, 158, 161
    prolonged absorption 161
    proteins 156

Platelet-derived growth factor (PDGF) 82
PLE techniques 33
*Polyalthia korintii* 110, 115
Polysaccharides 234
Post-translational 1, 10
   modification 10
   protein modifications 1
Power 225, 229
 anti-oxidant 229
 protective 225
Pressurized liquid extraction (PLE) 17, 19, 30, 31
Processes 5, 8, 17, 19, 21, 22, 27, 28, 32, 149, 181, 183, 185, 186, 188, 247, 248, 250, 286, 323, 326, 330
 behavioral 247
 biochemical 149, 183, 188
 biotechnological 330
 green 17
 isomerase 8
 neurological 248
 respiration 5
 synergistic 286
Production 42, 43, 44, 58, 59, 182, 184, 190, 268, 299, 300, 320, 322, 323, 325, 326, 327
 biopesticide 268
 biotechnology 320
Products 154, 177, 224, 266, 285, 287, 329
 organic 285, 287
 perfumery 329
 pharmaceutical 177, 224
 plant-derived 154
 pyrethrum-based 266
Properties 18, 28, 94, 120, 150, 159, 225, 228, 231, 232, 247, 266, 270, 288, 299, 313, 325
 analgesic 150
 antidiabetic 94
 anti-inflammatory 313
 flavoring 228
 healing 299
 medicinal 325
 regeneration 313
 sensory 231, 232

 synergistic 266
Propolis 300, 306, 313
 green 300, 306
 flavonoids 313
Prostate cancer cells 117
Protein(s) 2, 4, 10, 11, 18, 177, 188, 234, 287, 289, 300
 associated enzyme 289
 cytoplasmic 287
 cytosolic 4
 heat shock 11
 membrane-bounded 18
 prenylation 2, 10
 synthesis 188
*Pseudofusicoccum stromaticum* 45
*Pseudomonas aeruginosa* 281, 282, 286, 289, 302, 308, 309
*Pseuduvaria macrophylla* 110, 113
*Psidium* 70, 80, 82, 84, 85, 86, 88, 94, 109
 *cattleianum* 94
 *guajava* 70, 82, 84, 86, 88, 94, 109
 *guineense* 80, 85

# R

Radiation 19, 21, 182, 184, 328
 radio 19
 solar 182
 ultraviolet 328
Reactions 2, 3, 5, 6, 8, 9, 10, 12, 43, 51, 161, 171, 173, 174, 250, 251, 289, 321, 324
 acetylation 51
 condensation 2, 9, 171
 decarboxylative elimination 3
 dehydration 171
 enzyme-dependent 289
 metabolic 161
Regulation, cardiovascular disease 70
Renewable natural products 19
Repellency 129, 132, 136, 138, 141, 142
 process 136
 tests 138
 activity 129, 132, 136, 141, 142

Repellent(s) 68, 129, 130, 131, 136, 139, 140, 142, 181, 187, 248, 267
  commercial 130
  products 130
Resistance 41, 149, 157, 159, 251, 254, 257, 269, 313
  metabolic 251, 254
  microbial 313
Respiration, mitochondrial 7
Respiratory enzymes, inhibited 289
Response 159, 225
  anthelmintic 159
  immune 225
Rheumatisms 313
*Rhyzopertha dominica* 247
*Riphicephalus annulatus* 132
RNA transfection 285
Rocky mountain spotted fever (RMSF) 139
Rosemary 29, 31, 140, 203, 228, 253, 259, 263, 264, 265, 266, 286
  essential oils of 31, 263
Rosemary oil 259
*Rosmarinus officinalis* 29, 253, 259, 266, 286
Ruminal 153
  metabolism 153
  microflora 153

**S**

*Saccharomyces cerevisiae* 282, 322, 325
*Salmonella* 54, 282, 287
  *enterica* 287
  *setubal* 54
  *typhimurium* 282
*Salvia officinalis* 29, 30, 206, 207, 253
*Sambucus nigra* 25, 26
*Sarcina lutea* 302
*Satureja horvatii* 287
Scanning electron microscopy 151
Secondary metabolites 1, 39, 41, 42, 43, 54, 181, 183, 184, 185, 188, 189, 330
  production 42, 188, 189, 330
SFE process 28, 30
Signals 5, 176

anti-aggregating 176
*Sitophilus* 120, 247
  *oryzae* 247
  *zeamais* 120
Skin 56, 132, 267, 328
  care products 328
  infection 56
  irritations 132, 267
Soil 184, 187, 190, 191, 201
  alkaline 187
  decomposition process 191
  environment 190, 191
  fertility 184
  microbial ecology 191
  microorganisms 201
  nutrients 191
*Solanum* 206, 208, 231
  *lycopersicum* 206, 208
  *tuberosum* 231
*Solidago canadensis* 206
Solvent-free microwave extraction (SFME) 21, 22, 25
Sonication time 23
Species, oxygen-reactive 174
*Sphaeropsis sapinea* 211
Spoilage microbiota 287
*Staphylococcus* 114, 115, 302, 309
  *epidermidis* 114, 115, 302, 309
  *pyogenes* 309
*Staphylococcus aureus* 113, 114, 115, 152, 280, 282, 302, 308, 309, 311, 312, 326
  methicillin-resistant 113, 280
Steviol glycosides (SGs) 325, 329
*Streptococcus* 113, 115, 282, 302
  *mutans* 282, 302
  *pneumonia* 282
  *pyogenes* 113, 115, 282, 302
Stress 55, 81, 184, 229, 234, 320
  environmental 184
  herbivore 320
  inflammatory 234
  oxidative 81, 229
Subcritical water extraction (SWE) 17, 19, 30, 31, 32, 33

Substances 18, 80, 138, 142, 169, 184, 189, 224, 225, 229, 247, 253, 258, 262, 298, 299, 300, 301, 309
　anti-inflammatory 80
　anti-oxidant 229
　chemical 18, 169
　phenolic 229
　synergistic 262
　toxic 189
　triterpenic 309
　water-soluble 184
Sugarcane bagasse 32
Sunflower oil 231
Supercritical fluid extraction (SFE) 17, 19, 27, 29, 32, 33
Sustainable 85, 269, 323
　food 269
　mass production 323
　natural products 85
Symbiosis 169, 181, 190
　plant-microorganisms 169, 181
Synergism 82, 94, 152, 159, 246, 256, 259, 268
　pharmacodynamic 159
Synergistic 255, 256, 281
　antibacterial effect 281
　binary interactions 255
　combinations 256
Synthetic(s) 85, 89, 152, 224, 231, 249, 254
　anthelmintic combination 152
　drug doxorubicin 89
　insecticides 85, 249, 254
Systems 19, 23, 28, 82, 148, 169, 177, 250, 286, 290, 324, 327
　dermal fibroblast 82
　dynamic 169
　microbial 324
　microwave heating 19
　neuromuscular 148
　respiratory 250
　synergistic 286
　tricyclic 177
*Syzygium guineense* 79
*Syzygium* 73, 74, 78, 79, 82, 84, 88, 91, 93, 94, 232, 266

　*aromaticum* 78, 79, 93, 94, 232, 266
　*cumini* 73, 82, 84, 88, 91
　*samarangense* 74, 82

**T**

Targets, therapeutic 234
*Taxomyces andreanae* 41
Techniques 290, 322
　metabolic engineering 322
　microextraction 322
　natural food-processing 290
Technologies 17, 19, 33
　green 19
Terpenes 17, 21, 32, 43, 68, 129, 131, 132, 136, 151, 152, 159, 169, 191, 224, 225, 228, 301, 302, 322, 323, 324, 325, 327, 329
　aromatic 323
　based biopesticides 329
　cedarwood oil 225
　lemon 225
　lipophilic 151
　noncyclic 301
　production 228, 323, 324
　purify hydrophobic 322
　synthase products 325
　synthases 323
　volatile oil 21
Terpenoid 1, 2, 3, 4, 5, 6, 7, 8, 9, 33, 54, 139, 172, 173, 177, 184, 191, 255, 259, 280, 326
　biosynthesis 1, 8, 9, 191, 326
　biosynthetic pathways 2, 5
　extraction processes 33
　fraction 280
　pathways 54
　precursor biosynthesis 3, 4, 6, 7, 9
　repellent action 139
　reservoirs 184
　synergistic 255, 259
　synthesis 172, 173
　terpenoids 177
Thyme oils 140

*Thymus* 203, 206, 232, 253, 266
  *vulgaris* 203, 206, 253, 266
  *zygis* 232
  *zygis* oils 232
Tick(s) 132, 137, 142
  attacks 132
  climbing bioassay 137
  parasites 142
Tissue, parasite location 155
Toxicity 117, 136, 176, 247, 248, 254, 259, 260, 261, 262, 268
  residual 261
Toxin-induced neurotoxicity 234
Trafficking of terpenoids precursors 8
Transient receptor potential (TRP) 249
Transmission electron microscopy 151
Transport 154, 155, 183, 186
  mediated digoxin 155
  proteins 154, 155
Treatment, anthelmintic 161
*Tribolium castaneum* 247
*Trichoderma reesei* 115
*Tripterygium wilfordii* 5
Triterpenoid(s) 26, 281, 285
  pentacyclic 26, 281, 285
  saponin 285
*Triticum aestivum* 41, 206
*Trypanosoma cruzi* 90, 92
Tumor necrosis factor (TNF) 82

**U**

Ubiquinone synthesis 1
Ultrasound-assisted extraction (UAE) 17, 19, 22, 24, 25, 26, 33
Uterine cancers 225

**V**

*Verbena officinalis* 207
*Virola michelli* 56
Virulence factors 40
Viruses 283, 285, 327
  herpes simplex 283

vesicular stomatitis 285
Volatile organic compounds (VOCs) 45

**W**

Water 184
  deficiency 184
  stress 184

**X**

*Xanthomonas oryzae* 177
Xenobiotics 153, 155, 251, 257, 268
*Xylopia* 111, 115, 119, 120
  *aethiopica* 111, 115
  *frutescens* 119
  *sericea* 120
*Xylopica aethiopica* 233

**Y**

Yellow fever 130

**Z**

Zeaxanthin epoxidase 12
*Zingiber officinale* 203

www.ingramcontent.com/pod-product-compliance
Lightning Source LLC
Chambersburg PA
CBHW050804220326
41598CB00006B/111